MW01255858

Textbook of Animal Biotechnology

Textbook of Animal Biotechnology

Edited by **Carlos Wyatt**

New York

Published by Syrawood Publishing House,
750 Third Avenue, 9th Floor,
New York, NY 10017, USA
www.syrawoodpublishinghouse.com

Textbook of Animal Biotechnology
Edited by Carlos Wyatt

International Standard Book Number: 978-1-68286-067-0 (Hardback)

Printed in the United States of America.

Contents

Preface

Animal biotechnology is focused on animal health and livestock production. The applications of this field are spread across industries such as pharmaceutics, dairy farming, etc. This book aims to understand the innovative methods and techniques for animal breeding and genetics. It comprises of research on different aspects of feed processing technology and bioevaluation. This text will benefit students, researchers and professionals engaged in this field by keeping them updated with global trends.

This book is the end result of constructive efforts and intensive research done by experts in this field. The aim of this book is to enlighten the readers with recent information in this area of research. The information provided in this profound book would serve as a valuable reference to students and researchers in this field.

At the end, I would like to thank all the authors for devoting their precious time and providing their valuable contributions to this book. I would also like to express my gratitude to my fellow colleagues who encouraged me throughout the process.

Editor

Haplotype phasing after joint estimation of recombination and linkage disequilibrium in breeding populations

Luis Gomez-Raya[1*], Amanda M Hulse[2], David Thain[3] and Wendy M Rauw[1]

Abstract

A novel method for haplotype phasing in families after joint estimation of recombination fraction and linkage disequilibrium is developed. Results from Monte Carlo computer simulations show that the newly developed E.M. algorithm is accurate if true recombination fraction is 0 even for single families of relatively small sizes. Estimates of recombination fraction and linkage disequilibrium were 0.00 (SD 0.00) and 0.19 (SD 0.03) for simulated recombination fraction and linkage disequilibrium of 0.00 and 0.20, respectively. A genome fragmentation phasing strategy was developed and used for phasing haplotypes in a sire and 36 progeny using the 50 k Illumina BeadChip by: a) estimation of the recombination fraction and LD in consecutive SNPs using family information, b) linkage analyses between fragments, c) phasing of haplotypes in parents and progeny and in following generations. Homozygous SNPs in progeny allowed determination of paternal fragment inheritance, and deduction of SNP sequence information of haplotypes from dams. The strategy also allowed detection of genotyping errors. A total of 613 recombination events were detected after linkage analysis was carried out between fragments. Hot and cold spots were identified at the individual (sire level). SNPs for which the sire and calf were heterozygotes became informative (over 90%) after the phasing of haplotypes. Average of regions of identity between half-sibs when comparing its maternal inherited haplotypes (with at least 20 SNP) in common was 0.11 with a maximum of 0.29 and a minimum of 0.05. A Monte-Carlo simulation of BTA1 with the same linkage disequilibrium structure and genetic linkage as the cattle family yielded a 99.98 and 99.94% of correct phases for informative SNPs in sire and calves, respectively.

Keywords: Breeding, Haplotype phasing, Linkage disequilibrium, SNP

Background

Advances in molecular biology have allowed rapid and massive genotyping of Single Nucleotide Polymorphisms (SNP) in animal and plant species. SNP arrays have been implemented for genome-wide selection to enhance genetic improvement of farm animals [1]. In many instances, it has been done without consideration of the nature of the SNP information. That is, SNPs have been assumed to be unlinked in spite of providing low information (maximum PIC values of 0.375 at intermediate frequencies for two alleles). In reality, SNP polymorphisms are arranged in aligned sequences forming haplotypes where the order of the alleles for consecutive SNPs within each homologue chromosome contains relevant information. To make full use of SNP microarray technology, haplotype phasing must be estimated because genotyping only generates information for each single SNP. Haplotype phasing consists of arranging the order of allelic variants in a chromosomal segment within each of the two homologous chromosomes of diploid species. Phasing knowledge can be applied to trace SNP inheritance and to account for regions that are identical by descent in genomic evaluations aimed at the genetic improvement of agricultural species [2,3]. Haplotype phasing can be assessed in the laboratory or computationally [4]. Computational methods can make use of the population structure of linkage disequilibrium assuming no relationships

* Correspondence: gomez.luis@inia.es
[1]Departamento de Mejora Genética Animal, Instituto Nacional de Investigación y Tecnología Agraria y Alimentaria (INIA), Ctra. de La Coruña km 7, 28040, Madrid, Spain
Full list of author information is available at the end of the article

among individuals or may also establish inheritance of haplotypes within pedigrees in order to reconstruct haplotypes. In the latter, haplotypes can be traced up to ancestors in the top of the pedigree [5-7]. PHASE [7], FASTPHASE [6], and BEAGLE [5] implement a Bayesian statistical method for reconstructing using coalescent and hidden Markov models. FASTPHASE [6] cannot provide estimates of recombination rates but PHASE [7] does. BEAGLE [5] can infer haplotypes from unrelated individuals, parent-offspring pairs, and parent-offspring trios. Rohde and Fuerst [8] developed methods for haplotype inference based on maximum likelihood, which could be used with nuclear families. Their method was based on searching long genomic segments for which an allele was shared for two individuals and has been also used for phasing haplotypes. A linkage program like GENEHUNTER can extract multiple relative's information [9], although it is done by assuming linkage equilibrium between polymorphisms, which may lead to incorrect phasing. If the genotype information is from sibs and genotyping information on one of the parents is missing then linkage disequilibrium (from gametes from the other parent) may not be separated from genetic linkage (within the parent with genotype information). Another group of methods makes use of relative's information [10,11] together with laws of Mendelian inheritance to infer haplotypes. Haplotypes inferred from data sets from families are accurate across extended genomic distances if family sizes are large. Williams *et al.* [12] proposed a new haplotype inference method for nuclear families based on maximum likelihood and minimum recombination using linear programming (implemented in software Hapi). More recently, Lai *et al.* [13] proposed an algorithm for reconstructing haplotypes in a pedigree by assuming zero recombinants. Their method can be applied to general pedigrees but it is not designed to make use of nuclear families with one single parent.

All of the above methods for haplotype inference cannot be readily applied to the situation most commonly found in farm animals. This is, genotyping information is available from just one of the parents which in turn have large progeny groups (e.g., dairy bulls with up to one million progeny). Therefore, none of the above methods can estimate recombination and linkage disequilibrium simultaneously because they assume that there is linkage equilibrium. As stated by Browning and Bronwing [4]: "When sites are in LD, linkage programs that assume linkage equilibrium may falsely infer IBD in situations in which is not present".

In this study, we develop a computational method to estimate haplotype phases in breeding populations using closely linked biallelic markers densely distributed in the entire genome. The strategy is based on making use of large families of breeding populations in order to estimate recombination fraction and linkage disequilibrium (LD) simultaneously in order to phase parents. Then, progeny are phased using Mendelian laws of inheritance together with genotyping information from parents and progeny as it has been done elsewhere e.g., [9-11]. A new EM algorithm is developed to simultaneously estimate recombination fraction and linkage disequilibrium in half-sib families (required for accurate phasing). The method to phase haplotypes is tested in a cattle family with 36 calves. Montecarlo-computer simulations were carried out for joint estimation of recombination and linkage equilibrium in a family resembling recombination and linkage disequilibrium estimated from real data in BTA1.

Methods

Genome fragmentation phasing strategy

Assume SNP microarrays with a dense coverage of the animal genome are used for large-scale genotyping of animals. The situation considered is dairy cattle in which bull sires and bull dams are mated (usually by artificial insemination) and have usually one progeny from each dam (half-sibs). Among the progeny of those elite bulls are young bulls that are typed for microarrays in order to carry out pre-selection based on genomic information. Some of the bull dams are also progeny of elite bulls and this process repeats itself in successive generations. Therefore, all male selection candidates are typed for microarrays. A method named genome fragmentation phasing strategy (GFPS) is proposed to phase haplotypes when information is available in families as it is common in breeding populations. GFPS steps assumptions are: 1) families are large with progeny sharing at least one parent; 2) both linkage disequilibrium and recombination events are modeled; and 3) use is made of SNP arrays with high coverage of the genome and with known SNP location. The steps for GFPS are:

A. ***Estimation of the recombination fraction in consecutive SNPs using family information in the first generation***
There are two possibilities depending on the available genotype information from one or from two parents. If both parents are available then standard linkage analyses can be performed. Non-informative progeny can be ignored and the resulting recombination fraction estimates will be unbiased. If only one parent is known then joint estimation of estimation of recombination fraction and linkage disequilibrium (LD) is needed for each pair of consecutive closely linked SNPs. An EM algorithm is proposed to estimate recombination and LD (Appendix 1). Monte-Carlo computer simulation was carried out to validate joint

estimation of LD. If recombination fraction between two consecutive SNP, T/t and M/m, is zero then:

A1 *Establish linkage phase in the parent or parents by eq. A3:*

$$\Pr \textit{ of Phase } (TM/tm) = \frac{L_{TM/tm}(\delta, f_T, f_M, c|\text{data})}{L_{TM/tm}(\delta, f_T, f_M, c|\text{data}) + L_{Tm/tM}(\delta, f_T, f_M, c|\text{data})}$$

where $L_{TM/tm}(\delta, f_T, f_M, c|\text{data})$ and $L_{Tm/tM}(\delta, f_T, f_M, c|\text{data})$ are the likelihoods of phase (TM/tm) and phase (Tm/tM) under a model estimating linkage disequilibrium (δ), allele frequencies for one of the alleles at markers T/t and M/m (f_T, and f_M), and the recombination fraction (c). See Appendix 1 for a full description of the maximum likelihood method. Repeat this process for each pair of consecutive SNPs in which the last SNP in a pair is the first SNP of the following until an estimate of recombination fraction is greater than 0. If recombination fraction is greater than 0 then it means that either a recombination event has taken place or it is a genotyping error. This is a signal for the termination of one fragment and the starting of the next one. Genotyping errors can be of two kinds: failure of genotyping itself or incorrect alignment of SNPs in the array. At this point, all consecutive SNPs fully linked and aligned two by two are considered a fragment, or haplotype block.

A2 *Reconstruct phases of parent(s) for each fragment*

The phases for each pair of SNPs in the sire are reconstructed by aligning the most likely phases for each pair of SNPs until fragment phasing information is completed. Therefore, a fragment is a piece of a chromosome in which no recombination events are detected between any consecutive markers. If the sire is homozygote then the same allele would appear in the two homologous chromosomes. Figure 1 illustrates reconstruction of parent haploypes for a sire. The blue arrow indicates heterozygote SNPs that are concatenated according the linkage phases having the higher probability.

A3 *Reconstruct phases in progeny with paternal and maternal haplotypes*

The sire has two fragments and the progeny has one paternal fragment and another maternal fragment that it is deducted from the genotype information. Progeny phases can be reconstructed by the following rules. If a progeny is homozygous for one of the alleles for which the parent is heterozygous then the full haplotype of that fragment is inherited by that progeny (square in the Figure 1). The contribution from the mother completes the genotype information for each SNP within the fragment. Homozygous calf's genotypes from an heterozygous sire are represented in red in Figure 1. If the progeny is heterozygous at one SNP but the parent is homozygous then that paternal allele in the progeny is the sire allele while the maternal allele is the other allele (green in Figure 1).

B. ***Linkage analyses between fragments***

Fragments with no recombination events detected between SNPs can be used in linkage analyses considering long haplotypes as highly informative "super-alleles" (haplotypes of fragments whose inheritance can be traced to their paternal and maternal contributions), which allows detection of recombination events. Also changes in the haplotype configuration inherited in the calf when compared to

SNP	Location	Sire phase		Calf phase	
BFGL-NGS-115671	2202187	G	G	G	A
ARS-BFGL-NGS-71085	2270722	G	A	A	G
ARS-BFGL-NGS-103391	2333174	G	A	A	G
ARS-BFGL-NGS-22858	2391304	A	C	C	C
ARS-BFGL-NGS-92033	2485621	G	G	G	G
ARS-BFGL-NGS-79093	2522907	A	A	A	A
ARS-BFGL-NGS-58892	2558854	G	A	A	A
ARS-BFGL-NGS-98287	2580759	C	C	C	A
BTB-00001612	2741612	A	G	G	G
BTB-00001299	2775726	A	G	G	G
ARS-BFGL-NGS-37412	2809422	A	A	A	A
ARS-BFGL-NGS-16784	2877871	G	G	G	A
BTB-02081949	2914487	A	A	A	G
BTB-00001074	3015076	A	G	G	G
ARS-BFGL-NGS-30999	3083367	A	A	A	G
Hapmap57114-rs29012843	3211001	C	C	C	A
ARS-BFGL-NGS-108686	3239884	A	G	G	G

Figure 1 Scheme of the genome fragmentation phasing strategy using SNP information from a sire and one progeny of cattle.

sire's phased chromosomes may reveal a recombination event in that calf.

C. ***Phasing of haplotypes in following generations***
Once the phase is established in the parents and progeny, phases of individuals in the following generation is straightforward. Phases of progeny that become a parent and contributes to the next generation are used to reconstruct phases in the progeny of the progeny. Linkage analyses are carried out in the next generation using family information in order to establish haplotype phases in each of the subsequent generations.

Monte-Carlo computer simulation of a half-sib family for joint estimation of LD and genetic linkage

A Monte-Carlo computer simulation was carried out in order to validate methods to estimate recombination fraction and LD jointly in half-sib families. The sire was assumed to be a double heterozygote. A random generator from the uniform distribution was used to assign progeny with the genotypes *TTMM*, *TTMm*, *TTmm*, *TtMM*, *TtMm*, *Ttmm*, *ttMM*, *ttMm*, and *ttmm* according to their probability (frequency): $\phi_{TTMM} = {}^1/_2(1-c)f_{TM}$, $\phi_{TTMm} = {}^1/_2(1-c)f_{Tm} + {}^1/_2c\,f_{TM}$, $\phi_{TTmm} = {}^1/_2c\,f_{Tm}$, $\phi_{TtMM} = {}^1/_2(1-c)f_{tM} + {}^1/_2c\,f_{TM}$, $\phi_{TtMm} = {}^1/_2(1-c)(f_{tm} + f_{TM}) + {}^1/_2c\,(f_{tM} + f_{Tm})$, $\phi_{Ttmm} = {}^1/_2(1-c)f_{Tm} + {}^1/_2c\,f_{tm}$, $\phi_{ttMM} = {}^1/_2c\,f_{tM}$, $\phi_{ttMm} = {}^1/_2(1-c)f_{tM} + {}^1/_2c\,f_{tm}$, and $\phi_{ttmm} = {}^1/_2(1-c)f_{tm}$. If the drawing of the uniform distribution was between 0 and ϕ_{TTMM}, then the offspring had genotype *TTMM*. If the drawing of the uniform distribution was between ϕ_{TTMM} and $\phi_{TTMM} + \phi_{TTMm}$ then the offspring genotype was *TTMm*. Assigning other genotypes to offspring was done following the same rule.

Multi-family estimation of linkage disequilibrium

A total of six families with sizes 94, 77, 106, 81, 79, and 100 half-sib progeny were simulated resembling the sire Norwegian cattle population (after pooling selected and culled bulls in Table 1 of [14]). The allele frequencies were intermediate, recombination fraction was 0, 0.25, or 0.50, and linkage disequilibrium ranged from 0 to 0.25. The sires were simulated as if they were coming from a population with the same linkage disequilibrium and allele frequencies as used to generate the half-sib progeny. In order to do so, the two haplotypes of each sire were generated following the same principles as above, with probabilities according to the simulated frequencies: $f_{TM} = \delta + f_T f_M$, $f_{Tm} = -\delta + f_T f_m$, $f_{tM} = -\delta + f_t f_M$, and $f_{tm} = \delta + f_t f_m$, in which allele frequencies f_M, f_m, f_T, f_t, and δ were input parameters. Thus, the sire could be a double homozygote, homo-heterozygote or double heterozygote after assigning the two haplotypes. The half-sib progeny was generated as described in the previous section. Join estimation of linkage disequilibrium and recombination fraction was carried out using the developed E.M. algorithm for multiple families (Appendix 1). Each experiment was replicated 10,000 times.

Table 1 Average estimates of linkage disequilibrium in a double heterozygote sire with varying half-sib family size and varying recombination fraction (*c*) and linkage disequilibrium (δ)

		Family size					
Simulated *c*	**Simulated δ**	**36**	**100**	**300**	**500**	**1,000**	**2,000**
0.00	−0.10	−0.0905 (0.0401)	−0.0998 (0.0228)	−0.1001 (0.0131)	−0.1001 (0.0101)	−0.1000 (0.0072)	−0.1000 (0.0051)
	0.00	−0.0044 (0.0412)	−0.0014 (0.0250)	−0.0007 (0.0143)	−0.0003 (0.0110)	−0.0000 (0.078)	−0.0001 (0.0056)
	0.05	0.0434 (0.0409)	0.0479 (0.0245)	0.0492 (0.0140)	0.0496 (0.0109)	0.0499 (0.0077)	0.0499 (0.0055)
	0.10	0.0914 (0.0392)	0.0973 (0.0232)	0.0990 (0.0133)	0.0995 (0.0101)	0.0998 (0.0072)	0.0998 (0.0051)
	0.20	0.1873 (0.0304)	0.1957 (0.0163)	0.1985 (0.0089)	0.1992 (0.0068)	0.1996 (0.0048)	0.1998 (0.0034)
0.10	−0.10	−0.0958 (0.0540)	−0.0997 (0.0277)	−0.1003 (0.0153)	−0.1002 (0.0118)	−0.0999 (0.0084)	−0.1000 (0.0059)
	0.00	0.0076 (0.0069)	0.0042 (0.0422)	0.0002 (0.0192)	0.0001 (0.0134)	0.0001 (0.0099)	0.0000 (0.0000)
	0.05	0.0544 (0.0664)	0.0581 (0.0493)	0.0531 (0.0281)	0.0513 (0.0189)	0.0505 (0.0118)	0.0502 (0.0081)
	0.10	0.0925 (0.0576)	0.1081 (0.0493)	0.1074 (0.0346)	0.1059 (0.0281)	0.1032 (0.0197)	0.1013 (0.0125)
	0.20	0.1534 (0.0364)	0.1721 (0.294)	0.1856 (0.0262)	0.1882 (0.0224)	0.1891 (0.0178)	0.1909 (0.0144)
0.20	−0.10	−0.0889 (0.0705)	−0.0978 (0.0374)	−0.1001 (0.0184)	−0.1001 (0.0141)	−0.0999 0.0099	0.1000 (0.0071)
	0.00	0.0776 (0.0764)	0.0116 (0.0553)	0.0053 0.0340	0.0026 (0.0239)	0.0010 (0.0147)	0.0003 (0.0096)
	0.05	0.0485 (0.0723)	0.0572 (0.0535)	0.0580 (0.0384)	0.0569 (0.0322)	0.0549 (0.0246)	0.0525 (0.0167)
	0.10	0.0818 (0.0651)	0.0961 (0.0508)	0.1008 (0.0362)	0.1017 (0.0307)	0.1033 (0.0258)	0.1039 (0.0215)
	0.20	0.1237 (0.0461)	0.1474 (0.0425)	0.1526 (0.0316)	0.1536 (0.0274)	0.1538 (0.0222)	0.1539 (0.0184)

The simulated allele frequencies were $f_T = 0.5$ and $f_M = 0.5$. The values between parentheses are standard deviations among replicates.

Programming source code for GFPS

Subroutines were written in Fortran 90 to compute join estimates of LD and recombination fraction using the E.M. algorithm as well as to perform all steps of the GFPS algorithm. All source code is available on request to the authors (gomez.luis@inia.es).

Genome analyses of LD in a beef cattle half-sib family

A half-sib family consisting of 36 calves from commercial beef cattle at the Gund ranch in Nevada was used to illustrate and to compare alternative methods for estimation of linkage disequilibrium. The first step was to determine paternity of the calves at the ranch. A set of 25 microsatellites (BMS410, BMS499, BMS650, BMS1244, BMS1634, TGLA227, BMS601, BMS1789, BMS2005, ILSTS081, BMS1315, BMS1226, BMS2573, ILSTS058, TGLA126, CSSM66, SPS115, TGLA53, BM1824, BM2113, ETH3R, TGLA122, INRA023, ETH225, ETH10) was used to assign paternity that was carried out using Cervus software. The procedures are fully described in [15]. The Illumina bovine 50K BeadChip was used with bull #302 and his 36 calves in order to compare methods to estimate LD in half-sib families. Only SNPs with a call rate higher than 0.80 in at least 24 calves and with MAF of 0.10 or more were used. The data was also filtered for SNPs that were not consistent for inheritance from sire to progeny. If a SNP was not consistent for one progeny then the SNP information was discarded for the entire family. Only pairs of SNPs within the same chromosome and within a distance of 50Mb or less were used for estimating linkage disequilibrium and recombination fraction.

The GFPS algorithm was used to reconstruct haplotypes of fragments in the sire and its 36 calves. Recombination events between fragments were detected. In addition, regions of identity (ROI) were used as a measure of molecular relatedness. ROIs are a generalization of runs of homozygosity (ROH) and are the proportion (in respect to the total of the genome) of long haplotypes shared by two individuals. ROH have been thoroughly investigated in human [16,17] and animal populations [18], and in this study were estimated after reconstruction of haplotypes via GFPS. Thus, only haplotypes made up by 20 or more identical SNP alleles for each two individuals were used. ROIs were used to estimate patterns of Mendelian segregation as well as genetic relatedness between two individuals due to the inheritance of paternal or maternal gametes.

Monte-Carlo computer simulation of a phased chromosome

A Monte-Carlo computer simulation of BTA1 using phasing results of the cattle family was performed. A total number of progeny was 36. Only heterozygous SNPs (for the sire) were simulated (sire homozygous SNPs are trivial for phasing purposes). The chromosome coming from the sire to their calves was assumed to come from meiosis with a probability of 0, 1, 2, or 3 recombination events of 0.511494, 0.398467, 0.081418, and 0.00862, respectively. These are the genome-wide values found for recombination events per chromosome in our study. Once, the two resulting gametes from the sires were created then a drawing from the uniform distribution was used to assign either of the two chromosomes to a calf. The chromosome coming from the dams was generated using the haplotype frequencies (of chromosome 1) of the two first SNPs as estimated in this study. The next SNP was simulated using haplotype frequencies (for SNPs second and third) but conditional to the SNP allele at the second SNP. This process was repeated until the entire chromosome was terminated. The same procedure was followed to generate BTA1 for each of the 36 calves. The SNP data was randomly allocate to disturb the order of alleles and the resulting data was analyzed with the methods developed in this paper. The location of each (base pair) was the same as SNPs in the real data. Also, markers with MAF<0.10 were excluded. A total of 7,000 replicates were performed.

Results

Monte Carlo computer results from joint estimation of recombination fraction and linkage disequilibrium

An E.M. algorithm was implemented for the joint estimation of recombination fraction and linkage disequilibrium in both single and multiple family situations (Appendix 1). Tables 1 and 2 show the results for the joint estimation of recombination fraction in a half-sib family with varying family sizes (36 to 2,000). The results show that both recombination fraction and linkage disequilibrium are accurately estimated when either true recombination or true disequilibrium is 0 even for relative small family sizes (36). On the contrary, estimates tend to be biased when both recombination fraction and linkage disequilibrium (absolute value) are greater than 0. Substituting the observed counts in equation A1 by their expected values according to true parameters (recombination fraction, linkage disequilibrium, allele frequencies) was carried out to plot the log of the likelihood against those parameters in order to investigate if the maximum likelihood method can or cannot separate effects of recombination fraction and linkage disequilibrium (i.e., behavior of the likelihood when sample size is infinite). Figure 2 shows several scenarios (A to D) regarding true parameters. If the method would work properly then the highest peak (maximum value of the likelihood) should correspond to true parameters. If either true recombination fraction or linkage disequilibrium is 0 then there is one single maximum value, which corresponds to the true parameters (Figure 2A, C and D). If the true values of both recombination fraction and linkage disequilibrium are different from 0 then estimates of both parameters are biased (Figure 2B). This occurs because the estimation of both

Table 2 Average estimates of recombination fraction in a double heterozygote sire with varying half-sib family size and varying recombination fraction (*c*) and linkage disequilibrium (*δ*)

		Family size					
Simulated *c*	**Simulated *δ***	**36**	**100**	**300**	**500**	**1,000**	**2,000**
0.00	−0.10	0.0007 (0.1232)	0.0000 (0.0000)	0.0000 (0.0000)	0.0000 (0.0000)	0.0000 (0.0000)	0.0000 (0.0000)
	0.00	0.0000 (0.0031)	0.0000 (0.0000)	0.0000 (0.0000)	0.0000 (0.0000)	0.0000 (0.0000)	0.0000 (0.0000)
	0.05	0.0000 (0.0000)	0.0000 (0.0000)	0.0000 (0.0000)	0.0000 (0.0000)	0.0000 (0.0000)	0.0000 (0.0000)
	0.10	0.0000 (0.0000)	0.0000 (0.0000)	0.0000 (0.0000)	0.0000 (0.0000)	0.0000 (0.0000)	0.0000 (0.0000)
	0.20	0.0000 (0.0000)	0.0000 (0.0000)	0.0000 (0.0000)	0.0000 (0.0000)	0.0000 (0.0000)	0.0000 (0.0000)
0.10	−0.10	0.1046 (0.1080)	0.0996 (0.0518)	0.0998 (0.0284)	0.0997 (0.0156)	0.0997 (0.0156)	0.0999 (0.0111)
	0.00	0.1212 (0.1444)	0.1098 (0.0881)	0.1016 (0.0398)	0.1008 (0.2290)	0.1002 (0.0204)	0.1002 (0.0145)
	0.05	0.1157 (0.1425)	0.1184 (0.1033)	0.1078 (0.0603)	0.1034 (0.0412)	0.1011 (0.0259)	0.1007 (0.0176)
	0.10	0.0926 (0.1262)	0.1168 (0.1029)	0.1154 (0.0733)	0.1122 (0.0605)	0.1065 (0.0427)	0.1029 (0.0277)
	0.20	0.0227 (0.0550)	0.0459 (0.0639)	0.0740 (0.0541)	0.0732 (0.0462)	0.0770 (0.0369)	0.0808 (0.0299)
0.20	−0.10	0.2148 (0.1510)	0.2034 (0.0770)	0.2007 (0.0382)	0.2000 (0.0292)	0.2001 (0.0206)	0.2001 (0.0145)
	0.00	0.2167 (0.1670)	0.2212 (0.1204)	0.216 (0.0749)	0.2055 (0.0536)	0.2019 (0.0336)	0.2009 (0.0221)
	0.05	0.1970 (0.1607)	0.2110 (0.1172)	0.2152 (0.0841)	0.2131 (0.0703)	0.2094 (0.0541)	0.2052 (0.0373)
	0.10	0.1627 (0.1493)	0.1868 (0.1087)	0.1986 (0.0778)	0.2007 (0.0663)	0.2044 (0.0551)	0.2069 (0.0460)
	0.20	0.0568 (0.0959)	0.0944 (0.0905)	0.1040 (0.0663)	0.1061 (0.0573)	0.1068 (0.0467)	0.1075 (0.0388)

The simulated allele frequencies were $f_T = 0.5$ and $f_M = 0.5$. The number of replicates was 10^4. The values between parentheses are standard deviations among replicates.

parameters is confounded. Nevertheless, estimation of recombination fraction would be unbiased if true recombination is 0, which is necessary for phasing parents.

Table 3 shows the average of the estimate of the recombination fraction after assuming linkage equilibrium as described by [19] and as it has been assumed for all published QTL mapping experiments [20]. It is shown that estimates of the recombination fraction are generally biased even for true recombination fraction of 0 when family size is small. Table 4 shows computer simulation results for multi-family joint estimation of recombination fraction and linkage disequilibrium. The results show that estimates of both linkage disequilibrium and recombination fraction are unbiased in this setting regardless of the true (simulated) value of those parameters.

Genome fragmentation phasing strategy in a cattle half-sib family

A method named genome fragmentation phasing strategy was developed for phasing parents and progeny and used for phasing a cattle family comprising 36 calves to illustrate GFPS. The first step was to estimate recombination fraction between each of two consecutive SNPs. The genome-wide distribution of recombination events between each of two consecutive SNPs is depicted in Figure 3. Although the great majority (over 92%) of the estimates were 0.00, there were many estimates of recombination fraction too high for the physical distance separating them. It may be attributed to either miss location of SNPs during the sequencing and alignment or genotyping errors. A recombination fraction larger than 0 was used (as a signal) to terminate a fragment and to initiate the next one. The fragmentation yielded a distribution of fragment size across the genome (number of SNPs per fragment) shown in Figure 4. Most of the fragments were rather small but some relatively large fragments (more than 200 SNPs) allowed genome-wide identification of cold spots (Figure 5). These cold spots are for the sire producing meiosis and if tested in multiple families would allow distinguishing between cold spots at the individual and population levels.

The number of recombination events per chromosome per individual (gamete) is shown in Figure 6. There were some calves (189 and 284) with only single recombination event per chromosome. The distribution of all recombination events between fragments generated by GFPS along the 29 autosomes is given in Figure 7. There were 613 recombination events detected. In some situations, recombinations are evenly distributed in the genome. However, some areas have higher values suggesting either hot spots or miss location of whole fragments during the process of sequencing and assembling to produce Illumina's array. For example, recombination fraction between the first fragment in chromosome 17 (unlinked with nearby fragments) was genome-wide tested (all fragments in the entire genome) and resulted in a recombination fraction estimate of 0.07 with another fragment in chromosome 19 which suggests an error of assembling in the Illumina array.

The informativity of SNPs greatly increased after haplotype reconstruction using GFPS. More than 90%

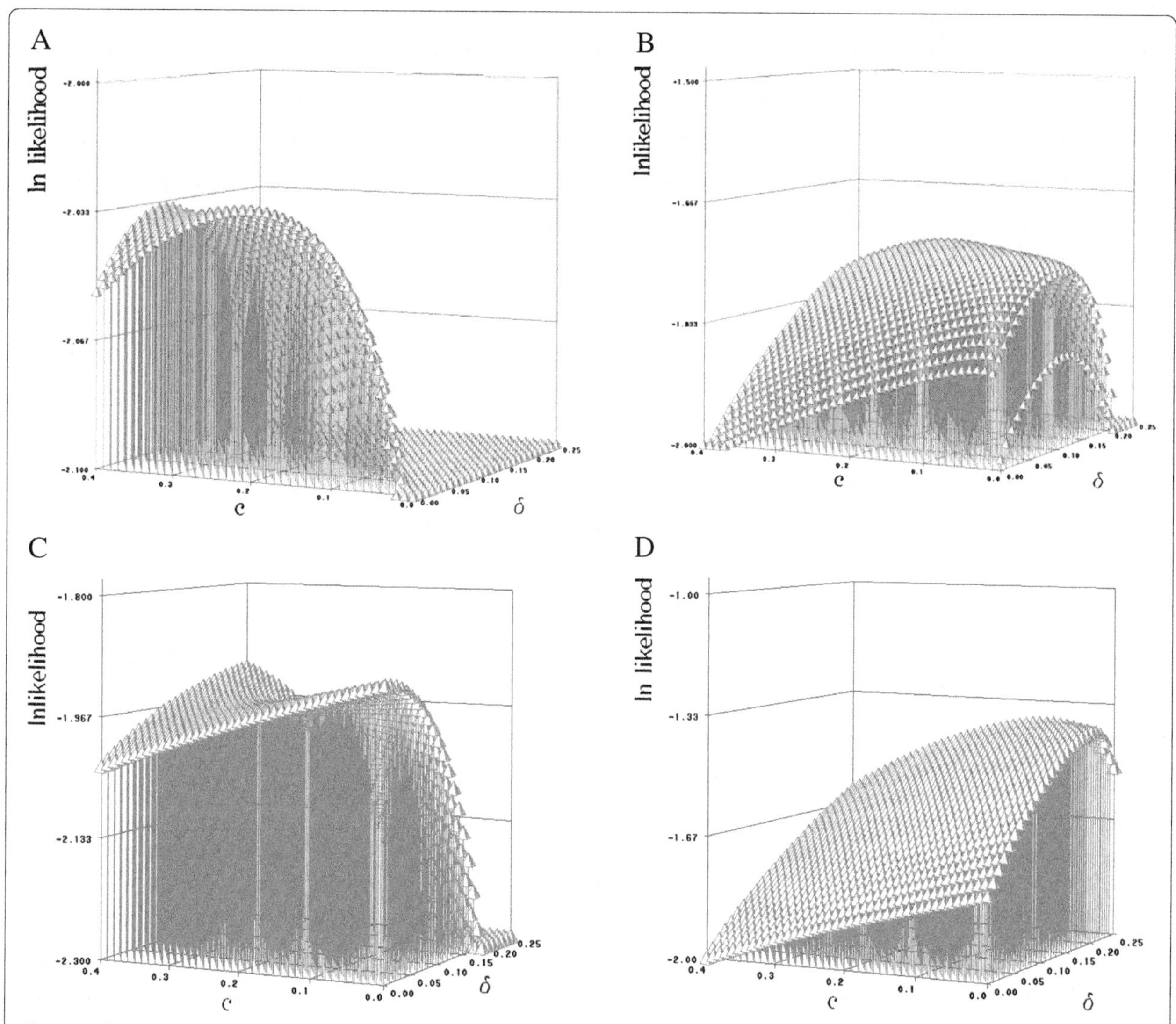

Figure 2 Three-dimensional plot of ln of the likelihood (equation A1) along recombination fraction (c) and linkage disequilibrium (δ) when testing a family of infinite size for different situations regarding c and δ. The allele frequencies were intermediate at the two loci. **A)** Situations simulated is c=0.20, δ=0. The maximum value was c=0.20, δ=0. **B)** Situations simulated is c=0.20, δ=0.20. The maximum value was at c=0.10, δ=0.15. **C)** Situations simulated is c=0.00, δ=0.00. The maximum value was c=0.00, δ=0.00. **D)** Situations simulated is c=0.00, δ=0.20. The maximum value was c=0.00, δ=0.20.

(range 86.1 to 95.1% for the 29 chromosomes) of non-informative SNPs become informative due to the use of the information of SNPs linked to them.

The analysis of genome regions of identity shared by two individuals is given in Figure 8. The analyses compare regions of identity (ROIs) inherited from sire or from dam for each of two individuals in a half-sib family. As expected for half-sibs, the regions shared or identical from paternal origin between two individuals is on average 0.52, with a range of values between 0.40 and 0.68. The maternally inherited ROIs between individuals ranged between 0.05 and 0.29 with a mean of 0.11. These results suggest that there is a significant amount of fragments with a relative large variation that are identical by descent among unrelated dams. Figure 9 shows the tail of the distributions of paternal and maternal ROIs. Fragments of more than 20 Mb of maternal autosomes are commonly shared by individuals with unrelated mothers.

The results of the simulation of BTA1 with the same LD and genetic linkage structure of our data are depicted in Table 5. The results of correct phasing are for informative segments. That is, homozygote calves allowed identification of origin of haplotype (paternal or maternal). Non informative areas are detectable and cannot be phased out. The phasing method was very accurate to identify haplotypes in both sires and calves. Recombination events were considered identified if they were less that 3Mb apart from the true (simulated)

Table 3 Average estimates of recombination fraction assuming linkage equilibrium [19] in a double heterozygote sire with varying half-sib family size and varying recombination fraction (*c*) and linkage disequilibrium (*δ*)

		Family size					
Simulated *c*	Simulated *δ*	36	100	300	500	1,000	2,000
0.00	−0.10	0.0634 (0.1460)	0.0053 (0.0325)	0.0000 (0.0009)	0.0000 (0.0000)	0.0000 (0.0000)	0.0000 (0.0000)
	0.00	0.0123 (0.0566)	0.0002 (0.0057)	0.0000 (0.0000)	0.0000 (0.0000)	0.0000 (0.0000)	0.0000 (0.0000)
	0.05	0.0044 (0.0286)	0.0001 (0.0026)	0.0000 (0.0000)	0.0000 (0.0000)	0.0000 (0.0000)	0.0000 (0.0000)
	0.10	0.0012 (0.0124)	0.0000 (0.0003)	0.0000 (0.0000)	0.0000 (0.0000)	0.0000 (0.0000)	0.0000 (0.0000)
	0.20	0.0000 (0.0012)	0.0000 (0.0000)	0.0000 (0.0000)	0.0000 (0.0000)	0.0000 (0.0000)	0.0000 (0.0000)
0.10	−0.10	0.2620 (0.1938)	0.2259 (0.1181)	0.2108 (0.0580)	0.2085 (0.0444)	0.2067 (0.0313)	0.2066 (0.0220)
	0.00	0.1250 (0.1360)	0.1042 (0.0636)	0.1017 (0.0350)	0.1009 (0.0190)	0.1002 (0.0190)	0.1002 (0.0135)
	0.05	0.0814 (0.1006)	0.0710 (0.0492)	0.0698 (0.0274)	0.0694 (0.0212)	0.0691 (0.0149)	0.0692 (0.0106)
	0.10	0.0510 (0.0721)	0.0463 (0.0373)	0.0457 (0.0210)	0.0457 (0.0163)	0.0455 (0.0116)	0.0456 (0.0082)
	0.20	0.0123 (0.0293)	0.0119 (0.0171)	0.0126 (0.0097)	0.0124 (0.0077)	0.0124 (0.0053)	0.0124 (0.0040)
0.20	−0.10	0.3788 (0.1609)	0.3889 (0.1139)	0.3793 (0.0731)	0.3727 (0.0541)	0.3692 (0.0352)	0.3689 (0.0245)
	0.00	0.2343 (0.0764)	0.2068 (0.0876)	0.2024 (0.0470)	0.2010 (0.0359)	0.2003 (0.0254)	0.2004 (0.0181)
	0.05	0.1652 (0.1410)	0.1459 (0.0691)	0.1442 (0.0387)	0.1433 (0.0296)	0.1429 (0.0210)	0.1431 (0.0150)
	0.10	0.1078 (0.1086)	0.0981 (0.0546)	0.0973 (0.0308)	0.0970 (0.0238)	0.0967 (0.0168)	0.0968 (0.0112)
	0.20	0.0285 (0.0486)	0.0270 (0.0270)	0.0272 (0.0152)	0.0268 (0.0123)	0.0270 (0.0085)	0.0271 (0.0061)

The simulated allele frequencies were $f_T = 0.5$ and $f_M = 0.5$.The number of replicates was 10^4. The values between parentheses are standard deviations among replicates.

event. The average distance between the true (simulated) and identified recombination events was 0.79 Mb.

Discussion

This paper describes a new method, GFPS, for reconstructing haplotypes after the use of SNP arrays in breeding populations in which large groups of progeny from at least one progenitor are available. The method permits reconstruction of long haplotypes utilizing consecutive SNPs within a fragment in a similar fashion to the process of sequencing and assembling small fragments of DNA that are aligned using common sequences in the extremes in order to generate larger fragments. In turn, linkage analyses of long fragments

Table 4 Average of the estimates of linkage disequilibrium (*δ*) and recombination fraction (*c*) using the E.M. algorithm for joint estimation in multiple half-sib families

	Simulated *c*					
Simulated *δ*	0		0.25		0.50	
	δ	*c*	*δ*	*c*	*δ*	*c*
0.000	0.0002 (0.0165)	0.0001 (0.0042)	0.0002 (0.0165)	0.2563 (0.0816)	0.0002 (0.0165)	0.5017 (0.0937)
0.010	0.0102 (0.0165)	0.0002 (0.0052)	0.0102 (0.0165)	0.2560 (0.0827)	0.0102 (0.0165)	0.5013 (0.0935)
0.020	0.0202 (0.0164)	0.0003 (0.0059)	0.0202 (0.164)	0.2557 (0.0835)	0.0202 (0.0164)	0.5002 (0.0930)
0.030	0.0301 (0.0164)	0.0007 (0.0093)	0.0301 (0.0165)	0.2554 (0.0841)	0.0301 (0.1646)	0.5000 (0.0922)
0.040	0.0400 (0.0163)	0.0011 (0.0112)	0.0400 (0.0163)	0.2550 (0.0840)	0.0400 (0.0163)	0.4993 (0.0905)
0.050	0.0499 (0.0163)	0.0016 (0.0133)	0.4999 (0.0163)	0.2547 (0.0827)	0.0499 (0.0163)	0.4991 (0.0887)
0.075	0.0748 (0.0160)	0.0043 (0.0201)	0.0748 (0.0160)	0.2539 (0.0787)	0.0748 (0.0160)	0.4990 (0.0812)
0.100	0.0998 (0.0155)	0.0079 (0.0256)	0.0997 (0.0154)	0.2529 (0.0730)	0.0997 (0.0155)	0.4989 (0.0732)
0.125	0.01245 (0.0146)	0.0107 (0.0267)	0.1245 (0.0146)	0.2507 (0.0658)	0.1245 (0.0146)	0.4985 (0.0646)
0.150	0.1494 (0.0136)	0.0135 (0.0269)	0.1494 (0.0136)	0.2500 (0.0585)	0.1494 (0.0136)	0.4991 (0.0577)
0.175	0.1743 (0.0122)	0.0145 (0.0249)	0.1743 (0.0122)	0.2500 (0.0508)	0.1743 (0.0122)	0.4994 (0.0507)
0.200	0.1991 (0.0102)	0.0133 (0.0203)	0.1991 (0.0102)	0.2497 (0.0430)	0.1991 (0.0102)	0.4996 (0.0445)
0.250	0.2489 (0.0020)	0.0000 (0.0000)	0.2488 (0.0020)	0.2498 (0.0289)	0.2489 (0.0020)	0.5001 (0.0339)

The simulated allele frequencies were $f_T = 0.5$ and $f_M = 0.5$. The simulation was carried out for varying recombination fractions, linkage disequilibria and resembling the Norwegian cattle population structure. The number of replicates was 10^4. The values between parentheses are the average of standard deviations of the estimates of *δ* and *c*.

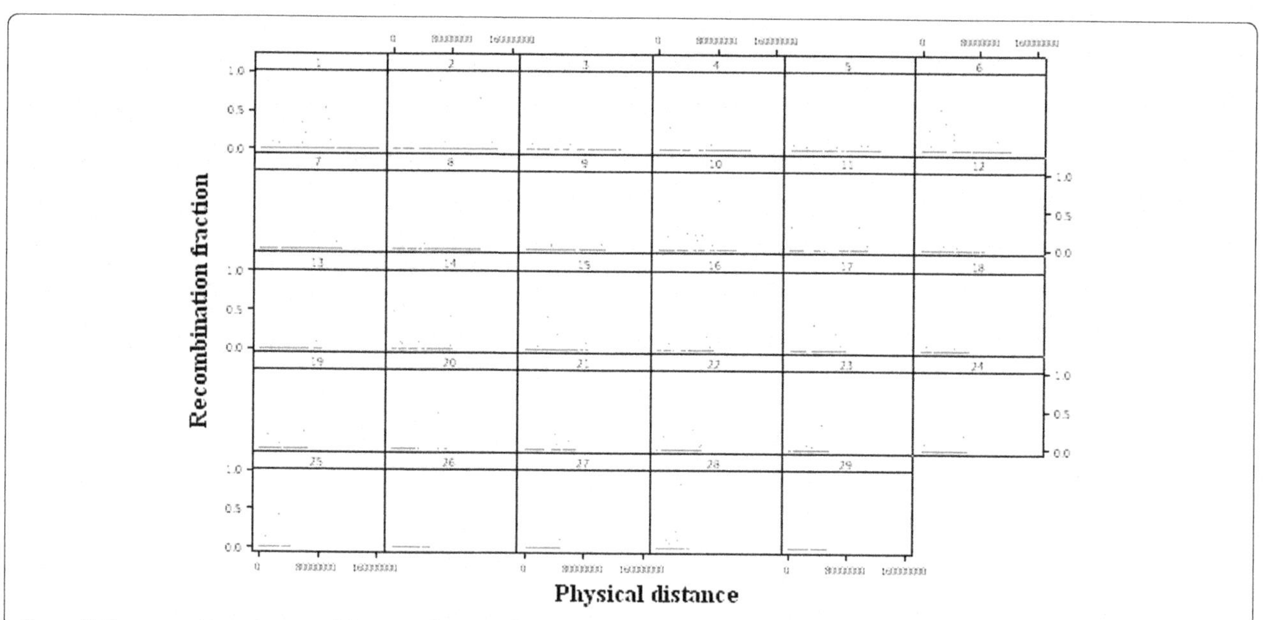

Figure 3 Genome-wide estimates of the recombination fraction between consecutive SNP during fragmentation along physical distance for the 29 autosomal chromosomes.

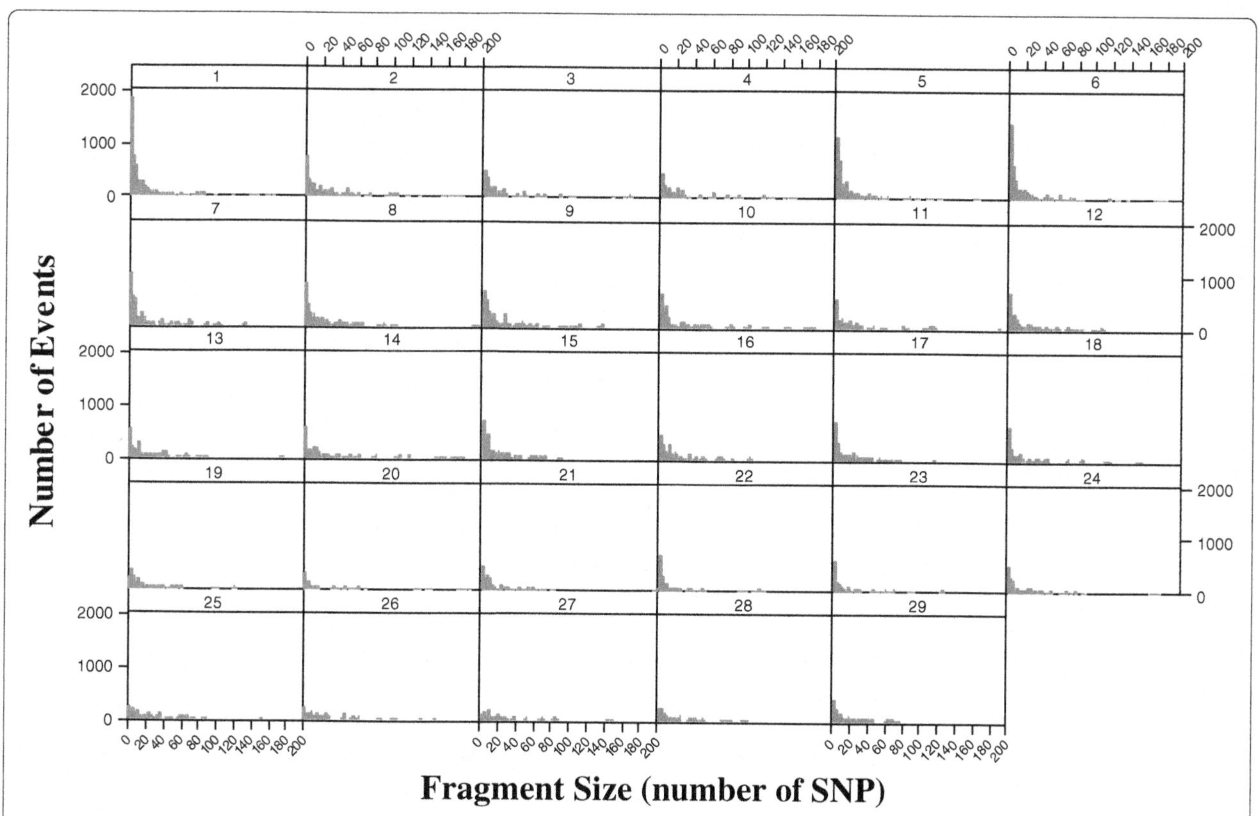

Figure 4 Histograms of the number of fragments for the 29 autosomal chromosomes according to fragment's size (number of SNPs).

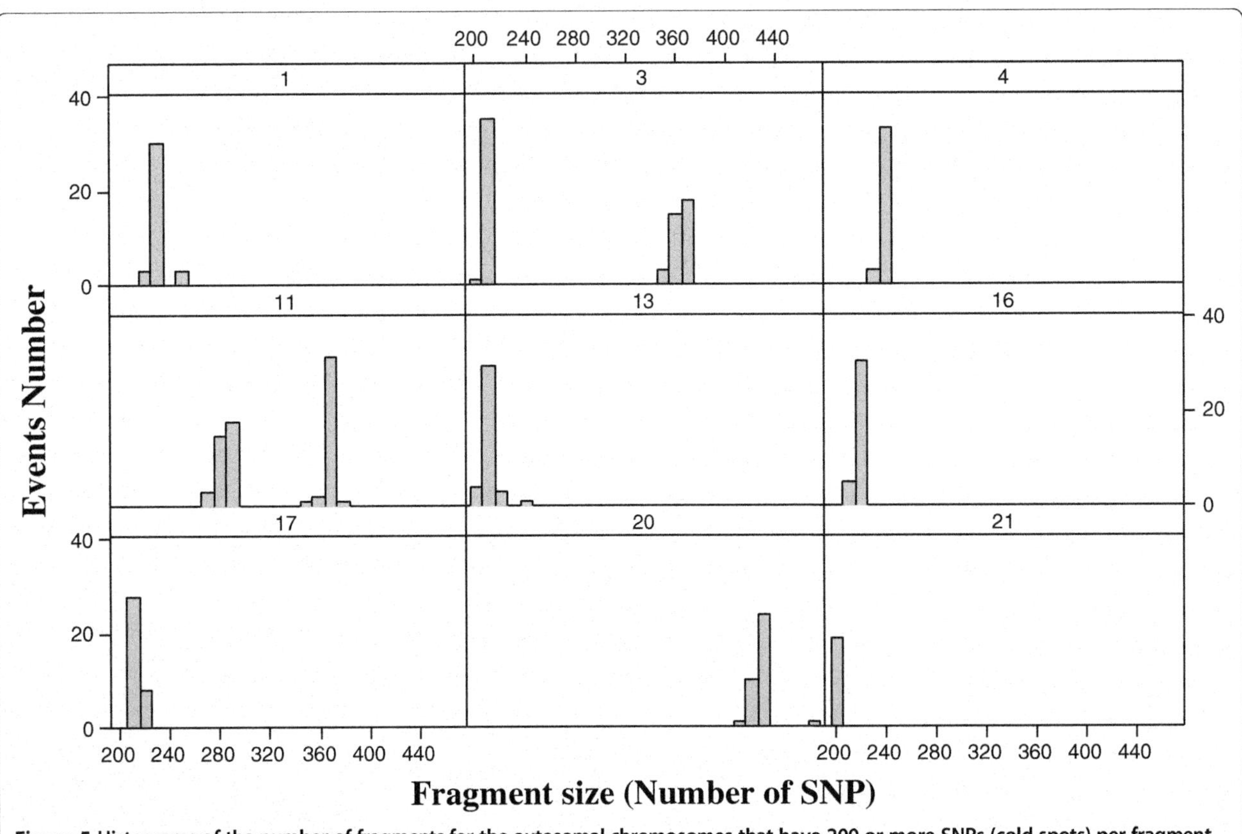

Figure 5 Histograms of the number of fragments for the autosomal chromosomes that have 200 or more SNPs (cold spots) per fragment for autosomes 1,3,4,11,13,16,17,20, and 21.

can identify recombination events and pinpoint genotyping or assembling errors. The application of this method provides a new insight into the genome of highly used sires or dams by allowing identification of individual hot and cold spots or identifying chromosomes with non Mendelian segregation which might be caused by chromosomal abnormalities. The application of this method allows a high control of the haplotypes passing from generation to generation and can provide a better understanding of the genetic basis of production and diseases in breeding populations.

The alternative to the proposed GFPS is long-range phasing and imputation methods [21,22] which make use of surrogate fathers and mothers in order to identify haplotypes. In general, imputation methods utilize allele frequencies in order to reconstruct haplotypes by assuming that the larger the allele frequencies, the larger the chance of being represented in the haplotypes. Consequently, there is always an error associated to imputed haplotypes and rare alleles are missed. The advantage of the method presented here and described elsewhere is that information on sibs is used together with the laws of inheritance [5,10]. The novelty of our method is that it includes joint estimation of recombination and LD for reconstruction of parents and progeny haplotypes. As shown in this paper, if only one parent is available then both parameters can be accurately estimated if at least one of them is 0 as has been shown here with Monte Carlo computer simulations. Homozygous progeny from heterozygous sires flag which allele and haplotype is inhered from the sire, and consequently, allow identification of haplotypes of paternal and maternal origin. It dramatically increases informativity of heterozygous SNPs between parent and progeny (over 90%).

Many of the applications of SNP arrays in farm animals ignore the sequential nature of polymorphism in haplotypes. For example, in many instances genomic selection [1] assumes a large number of unlinked loci. As discussed in previous work it has implications in genomic evaluations that are systematically ignored [2,3,23]: a) progeny from the same parent share large haplotypes (not just single SNP) as shown with ROIs in this paper, and b) single heterozygous SNPs for both sire and half-sib progeny are non-informative, and consequently information is reduced or lost in genomic evaluations. Other advantages and potential applications of the proposed method are: 1) requires only currently available genotypic information on progeny as required for genomic selection when using juveniles for shortening generation intervals, 2) can be applied to any farm animal

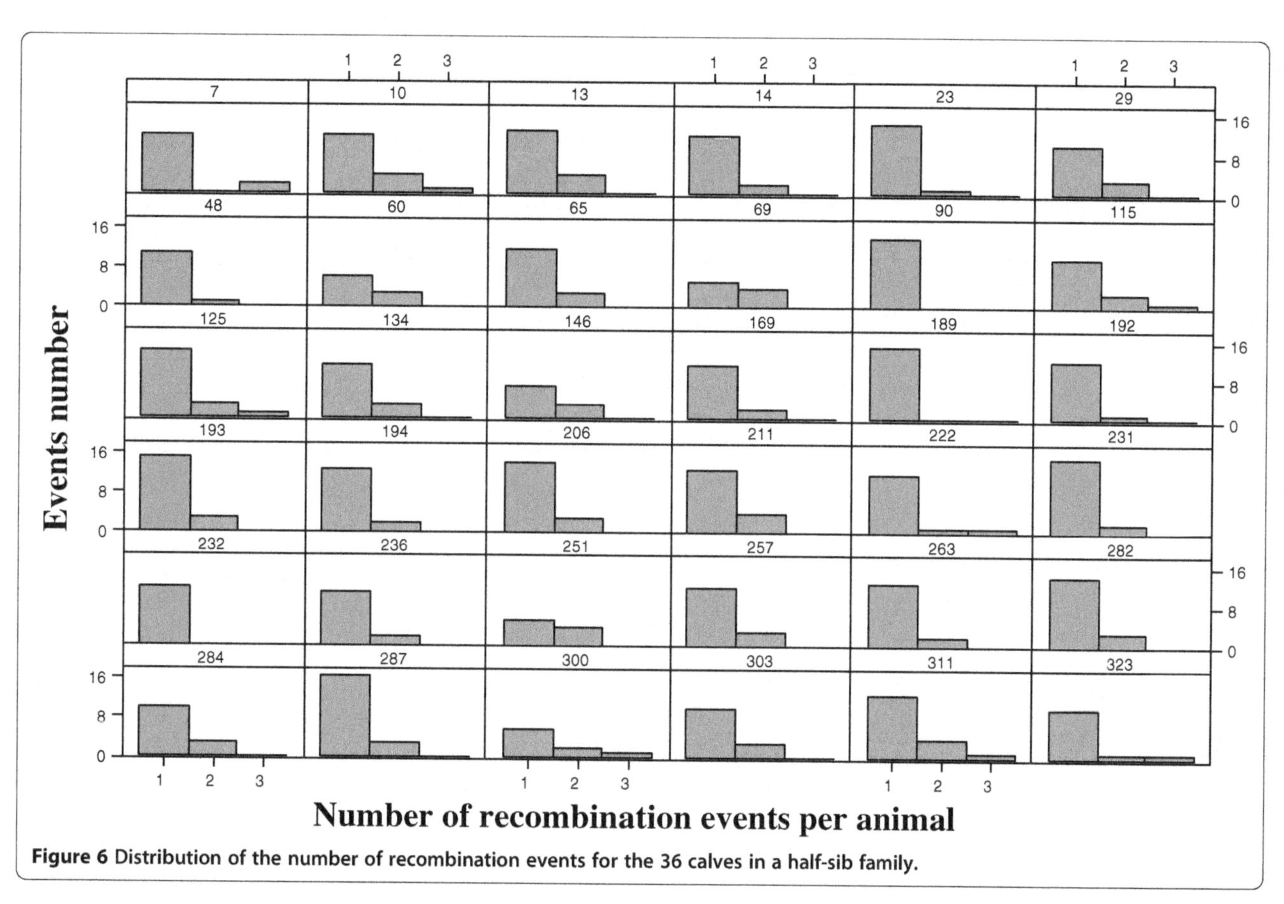

Figure 6 Distribution of the number of recombination events for the 36 calves in a half-sib family.

with large families even if only one parent is common to a group of progeny, 3) facilitates haplotyping the entire breeding population for new investigations such as signatures of selection or high order linkage disequilibrium, 4) can be used to estimate molecular relatedness more precisely by using haplotypes of given length rather than the sum of single SNPs (large fragments are likely identical by descent), and 5) can help to tracing up allele and haplotypes through generations which may facilitate the detection of genes involved in diseases or production.

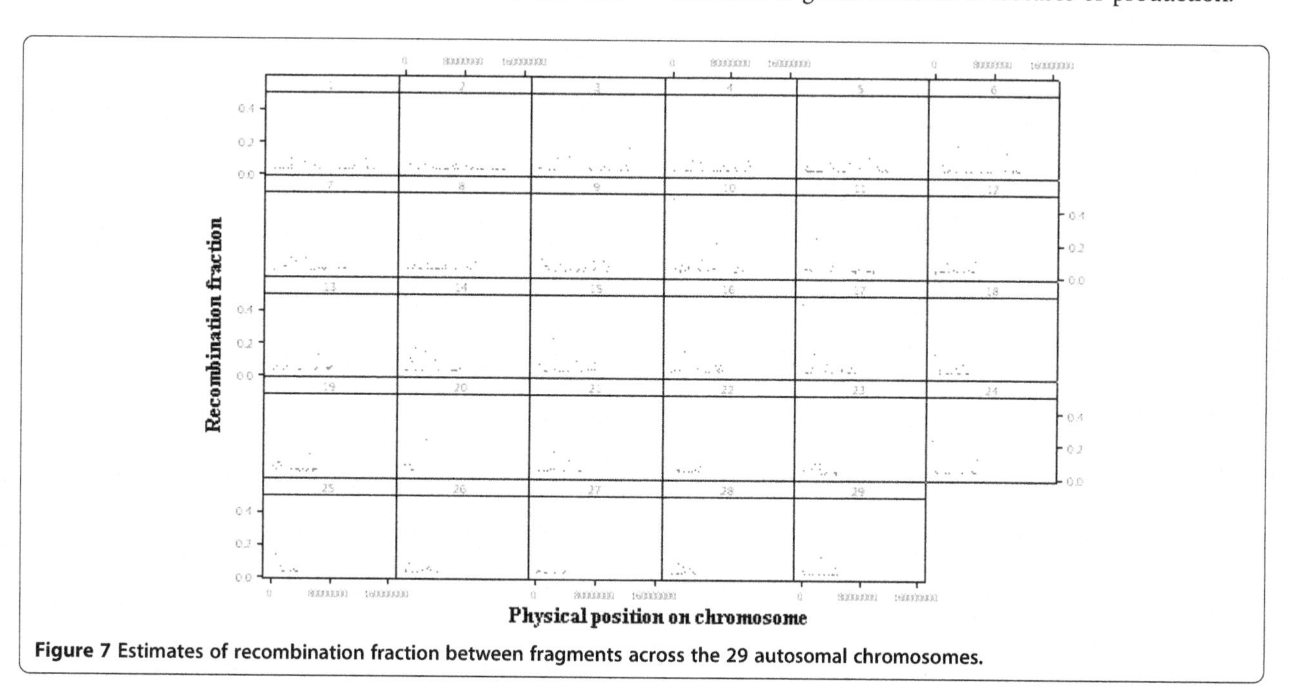

Figure 7 Estimates of recombination fraction between fragments across the 29 autosomal chromosomes.

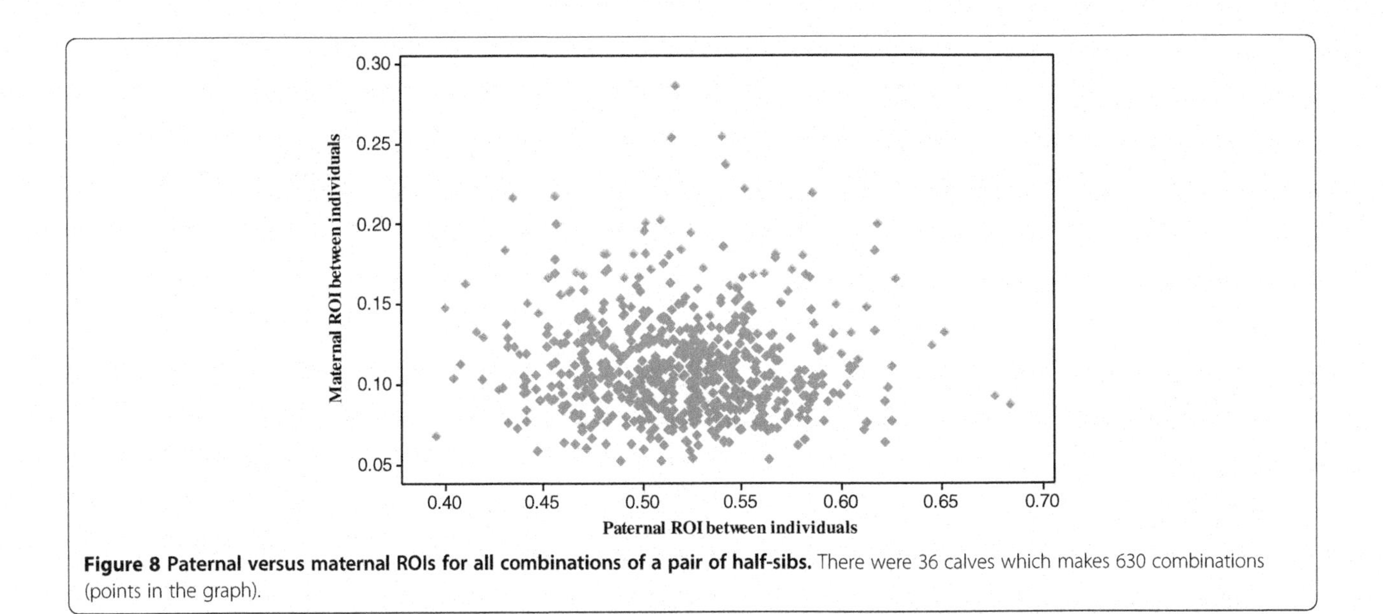

Figure 8 Paternal versus maternal ROIs for all combinations of a pair of half-sibs. There were 36 calves which makes 630 combinations (points in the graph).

The methods developed are designed for haplotyping farm animals with large families. However, they can also be applied to wild animals as long as large families and dense SNP arrays are available in those species. After all, the cattle used in our study came from a free range production system where the only samples or records taken were from DNA which was first used for paternity testing and then for phasing haplotypes.

Conclusions

Haplotype phasing is possible and highly accurate when estimating jointly linkage disequilibrium and genetic linkage in animal breeding populations as long as large families are available.

Appendix 1

Joint estimation of genetic linkage and linkage disequilibrium in half-sib families

Let the sire have genotype *TtMm* at two SNPs, *T/t*, and *M/m* and linkage phase (*TM/tm*) and $n_{j,i}$ be the genotype counts from offspring from the *i-th* sire family (*j=TTMM, TTMm, TTmm, TtMM, TtMm, Ttmm, ttMM, ttMm*, and *ttmm*). The recombination fraction, *c*, is estimated simultaneously with linkage disequilibrium (δ), and allele

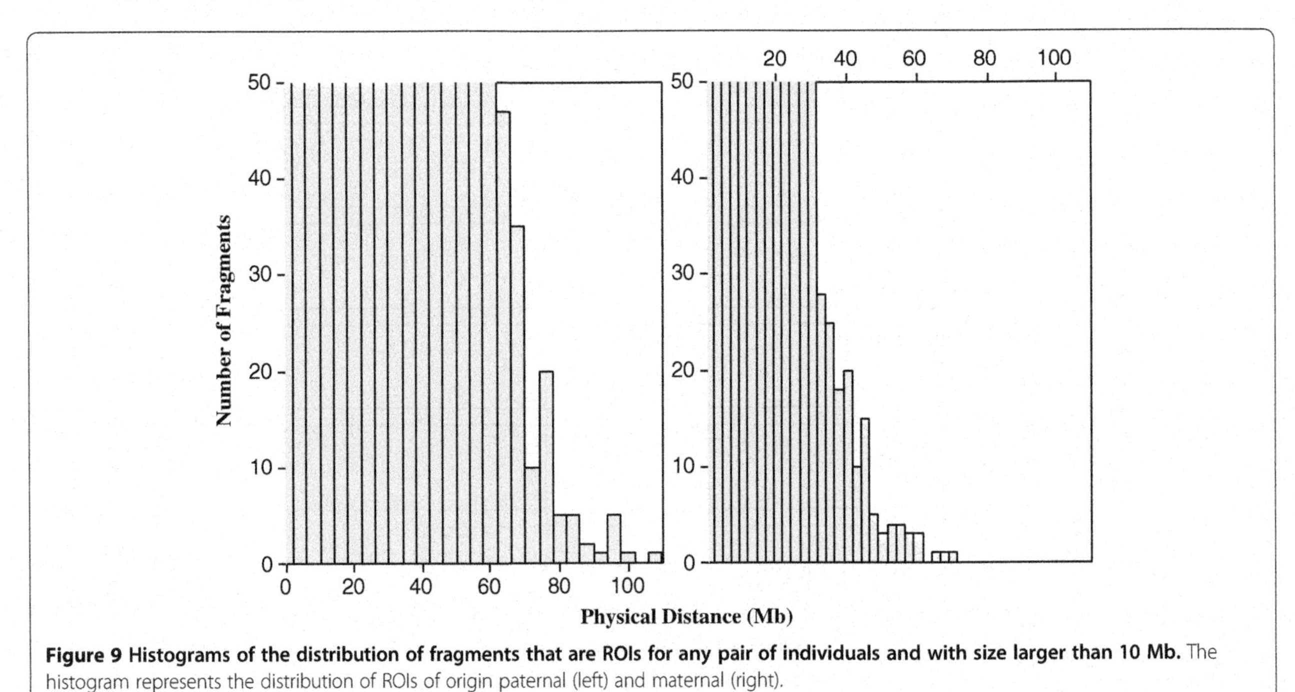

Figure 9 Histograms of the distribution of fragments that are ROIs for any pair of individuals and with size larger than 10 Mb. The histogram represents the distribution of ROIs of origin paternal (left) and maternal (right).

Table 5 Monte-Carlo simulation results of BTA1 with the same LD and genetic linkage observed in the real data

Average over replicates	
Total number of SNPs in BTA1	3,343
Total number of SNPs bull heterozygote	751
Percentage of correct phases in informative segments in sire	99.98%
Percentage of correct identification of recombination events	99.87%
Average distance between true and estimated position for recombination events	0.793 Mb (0.002)
Percentage of correct phases in informative segments in calves	99.94%

Informative SNPs are those that can be traced up to sire or dam.

frequencies (f_T, f_M). The likelihood equation for linkage phase TM/tm for data of the *i-th* half-sib family is:

$$L_{TM/tm,i}(\delta,f_T,f_M,c|\ nG) = K(\phi_{TTMM})^{n_{TTMM,i}}(\phi_{TTMm})^{n_{TTMm,i}} (\phi_{TTmm})^{n_{TTmm,i}}(\phi_{TtMM})^{n_{TtMM,i}}(\phi_{TtMm})^{n_{TtMm,i}} (\phi_{Ttmm})^{n_{Ttmm,i}}(\phi_{ttMM})^{n_{ttMM,i}}(\phi_{ttMm})^{n_{ttMm,i}}(\phi_{ttmm})^{n_{ttmm,i}} \quad \text{(A1)}$$

where the probabilities of offspring genotypes among half-sib offspring are obtained from Table 6: $\phi_{TTMM} = {}^1/_2(1{-}c)f_{TM}$; $\phi_{TTMm} = {}^1/_2(1{-}c)f_{Tm} + {}^1/_2c\ f_{TM}$; $\phi_{TTmm} = {}^1/_2c\ f_{Tm}$; $\phi_{TtMM} = {}^1/_2(1{-}c)f_{tM} + {}^1/_2c\ f_{TM}$; $\phi_{TtMm} = {}^1/_2(1{-}c)(f_{tm} + f_{TM}) + {}^1/_2c\ (f_{tM} + f_{Tm})$; $\phi_{Ttmm} = {}^1/_2(1{-}c)f_{Tm} + {}^1/_2c\ f_{tm}$; $\phi_{ttMM} = {}^1/_2c\ f_{tM}$ $\phi_{ttMm} = {}^1/_2(1{-}c)f_{tM} + {}^1/_2c\ f_{tm}$; and $\phi_{ttmm} = {}^1/_2(1{-}c)f_{tm}$.

Likelihood equation [A1] can be solved by applying the E.M algorithm:

$$\hat{f}^i_{TM} = \frac{1}{N_i}\left(n_{TTMM,i} + \frac{c\hat{f}^i_{TM}n_{TTMm,i}}{c\hat{f}^i_{TM} + (1{-}c)\hat{f}^i_{Tm}} + \frac{c\hat{f}^i_{TM}n_{TtMM,i}}{c\hat{f}^i_{TM} + (1{-}c)\hat{f}^i_{tM}} + \frac{(1{-}c)\hat{f}^i_{TM}n_{TtMm,i}}{\left[c\left(\hat{f}^i_{Tm} + \hat{f}^i_{tM}\right) + (1{-}c)\left(\hat{f}^i_{TM} + \hat{f}^i_{tm}\right)\right]}\right)$$

$$\hat{f}^i_{Tm} = \frac{1}{N_i}\left(n_{TTmm,i} + \frac{(1{-}c)\hat{f}^i_{Tm}n_{TTMm,i}}{c\hat{f}^i_{TM} + (1{-}c)\hat{f}^i_{Tm}} + \frac{(1{-}c)\hat{f}^i_{Tm}n_{Ttmm,i}}{c\hat{f}^i_{tm} + (1{-}c)\hat{f}^i_{Tm}} + \frac{c\hat{f}^i_{Tm}n_{TtMm,i}}{\left[c\left(\hat{f}^i_{Tm} + \hat{f}^i_{tM}\right) + (1{-}c)\left(\hat{f}^i_{TM} + \hat{f}^i_{tm}\right)\right]}\right)$$

$$\hat{f}^i_{tM} = \frac{1}{N_i}\left(n_{ttMM,i} + \frac{(1{-}c)\hat{f}^i_{tM}n_{TtMM,i}}{c\hat{f}^i_{TM} + (1{-}c)\hat{f}^i_{tM}} + \frac{(1{-}c)\hat{f}^i_{tM}n_{ttMm,i}}{c\hat{f}^i_{tm} + (1{-}c)\hat{f}^i_{tM}} + \frac{c\hat{f}^i_{tM}n_{TtMm,i}}{\left[c\left(\hat{f}^i_{Tm} + \hat{f}^i_{tM}\right) + (1{-}c)\left(\hat{f}^i_{TM} + c\hat{f}^i_{tm}\right)\right]}\right)$$

$$\hat{f}^i_{tm} = \frac{1}{N_i}\left(n_{ttmm,i} + \frac{c\hat{f}^i_{tm}n_{Ttmm,i}}{c\hat{f}^i_{tm} + (1{-}c)\hat{f}^i_{Tm}} + \frac{c\hat{f}^i_{tm}n_{ttMm,i}}{c\hat{f}^i_{tm} + (1{-}c)\hat{f}_{tM}} + \frac{(1{-}c)\hat{f}^i_{tm}n_{TtMm,i}}{\left[c\left(\hat{f}^i_{Tm} + \hat{f}^i_{tM}\right) + (1{-}c)\left(\hat{f}^i_{TM} + \hat{f}^i_{tm}\right)\right]}\right)$$

$$\hat{c}^i = \frac{1}{N_i}\left(n_{TTmm,i} + n_{ttMM,i} + \frac{c\hat{f}^i_{TM}n_{TTMm,i}}{(1{-}c)\hat{f}^i_{TM} + c\hat{f}^i_{Tm}} + \frac{c\hat{f}^i_{TM}n_{TtMM,i}}{c\hat{f}^i_{TM} + (1{-}c)\hat{f}^i_{tM}} + \frac{c\left(\hat{f}^i_{Tm} + \hat{f}^i_{tM}\right)n_{TtMm,i}}{\left[c\left(\hat{f}^i_{Tm} + \hat{f}^i_{tM}\right) + (1{-}c)\left(\hat{f}^i_{TM} + \hat{f}^i_{tm}\right)\right]} + \frac{c\hat{f}^i_{tm}n_{Ttmm,i}}{c\hat{f}^i_{tm} + (1{-}c)\hat{f}^i_{Tm}} + \frac{c\hat{f}^i_{tm}n_{ttMm,i}}{c\hat{f}^i_{tm} + (1{-}c)\hat{f}^i_{tM}}\right) \quad \text{(A2)}$$

where N_i is the size of the *i-th* half-sib family. Using initial values of the haplotype frequencies and iterating over equation A2 will converge to ML estimates of

Table 6 Genotypes and their frequencies among half-sib progeny from a double heterozygote sire

	Sire (phase TM/tm)			
Dam	**TM**	**Tm**	**tM**	**tm**
G freq	½ (1-c)	½ c	½ c	½ (1-c)
TM f_{TM}	TTMM ½ (1-c) f_{TM}	TTMm ½ c f_{TM}	TtMM ½ c f_{TM}	TtMm ½ (1-c) f_{TM}
Tm f_{Tm}	TTMm ½ (1-c) f_{Tm}	TTmm ½ c f_{Tm}	TtMm ½ c f_{Tm}	Ttmm ½ (1-c) f_{Tm}
tM f_{tM}	TtMM ½ (1-c) f_{tM}	TtMm ½ c f_{tM}	ttMM ½ c f_{tM}	ttMm ½ (1-c) f_{tM}
tm f_{tm}	TtMm ½ (1-c) f_{tm}	Ttmm ½ c f_{tm}	ttMm ½ c f_{tm}	ttmm ½ (1-c) f_{tm}
	Sire (phase Tm/tM)			
Dam	**TM**	**Tm**	**tM**	**tm**
G freq	½ c	½ (1-c)	½ (1-c)	½ c
TM f_{TM}	TTMM ½ c f_{TM}	TTMm ½ (1-c)f_{TM}	TtMM ½ (1-c)f_{TM}	TtMm ½ c f_{TM}
Tm f_{Tm}	TTMm ½ c f_{Tm}	TTmm ½ (1-c)f_{Tm}	TtMm ½ (1-c)f_{Tm}	Ttmm ½ c f_{Tm}
tM f_{tM}	TtMM ½ c f_{tM}	TtMm ½ (1-c)f_{tM}	ttMM ½ (1-c)f_{tM}	ttMm ½ c f_{tM}
tm f_{tm}	TtMm ½ c f_{tm}	Ttmm ½ (1-c)f_{tm}	ttMm ½ (1-c)f_{tm}	ttmm ½ c f_{tm}

G gametes, *freq* frequency.

haplotype frequencies. Linkage disequilibrium is estimated by $\hat{\delta} = \hat{f}^i_{TM}\hat{f}^i_{tm} - \hat{f}^i_{Tm}\hat{f}^i_{tM}$.

If the linkage phase of the sire is Tm/tM then the E.M. equations are:

$$\hat{f}^i_{TM} = \frac{1}{N_i}\Bigg(n_{TTMM,i} + \frac{(1-c)\hat{f}^i_{TM} n_{TTMm,i}}{(1-c)\hat{f}^i_{TM} + c\hat{f}^i_{Tm}} + \frac{(1-c)\hat{f}^i_{TM} n_{TtMM,i}}{(1-c)\hat{f}^i_{TM} + c\hat{f}^i_{tM}} + \frac{c\hat{f}^i_{TM} n_{TtMm,i}}{\left[(1-c)\left(\hat{f}^i_{Tm} + \hat{f}^i_{tM}\right) + c\left(\hat{f}^i_{TM} + \hat{f}^i_{tm}\right)\right]}\Bigg)$$

$$\hat{f}^i_{Tm} = \frac{1}{N_i}\Bigg(n_{TTmm,i} + \frac{c\hat{f}^i_{Tm} n_{TTMm,i}}{(1-c)\hat{f}^i_{TM} + c\hat{f}^i_{Tm}} + \frac{c\hat{f}^i_{Tm} n_{Ttmm,i}}{(1-c)\hat{f}^i_{tm} + c\hat{f}^i_{Tm}} + \frac{(1-c)\hat{f}^i_{Tm} n_{TtMm,i}}{\left[(1-c)\left(\hat{f}^i_{Tm} + \hat{f}^i_{tM}\right) + c\left(\hat{f}^i_{TM} + \hat{f}^i_{tm}\right)\right]}\Bigg)$$

$$\hat{f}^i_{tM} = \frac{1}{N_i}\Bigg(n_{ttMM,i} + \frac{c\hat{f}^i_{tM} n_{TtMM,i}}{(1-c)\hat{f}^i_{TM} + c\hat{f}^i_{tM}} + \frac{c\hat{f}^i_{tM} n_{ttMm,i}}{(1-c)\hat{f}^i_{tm} + c\hat{f}^i_{tM}} + \frac{(1-c)\hat{f}^i_{tM} n_{TtMm,i}}{\left[(1-c)\left(\hat{f}^i_{Tm} + \hat{f}^i_{tM}\right) + c\left(\hat{f}^i_{TM} + \hat{f}^i_{tm}\right)\right]}\Bigg)$$

$$\hat{f}^i_{tm} = \frac{1}{N_i}\Bigg(n_{ttmm,i} + \frac{(1-c)\hat{f}^i_{tm} n_{Ttmm,i}}{(1-c)\hat{f}^i_{tm} + c\hat{f}^i_{Tm}} + \frac{(1-c)\hat{f}^i_{tm} n_{ttMm,i}}{(1-c)\hat{f}^i_{tm} + c\hat{f}^i_{tM}} + \frac{c\hat{f}^i_{tm} n_{TtMm,i}}{\left[(1-c)\left(\hat{f}^i_{Tm} + \hat{f}^i_{tM}\right) + c\left(\hat{f}^i_{TM} + \hat{f}^i_{tm}\right)\right]}\Bigg)$$

$$\hat{c}^i = \frac{1}{N_i}\Bigg(n_{TTMM,i} + n_{ttmm,i} + \frac{c\hat{f}^i_{Tm} n_{TTMm,i}}{(1-c)\hat{f}^i_{TM} + c\hat{f}^i_{Tm}} + \frac{c\hat{f}^i_{tM} n_{TtMM,i}}{c\hat{f}^i_{tM} + (1-c)\hat{f}^i_{TM}} + \frac{c\left(\hat{f}^i_{tm} + \hat{f}^i_{TM}\right) n_{TtMm,i}}{\left[c\left(\hat{f}^i_{tm} + \hat{f}^i_{TM}\right) + (1-c)\left(\hat{f}^i_{tM} + \hat{f}^i_{Tm}\right)\right]} + \frac{c\hat{f}^i_{Tm} n_{Ttmm,i}}{c\hat{f}^i_{Tm} + (1-c)\hat{f}^i_{tm}} + \frac{c\hat{f}^i_{tM} n_{ttMm,i}}{c\hat{f}^i_{tM} + (1-c)\hat{f}^i_{tm}}\Bigg)$$

The probability of phase TM/tm is:

$$\Pr\ of\ Phase\ (TM/tm) = \frac{L_{TM/tm}(\delta, f_T, f_M, c|\text{data})}{L_{TM/tm}(\delta, f_T, f_M, c|\text{data}) + L_{Tm/tM}(\delta, f_T, f_M, c|\text{data})} \tag{A3}$$

where $L_{TM/tm}(\delta, f_T, f_M, c|\text{data})$ and $L_{Tm/tM}(\delta, f_T, f_M, c|\text{data})$ are the likelihoods of phase (*TM/tm*) and phase (*Tm/tM*) under a model estimating linkage disequilibrium (δ), allele frequencies for one of the alleles at markers *T/t* and *M/m* (f_T, and f_M), and the recombination fraction (c).

Estimation of LD across multiple half-sib families

The likelihood equation to estimate LD across half-sib families is:

$$L(\delta, f_T, f_M, c|\ nG) = \prod_{i=1}^{nf} L_i(\delta, f_T, f_M, c|\ nG)$$

where $L_i(\delta, f_T, f_M, c\ |nG)$ is the likelihood for the *i-th* half-sib family conditional to genotype marker information (nG) and nf is the number of families. Note that depending on the sire genotype, allele frequencies for T and M (double homozygote) or M (homo-heterozygote) do not need to be estimated. The E. M. algorithm can be applied to multiple families by iterating on the four haplotype frequencies and recombination fraction:

$$\hat{f}_{TM} = \frac{\sum_{i=1}^{nf}\left(N_i\hat{f}^i_{TM}\right)}{\sum_{i=1}^{nf} N_i}, \quad \hat{f}_{Tm} = \frac{\sum_{i=1}^{nf}\left(N_i\hat{f}^i_{Tm}\right)}{\sum_{i=1}^{nf} N_i}, \quad \hat{f}_{tM} = \frac{\sum_{i=1}^{nf}\left(N_i\hat{f}^i_{tM}\right)}{\sum_{i=1}^{nf} N_i}, \quad \hat{f}_{tm} = \frac{\sum_{i=1}^{nf}\left(N_i\hat{f}^i_{tm}\right)}{\sum_{i=1}^{nf} N_i} \tag{A4}$$

$$\hat{c} = \frac{\sum_{i=1}^{nf}\left(N_i\hat{c}^i\right)}{\sum_{i=1}^{nf} N_i}$$

where equations for haplotype frequencies for each single family varies depending on the sire genotype (see [15] for a full description of all possible situations). Equation [A4] was solved iteratively after giving a starting value to the haplotype frequencies and by estimating in each iteration $\hat{f}_T = \hat{f}^i_{Tm} + \hat{f}^i_{TM}$ and $\hat{f}_M = \hat{f}^i_{TM} + \hat{f}^i_{tM}$.

Abbreviations

SNP: Single Nucleotide Polymorphism; LD: Linkage disequilibrium; ROI: regions of identity; GFPS: Genome fragmentation phasing strategy.

Competing interests

The authors declare that they have no competing interests.

Authors' contributions

LGR contributed to the developed the idea and carried out the mathematical modeling for the joint estimation of LD and linkage disequilibrium. He drafted the first version of the manuscript. AMH participated in the design of the study, contributed with preparation of DNA samples and writing of several parts of the manuscript. DT carried out the tissue sampling, and contributed to the writing of the manuscript. WMR contributed to the development of the idea, helped to carry out tissue sampling and contributed to the writing of the manuscript. All authors read and approved the final manuscript.

Acknowledgements

WMR acknowledges support from a Marie Curie International Reintegration Grant from the European Union, project no. PIRG08-GA-2010-277031 "SelectionForWelfare". LGR and WMR acknowledge support from project AGL2012-39137.

Author details

[1]Departamento de Mejora Genética Animal, Instituto Nacional de Investigación y Tecnología Agraria y Alimentaria (INIA), Ctra. de La Coruña km 7, 28040, Madrid, Spain. [2]Interdisciplinary Program in Genetics, Texas A&M University, College Station, TX 77843, USA. [3]Department of Animal Biotechnology, University of Nevada, 1664 North Virginia Street, Reno, NV 89557, USA.

References

1. Meuwissen THE, Hayes BJ, Goddard ME: **Prediction of total genetic value using genome wide dense marker maps.** *Genetics* 2001, **157:**1819–1829.
2. Edriss V, Fernando RL, Su G, Lund MS, Guldbrandtsen B: **The effect of using genealogy-based haplotypes for genomic prediction.** *Genet Sel Evol* 2013, **45:**5. doi:10.1186/1297-9686-45-5.
3. Mulder HA, Calus MP, Veerkamp RF: **Prediction of haplotypes for ungenotyped animals and its effect on marker-assisted breeding value estimation.** *Genet Sel Evol* 2010, **42:**10. doi:10.1186/1297-9686-42-10.
4. Browning SR, Browning BL: **Haplotype phasing: existing methods and new developments.** *Nat Rev Genet* 2011, **12:**703–714.
5. Browning SR, Browning BL: **Rapid and accurate haplotype phasing and missing data inference for whole genome association studies using localized haplotype clustering.** *Am J Hum Genet* 2007, **81:**1084–1097.
6. Scheet P, Stephens M: **A fast and flexible statistical model for large-scale population genotype data: applications to inferring missing genotypes and haplotypic phase.** *Am J Hum Genet* 2006, **78:**629–644.
7. Stephens M, Scheet P: **Accounting for decay of linkage disequilibrium in haplotype inference and missing-data imputation.** *Am J Hum Genet* 2005, **76:**449–462.
8. Rohde K, Fuerst R: **Haplotyping and estimation of haplotype frequencies for closely linked biallelic multilocus genetic phenotypes including nuclear family information.** *Hum Mutat* 2001, **17:**289–295.
9. Kruglyak L, Daly MJ, Reeve-Daly MP, Lander ES: **Parametric and nonparametric linkage analysis: a unified multipoint approach.** *Am. J. H. Genet* 1996, **58:**1347–1363.
10. Abecasis GR, Cherny SS, Cookson WO, Cardon LR: **Merlin - rapid analysis of dense genetic maps using sparse gene flow trees.** *Nat Genet* 2002, **30:**97–101.
11. Lander ES, Green P: **Construction of multilocus genetic linkage maps in humans.** *Proc Natl Acad Sci USA* 1987, **84:**2363–2367.
12. Williams AL, Housman DE, Rinard MC, Gifford DK: **Rapid haplotype inference for nuclear families.** *Genome Biol* 2010, **11:**R108.
13. Lai E-Y, Wang W-B, Jiang T, Wu K-P: **A linear-time algorithm for reconstructing zero-recombinant haplotype configuration on a pedigree.** *BMC Bioinforma* 2012, **13**(Suppl 17)**:**S19.
14. Gomez-Raya L, Olsen HG, Klungland H, Våge DI, Olsaker I, Talle SB, Aasland M, Lien S: **The use of genetic markers to measure genomics response to selection in livestock.** *Genetics* 2002, **162:**1381–1388.
15. Gomez-Raya L: **Maximum likelihood estimation of linkage disequilibrium in half-sib families.** *Genetics* 2012, **191:**195–213.
16. Pemberton TJ, Absher D, Feldman MW, Myers RM, Rosenberg NA, Li JZ: **Genomic patterns of homozygosity in worldwide human populations.** *Am J Hum Genet* 2012, **91:**275–292. doi:10.1016/j.ajhg.2012.06.014.
17. Szpiech ZA, Xu J, Pemberton TJ, Peng W, Zöllner S, Rosenberg NA, Li JZ: **Long Runs of Homozygosity Are Enriched for Deleterious Variation.** *Am J Hum Genet* 2013. doi:10.1016/j.ajhg.2013.05.003.
18. Purfield DC, Berry DP, McParland S, Bradley DG: **Runs of homozygosity and population history in cattle.** *BMC Genet* 2012:. doi:10.1186/1471-2156-13-70.
19. Gomez-Raya L: **Biased estimation of the recombination fraction using half-sib families and informative offspring.** *Genetics* 2001, **157:**1357–1367.
20. Georges M, Nielsen D, Mackinnon M, Mishra A, Okimoto R, *et al*: **Mapping quantitative trait loci controlling milk production in dairy cattle by exploting progeny testing.** *Genetics* 1995, **139:**907–920.
21. Hickey JM, Kinghorn BP, Tier B, Wilson JF, Dunstan N, van der Werf JHJ: **A combined long-range phasing and long haplotype imputation method to impute phase for SNP genotypes.** *Genet Sel Evol* 2011, **43:**12.
22. Kong A, Masson G, Frigge ML, Gylfason A, Zusmanovich P, Thorleifsson G, Olason PI, Ingason A, Steinberg S, Rafnar T, Sulem P, Mouy M, Jonsson F, Thorsteinsdottir U, Gudbjartsson DF, Stefansson H, Stefansson K: **Detection of sharing by descent, long-range phasing and haplotype imputation.** *Nature Genet* 2008, **40:**1068–1075.
23. Meuwissen THE, Goddard ME: **Prediction of identity by descent probabilities from marker-haplotypes.** *Genet Sel Evol* 2001, **33:**605–634.

2

Identification and characterization of genes that control fat deposition in chickens

Hirwa Claire D'Andre[1,3*], Wallace Paul[2], Xu Shen[3], Xinzheng Jia[3], Rong Zhang[3], Liang Sun[3] and Xiquan Zhang[3]

Abstract

Background: Fat deposits in chickens contribute significantly to meat quality attributes such as juiciness, flavor, taste and other organoleptic properties. The quantity of fat deposited increases faster and earlier in the fast-growing chickens than in slow-growing chickens. In this study, Affymetrix Genechip® Chicken Genome Arrays 32773 transcripts were used to compare gene expression profiles in liver and hypothalamus tissues of fast-growing and slow-growing chicken at 8 wk of age. Real-time RT-PCR was used to validate the differential expression of genes selected from the microarray analysis. The mRNA expression of the genes was further examined in fat tissues. The association of single nucleotide polymorphisms of four lipid-related genes with fat traits was examined in a F_2 resource population.

Results: Four hundred genes in the liver tissues and 220 genes hypothalamus tissues, respectively, were identified to be differentially expressed in fast-growing chickens and slow-growing chickens. Expression levels of genes for lipid metabolism (*SULT1B1, ACSBG2, PNPLA3, LPL, AOAH*) carbohydrate metabolism (*MGAT4B, XYLB, GBE1, PGM1, HKDC1*)cholesttrol biosynthesis (*FDPS, LSS, HMGCR, NSDHL, DHCR24, IDI1, ME1*) *HSD17B7* and other reaction or processes (*CYP1A4, CYP1A1, AKR1B1, CYP4V2, DDO*) were higher in the fast-growing White Recessive Rock chickens than in the slow-growing Xinghua chickens. On the other hand, expression levels of genes associated with multicellular organism development, immune response, DNA integration, melanin biosynthetic process, muscle organ development and oxidation-reduction (*FRZB, DMD, FUT8, CYP2C45, DHRSX*, and *CYP2C18*) and with glycol-metabolism (*GCNT2, ELOVL 6*, and *FASN*), were higher in the XH chickens than in the fast-growing chickens. RT-PCR validated high expression levels of nine out of 12 genes in fat tissues. The G1257069A and T1247123C of the *ACSBG2* gene were significantly associated with abdominal fat weight. The G4928024A of the *FASN* gene were significantly associated with fat bandwidth, and abdominal fat percentage. The C4930169T of the *FASN* gene was associated with abdominal fat weight while the A59539099G of the *ELOVL 6* was significantly associated with subcutaneous fat. The A8378815G of the *DDT* was associated with fat band width.

Conclusion: The differences in fat deposition were reflected with differential gene expressions in fast and slow growing chickens.

Keywords: Chicken, Fat deposition, Genes

Background

Fat deposition is a crucial aspect in modern chicken breeding schemes because it is associated with selection for increased body weight in broilers [1-7]. The growth of broiler chicken is accompanied by an increased percentage of body fat with a concomitant increase in the mass of abdominal and visceral fat [8]. The quantity of fat deposited increases faster and earlier in fast-growing chickens than in slow-growing chickens [9-12]. Excessive adiposity is a problem in modern broiler industry [13]; and needs to be controlled to reduce negative effects on productivity, acceptability, and health of consumers. In meat-type chickens, excessive adipose tissue decreases both feed efficiency during rearing and the yield of lean meat after processing. However, fat is the major contributor to meat flavor; and the presence of

* Correspondence: chirwa02@yahoo.fr
[1]Rwanda Agriculture Board, Research Department, P. O Box 5016, Kigali, Rwanda
[3]Department of Animal Genetics, Breeding and Reproduction, College of Animal Science, South China Agricultural University, Guangzhou, Guangdong 510642, China
Full list of author information is available at the end of the article

intramuscular fat confers high eating quality of meat. Therefore, regulating fat deposition plays an important role in broiler chicken production.

In birds, lipogenesis, takes place primarily in the liver whereas adipocyte serves as the storage site for triglycerides [14]. Hepatic lipogenesis contributes 80 to 85% of the fatty acids stored in adipose tissue [15] because lipogenic activity in chickens is much greater in the liver than in adipose tissue [16-18].

In the past decade, genetic mechanisms underlying chicken fat deposition were widely studied but few studies were conducted to determine the gene expression involved in pathways as well as mechanisms that lead to adiposity in chickens [19]. In the present study, fast-growing White Recessive Rock chickens (WRR) and slow-growing Xinghua chickens (XH) were used to characterize specific genes for fat deposition in chickens. Global gene expression patterns within the liver and hypothalamus tissue of WRR and XH chickens were determined using Partek GS 6.4 Affymetrix Genechip® Chicken Genome Arrays and the differentially expressed genes were identified. Some of the differentially expressed genes were validated by determining their mRNA expression in liver, hypothalamus and fat tissues. The association of single nucleotide polymorphisms of the genes with chicken fat traits was also investigated.

Materials and methods

Chicken populations

Eight WRR (4♂ + 4♀, Institute of Animal Science, Guangdong Academy of Agricultural Sciences, Guangzhou, China), and 8 XH chickens (4♂ + 4♀, Fengkai Zhicheng Poultry Breeding Company, Guangdong, China), were used for differential expression observation with microarray hybridization. All the birds were fed a nutritionally balanced corn-soybean diet [20]. The birds had free access to water. They were slaughtered at 8 wk of age, and the liver and hypothalamus were excised, snapped frozen in liquid nitrogen and stored at −80°C until required for further analyses.

Six sets of WRR (3♂ + 3♀), and another six of XH (3♂ + 3♀), were used to study mRNA expression of the *SULT1B1, PNPLA3, GPAM, ELOVL6, LPL, FASN, ACSBG2, FDPS*, and *FRZB* genes in abdominal fat, subcutaneous fat, breast muscle, and pituitary tissues in the liver and hypothalamus tissues.

For association analysis, an F_2 resource population was constructed by crossing WRR with XH chickens [21]. The fat traits such as abdominal fat weight, subcutaneous fat thickness, fat band width, abdominal fat percentage were recorded in all F_2 full-sib individuals.

Ethics statement

The study was approved by the Animal Care Committee of South China Agricultural University (Guangzhou, People's Republic of China). Animals involved in this study were humanely sacrificed as necessary to ameliorate their suffering.

Microarray hybridization and data preprocessing

Total RNA was isolated from frozen tissues (50 mg) using TRIzol reagent (Invitrogen, CA, USA) according to the manufacturer's instructions. Total RNA concentration was determined by spectrophotometry. The RNA labelling and microarray hybridization were carried out according to the Affymetrix Expression Analysis Technical Manual (Biochip Corporation, Shanghai, China). The arrays were scanned using the Affymetrix Scanner 3000.

The GeneChip Chicken Genome Array used in the present study was created by Affymetrix Inc. (Santa Clara, USA) at the end of 2006, with comprehensive coverage of over 38,000 probe sets representing 32,773 transcripts corresponding to over 28,000 chicken genes (Chicken Genome Sequencing Consortium 2.1). Sequence information for this array was selected from the following public data sources: GenBank, UniGene and Ensembl.

Data normalization was used to eliminate dye-related artifacts. Consecutive filtering procedures were performed to normalize the data, and to remove noise derived from absent genes, background, and nonspecific hybridizations. Comparisons of expression levels were performed for each gene, and genes with the most significant differential expression ($P < 0.05$) were retained. Raw data sets were normalized to total fluorescence, which represents the total amount of RNA hybridized to a microarray, using the Partek GS 6.4 (Affymetrix Genechip® Chicken Genome Arrays, USA). QVALUE was used to obtain false-discovery rates (FDR).

The data obtained were subjected to Partek GS 6.4 for comparison using Affymetrix Expression Console Software, for expression algorithm robust multi-array (RMA) analysis. Multivariate ANOVA was used to determine significant differences among the replicates. Differentially expressed genes between WRR and XH chickens were identified by cutoff of fold-change (fold change) ≥ 2 and $P < 0.05$. Molecular functions of differentially expressed genes were classified according to molecule annotation system (MAS) 3.0 (http://bioinfo.capitalbio.com/mas3/). Database from the Kyoto Encyclopedia of Genes and Genomes (KEGG) were used for pathway analysis on differentially expressed genes using AgriGO (GO Analysis Toolkit and Database for Agricultural Community) http://bioinfo.cau.edu.cn/agriGO/) and Database for Annotation, Visualization and Integrated Discovery

(DAVID) Bioinformatics Resources (http://david.abcc.ncifcrf.gov/).

Validation of the differential expression with real-time RT-PCR

The primers were designed based on the published cDNA sequences of *SULT1B1*, the *LPL*, *ELOVL6*, *ACSBG2*, *SCD5*, *FADS1*, *PNPLA3*, *GAPDH*, *BEAN*, *SLC22A2*, *DDT*, *PLA2G12A*, and *18S* genes (http://www.ncbi.nlm.nih.gov) using GENETOOL software (BioTools, Alberta, Canada). The RNA was reverse-transcribed using the RevertAid Fist Strand cDNA Synthesis (Toyobo, Japan). After reverse transcription, the cDNA of the selected genes were amplified by real-time reverse transcription PCR. The relative level of each mRNA normalized to the *18 s* gene was calculated using the following equation: fold change = $2^{\text{Ct target (WRR)}-\text{Ct target (XH)}}/2^{\text{Ct } 18S \text{ (WRR)} - \text{Ct 18S (XH)}}$. The linear amount of target molecules relative to the calibrator was calculated by $2^{-\Delta\Delta CT}$. Therefore, all gene transcription results are reported as the *n*-fold difference relative to the calibrator. Specificity of the amplification product was verified by electrophoresis on a 0.8% agarose-gel. The results were expressed as mean ± SE.

Fat tissue expression of the differential expression genes with real-time RT-PCR studies

The same primers as those used in validation were used for determining fat tissue expression. The real-time RT-PCR reactions were performed using the iCycler Real-Time PCR detection System. Each sample reaction was ran in triplicate and the expression quantified as the number of cycles (CT) after which fluorescence exceeds the background threshold minus the CT for the housekeeping control (ΔCT). The calculation of absolute mRNA levels was based on the PCR efficiency and the threshold cycle (Ct) deviation of unknown cDNA versus the control cDNA. The quantitative values were obtained from the Ct values, which were the inverse ratios relative to the starting PCR product. The linear amount of target molecules relative to the calibrator was calculated by $2^{-\Delta\Delta CT}$. Briefly, the relative levels of each mRNA were expressed as the same as above.

SNP identification and association analysis

Tree variation sites were identified in intronic of chicken genes *ACSBG2, FASN and ELOVL6;* and one variation site was identified as non synonymous of chicken *ACSBG2 and* synonymous coding region of chicken *DDT* gene by using GENBANK (Table 1).

The data for association study were analyzed by ANOVA (SAS 8.1). The statistical significance threshold was set at $P < 0.05$. Values were expressed as the mean ± SEM, and the differences in the means were compared using Duncan's Multiple Range Test at 5% level of significance.

Results

Differentially expressed genes in fast-growing WRR and slow-growing XH chickens at 8 wk of age

After normalization and statistical analyses, 400 and 220 genes with at least 2-fold differences were identified ($P < 0.05$, FC ≥ 2) in liver and hypothalamus tissues of WRR and XH chickens, respectively. When fast-growing WRR chickens were compared with slow-growing XH chickens, 214 and 91 genes were up-regulated, and 186 and 129 genes were down-regulated in liver and hypothalamus tissues (Figure 1A and B; Tables 2 and 3).

In the liver, lipid metabolism genes viz *SULT1B1*, *ACSBG2*, *LPL*, *AACS*, *PNPLA3*, were up-regulated while *AOAH* gene was down-regulated. The carbohydrate metabolism genes: *MGAT4B*, *XYLB*, *GBE1*, *PGM1*, and *HKDC1*, were up-regulated (Table 2; Figure 1A). The fatty acid biosynthesis genes, *ELOVL6* and *FASN*, cholesterol biosynthesis genes, *LSS*, *HMGCR*, *FDPS*, *DHCR24*, malate metabolism process gene, *ME1*, proline biosynthesis process genes, *PYCR2* and *ALDH18A1*, oxidation-reduction reactions genes, *CYP1A4*, *CYP1A1 similar to aldose reductase*, *AKR1B1*, *CYP4V2*, and *DDO*, cyclic nucleotide catabolic process gene, *N4BP2L1*, and multicellular organism development genes, *SEMA5A* and

Table 1 The identified SNPs of the 4 fat deposition related genes

Variation ID	Genes name	Chr.	Position on chromosome (bp)	Consequence to transcript	Allele
rs10731268	*ACSBG2*	28	1257069	NON_SYNONYMOUS_CODING	G/A
rs15248801	*ACSBG2*	28	1247123	INTRONIC	T/C
rs15822158	*FASN*	18	4928024	INTRONIC	G/A
rs15822181	*FASN*	18	4930169	INTRONIC	C/T
rs15822181	*ELOVL6*	4	59539099	INTRONIC	A/G
rs14092745	*DDT*	15	8378815	SYNONYMOUS_CODING	A/G

SNP position was determined based on the reported SNP in ensembl http://www.ensembl.org/biomart/martview.
ACSBG2, acyl-CoA synthetase bubblegum family member 2; *FASN*, fatty acid synthase; *ELOVL6*, elongation of long chain fatty acids; *DDT*, D-dopachrome tautomerase.

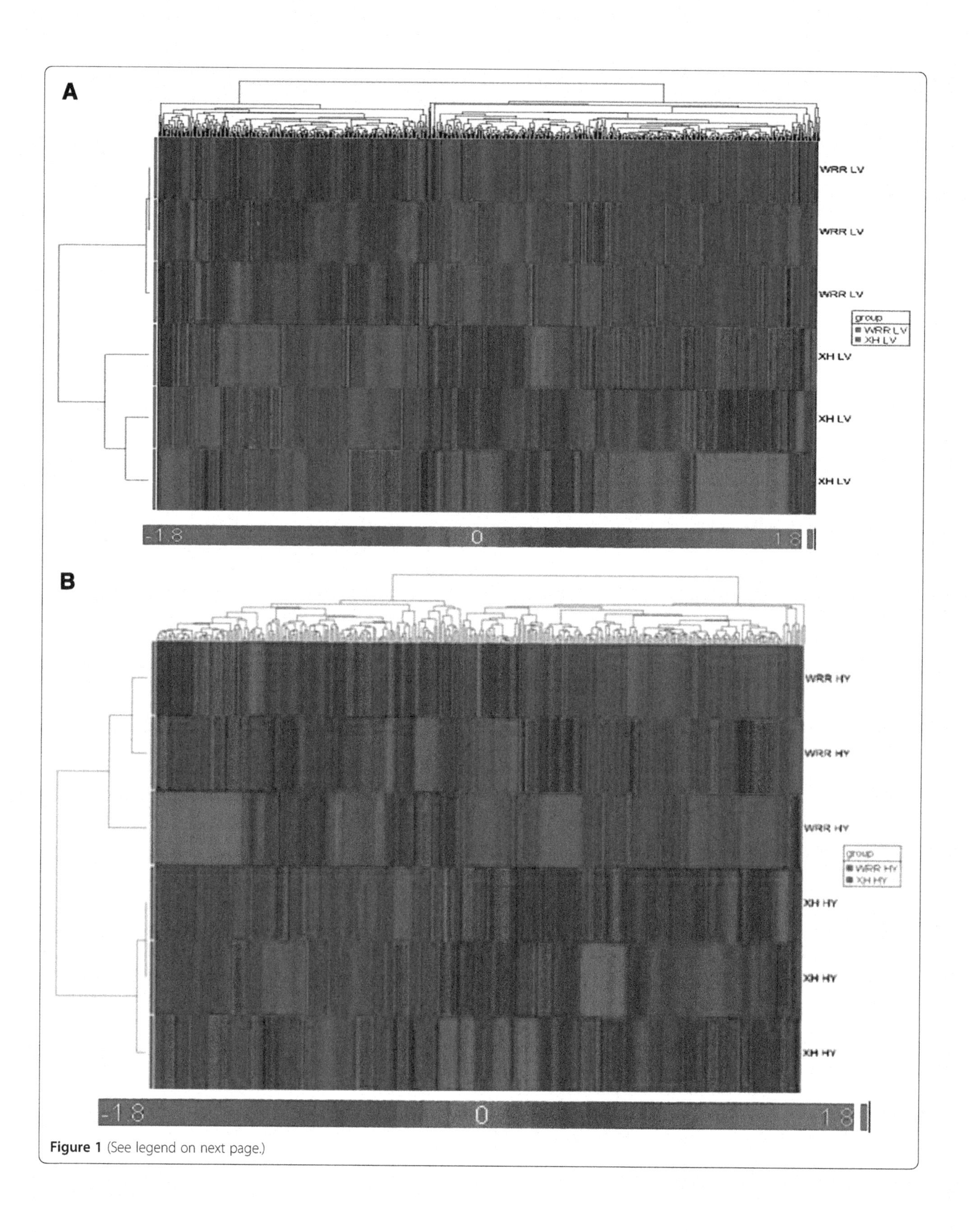

Figure 1 (See legend on next page.)

(See figure on previous page.)

Figure 1 Heat maps of differentially expressed genes of FG and SG chicken during developmental stages of liver and hypothalamus tissue. A Heat map of differentially expressed genes of FG and SG chicken during developmental stages of liver tissue. The red color represents fast growing chicken (WRR) genes while the blue color represents the slow growing chicken (XH) genes. The fold changes were arranged from −1.8 up to 1.8 ($P < 0.05$). WRRLV means liver tissue from White Recessive chickens. XHLV means liver tissue from Xinghua chickens. **B** Heat map of differentially expressed genes of FG and SG chicken during developmental stages of Hypothalamus Tissue . The red color represents fast growing chicken (WRR) genes while the blue color represents the slow growing chicken (XH) genes. The all identified gene, fold changes were arranged from −1.8 up to 1.8 ($P < 0.05$). WRRHY means Hypothalamus tissue from White Recessive chickens. XHHY means Hypothalamus tissue from Xinghua chickens.

C1orf107, were identified highly expressed in WRR chickens. In contrast, genes highly expressed in XH chickens were associated with multicellular development, *FRZB*, immune response, DNA integration, melanin biosynthetic process, *DDT*, muscle organ development, *DMD*, transforming growth factor beta receptor signaling pathway, *FUT8*, and oxidation-reduction, *CYP2C45*, *DHRSX*, *MICAL1*, and *CYP2C18*. In addition, the genes for the biosynthesis of steroids and fatty acid, *ELOVL6*, and *FASN* were also observed highly expressed in XH chickens (Table 2; Figure 1). The metabolic process genes, *ACSM5* (hypothetical protein), were down- regulated by 5- fold, while another metabolic process genes, *ENPEP*, were up-regulated by 5- fold (Table 2).

In the hypothalamus, the cyclic nucleotide catabolism gene, *N4BP2L1*, was up-regulated in fast growing WRR chickens by a 3.7-fold change. The negative regulation of endothelial cell proliferation gene, *TNFSF15*, was up-regulated by a 2.5-fold change. The proteolysis gene, *ITGBL1*, the protein amino acid phosphorylation genes, *SGK1* and *RIPK2*, are up-regulated in the WRR chickens. The copper ion transport gene, *SLC31A1*, was localized on chr17:1874758–1884555, and was up-regulated in the WRR chickens by a 7.3-fold change. The *PODXL*, *RAD54B*, *PODXL*, *PMP2* and *TMSB10*, were up-regulated in the WRR chickens. The melanin biosynthesis gene, *DDT*, ion transport genes, *SLC22A2*, and *GLRA1*, lipid metabolism process gene, *P20K* (also known as *EX-FABP*), cellular amino acid metabolism gene, *LOC772201*, protein complex gene, *ATPAF1*, proteolysis genes, *FOLH1*, *C1R*, and *VSP13B*, striated muscle contraction gene, *MYBPC2*, nitrogen compound metabolism process gene, *Vanin1*, porphyrin biosynthesis process gene, *FECH*, and response to stress genes, *HSP70*, *HSP25*, and *HSPB1*, were down-regulated in slow-growing XH chickens. In addition, the signal transduction genes, similar *to KIAA0712* protein, and *ANK2*, small GTPase mediated signal transduction gene, *RAB30*, DNA integration gene, *LOC770705*, amino acid phosphorylation gene, *PRKD3*, carbohydrate metabolic process gene, *CBR1*, and *NAT13*, neuron migration gene, *MDGA1*, hemophilic cell adhesion gene, *PCDH24*, sodium transport gene, *SLC13A5*, regulation of transcript DNA-dependent genes, *CREB3L2*, and *MLL3*, were also down-regulated in slow-growing XH chickens (Table 3).

Different gene ontology (GO) terms for biological process were identified in the livers of WRR and XH chickens. The highest GO clustered was in lipid biosynthesis process and fatty acid metabolism process (Figure 2).

In hypothalamus tissue, the GO terms for biological process in the WRR and XH chickens were mostly observed in response to stimulus, response to stress, and response to abiotic stimulus. Pigment metabolic process, melanin metabolic process, response to radiation, response to heat, response to temperature stimulus, leucocyte proliferation, pigment biosynthesis process, lymphocyte proliferation, mononuclear cell proliferation and response to ionizing radiation were also observed (Figure 3).

In the pathway study, a number of lipid-related genes: *ACSBG2*, *FASN*, *LPL*, *GPAM*, *FDPS*, and others were identified. The cicardian clock gene, *ARNTL* also known as *Bmal1*, was observed.

Based on the pathways, differentially expressed genes participated in several function related to lipid (Tables 2 and 4). The lipid related genes were *ACSBG2*, *SULT1B1*, and *LDLR* of lipid metabolism, *LPL* of glycerolipid metabolism, and *MTTP* of lipid transporter activity, *FASN* and *ELOVL6* of biosynthesis of unsaturated fatty acids, *LSS*, *HMGCR*, *NSDHL*, *DHCR24*, *IDI1*, of *HSD17B7* of biosynthesis of steroid, *AGPAT4* and *FRZB* of triacylglyceride synthesis, *GPAM* of glycerolipid metabolism, *PHOSPHO1* and *PTDSS1* of glycerophospholipid metabolism, *ATP6V1C2* of oxidative phosphorylation, *ACSS2* of glycolysis, *GCNT2* of glycosphingolipid biosynthesis – lactoseries, and *ME1* of pyruvate metabolism (Figure 3).

In hypothalamus tissue, three genes related to VEGF signaling pathway, four genes related to MAPK signaling pathway, one gene each related to alpha-linolenic acid metabolism, nitrogen metabolism, linoleic acid metabolism, porphyrin and chlorophyll metabolism were identified. Then a homologous recombination, heparan sulfate biosynthesis, ether lipid metabolism, arginine and proline metabolism, arachidonic acid metabolism, N-glycan biosynthesis, glycerophospholipid metabolism, ErbB signaling

Table 2 Fold-changes of significantly differential expressed genes in WRR and XH chickens

Gene symbol	Gene title	*P* value	Fold change	Chromosome alignment s
Lipid metabolic process				
SULT1B1	sulfotransferase family, cytosolic, 1B, member 1	0.0001	7.689	chr4:53309684-53311980
ACSBG2	acyl-CoA synthetase bubblegum family member 2	0.004	5.382	chr28:1247898-1259038
LPL	lipoprotein lipase	0.018	2.528	chrZ:53399697-53408327
AACS	acetoacetyl-CoA synthetase	0.021	2.507	chr15:4477440-4512637
PNPLA3	Patatin-like phospholipase domain containing 3	0.024	3.028	chr1:71256654-71270462
AOAH	acyloxyacyl hydrolase (neutrophil)	0.043	−2.516	chr2:46723433-46778195
Carbohydrate metabolic process				
MGAT4B	mannosyl (alpha-1,3-)-glycoprotein beta-1,4-N-acetylglucosaminyltransferase, iso	8.33E-05	2.178	chr13:13578206-13590970
XYLB	xylulokinase homolog (H. influenzae)	0.0001	2.603	chr2:6032066-6115406
GBE1	glucan (1,4-alpha-), branching enzyme 1 (glycogen branching enzyme, Andersen dis	0.0008	2.119	chr1:98522850-98669948
PGM1	phosphoglucomutase 1	0.002	2.179	chr8:28644700-28665874
HKDC1	hexokinase domain containing 1	0.038	7.368	chr6:11960338-11966483
Fatty acid biosynthetic process				
ELOVL6	ELOVL family member 6, elongation of long chain fatty acids	0.002	2.181	chr4:59493262-59560594
FASN	fatty acid synthase	0.029	2.840	chr18:4906222-4942593
Cholesterol biosynthetic process				
LSS	lanosterol synthase (2,3-oxidosqualene-lanosterol cyclase)	0,001	2,186	chr7:6878402-6888484
HMGCR	3-hydroxy-3-methylglutaryl-Coenzyme A reductase	0,005	3,236	chrZ:23472632-23474241
FDPS	farnesyl diphosphate synthase (farnesyl pyrophosphate synthetase, dimethylallylt	0,021	2,167	chrUn_random:7545445-7546725
DHCR24	24-dehydrocholesterol reductase	0,026	2,587	chr8:26011324-26019531
HMGCR	3-hydroxy-3-methylglutaryl-Coenzyme A reductase	0,027	2,805	chrZ:23472597-23491333
Oxidation reduction				
CYP1A4	cytochrome P450 1A4	0,001	9,342	chr10:1822784-1826314
CYP1A1	cytochrome P450, family 1, subfamily A, polypeptide 1	0,003	6,485	chr10:1806680-1809495
DHRSX	dehydrogenase/reductase (SDR family) X-linked	0,004	−2,1001	chr1:132739051-132944192
LOC418170	similar to aldose reductase	0,014	2,042	chr1:64269892-64273020
CYP2C45	cytochrome P-450 2C45	0,019	−5,673	chr6:17648418-17654233
AKR1B1	aldo-keto reductase family 1, member B1 (aldose reductase)	0,028	2,788	chr1:64293981-64312331
MICAL1	microtubule associated monoxygenase, calponin and LIM domain containing 1	0,029	−2,186	chr26:25422-27136
CYP2C18	cytochrome P450, family 2, subfamily C, polypeptide 18	0,040	−3,214	chr6:18655324-18664396
CYP4V2	cytochrome P450, family 4, subfamily V, polypeptide 2	0,048	2,426	chr4:63195381-63202122
DDO	D-aspartate oxidase	0,049	2,219	chr3:69194822-69198140
Cicardian clock genes				
ARNTL	aryl hydrocarbon receptor nuclear translocator-like	0,002	−2,043	chr5:8501344-8546127
Transforming growth factor beta receptor signaling pathway				
FUT8	fucosyltransferase 8 (alpha (1,6) fucosyltransferase)	0,028	−2,473	chr5:24711230-24725772

Table 2 Fold-changes of significantly differential expressed genes in WRR and XH chickens *(Continued)*

Asparagine biosynthetic process				
ASNS	asparagine synthetase	1,48E-05	9,945	chr2:24628018-24641745
Melanin biosynthetic process				
DDT	D-dopachrome tautomerase	3,50E-05	−13,908	chr15:8372896-8375331

Positive values indicated that the genes were up-regulated when fast growing WRR chickens are compared with slow growing XH chickens.
Negative values meant down-regulation when comparison between WRR and XH chickens are made, Data were significantly different (*P* > 0,05), and fold changes were not smaller than 2.
"—" meant unknown.

pathway, Wnt signaling pathway were also observed in our present study (Table 5).

Validation of differential expression by real-time RT-PCR

The mRNA levels of 9 genes involved in fat deposition were further quantified using real-time RT-PCR (Table 6). The level of *18S rRNA* was chosen as reference and confirmed to be invariable. The expression levels (normalized to 18S) of the 9 genes were determined. Fold changes of gene expression determined by real-time RT-PCR were compared with the fold changes obtained from microarray analysis (Table 6). The highest fold changes in WRR chickens compared with XH chickens were confirmed in the *SULT1B1, ACSBG2, ELOVL6, SLC31A1,* and *PNPLA3* genes. The lowest fold-changes were observed in the *DDT* and *BEAN* genes.

Expression levels of the Fat deposition related genes in the Fat tissues of WRR and XH chickens

When WRR males were compared with XH males, the expression of the *LPL, FDPS, PNPLA3, GPAM,* and *SULT1B1* genes were up-regulated, and the *FASN, ACSBG2,* and *FRZB* were down-regulated in the abdominal fat tissue (Figure 4). In the subcutaneous fat tissue, the *LPL, FDPS, PNPLA3,* and *SULT1B1* were up-regulated, and the *FASN, GPAM, ACSBG2,* and *FRZB* genes were down-regulated. In the breast muscle tissues, the *FDPS, PNPLA3, GPAM,* and *FRZB* were up-regulated, and the *LPL, FASN, ACSBG2, SULT1B1,* and *ELOVL6* genes were down-regulated (Figure 4). In the pituitary tissues, the *LPL, FASN, SULT1B1,* and *ELOVL6* genes were up-regulated, and the *FDPS, PNPLA3, GPAM, ACSBG2,* and *FRZB* genes were up-regulated (Tables 7, 4 and 5).

Polymorphisms of fat deposition genes associated with fat trait in chickens

The SNP rs10731268 of the *ACSBG2* gene was associated with abdominal fat weight ($P = 0.005$), and abdominal fat percentage ($P = 0.022$). The SNP rs15248801 of the *ACSBG2* gene was associated with abdominal fat weight ($P = 0.039$) [Table 8]. The SNP rs15822158 of the *FASN* gene was associated with fat band width ($P = 0.0003$), abdominal fat percentage ($P = 0.001$), and abdominal fat percentage ($P = 0.005$) [Table 9]. The SNP rs15822181 of the *FASN* gene was associated with abdominal fat weight ($P = 0.049$) while the SNP rs16418687 of the *ELOVL6* gene was associated with subcutaneous fat ($P = 0.034$). The SNP rs14092745 of the *DDT* gene was associated with fat band width ($P = 0.048$) [Table 9, 10].

Discussion

The approach of selective-fat-deposition-related-genes in animals is a relatively new strategy aimed at improving production efficiency while enhancing meat quality. Efforts to reduce fat deposition in animals include genetic selection, feeding strategies, housing and environmental strategies as well as hormone supplementation. While these efforts have improved production efficiency and reduced carcass lipid deposition, negatively impacts on meat quality were due to reduced intramuscular fat deposition [22]. Based on the comparison of two types of breeds of chicken whose fat deposition and growth rate are exceptionally varied, a functional genomics approach was chosen in order to identify chicken fat-deposition-related-genes. In this genomic approach, liver tissue was used. The liver is the site of fat synthesis, and hypothalamus, which is a major gland for the endocrine system. Few studies focused on global gene expression surveys in chickens. Wang *et al.* [19] provided analysis of chicken adipose tissue gene expression profile. Other hepatic transcriptional analyses had been reported, using dedicated chicken 3.2 K liver-specific microarray [14,23] or a 323 cDNA microarray [24].

Differential gene expressions in liver during the fat developmental stage in fast growing WRR and slow growing XH chickens were related to lipid metabolism in our study. It has been reported that some genes, e.g. *3-hydroxyacyl-CoA dehydrogenase, long chain acyl-CoA thioesterase, fatty-acid elongation enzymes* and *cytosolic fatty-acid- and acyl-CoA-binding proteins,* are known to play key roles in mammalian fat or lipid metabolism [25]. Glyco-metabolism such as glycol-sphingolipids (*GCNT2*), biosynthesis of steroids, fatty acid biosynthesis (*ELOVL6* and *FASN*) was observed in this study. Collin *et al.*[26] reported that fast growing chickens developed excessive adiposity besides the high muscle mass

Table 3 Differentially expressed genes in hypothalamus of WRR and XH chickens

Gene symbol	Gene title	*P* value	Fold-change	Chromosomes alignment
Lipid metabolic process				
P20K	quiescence-specific protein	0,0109	−2,402	chr17:881078-883996
Porphyrin biosynthetic process				
FECH	ferrochelatase (protoporphyria)	0,009	−2,155	chrZ:267090-278253
Nitrogen compound metabolic process				
RCJMB04_35g11	vanin 1	0,0406	−2,312	chr3:58745711-58758866
Transport				
PMP2	peripheral myelin protein 2	0,031	3,405	chr2:126148069-126152515
Ion transport				
SLC22A2	solute carrier family 22 (organic cation transporter), member 2	0,004	−2,123	chr3:47342914-47355320
GLRA1	glycine receptor, alpha 1 (startle disease/hyperekplexia)	0,035	−2,146	chr13:12903161-12940509
sodium ion transport				
SLC13A5	solute carrier family 13 (sodium-dependent citrate transporter), member 5	0,003	−2,046	chr19:9754843-9769124
Copper ion transport				
SLC31A1	solute carrier family 31 (copper transporters), member 1	0,017	7,335	chr17:1874758-1884555
Striated muscle contraction				
MYBPC2	myosin binding protein C, fast type	0,034	−3,717	un
Response to stress				
HSP70	heat shock protein 70	0,001	−2,093	chr5:55409752-55412248
HSP25	heat shock protein 25	0,006	−2,759	chr27:4486260-4487251
Signal transduction				
LOC419724	similar to KIAA0712 protein	0,009	−2,393	chr24:1172171-1264246
cyclic nucleotide catabolic process				
N4BP2L1	NEDD4 binding protein 2-like 1	0,006	3,732	chr1:178835487-178837064
Negative regulation of endothelial cell proliferation				
TNFSF15	tumor necrosis factor (ligand) superfamily, member 15	0,027	2,553	chr17:2943951-2959613
Regulation of transcription, DNA-dependent				
CREB3L2	cAMP responsive element binding protein 3-like 2	0,001	−2,197	chr1:59488049-59500378
MLL3	myeloid/lymphoid or mixed-lineage leukemia 3	0,008	−4,110	chr2:6484781-6486027
Protein complex assembly				
ATPAF1	ATP synthase mitochondrial F1 complex assembly factor 1	0,003	−3,241	chr8:22570693-22574048
Protein amino acid phosphorylation				
PRKD3	protein kinase D3	0,0005	−2,745	chr3:34793675-34819927
SGK1	serum/glucocorticoid regulated kinase 1	0,004	2,419	chr3:58130872-58134430
RIPK2	receptor-interacting serine-threonine kinase 2	0,013	2,532	chr2:129010265-129029220
Proteolysis				
FOLH1	folate hydrolase (prostate-specific membrane antigen) 1	0,005	−2,344	chr1:191872775-191933212
C1R	complement component 1, r subcomponent	0,011	−2,014	chr1:80553474-80558910
ITGBL1	integrin, beta-like 1 (with EGF-like repeat domains)	0,024	2,43	chr1:147579457-147711379
VPS13B	vacuolar protein sorting 13 homolog B (yeast)	0,041	−2,155	chr2:133389475-133390863

Positive values simply mean that the genes were up-regulated when WRR chickens are compared with XH chickens.
Similarly, negative values mean down-regulation when comparison between WRR and XH chickens is made, Data were significantly different [$P > 0, 05$ (fold change ≥ 2)].

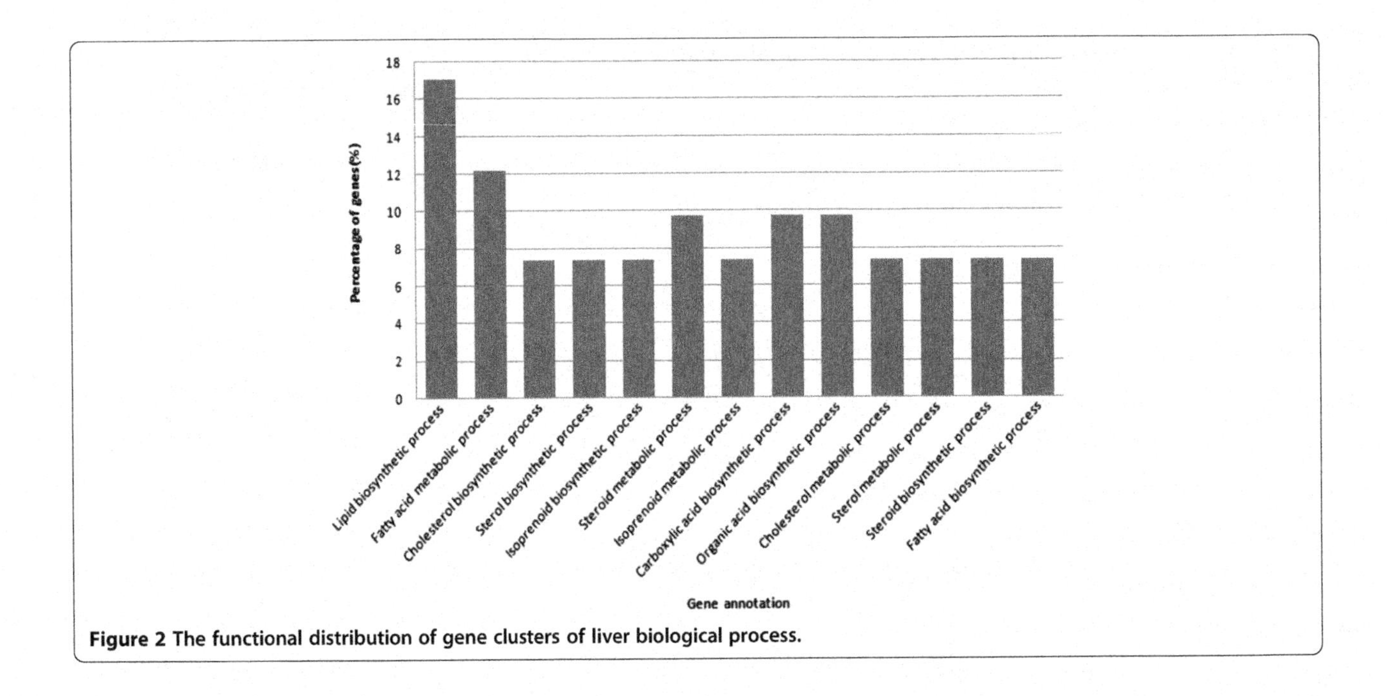

Figure 2 The functional distribution of gene clusters of liver biological process.

resulting from selection. The suggestion is that differential expression of the lipid metabolism related genes might be one of the factors in the differences of fat deposition between fast growing and slow growing chickens at the developmental stage.

The liver is the main site for fatty acid biosynthesis and the fatty acids are then transported to the adipose tissue for storage. The tasks are accomplished through the generation of triglycerides by the liver from fatty acids and L-a-glycerophosphate, packaged into very low density lipoproteins (VLDL), and then, secreted into the blood. The triglycerides in VLDL are processed by the adipose tissue and finally deposited in the central vacuole of the adipocyte. It was suggested that several mechanisms regulate intracellular non-esterified fatty acids composition, including fatty acid transport, acyl CoA synthetases, fatty acid elongases, desaturases, neutral and polar lipid lipases and fatty acid oxidation. Most of these mechanisms are regulated by PPAR alpha or SREBP-1c. Together, these mechanisms control hepatic lipid composition and affect whole-body lipid composition [27]. LPL catalyzes the hydrolysis of plasma lipoproteins, which is a rate-limiting step in

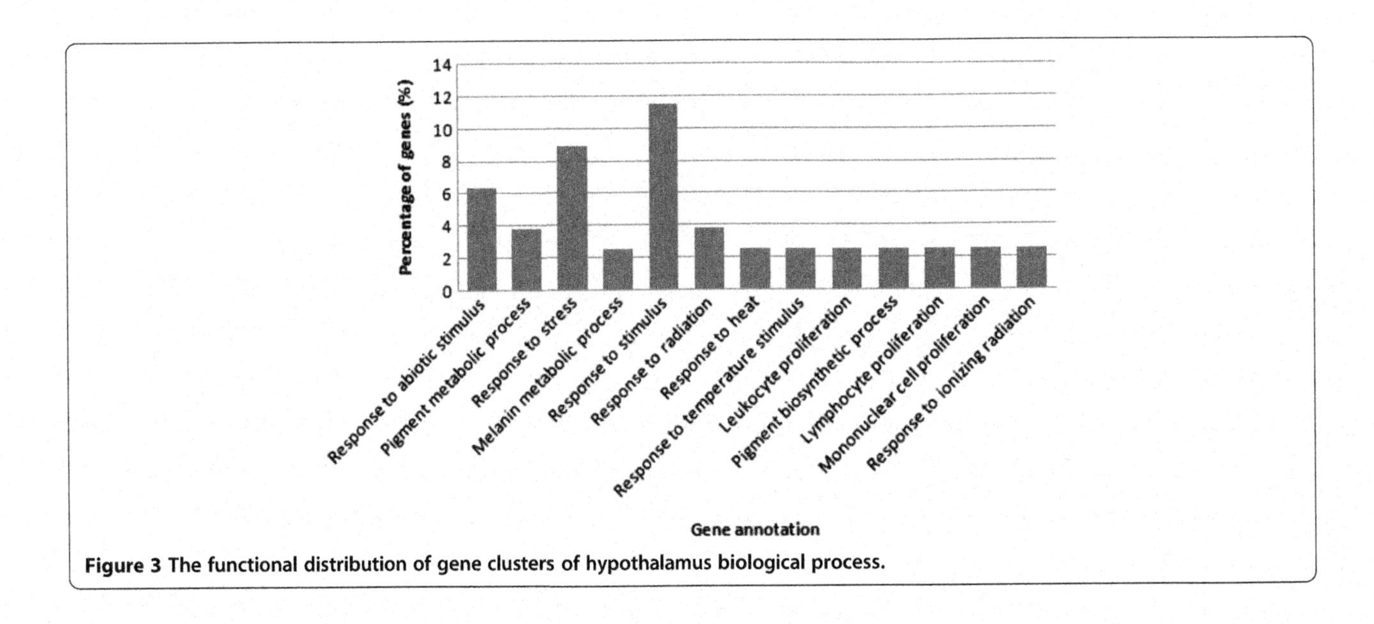

Figure 3 The functional distribution of gene clusters of hypothalamus biological process.

Table 4 Pathways of the fat-deposition-related genes in the liver of WRR and XH chickens

Probeset ID	Gene	Pathway
GgaAffx.12964.1.S1_s_at	*LSS*	Biosynthesis of steroids
Gga.13365.1.S1_at	*AGPAT4*	Triacylglyceride_Synthesis_BiGCaT
GgaAffx.21515.1.S1_s_at	*PTDSS1*	Glycerophospholipid metabolism
GgaAffx.12469.1.S1_at	*ELOVL6*	Biosynthesis of unsaturated fatty acids
Gga.7215.2.S1_a_at	*HSD17B7*	Biosynthesis of steroids
Gga.2334.1.S2_at	*PHOSPHO1*	Glycerophospholipid metabolism
Gga.2298.1.S1_at	*ATP6V1C2*	Oxidative phosphorylation
GgaAffx.5529.1.S1_at	*GPAM*	Glycerolipid metabolism
Gga.8851.1.S1_a_at	*IDI1*	Biosynthesis of steroid
GgaAffx.2094.4.S1_s_at	*ACSS2*	Glycolysis/Gluconeogenesis
Gga.1132.1.S1_at	*ME1*	Pyruvate metabolism
GgaAffx.21769.1.S1_s_at	*LPL*	Glycerolipid metabolism
Gga.9630.1.S1_s_at	*LDLR*	Lipid metabolism
Gga.9949.1.S1_at	*NSDHL*	Biosynthesis of steroids
GgaAffx.12935.1.S1_s_at	*DHCR24*	Biosynthesis of steroids
Gga.2785.1.S1_s_at	*HMGCR*	Biosynthesis of steroids
Gga.2448.1.S2_at	*FASN*	Fatty acid biosynthesis
GgaAffx.8101.1.S1_at	*GCNT2; LOC428479*	Glycosphingolipid biosynthesis - lactoseries
GgaAffx.23852.1.S1_at	*MTTP*	lipid transporter activity
Gga.7792.1.S1_s_at	*ACSBG2*	Lipid metabolism
Gga.8853.2.S1_a_at	*SULT1B1*	Molecular_function–transferase_activity
Gga.4955.1.S1_at	*FRZB*	Adipogenesis; Cellular_component

the transportation of lipids into peripheral tissues [28,29]. The *LPL* gene expression in fast growing chicken was 2.5-fold greater than that in the slow growing type at the developmental stage in this study. In mammals, increased LPL activity is strongly associated with fat deposition and obesity, and these are regulated by both insulin and glucocorticoids according to Fried *et al.* [30]. The major site of lipogenesis in birds, however, is the liver rather than the adipose tissue [31]. The role of fatty acid-binding protein in the intramuscular trafficking of long-chain fatty acids within intramuscular adipocytes has been studied and found to be related to intramuscular levels in different species [32,33].

Table 5 Pathway of the fat-deposition-related genes expressed in hypothalamus tissue of WRR and XH chickens

Pathway	Count	*P*-Value	*Q*-Value	Gene
VEGF signaling pathway	3	4,32E-05	2,16E-05	*HSPB1;PLA2G12A;KRAS*
MAPK signaling pathway	4	1,02E-04	3,40E-05	*HSPB1;HSP70;PLA2G12A;KRAS*
alpha-Linolenic acid metabolism	1	0,015	0,001	*PLA2G12A*
Nitrogen metabolism	1	0,019	0,002	*CA3*
Linoleic acid metabolism	1	0,019	0,002	*PLA2G12A*
Porphyrin and chlorophyll metabolism	1	0,023	0,002	*FECH*
Homologous recombination	1	0,026	0,002	*RAD54B*
Heparan sulfate biosynthesis	1	0,026	0,002	*HS6ST2*
Ether lipid metabolism	1	0,030	0,002	*PLA2G12A*
Arginine and proline metabolism	1	0,032	0,002	*LOC396507*
Arachidonic acid metabolism	1	0,033	0,002	*PLA2G12A*
N-Glycan biosynthesis	1	0,043	0,002	*ALG13*
Inositol phosphate metabolism	1	0,052	0,0027	*IPMK*
Glycerophospholipid metabolism	1	0,056	0,002	*PLA2G12A*
ErbB signaling pathway	1	0,086	0,004	*KRAS*
Wnt signaling pathway	1	0,138	0,005	*TCF7L2*

Fatty acid synthesis (FAS) occurs during periods of energy surplus and concomitantly its gene expression is down-regulated during starvation in the liver [34], which is the major site of lipogenesis in avian species [35-37]. The regulation of hypothalamic fatty acid synthesis gene expression in response to starvation is similar to that of liver fatty acid synthesis. In birds, like in humans, fatty acid synthesis primarily occurs in the liver. Demeure *et al.* [38] reported that chicken *FASN* gene is directly the target of liver cross receptor (LxR) alpha and therefore, expands the role of LxR alpha as a regulator of lipid metabolism. *FASN* and *GPAM* are two enzymes that play central roles in *de novo* lipogenesis. The G4928024A of the *FASN* gene is significantly associated with fat band width, abdominal fat percentage, and abdominal fat percentage.

The *DDT* gene was observed down-regulated in both tissues when fast growing WRR chickens were compared with slow growing XH chickens. This gene

Table 6 Comparison of liver tissue gene expression levels between microarray and qRT-PCR

Genes	Microarray Fold changes in WRR vs. XH	Real-time PCR Fold changes in WRR vs. XH
SULT1B1	4,13	3,28
LPL	2,5	2,48
ELOVL6	2,18	1,76
ACSBG2	5,3	2,29
PNPLA3	3,03	4,6
BEAN	−4,2	−5,76
SLC31A1	7,3	1,09
DDT	−6,59	−0,4
PLA2G12A	−2,8	−2,6

Validation of differentially expressed genes between WRR and XH chickens by RT-PCR.
The data presented indicate the relative mRNA expression of both microarray and qRT-PCR.
Positive values mean that the gene was up-regulated when WRR chickens were compared with XH chickens. Similarly, a negative number means that the gene was down-regulated.

has function in melanization which can play a role in the pigmentation of abdominal fat. It also, has a high correlation with the accumulation of melanin in the skin of the shanks. Melanization of abdominal fascia is not harmful but it may cause severe economic losses to the producer. It was surprising to observe that the *FDPS, LSS, HMGCR, NSDHL, DHCR24, IDI1,* and *HSD17B7* were up-regulated in fast growing WRR chickens. These genes are considered as the ones which has some functions in cholesterol biosynthesis. The glycolytic genes (*ACSS2*), carbohydrate metabolic and fatty acid biosynthesis were also up-regulated in the WRR chickens. It is suggested that the genes related to cholesterol biosynthesis, carbohydrate metabolic and fatty acid biosynthesis may have influence on fat development.

This study also showed that the genes related to proline biosynthetic process, *member 2* of *pyrroline-5-carboxylate reductase* family, *member A1* of *aldehyde dehydrogenase 18* families, and oxidation reduction, *CYP1A4, CYP1A1, AKR1B1, CYP4V2, DDO,* and *similar to aldose reductase,* were differently expressed between the WRR and XH chickens. The *CYP2H1, CYP2C45, CYP2C18, MICAL1* and *CYP3A37* genes were significantly different in between the WRR and XH chickens. In this study, many lipid-related genes were identified, *ACSBG2, FASN, LPL, GPAM,* and *FDPS.* The cicardian clock gene (*ARNTL)* was observed, it plays a role in glucose, lipid metabolism and adipogenesis [39-41]. Moreover, a network of 11 genes, *LPL, ACSBG2, AACS, FASN, LSS, FDPS, SULT1B1, HMGCR, DPP4, FUT8,* and *PLAU,* was observed. Parallel expression patterns of these functionally relevant genes provided strong evidence for their coordinated involvement in lipid biosynthesis, cholesterol biosynthesis and fatty acid degradation in chickens. In chickens, the *ACSBG2* gene has been found to play a significant role in lipid metabolism. The present study confirmed this conclusion.

In order to support the results of the microarray study, all the genes used for the mRNA assay were found to have good relationship with fat-related genes as their functions related to lipid metabolism, cholesterol biosynthesis and fatty acid metabolism. Interestingly, the *SULT1B1, PNPLA3, GPAM, ELOVL6, LPL, FASN, ACSBG2, FDPS,* and *FRZB* genes were preferentially expressed in 4 fatty tissues of abdominal fat, subcutaneous fat, breast muscle and pituitary gland when WRR were compared with XH chickens.

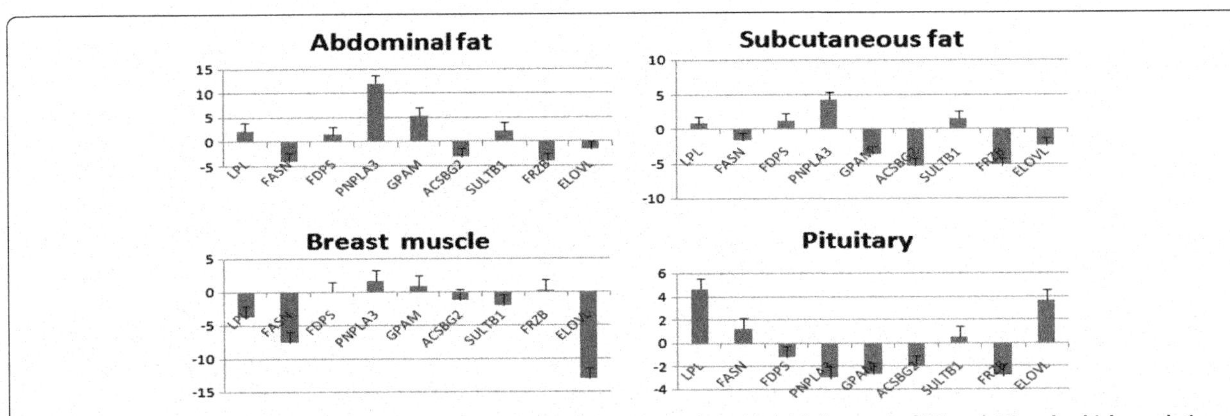

Figure 4 Data presented indicate the different mRNA gene relative expressions (as fold changes) of FG and SG male chicken relative to different fat tissues. Error bars represent the standard errors. Positive values imply genes were up-regulated in fast-growing chicken compared to slow-growing ones.

Table 7 The localization of differentially expressed fat-related genes

Alignments	Gene Symbol	Transcript ID	Gene ontology
chrZ:53399697-53408327	*LPL, ENSGALG00000015425*	NM_205282	GO:0004465 lipoprotein lipase activity;
			GO:0004806 triacylglycerol lipase activity;
			GO:0006629 lipid metabolism;
chr4:53309684-53311980	*SULT1B1, SULT1B1*	NM_204545	GO:0006629 lipid metabolism;
			GO:0008202 steroid metabolism;
chr28:1247898-1259038	*ACSBG2(RCJMB04_9i11)*	NM_001012846	006629 lipid metabolism;
			GO:0006631 fatty acid metabolism;
chr15:4477440-4512637	*AACS, ENSGALG00000002899*	NM_001006184	GO:0006629 lipid metabolism;
			GO:0006631 fatty acid metabolism;
			GO:0005829 cytosol
chr1:71256654-71270462	*PNPLA3*	XM_416457	GO:0006629 lipid metabolism;
			GO:0006629 lipid metabolism;
			GO:0016042 lipid catabolism
chr7:2296059-2312305	*FRZB, ENSGALG00000002763*	NM_204772	GO:0017147 Wnt-protein binding;
			GO:0007275 development;
			GO:0016055 Wnt receptor signaling pathway
chr15:8372896-8375331	*DDT(RCJMB04_2c16)*	NM_001030667	GO:0006583 melanin biosynthesis from tyrosine;
chr1:132739051132944192	*DHRSX*	XM_001232713	GO:0055114 oxidation reduction
chr7:6879282-6888069	*LSS*	NM_001006514	GO:0006695 cholesterol biosynthesis
chr7:22713218-22757360	*DPP4*	NM_001031255	GO:0005515 protein binding;
			GO:0008239 dipeptidyl-peptidase activity;
			GO:0042803 protein homodimerization activity;
chr13:13578206-13590970	*MGAT4B*	XM_414605	GO:0005975 carbohydrate metabolism;
chrZ:23472632-23474241	*HMGCR, ENSGALG00000014948*	NM_204485	GO:0004420 hydroxymethylglutaryl-CoA reductase (NADPH) activity;
			GO:0016491 oxidoreductase activity;
			GO:0050661 NADP binding;
chr5:24711230-24725772	*ENSGALG00000008078, FUT8*	NM_001004766	GO:0008424 glycoprotein 6-alpha-L-fucosyltransferase activity;
			GO:0046921 alpha(1,6)-fucosyltransferase activity;
			GO:0007179 transforming growth factor beta receptor signaling pathway
chr6:28077530-28103087	*GPAM*	XM_421757	GO:0006631 fatty acid metabolism;
			GO:0019432 triacylglycerol biosynthesis;
			GO:0040018 positive regulation of body size;
chr20:3629311-3638107	*SGK2*	XM_417346;CR387909	GO:0004713 protein-tyrosine kinase activity;
			GO:0005524 ATP binding;
chr3:34793675-34819927	*PRKD3*	XM_419526	GO:0004713 protein-tyrosine kinase activity;
			GO:0005524 ATP binding;
chr20:2633184-2647420	*ACSS2*	XM_417342	GO:0006085 acetyl-CoA biosynthesis;
Chr1: 64269656-64276867	*LOC418170*	XM_416401	GO:0055114 oxidation reduction
chr6:17648418-17654233	*CYP2C45*	SNM_001001752	GO:0055114 oxidation reduction
chr4:11367346-11377074	*NSDHL*	XM_420279	GO:0005975 carbohydrate metabolism;
			GO:0008203 cholesterol metabolism;
chrUn_random:7545445-7546725	*FDPS*	XM_422855	GO:0006695 cholesterol biosynthesis

Table 7 The localization of differentially expressed fat-related genes *(Continued)*

chr2:144322566-144334593	*SQLE*	NM_001030953	GO:0055114 oxidation reduction;
chr8:26011324-26019531	*DHCR24*	NM_001031288	GO:0016125 sterol metabolism;
chr1:64293981-64312331	*AKR1B1*	XR_026805	GO:0055114 oxidation reduction
chr1:118069255-119072613	*DMD*	NM_205299	GO:0007519 striated muscle development; cytoplasm;
			GO:0016010 dystrophin-associated glycoprotein complex;
			GO:0045121 lipid raft;
chr18:4906222-4942593	*FASN, ENSGALG00000002747*	NM_205155;J02839	GO:0000036 acyl carrier activity;
			GO:0004312 fatty-acid synthase activity;
			GO:0006633 fatty acid biosynthesis;
chr4:59493262-59560594	*ELOVL6*		GO:0006633 fatty acid biosynthesis
chr6:11960338-11966483	*HKDC1*	XM_421579	GO:0005524 ATP binding;
			GO:0005975 carbohydrate metabolism;
			GO:0006096 glycolysis;
chr2:64406040-64411831	*GCNT2*	XM_426036;XM_418950	GO:0008375 acetylglucosaminyltransferase activity;
chr1:98522850-98669948	*GBE1*	XM_425536	GO:0005975 carbohydrate metabolism

However, *PNPLA3* mRNA level was higher in all tissues in WRR except pituitary tissue, where it expressed lower levels in XH chicken. *PNPLA3,* also referred to as adiponutrin, was originally identified as a highly adipose–specific transcript that rapidly responds to nutritional status [42]. The microarray assay demonstrated that there was a 3 times higher expression of the *PNPLA3* gene in liver tissue at 8 wk of age in WRR than in XH chickens. It could be concluded that the *PNPLA3* gene is involved in fat deposition.

The microarray data showed that the *SULT1B1* is abundantly expressed in liver tissue with 7 fold change in WRR with XH chickens. The gene was reported to be expressed in liver and other numerous extra-hepatic tissues [43]. *FDPS* is an important intermediate in cholesterol and sterol biosynthesis, a substrate for protein farnesylation and geranylgeranylation, and a ligand or agonist for certain hormone receptors and growth receptors. In this study, the *FDPS* was found to belong to the cholesterol biosythetic group. The *FDPS* mRNA level was higher in subcutaneous fat and pituitary tissue of WRR female chicken against XH counterpart.

The *GPAM* gene plays a vital role in the regulation of cellular triacylglycerol and phospholipid levels [44,45]. In this study, adipose tissues such as abdominal fat and subcutaneous fat were found to have the highest levels of *GPAM* mRNA expression whereas it was rarely detectable in the liver in the microarray assay. The *FRZB* gene (also known as *SFRP3*) is a member of the secreted frizzled receptor family of soluble proteins which binds to and antagonises Wnt receptor [46]. Wnts are secreted lipid-modified signaling proteins that influence multiple processes in the development of animals. The *FRZB* was shown to play a major role in adipogenesis in the microarray analysis of WRR and XH at 8 wk of age. *ELOVL6* is involved in *de novo* lipogenesis and is regulated by dietary, hormonal and developmental factors

Table 8 Association of the G127069A, T1247123C in the ACSBG2 gene with chicken fat traits

Traits	rs10731268 = G1257069A				rs15248801 = T1247123C			
	P value	Genotypes			*P* value	Genotypes		
		GG	GA	AA		TT	TC	CC
Fat Bandwidth (mm)	0,587	0	10,05 ± 6,76(33)	13,85 ± 1,80(462)	0,66	11,65 ± 7,34(28)	11,34 ± 3,26(142)	14,75 ± 2,16(325)
Abdominal fat weight (g)	0,005**	0	18,58 ± 2,27(33)b	27,21 ± 0,79(463)a	0,0396*	26,07 ± 3,23(28)	23,68 ± 1,43(142)	28,06 ± 0,94(325)
Abdominal fat percentage	0,3331	0	12,16 ± 9,83(33)	22,02 ± 2,62(463)	0,3627	18,07 ± 57,75(28)	15,61 ± 3,44(142)	21,43 ± 2,27(326)

Means with different letter are significantly different ** ($P > 0.01$); * ($P > 0.05$).
Data are presented at least square means ± SE.
The number shown in parentheses stands for the selected individuals.

Table 9 Association of the G4928024A, C4930169T in the *FASN* gene with chicken fat traits

rs15822158 = G4928024A					rs15822181 = C4930169T			
Traits	P,Value	Genotypes			P.Value	Genotypes		
		AA	AB	BB		CC	CT	TT
Subcutaneous fat thickness (mm)	0,7051	4,34 ± 0,39(14)	3,94 ± 0,29(26)	4,03 ± 0,07(438)	0,8198	4,02 ± 0,07 (352)	4,12 ± 0,14 (107)	4,03 ± 0,33 (20)
Fat Bandwidth (mm)	0,0003**	10,52 ± 10,4(14) b	44,02 ± 7,6(26) a	11,98 ± 1,86 (437)b	0,8934	14,21 ± 2,11 (351)	12,32 ± 3,82 (107)	11,96 ± 8,84 (20)
Abdominal fat weight (g)	0,2155	21,98 ± 4,60(14)	22,47 ± 3,37 (26)	27,28 ± 0,82 (437)	0,0491*	25,97 ± 0,92 (351)	30,54 ± 1,66 (107)	24,85 ± 3,85 (20)
Abdominal fat percentage	0,0016**	15,52 ± 11,02 (14)b	48,12 ± 0,09 (26)a	18,18 ± 1,97 (438)b	0,9755	19,71 ± 2,22 (352)	20,34 ± 4,04 (107)	18,18 ± 9,34 (20)

Means with different letter are significantly different ** (*P* > 0, 01); * (*P* > 0,05).
Data are presented at least square means ± SE.
The number shown in parentheses stands for the selected individuals.

[47]. In this study, *ELOVL6* mRNA level was higher in all tissues of XH chickens than of WRR chickens.

LPL is a glycoprotein enzyme that is produced in several tissues of mammals such as adipose tissue, skeletal muscle, heart, macrophages and lactating mammary gland, but not in the liver of adults [48,49]. In chickens, LPL hydrolyzes lipids in lipoproteins, such as those found in chylomicrons and very low-density lipoproteins (VLDL) into three free fatty acid molecules and one glycerol molecule [29,50-52]. In studying the deposition of fat in the abdominal fat pads of chicken, it has become clearer that LPL-catalyzed hydrolysis of triacylglycerol in adipose tissue is a rate-limiting step in fat accumulation [28]. Therefore, the transport and incorporation of exogenous lipids, i.e. plasma VLDL lipoprotein and portomicron, are essential for the deposition of cytoplasmic triglycerides in abdominal adipose tissue. These are characteristics of lipid metabolism in avian species since lipogenic activity is much greater in the liver than in adipose tissue [28,53,54]. This study showed that the *LPL* gene expression was significantly higher in fast-growing chickens than in slow-growing chickens.

The association study provides direct evidence of genes related to fat deposition. In our association study, the A59539099G of the *ELOVL6* gene was significantly associated with subcutaneous fat. The A8378815G of the *DDT* gene was associated with fat band width. The C4930169T of the *FASN* gene was also found to be associated with abdominal fat weight. G1257069A and T1247123C of the *ACSBG2* gene were significantly associated with fat traits. The above results further confirmed that the *ELOVL6, DDT, FASN,* and *ACSBG2* genes are related to chicken fat deposition.

Conclusion

The differential genes expressions in fast and slow growing chickens show differences in fat developmental stage which is supported by lipid-related genes identified and characterized in these two types of chicken. The findings indicate that the variation of the *ACSBG2, FASN, ELOVL 6,* and *DDT* genes were significantly associated with fat deposition.

Table 10 Association of the A59539099G in the *ELOVL 6* and the A8378815G in the *DDT* gene on chicken fat traits

	chr4/ELOVL 6				chr, 15/DDT			
	rs16418687 = A59539099G				rs14092745 = A8378815G			
Traits	*P* value	Genotypes			*P* value	Genotypes		
		AA	AG	GG		AA	AG	GG
Subcutaneous fat thickness (mm)	0,033*	4,05 ± 0,06(475)a	1,820 ± 1,04(2)	0	0,100	4,31 ± 0,18(60)	4,08 ± 0,09(229)	3,96 ± 0,11(185)
Fat Bandwidth (mm)	0,867	16,34 ± 3,19(475)	8,08 ± 49,23(2)	0	0,048*	27,04 ± 5,07(60)	12,02 ± 2,60(228)	11,55 ± 2,89(185)
Abdominal fat weight (g)	0,338	26,84 ± 0,79(474)	15,15 ± 12,16(2)	0	0,649	28,18 ± 5,54(60)	31,17 ± 2,83(229)	26,33 ± 3,16(185)
Abdominal fat percentage	0,811	23,42 ± 3,22(475)	11,47 ± 49,75(2)	0	0,254	34,17 ± 7,41(60)	22,17 ± 3,79(229)	17,25 ± 4,22(185)

Means with different letter are significantly different ** (*P* > 0.01); * (*P* > 0.05).
Data are presented at least square means ± SE.
The number shown in parentheses stands for the selected individuals.

Abbreviations

WRR: White recessive rock; XH: Xinghua; RMA: Robust multi-array; MAS: Molecule annotation system; KEGG: Kyoto encyclopedia of genes and genomes; AgriGO: GO analysis toolkit and database for agricultural community; DAVID: Database for annotation, visualization and integrated discovery.

Competing interests

The authors declare that they have no competing interests.

Authors' contributions

HCA is a correspondence author, conducted all the experiments and written and approved the final manuscript. WP participated in data analysis and approved the final manuscript. SX participated in data collection, laboratory experiment and approved final manuscript. JX participated in data collection, laboratory experiment and approved final manuscript. ZR participated in data collection, laboratory experiment and approved final manuscript. SL corried out the data analysis and approved final manuscript. ZX guided in gene expression analysis and approved final manuscript.

Acknowledgments

Funds for this work were partly provided by The Ministry of Higher Education Student Financing Agency of Rwanda, China Scholarship Council and South China Agricultural University, Guangzhou, China. Further support was provided by the Major State Basic Research Development Program, China (project no. 2006CB102107), and the National High Technology Research and Development Program of China (863 Program, project no. 2007AA10Z163).

Author details

[1]Rwanda Agriculture Board, Research Department, P. O Box 5016, Kigali, Rwanda. [2]Council for Scientific and Industrial Research (CSIR), Animal Research Institute, P. O. Box AH 20, Accra, Achimota, Ghana. [3]Department of Animal Genetics, Breeding and Reproduction, College of Animal Science, South China Agricultural University, Guangzhou, Guangdong 510642, China.

References

1. Ricard FH, Rouvier R: **Etude de la composition anatomique du poulet. 1. Variabilite de la repartition des differentes parties corporelle chez de coquelets "Bresse-Pile".** *Anri Zoorerh* 1967, **16**:23.
2. Proudman W, Mellen J, Anderson DL: **Utilization of feed in fast- and slow-growing lines of chickens.** *Poult Sci* 1970, **49**:961–972.
3. Leclercq B, Blum JC, Boyer JP: **Selecting broilers for low or high abdominal fat: initial observations.** *Br Poult Sci* 1980, **21**:107–113.
4. Becker WA: *Genotypic and phenotypic relations of abdominal fat in chickens.* Kansas City, Missouri: Presented at the 27th Annual National Breeder's roundtable; 1978.
5. Huan-Xian C, Ran-Ran L, Gui-Ping Z, Mai-Qing Z, Ji-Lan C, Jie W: **Identification of differentially expressed genes and pathways for intramuscular fat deposition in *pectoralis major* tissues of fast-and slow-growing chickens.** *BMC Genomics* 2012, **13**:213.
6. Dunnington EA, Siegel PB: **Long-term divergent selection for eight-week body weight in white Plymouth rock chickens.** *Poultry Sci* 1996, **75**:1168–1179.
7. Zhao SM, Ma HT, Zou SX, Chen WH: **Effects of in ovo administration of DHEA on lipid metabolism and hepatic lipogenetic genes expression in broiler chickens during embryonic development.** *Lipids* 2007, **42**:749–757.
8. Zhao X, Mo DL, Li AN, Gong W, Xiao SQ, Zhang Y, Qin LM, Niu YN, Guo YX, Liu XH, Cong PQ, He ZY, Wang C, Li JQ, Chen YS: **Comparative analyses by sequencing of transcriptomes during skeletal muscle development between pig breeds differing in muscle growth rate and fatness.** *PLoS One* 2011, **6**(5):e19774. 1–18.
9. Havenstein GB, Ferket PR, Qureshi MA: **Carcass composition and yield of 1957 versus 2001 broilers when fed representative 1957 and 2001 broiler diets.** *Poult Sci* 2003, **82**:1509–1518.
10. Jakobsson A, Westerberg R, Jacobsson A: **Fatty acid elongases in mammals: their regulation and roles in metabolism.** *Prog Lipid Res* 2005, **45**:237–249.
11. Li WJ, Li HB, Chen JL, Zhao GP, Zheng MQ, Wen J: **Gene expression of heart and adipocyte-fatty acid-binding protein and correlation with intramuscular fat in Chinese chickens.** *Anim Biotechnol* 2008, **19**(3):189–193.
12. Carlborg O, Kerje S, Schütz K, Jacobsson L, Jensen P, Andersson L: **A global search reveals epistatic interaction between QTL for early growth in the chicken.** *Genome Res* 2003, **13**(3):413–421.
13. Zhao SM, Ma HT, Zou SX, Chen WH, Zhao RQ: **Hepatic lipogenesis gene expression in broiler chicken with different fat deposition during embryonic development.** *J Vet Med* 2007, **54**:1–6.
14. Cogburn LA, Wang X, Carre W, Rejto L, Aggrey SE, Duclos MJ, Simon J, Porter TE: **Functional genomics in chickens: development of integrated systems microarrays for transcriptional profiling and discovery of regulatory pathways.** *Comp Funct Genomics* 2004, **5**:253–261.
15. Richards MP, Poch SM, Coon CN, Rosebrough RW, Ashwell CM, McMurtry JP: **Feed restriction significantly alters lipogenic gene expression in broiler breeder chickens.** *J Nutr* 2003, **133**:707–715.
16. Goodridge AG, Ball EG: **Lipogenesis in the pigeon: in vivo studies.** *Am J Physiol* 1967, **213**:245–249.
17. O'Hea EK, Leveille GA: **Lipogenesis in isolated adipose tissue of the domestic chick (gallus domesticus).** *Comp Biochem Physiol* 1968, **26**:111–120.
18. Cai Y, Song Z, Zhang X, Wang X, Jiao H, Lin H: **Increased *de novo* lipogenesis in liver contributes to the augmented fat deposition in dexamethasone exposed broiler chickens (*Gallus gallus domesticus*).** *Comp Biochem Physiol C Toxicol Pharmacol* 2009, **150**(2):164–169.
19. Wang HB, Li H, Wang QG, Zhang XY, Wang SZ, Wang YX, Wang XP: **Profiling of chicken adipose tissue gene expression by genome array.** *BMC Genomics* 2007, **27**(8):193–207.
20. NRC: *Nutrient requirements of poultry.* 9th edition. Washington, DC: National Academy Press; 1994.
21. Lei MM, Nie QH, Peng X, Zhang DX, Zhang XQ: **Single nucleotide polymorphisms of the chicken insulin-like factor binding protein 2 gene associated with chicken growth and carcass traits.** *Poult Sci* 2005, **84**(8):1191–1198.
22. Dodson MV, Hausman GJ, Guan L, Du M, Rasmussen TP, Poulos SP, Mir P, Bergen WG, Fernyhough ME, McFarland DC, Rhoads RP, Soret B, Reecy JM, Velleman SG, Jiang Z: **Lipid metabolism, adipocyte depot physiology and utilization of meat animals as experimental models for metabolic research.** *Int J Biol Sci* 2010, **6**:691–699.
23. Duclos MJ, Wang X, Carre W, Rejto L, Simon J, Cogburn LA: *Nutritional regulation of global gene expression in chicken liver during fasting and re-feeding.* San Diego, CA: Plant Animal Genom XII Conf; 2004.
24. Bourneuf E, Herault F, Chicault C, Carre W, Assaf S, Monnier A, Mottier S, Lagarrigue S, Douaire M, Mosser J, *et al*: **Microarray analysis of differential gene expression in the liver of lean and fat chickens.** *Gene* 2006, **372**:162–170.
25. Ashrafi K, Chang FY, Watts JL, Fraser AG, Kamath RS, Julie A, Gary R: **Genome-wide RNAi analysis of Caenorhabditis elegans fat regulatory genes.** *Nature* 2003, **421**:268–272.
26. Collin A, Swennen Q, Skiba-Cassy S, Buyse J, Chartrin P, Le Bihan-Duval E, Crochet S, Duclos MJ, Joubert R, Decuypere E, Tesseraud S: **Regulation of fatty acid oxidation in chicken (Gallus gallus): Interactions between genotype and diet composition.** *Comp Biochem Physiol* 2009, **Part B 153**:171–177.
27. Jump DB, Botolin D, Wang Y, Xu J, Christian B, Demeure O: **Fatty acid regulation of hepatic gene transcription.** *J Nutr* 2005, **135**(11):2503–2506.
28. Sato K, Akiba Y, Chida Y, Takahashi K: **Lipoprotein hydrolysis and fat accumulation in Chicken adipose tissues are reduced by chronic administration of lipoprotein lipase monoclonal antibodies.** *Poult Sci* 1999, **78**:1286–1291.
29. Sato K, Sook SH, Kamada T: **Tissue distribution of lipase genes related to triglyceride metabolism in laying hens (*Gallus gallus*).** *Comp Biochem Physiol* 2010, **155**(1):62–66.
30. Fried SK, Russell CD, Grauso NL, Brolin RE: **Lipoprotein lipase regulation by insulin and glucocorticoid in subcutaneous and omental adipose tissues of obese women and men.** *J Clin Invest* 1993, **92**:2191–2198.
31. Leveille GA: **In vitro hepatic lipogenesis in the hen and chick.** *Comp Biochem Physiol* 1969, **28**:431–435.
32. Benistant C, Duchamps C, Cohen-Adad F, Rouanet JL, Barre H: **Increased in vitro fatty acid supply and cellular transport capacities in cold-acclimated ducklings (*Cairina moschata*).** *Am J Physiol* 1998, **275**:R683–R690.
33. Saez G, Davail S, Gentès G, Hocquette JF, Jourdan T, Degrace P, Baéza E: **Gene expression and protein content in relation to intramuscular fat content in Muscovy and Peki ducks.** *Poult Sci* 2009, **88**:2382–2391.

34. Back DW, Goldman MJ, Fisch JE, Ochs RS, Goodridge AG: **The fatty acid synthase gene in avian liver. Two mRNAs are expressed and regulated in parallel by feeding, primarily at the level of transcription.** *J Biol Chem* 1986, **261**:4190–4197.
35. Goodridge AG, Ball EG: **The Effect of Prolactin on Lipogenesis in the Pigeon. In Vitro Studies.** *Biochemistry* 1967, **6**(8):2335–2343.
36. Leveille GA, O'Hea EK, Chakrabarty K: **In vivo lipogenesis in the domestic chicken.** *Proc Soc Exp Biol Med* 1968, **128**:398–401.
37. Leveille GA, Romsos DR, Yeh YY, O'Hea EK: **Lipid biosynthesis in the chick. A consideration of site of synthesis, influence of diet and possible regulating mechanisms.** *Poult Sci* 1975, **54**:1075–1093.
38. Demeure O, Duby C, Desert C, Assaf S, Hazard D, Guillou H, Lagarrigue S: **Liver X receptor regulates fatty acid synthase expression in chicken.** *Poult Sci* 2009, **88**:2628–2635.
39. Shimba S, Ishii N, Ohta Y, Ohno T, Watabe Y, Hayashi M, *et al*: **Brain and muscle arnt-like protein-1 (BMAL1), a component of the molecular clock, regulates adipogenesis.** *Proc Natl Acad Sci U S A* 2005, **102**:12071–12076.
40. Gómez-Santos C, Gómez-Abellán P, Madrid JA, Hernández-Morante JJ, Lujan JA, Ordovas JM, Garaulet M: **Circadian rhythm of clock genes in human adipose explants.** *Obesity* 2009, **17**(8):1481–5.30.
41. Wu X, Zvonic S, Floyd ZE, Kilroy G, Goh BC, Hernandez TL, *et al*: **Induction of circadian gene expression in human subcutaneous adipose-derived stem cells.** *Obesity* 2009, **15**(11):2560–2570.
42. Baulande S, Lasnier F, Lucas M, Pairault J: **Adiponutrin, a transmembrane protein corresponding to a novel dietary and obesity-linked mRNA specifically expressed in the adipose lineage.** *J Biol Chem* 2001, **276**:33336–33344.
43. Dooley TP, Haldeman-Cahill R, Joiner J, Wilborn TW: **Expression profiling of human sulfotransferase and sulfatase gene superfamilies in epithelial tissues and cultured cells.** *Biochem Biophys Res Commun* 2000, **277**:236–245.
44. Yazdi M, Ahnmark A, William-Olsson L, Snaith M, Turner N, Osla F, Wedin M, Asztély AK, Elmgren A, Bohlooly-Y M, Schreyer S, Lindén D: **The role of mitochondrial glycerol-3 phosphate acyltransferase-1 in regulating lipid and glucose homeostasis in high-fat diet fed mice.** *Biochem Biophys Res Commun* 2008, **369**:1065–1070.
45. Palou M, Priego T, Sánchez J, Villegas E, Rodríguez AM, Palou A, Picó C: **Sequential changes in the expression of genes involved in lipid metabolism in adipose tissue and liver in response to fasting.** *Pflugers Arch* 2008, **456**(5):825–836.
46. Marsit CJ, Houseman EA, Christensen BC, Gagne L, Wrensch MR, *et al*: **Identification of methylated genes associated with aggressive bladder cancer.** *PLoS One* 2010, **5**(8):e12334.
47. Kan S, Abe H, Kono T, Yamazaki M, Nakashima K, Kamada T, Akiba Y: **Changes in peroxisome proliferator-activated receptor gamma gene expression of chicken abdominal adipose tissue with different age, sex and genotype.** *Anim Sci J* 2009, **80**:322–327.
48. Hoenig M, McGoldrick JB, De Beer M, Demacker PNM, Ferguson DC: **Activity and tissue-specific expression of lipases and tumor-necrosis factor a in lean and obese cats.** *Domest Anim Endocrinol* 2006, **30**:333–344.
49. Albalat A, Saera-Vila A, Capilla E, Gutierrez J, Perez-Sanchez J, Navarro I: **Insulin regulation of lipoprotein lipase (LPL) activity and expression in gilthead Sea bream (sparus aurata).** *Comp Biochem Physiol B* 2007, **148**:151–159.
50. Sato K, Akiba Y: **Lipoprotein lipase mRNA expression in abdominal adipose tissue is little modified by age and nutritional state in broiler chickens.** *Poultry Sci* 2002, **81**:846–885.
51. Nie Q, Fang M, Xie L, Shi J, Zhang X: **cDNA cloning, characterization, and variation analysis of chicken adipose triglyceride lipase (ATGL) gene.** *Mol Cell Biochem* 2009, **320**:67–74.
52. D'André Hirwa C, Yan W, Wallace P, Nie Q, Luo C, Li H, Shen X, Sun L, Tang J, Li W, Zhu X, Yang G, Zhang X: **Effects of the thyroid hormone responsive spot 14 {alpha} gene on chicken growth and fat traits.** *Poult Sc* 2010, **89**(9):1981–1991.
53. Griffin HD, Hermier D: **Plasma lipoprotein metabolism and fattening on poultry.** In *Leanness in domestic birds.* Edited by Leclercq B, Whitehead CC. London: Butterworths; 1988:175–201.
54. Griffin HD, Guo K, Windsor D, Butterwith SC: **Adipose tissue lipogenesis and fat deposition in leaner broiler chickens.** *J Nutr* 1996, **122**:363–368.

Hematologic and biochemical reference intervals for specific pathogen free 6-week-old Hampshire-Yorkshire crossbred pigs

Caitlin A Cooper[1], Luis E Moraes[1], James D Murray[1,2] and Sean D Owens[3*]

Abstract

Background: Hematologic and biochemical reference intervals depend on many factors, including age. A review of the literature highlights the lack of reference intervals for 6-wk-old specific pathogen free (SPF) Hampshire-Yorkshire crossbred pigs. For translational research, 6-wk-old pigs represent an important animal model for both human juvenile colitis and diabetes mellitus type 2 given the similarities between the porcine and human gastrointestinal maturation process. The aim of this study was to determine reference intervals for hematological and biochemical parameters in healthy 6-wk-old crossbred pigs. Blood samples were collected from 66 clinically healthy Hampshire-Yorkshire pigs. The pigs were 6 wks old, represented both sexes, and were housed in a SPF facility. Automated hematological and biochemical analysis were performed using an ADVIA 120 Hematology System and a Cobas 6000 C501 Clinical Chemistry Analyzer.

Results: Reference intervals were calculated using both parametric and nonparametric methods. The mean, median, minimum, and maximum values were calculated.

Conclusion: As pigs are used more frequently as medical models of human disease, having reference intervals for commonly measured hematological and biochemical parameters in 6-wk-old pigs will be useful. The reference intervals calculated in this study will aid in the diagnosis and monitoring of both naturally occurring and experimentally induced disease. In comparison to published reference intervals for older non SPF pigs, notable differences in leukocyte populations, and in levels of sodium, potassium, glucose, protein, and alkaline phosphatase were observed.

Keywords: Biochemical analytes, Hematology, Pigs, Reference interval, SPF

Introduction

Pigs are emerging as a useful model for studying gastrointestinal (GI) tract and metabolic development and dysfunction [1,2], and may prove to be a particularly good model for investigating the role of GI tract disturbances in inflammatory disease such as inflammatory bowel disease [3] and type 2 diabetes [4]. To better utilize young pigs as a model and understand changes in circulating leukocyte populations and blood chemistry during disease states, reference intervals for clinically healthy, specific pathogen free (SPF), young post-weaning pigs must be established. Reference intervals exist for different ages of pigs including 3-wk-old pigs [5], twelve-wk old pigs [6], and adult pigs [7]. However these pigs are from commercial, non-SPF populations which may be harboring common porcine pathogens [8-10], which could affect the reference intervals. For biomedical research SPF pigs are often used [11,12], and while there are reports of hematological and chemical parameters in SPF mini-pigs [13], there is no reference interval for 6-wk-old SPF pigs that are not derived from miniature pig lines. The aim of this study was to determine reference intervals for hematological and biochemical parameters in healthy 6-wk-old Hampshire-Yorkshire crossbred pigs raised in a SPF facility. The guidelines established by the American Society for Veterinary Clinical Pathology (ASVCP) were utilized to determine the number of animals needed and the correct procedures for determining

* Correspondence: sdowens@ucdavis.edu
[3]Department of Pathology, Microbiology and Immunology, School of Veterinary Medicine, University of California, One Shields Avenue, Davis, CA 95616, USA
Full list of author information is available at the end of the article

the reference intervals based on the distribution of each parameter.

Materials and methods

Animals

Blood samples were obtained from 66 Hampshire Yorkshire crossbred pigs. The pigs were 6-wks of age and represented both sexes: male (n = 40) and female (n = 26), and weighed between 10–20 kg. All pigs used in this study were examined and considered clinically healthy by a veterinarian. They had normal skin color, body condition and activity.

Housing

Pigs (*Sus scrofa*) originated from a closed herd and were bred, born, and raised at the University of California swine facility, which is a SPF facility for *Mycoplasma hyopneumoniae, Actinobacillus pleuropnemoniae*, porcine reproductive and respiratory syndrome (PRRS) virus, atrophic rhinitis (toxigenic *Pasteurella multocida*), influenza, *Brachyspira hyodysenteriae*, transmittable gastro-enteritis (TGE), *Salmonella typhimurium* and *S. choleraesuis*, internal and external parasites, brucellosis, and pseudorabies virus (PRV). Disease monitoring consists of routine slaughter checks performed by a licensed veterinarian on animals originating from the facility, including lung evaluation and inspection of the nasal passages for signs of atrophic rhinitis. At least four times a yr blood samples collected from adults within the herd undergo serology and PCR analysis at the University of California, Davis, Veterinary Teaching Hospital (VMTH) Clinical Laboratory to screen for all excluded pathogens. Once pigs are weaned, a full necropsy, including screening of feces for pathogens, is conducted on any pig that dies unexpectedly. The necropsies are performed by American College of Veterinary Pathologist (ACVP) board-certified pathologists at the California Animal Health and Food Safety (CAFHS) laboratory (UC Davis, Davis, CA, USA).

Husbandry

All 66 pigs had their incisor teeth clipped, ears notched, tails docked, and were dosed with 1 mL of oral antibiotic (Spectogard, Bimeda Inc., LeSueur, MN), at 1 d old. At d 3 of age all pigs received an intra-muscular injection of 100 mg iron dextran-200 (Durvet, INC., Blue Springs, MO) and male pigs were castrated. At d 21 of age the pigs were weaned and vaccinated with Fostera (Pfizer Animal Health, New York, NY) for porcine circovirus, then cohoused in mixed litter pens. Once weaned, pigs started to consume Pig A2000 Pellet Denagard/CTC starter diet (Akey, Brookville, OH) containing lactose, cereal food fines, soybean meal, oat groats, ground corn, animal plasma, poultry meal, fishmeal, cheese meal, vegetable and animal fat, and 0.0005% of Lincomix (Pfizer Animal Health, New York, NY) as an antibiotic growth promoter. This diet provided 21% crude protein, 8% crude fat, and 2% crude fiber. Pigs were switched to a standard grower diet (Associated Feed, Turlock, CA) after 2 wk. The grower diet contained wheat millrun, fat mixer, ground corn, blood meal, whole dried whey, soybean meal, Swine Micro 4 mix (Akey, Brookville, OH), and Tylan 40 antibiotic (Elanco Animal Health, Indianapolis, IN) at 0.00004%. This diet provided 20% protein, 7% crude fat, 2% crude fiber, and metabolizable energy of 13.6 MJ/kg. By 6 wks of age pigs weighed between 10 and 20 kg.

Blood collection

Pigs were placed in a recumbent position on a V shaped table to restrict their movement and blood was collected from the cranial vena cava. Samples for hematologic analysis were collected into 10 mL tubes containing EDTA (Becton Dickinson Company, Franklin Lakes, NJ); samples for biochemical analysis were collected into 5 mL empty serum collection tubes (Becton Dickinson Company, Franklin Lakes, NJ) The use of all animals in this study was approved by the UC Davis Institutional Animal Care and Use Committee, and study subjects were

Table 1 Methods of clinical analysis

Analyte	Method/Principle	Reaction type
Anion gap	Calculated	
Sodium	ISE, diluted	Potentiometry*
Potassium	ISE, diluted	Potentiometry*
Chloride	ISE, diluted	Potentiometry*
Bicarbonate	PEPC/NADH-NAD+	Zero-order kinetic*
Inorganic phosphate	Phosphomolybdate/ UV 340 nm	Endpoint*
Calcium	Schwarzenbach/UV 600 nm	Endpoint*
Urea nitrogen	Enzymatic: urease with GLDH	First-order kinetic*
Creatinine	Jaffe	First-order kinetic*
Glucose	Hexokinase	Endpoint*
Total protein	Biuret	Endpoint*
Albumin	Bromocresol green	Endpoint*
Globulin	Calculated	
AST	Modified IFCC	Zero-order kinetic*
Creatine kinase	Modified IFCC	Zero-order kinetic*
Alkaline phosphatase	Modified IFCC	Zero-order kinetic*
GGT	Modified IFCC	Zero-order kinetic*
Bilirubin total	Diazo	Endpoint*
SDH-37	D-Fructose to D-Sorbitol/ NADH/UV 340 nm	Zero-order kinetic

All analysis were carried out at 37°C.
AST, aspartate aminotransferase; GGT, γ-glutamyltransferase; SDH-37, sorbitol dehydrogenase; IFCC, International Federation of Clinical Chemistry; ISE, Ion specific electrode.
*Roche Diagnostics GmbH.

raised under an Association for Assessment and Accreditation of Laboratory Animal Care International (AAALAC) approved animal care program.

Hematology and blood chemistry

Following collection, blood samples were stored at 4°C before being delivered to the University of California, Davis, Veterinary Teaching Hospital (VMTH) Clinical Laboratory. Samples were analyzed within 4 h of collection. Hematological parameters were analyzed using an ADVIA® 120 Hematology System (Siemens Healthcare Diagnostics Inc., Tarrytown, NY) with a species-specific setting for pigs in the MultiSpecies System Software (Siemens Medical Solutions Diagnostics Inc., Tarrytown,

Table 2 Hematologic and clinical biochemical reference intervals for 6-week-old Hampshire-Yorkshire pigs between 10–20 kg

Analyte	Unit	N	Mean	Median	Min	Max	Reference interval	Data distribution
Hematology analytes								
RBC	m/uL	66	7.31	7.33	5.84	9.04	5.52 - 9.11	Non-Gaussian
HGB	gm/dL	65	10.7	10.8	8.6	12.8	8.8 - 12.7	Gaussian
HTC	%	66	35.5	35.7	25.4	43.8	28.3 - 42.7	Gaussian
MCV	fL	66	48.9	49.8	37.7	59.1	38.4 - 59.3	Gaussian
MCH	pgm	66	14.8	15.3	11.1	18.3	11.1 - 18.4	Non-Gaussian
MCHC	gm/dL	66	30.2	30.3	27.3	32.4	27.9 - 32.4	Gaussian
RDW	%	54	24.4	24.5	16.1	33.3	16.4 - 32.3	Gaussian
WBC	/µL	66	15,315	14,735	5,650	26,470	5,443 - 25,186	Gaussian
Neutrophils	/µL	66	5,668	5,266.5	1,192	15,745	810 - 13,397	Gaussian with square root transformation
Lymphocytes	/µL	66	8,677	8,239	4,045	15,856	3,810 - 14,919	Gaussian with square root transformation
Monocytes	/µL	66	692	619	237	1,460	219 - 1,705	Gaussian with logarithmic transformation
Eosinophils	/µL	64	219	201.5	58	574	45 - 481	Gaussian with square root transformation
Basophils	/µL	64	53	43	11	151	14 - 146	Gaussian with logarithmic transformation
Platelets	/µL	66	540,773	545,500	138,000	909,000	208,588 - 872,957	Gaussian
Biochemical analytes								
Anion gap	mmol/L	63	21	20	13	31	14 - 29	Gaussian with logarithmic transformation
Sodium	mmol/L	63	141	140	125	159	131 - 151	Non-Gaussian
Potassium	mmol/L	62	4.9	4.8	3.7	6.3	3.7 - 6.1	Gaussian with square root transformation
Chloride	mmol/L	63	100	100	90	112	93 - 108	Non-Gaussian
Bicarbonate	mmol/L	63	25	25	17	31	19 - 31	Gaussian
Phosphorus	mg/dL	63	8.9	8.8	6.1	12.3	6.3 - 11.5	Gaussian
Calcium	mg/dL	63	11.2	11.2	9.7	12.8	9.9 - 12.5	Gaussian
BUN	mg/dL	63	10	10	4	21	4 - 18	Gaussian with square root transformation
Creatinine	mg/dL	63	0.8	0.8	0.5	1.1	0.5 - 1.1	Non-Gaussian
Glucose	mg/dL	61	106	105	70	139	75 - 136	Gaussian
Total protein	g/dL	63	4.9	4.9	4.1	5.9	4.0 - 5.8	Gaussian
Albumin	g/dL	63	3.9	3.9	3.2	4.7	3.1 - 4.8	Non-Gaussian
Globulin	g/dL	63	1.0	0.9	0.2	1.8	0.3 - 1.7	Gaussian
AST	IU/L	63	44	36	13	141	13 - 111	Gaussian with logarithmic transformation
Creatine kinase	IU/L	62	1,358	921	170	6,823	153 - 5,427	Gaussian with logarithmic transformation
Alkaline phosphatase	IU/L	63	297	280	135	603	130 - 513	Gaussian with square root transformation
GGT	IU/L	63	57	55	34	112	33 - 94	Gaussian with logarithmic transformation
Total bilirubin	mg/dL	63	0.1	0.1	0	0.4	0 - 0.2	Non-Gaussian
SDH-37	IU/L	63	0.5	0	0	4	0 - 1.7	Non-Gaussian

RBC, red blood cells; HGB, hemoglobin; HTC, hematocrit; MCV, mean corpuscular volume; MCHC, mean corpuscular hemoglobin concentration; RDW, erythrocyte distribution width; WBC, white blood cells; BUN, blood urea nitrogen; AST, aspartate aminotransferase; GGT, γ-glutamyltransferase; SDH-37,sorbitol dehydrogenase.

NY, USA). The within-laboratory imprecision for the automated differentials (coefficient of variation, CV) for each variable, as determined by the VMTH Hematology Laboratory, is as follows: RBC 1.0%, HGB 0.8%, MCV 0.4%, RDW 0.6%, WBC 2.7%, and absolute counts for neutrophils 1.6%, lymphocytes 2.9%, monocytes 6.9%, eosinophils 8.8%, basophils 20%, and platelets 2.7%.

Blood chemistry analysis was performed using a Cobas® 6000 C501 Clinical Chemistry Analyzer (Roche Diagnostics, Indianapolis, IN) (Table 1). The CVs for each variable, as determined by the VMTH Clinical Chemistry Laboratory, are as follows: sodium 0.4%, potassium 0.0%, chloride 0.5%, bicarbonate 2.5%, phosphorus 1.3%, calcium 1.1%, BUN 0.7%, creatinine 1.8%, glucose 1.2%, total protein 0.9%, albumin 1.7%, AST 0.8%, creatine kinase 0.8%, alkaline phosphatase 0.5%, GGT 1.0%, total bilirubin 1.7% and SDH-37 2.2%.

Statistical analysis

The identification of outliers was conducted according to Grubbs [14]. Outliers were removed from the data and all variables were tested for Gaussian distribution using the Shapiro-Wilk test with a significance level of 5%. Variables that were not normally distributed were transformed with square root or log transformations. Reference intervals for such variables were calculated according to the ASVCP guidelines as the sample mean ± 2 standard deviations [15]. For variables that could not be transformed to Gaussian distribution, reference intervals were calculated through a bootstrap procedure. To accomplish this 10,000 bootstrap samples were generated though random sampling and replacement of values in the original dataset. The bootstrap generated standard error and standard deviation were then calculated according to Dimauro et al. [16]. Reference intervals were estimated using the bootstrap mean ± 2 bootstrap standard deviations.

Results

A total of 66 blood samples for hematologic analysis and 63 blood samples for biochemical analysis were collected. Outliers were identified and removed from the following datasets: hemoglobin, eosinophils, basophils, potassium, glucose, and creatine kinase. Due to a laboratory error the RDW was not measured for 12 samples. The means, medians, minimum and maximum values, reference intervals, and data distributions for the hematological and biochemical parameters are presented in Table 2.

Discussion

Hematological and biochemical parameters are affected by a variety of factors including age, sex, nutritional and health status, breed, season, and stress [17]. When evaluating results from hematological and biochemical tests these factors must be considered. The 6-wk-old pigs used in this study were all healthy, reared in the same conditions at a SPF facility, fed the same diet, and from a similar genetic background. These genetic, environmental, and nutritional factors should be considered when interpreting the hematological data presented.

The reference intervals of many hematological parameters including HCT, neutrophils, lymphocytes, monocytes, eosinophils, platelets, BUN, glucose, AST, and creatine have a wide range. This high level of variability in the circulating leukocytes is expected because in 6-wk-old pigs those populations are still expanding. The population of pig's sampled weight ranged from 10 to 20 kg, so parameters such as glucose and creatine which are respectively correlated to adiposity [18] and muscle growth [19], also display a high level of variability.

Pigs are becoming a more common animal model for biomedical research and using SPF pigs helps reduce confounding factors, such as sub-clinical disease, from skewing research results. The results from our study establish reference intervals for both hematological and biochemical parameters in six-wk-old SPF pigs. Six-wk-old pigs are a good animal model because at that age, six-wk old pigs are post-weaning, are undergoing rapid growth, and their immune systems are still maturing. Similarities between porcine and human gastrointestinal and immune system development highlight how the growing pig could represent an important animal model for the study of gastrointestinal and metabolic disease in growing children.

Competing interests

The authors declare that they have no competing interests.

Authors' contributions

CAC: Participated in the design of the study, performed sample collection, and was responsible for drafting the manuscript. LEM: Performed statistical analysis. JDM: Participated in the design of the study. SDO: Conceived of the study, participated in the study design and oversaw sample analysis. All authors read and approved the final manuscript.

Acknowledgements

We would like to thank Kent Parker and the staff of the UC-Davis Swine facility as well as Elizabeth Maga, Lydia Garas Klobas, Leslie Stewart, Justin Nunes, Erica Scott, Merritt Clark, and Sammi Lotti for their technical assistance.

Author details

[1]Department of Animal Science, University of California, Davis, USA. [2]Department of Population Health and Reproduction, University of California, Davis, USA. [3]Department of Pathology, Microbiology and Immunology, School of Veterinary Medicine, University of California, One Shields Avenue, Davis, CA 95616, USA.

References

1. Ji MH, Yang JJ, Wu J, Li RQ, Li GM, Fan YX, Li WY: **Experimental sepsis in pigs-effects of vasopressin on renal, hepatic, and intestinal dysfunction.** *Ups J Med Sci* 2012, **117**:257–263.
2. Spurlock ME, Gabler NK: **The development of porcine models of obesity and the metabolic syndrome.** *J Nutr* 2008, **138**:397–402.

3. Aloi M, Tromba L, Di Nardo G, Dilillo A, Del Giudice E, Marocchi E, Viola F, Civitelli F, Berni A, Cucchiara S: **Premature subclinical atherosclerosis in pediatric inflammatory bowel disease.** *J Pediatr* 2012, **161**:589–594.
4. Yang Y, Hayden MR, Sowers S, Bagree SV, Sowers JR: **Retinal redox stress and remodeling in cardiometabolic syndrome and diabetes.** *Oxid Med Cell Longev* 2010, **3**:392–403.
5. Egeli AK, Framstad T, Morberg H: **Clinical biochemistry, haematology and body weight in piglets.** *Acat Vet Scand* 1998, **39**:381–393.
6. Klem TB, Bleken E, Morberg H, Thoresen SI, Framstad T: **Hematologic and biochemical reference intervals for Norwegian crossbreed grower pigs.** *Vet Clin Pathol* 2010, **39**:221–226.
7. Framstad T, Morberg H, Aas RA: **Biochemical analysis of blood. Reference values for sows.** *Nor Vet Tidsskr* 1991, **103**:807–815.
8. Opriessnig T, Giménez-Lirola LG, Halbur PG: **Polymicrobial respiratory disease in pigs.** *Anim Health Res Rev* 2011, **12**:133–148.
9. Pontes NE, Barbosa CN, Jesus AL, Silva JG, Freitas AC: **Development and Evaluation of Single-tube nested PCR (STNPCR) for the detection of Porcine Circovirus type 2 (PCV2).** *Transbound Emerg Dis* 2012. PMID:23078249 [e-published ahead of print].
10. Wang R, Qiu S, Jian F, Zhang S, Shen Y, Zhang L, Ning C, Cao J, Qi M, Xiao L: **Prevalence and molecular identification of Cryptosporidium spp. in pigs in Henan, China.** *Parasitol Res* 2010, **107**:1489–1494.
11. Kim HI, Lee SY, Jin SM, Kim KS, Yu JE, Yeom SC, Yoon TW, Kim JH, Ha J, Park CG, Kim SJ: **Parameters for successful pig islet isolation as determined using 68 specific-pathogen-free miniature pigs.** *Xenotransplantation* 2009, **16**:11–18.
12. Takahashi J, Waki S, Matsumoto R, Odake J, Miyaji T, Tottori J, Iwanaga T, Iwahashi H: **Oligonucleotide microarray analysis of dietary-induced hyperlipidemia gene expression profiles in miniature pigs.** *PLoS One* 2012, **7**:e37581.
13. Yeom SC, Cho SY, Park CG, Lee WJ: **Analysis of reference interval and age-related changes in serum biochemistry and hematology in the specific pathogen free miniature pig.** *Lab Anim Res* 2012, **28**:245–253.
14. Grubbs FE: **Sample criteria for testing outlying observations.** *Ann Math Stat* 1950, **21**:27–58.
15. Friedrichs K, Barnhart K, Blanco J: **ASVCP Quality Assurance and Laboratory Standards Committee (QALS) Guidelines for the determination of reference intervals in veterinary species and other related topics.** [www.asvcp.org/membersonly/ReferenceInterval.cfm]
16. Dimauro C, Macciota NPP, Rassu SPG, Patta C, Pulina G: **A bootstrap approach to estimate reference intervals of biochemical variables in sheep using reduced sample sizes.** *Small Ruminant Res* 2009, **83**:34–41.
17. Thorn CE: **Normal hematology of the pig.** In *Schalm's veterinary hematology.* 5th edition. Edited by Feldman BF, Zinkl JG, Jain NC. Baltimore MD: Lippincott Williams & Wilkins; 2000:1089–1095.
18. Going SB, Lohman TG, Cussler EC, Williams DP, Morrison JA, Horn PS: **Percent body fat and chronic disease risk factors in U.S. children and youth.** *Am J Prev Med* 2011, **41**:S77–S86.
19. Dubreuil P, Lapierre H: **Biochemistry reference values for Quebec lactating dairy cows, nursing sows, growing pigs and calves.** *Can J Vet Res* 1997, **61**:235–239.

4

GMOs in animal agriculture: time to consider both costs and benefits in regulatory evaluations

Alison L Van Eenennaam

Abstract

In 2012, genetically engineered (GE) crops were grown by 17.3 million farmers on over 170 million hectares. Over 70% of harvested GE biomass is fed to food producing animals, making them the major consumers of GE crops for the past 15 plus years. Prior to commercialization, GE crops go through an extensive regulatory evaluation. Over one hundred regulatory submissions have shown compositional equivalence, and comparable levels of safety, between GE crops and their conventional counterparts. One component of regulatory compliance is whole GE food/feed animal feeding studies. Both regulatory studies and independent peer-reviewed studies have shown that GE crops can be safely used in animal feed, and rDNA fragments have never been detected in products (e.g. milk, meat, eggs) derived from animals that consumed GE feed. Despite the fact that the scientific weight of evidence from these hundreds of studies have not revealed unique risks associated with GE feed, some groups are calling for more animal feeding studies, including long-term rodent studies and studies in target livestock species for the approval of GE crops. It is an opportune time to review the results of such studies as have been done to date to evaluate the value of the additional information obtained. Requiring long-term and target animal feeding studies would sharply increase regulatory compliance costs and prolong the regulatory process associated with the commercialization of GE crops. Such costs may impede the development of feed crops with enhanced nutritional characteristics and durability, particularly in the local varieties in small and poor developing countries. More generally it is time for regulatory evaluations to more explicitly consider both the reasonable and unique risks and benefits associated with the use of both GE plants and animals in agricultural systems, and weigh them against those associated with existing systems, and those of regulatory inaction. This would represent a shift away from a GE evaluation process that currently focuses only on risk assessment and identifying ever diminishing marginal hazards, to a regulatory approach that more objectively evaluates and communicates the likely impact of approving a new GE plant or animal on agricultural production systems.

Keywords: Cost: benefit analysis, Genetic engineering, GMO, Regulation, Risk assessment, Safety

Introduction

A high proportion of soybean (81%), cotton (81%), corn (35%), and canola (30%) crops grown globally are genetically engineered (GE) varieties (Figure 1) [1]. It has been estimated that over 70-90% of harvested GE biomass is fed to food producing animals [2], making the world's livestock populations the largest consumers of the current generation of GE crops. Crops that are produced using GE are likely to become even more important to animal agriculture as the global livestock population grows in response to increased demand for animal protein products. Prior to commercialization, GE crops must go through an extensive safety evaluation. The Organisation for Economic Co-operation and Development (OECD) has established safety assessment processes based on the principle of "substantial equivalence" to assure that foods derived from GE crops are as safe and nutritious as those from plants derived through conventional breeding [3]. The concept is based on the principle that "*if a new food is found to be substantially equivalent in composition and nutritional characteristics to an existing food, it can be regarded as being as safe as the conventional food*" [4]. For GE crops, this comparison entails an extensive chemical analysis of key macronutrients, micronutrients, antinutrients and toxins. Most

Correspondence: alvaneenennaam@ucdavis.edu
Department of Animal Science, 2113 Meyer Hall, University of California, One Shields Avenue, Davis, CA 95616, USA

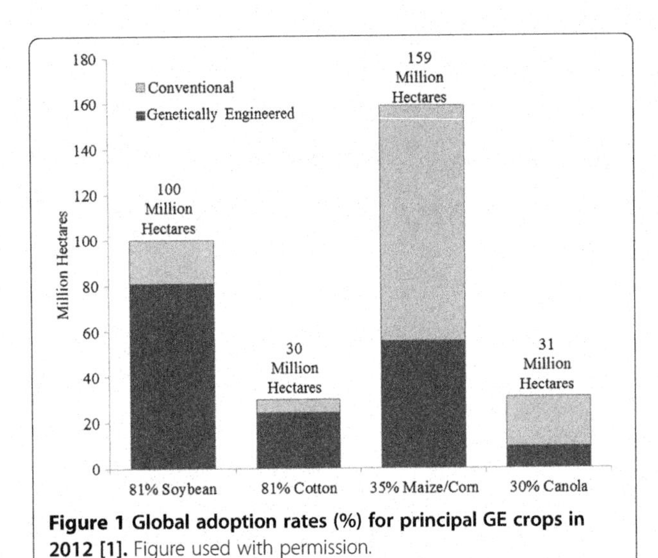

Figure 1 Global adoption rates (%) for principal GE crops in 2012 [1]. Figure used with permission.

conventionally-bred crops that are on the market have not ever been tested for their safety in animals, but they are known to be safe based on their history of safe use. Likewise, few foods have been subject to toxicological testing.

Inclusion in a balanced diet has often been a component of the regulatory approval package for a new GE crop. These studies are typically performed with rodents for a duration of 90 d. Although this is technically only required when the composition of the GE plant has been modified substantially, or there are indications for the potential occurrence of unintended effects [5], some are calling for an increase in the number, length, and target species (i.e. livestock) involved in animal feeding studies required for the regulatory approval of GE crops.

On June 2013, the European Commission published a Regulation (EU) No 503/2013 requiring an obligatory 90-d whole food/feed rodent feeding study for regulatory approval of each GE crop event [6]. Depending on the outcome of this study, a 2-year long-term GE feeding study in rats may also be requested, on a case-by-case basis. This Regulation passed despite the fact that the European Food Safety Authority (EFSA) questioned the need to provide such studies for the risk evaluation of each GE crop application as follows: "*When 'molecular, compositional phenotypic, agronomic and other analyses have demonstrated equivalence of the GM food/feed, animal feeding trials do not add to the safety assessment*" [5]. There are considerable costs involved in performing animal feeding studies. In 2007 it was estimated that the range of costs involved with animal performance and safety studies (typically a 90-d whole food/feed rodent feeding study) for approval of a GE crop ranged from $USD 300,000–845,000 [7]. Presumably costs have increased since that time, and if longer studies are required costs would likewise be increased. A recent two-year rat feeding study involving 200 rats was purported to cost 3.2 million Euro (~ $USD 4 million) [8]. Calls to do long-term or mutigenerational GE feeding studies on long-lived target species such as cattle would be orders of magnitude more expensive; assuming sufficient feed from the GE crop and its isogenic comparator was available to perform such work. Additionally the cost of rendering the animals would need to be factored into regulatory evaluations as animals would not be able to enter the food supply if fed an as yet unapproved GE crop variety.

The purpose of this paper is to review the rationale and results of peer-reviewed animal feeding studies using GE crops. Results from short-term and long-term studies are evaluated to determine if additional information was identified in these long-term studies that would not have been picked up in the short-term study, and the details of some highly controversial studies are reviewed. It is suggested that animal feeding studies should only be required if there is some reasonable food safety concern indicated during the regulatory evaluation of GE crops that has not been adequately addressed by *in silico* and *in vitro* analyses. Further, the need to evaluate both the risks and benefits in regulatory evaluations is discussed given the weight of scientific evidence on the safety and performance of GE crops that have been commercialized to date.

Short-term rodent feeding studies

The protocols for 90-d rodent studies were adapted from those for *in vivo* toxicological studies [9] and are intended to assess feed safety. This protocol recommends 10 animals per sex and per group, with three doses of the test substance and a control group. It was developed to test the toxicology of a chemically defined molecule (e.g. a drug), not complex materials like GE feed. It becomes somewhat problematic to appropriately "dose" the GE feed because diets must be balanced to meet the nutritional requirements of the rodents. Too much of a single crop or species in the diet may result in deleterious nutritional effects and associated phenotypes, independent of the GE status of the crop. GE feeding studies typically incorporate 33% GE animal feed in the test diet. Ideally, the GE line is compared to its near isogenic counterpart grown in the same location and environment, and possibly also a non-GE line (conventional comparator) considered to be safe. The latter is included to estimate the natural variability of analytes seen within the crop species. Several studies have revealed that environmental factors (such as field location, planting, sampling time, crop management practices), and genetic factors like line/breed and mutagenesis can result in more variability in gene expression between samples than is observed resulting from GE [10-12]. The failure of many researchers to appropriately match their experimental GE diets to appropriate isogenic and nutritionally equivalent control diets has resulted in some of the most controversial, and highly criticized, GE feed safety studies.

Long-term and multigenerational animal feeding studies

Ninety-day rodent toxicology feeding studies are not designed to measure effects on reproduction or development. Likewise, they are not designed to detect long term effects in animals, or the effect that eating a GE-based diet has on the next generation. This has resulted in a call for more long term and multigenerational animal feeding studies. Although, it should be noted that analyses of available data indicate that, for a wide range of substances, reproductive and developmental effects are not potentially more sensitive endpoints than those examined in subchronic toxicity tests [13]. Several review papers that summarize the results of long-term and multigenerational feeding studies in a variety of species have been published recently [2,14-16]. The duration of published long-term feeding studies using a GE-based diet ranged from 110 d [17-19] to 728 d [20]. The longest multigenerational study involved ten generations of quail fed up to 50% GE corn [21].

In a comprehensive review of the health effects of GE plants, Snell *et al.* [15] focused on 12 long-term and 12 multigenerational feeding trials with GE crops that also had a 90-d rodent study GE feeding study comparator [22-29]. It is important to note that these studies were financially supported by public funds. The question they specifically asked was, "Do long-term and multigenerational GE feeding studies provide any new evidence indicative of some adverse effect(s) that were not previously identified in the 90-d rat study"? The authors concluded that while none of the long-term or multigenerational studies they evaluated revealed any new effect that had not been found in the 90-d rodent toxicology study, there was a need to develop reproducible and standardized protocols for conducting and analyzing complementary fundamental research using different animal models on long-term and multigenerational studies. Some of the long-term and multigenerational studies examined did not use isogenic lines as controls, and the organs and parameters that were measured varied greatly among the studies. Few studies have been conducted using the same GE line and species, and even when they were conducted in the same species, different parameters were measured making a meta-analysis of the data problematic. The authors suggested that while a more standardized protocol for long-term and multigenerational studies would be useful for exploratory fundamental research projects, such studies should be conducted on a case-by-case basis for GE food safety only if some reasonable doubt remained after a 90-d rodent feeding trial.

Another review examined 60 high-throughput "-omics" comparisons between GE and non-GE crop lines, including 17 long-term and 16 multigenerational animal feeding studies, to determine if these additional tests raised new safety concerns [14]. High-throughput "-omics" – transcriptomics, proteomics, and metabolomics - methods have been suggested as a nontargeted approach to detect unintended effects in GE plants. Long-term studies included rats [20,30-32], mice [33-37], salmon [38,39], beef cattle [40], dairy cows [41], macaques [42], pigs [19], and quail [43]. Multigenerational studies included rats [44-48], mice [49-53], pigs [54-56], bulls [56], dairy cows [56], goats [57], sheep [56,58], broilers [56,59], laying hens [56,60], and quail [21,61]. These powerful studies consistently revealed that GE had fewer unintended effects than conventional breeding techniques. The authors suggested that the small number of unintended effects observed, including changes in the level of lactate dehydrogenase enzyme in goats fed GE soybean [57], and immune responses in mice fed GE triticale in the fifth generation of mice [49], fell within the normal range of variation, and did not suggest that they represented a health hazard. Even when GE crops were designed to intentionally have altered metabolic traits, "-omics" expression profiling technologies revealed few unintended effects [14]. The authors concluded that *"none of the "-omics" comparisons has raised new safety concerns about (marketed) GE varieties; neither did the long-term and multigenerational studies on animals"*. They further proposed that the data collected to date suggest that the risk assessment should actually be lowered for GE crops.

A highly controversial study by Séralini claimed that feeding GE glyphosate tolerant corn and a related herbicide formulation over a two year period caused organ damage, tumors, and early death among Sprague–Dawley rats [8]. The authors used a 90-d rodent toxicology feeding study design to study long-term carcinogenicity while failing to consider that 2-yr old rats of the Sprague–Dawley strain are known to be highly susceptible to developing tumors [62]. Independent scientists have noted numerous design flaws in the Séralini study [63,64] including too few animals per treatment group, too few controls (20 control animals (10 male and 10 female) versus 180 "treated" animals), inappropriate histological and statistical analysis of mortality and tumor rates, and ignoring the fact that many other peer-reviewed long-term studies with contradictory results have been conducted by independent scientists from around the world. This includes a two-yr rat feeding study, funded by the Japanese government, which found no deleterious effects of feeding GE feed in their long-term feeding trial [20]. In that study the researchers followed the suggested experimental design for a 104 wk carcinogenicity study [65,66] which includes the use of 50 animals per treatment group, use of a rat strain that has an acceptable survival rate for the long-term study, and appropriate statistical analysis of their data. The highly-publicized but poorly-executed Séralini investigation has since been thoroughly debunked by regulatory agencies throughout the world [67-75].

Another infamous study conducted by Ewen and Pusztai [76] in 1999 reported several injurious effects in the gastrointestinal tract of rats that had been fed GE potatoes expressing the antinutritive lectin *Galanthus nivalis* agglutinin (GNA), a compound with insecticide activity. It was claimed that the consumption of GE potatoes had significant effects on the immune system of rats in the feeding trials, because of some effect of GE itself rather than because of the particular gene inserted. However a report by the Royal Society concluded that the data reviewed *"provide no reliable or convincing evidence of adverse (or beneficial) effects, either of lectins added to unmodified potatoes or of potatoes genetically modified to contain a lectin gene, on the growth of rats or on their immunological function"* [77]. That report criticized the rat feeding study for the common trial design errors of too few animals per diet group and the lack of controls such as a standard rodent diet containing about 15% protein (the test diet was severely protein deficient at ~ 6%) [77]. Irrespective, given that lectin has been widely documented to be toxic and/or allergenic, GE crops expressing such a substance would be highly unlikely to ever obtain regulatory approval [78].

A critical review of the other published studies where change(s) in some parameters are reported to result from GE feed also reveal deviations from standard protocols [14,15,79]. These include control feed that was not derived from near isogenic lines, insufficient animal numbers for statistical power, over interpretation of differences which lie within the normal range of variation and hence are not biologically significant, and/or poor toxicological interpretation of the data. This emphasizes the need to follow required standard protocols in animal feeding studies. This lack of compliance with international protocols by some research groups, and the highly sensational presentation of their results in public settings have led to the unfortunate situation where companies are reticent to provide plant material for independent feeding studies [15]. This is particularly problematic for researchers who are interested in pursuing feeding trials in livestock species which typically require larger amounts of GE feed.

Animal reproduction

The reproductive effects of GE crops are another area that has generated debate [16]. In this regard several controversial studies are often cited. Some of these studies were not published in the peer-reviewed literature but rather were posted only on the internet and publicized at press conferences [80,81]. The Ermakova study [80] claimed that transgenic soybeans compromised the fertility of rats and dramatically decreased the survival and growth of their offspring. However the study was criticized for numerous design flaws by academic scientists [82]. The other internet study [81] housed male and female mice as breeding pairs for approximately 20 wk during which time they were allowed to produce litters continuously. The authors identified differences in reproductive parameters between mice fed with GE maize and the controls. They reported that there were statistically significantly fewer pups born in the GE group in the 3rd and 4th litters, and that there were fewer pups weaned in the 4th litter compared with the control group. The study was withdrawn from the internet by Austrian officials because of weaknesses in experimental design, calculation errors and deficiencies in the statistical analysis [83].

The fact that studies which did not even reach the accepted standard of peer-review publication can receive such wide publicity and be uncritically cited as evidence of the risks of GE crops by some authors [84] is unfortunate given a large number of less controversial, and hence less famous, carefully controlled peer-review studies that revealed no negative effects of GE-feed on various attributes (e.g. gonad weight, fecundity, fertility, gonadal histopathology) of female [23-26,46,47,85-96] and male [23-26,46,47,50,51,85,87,89-96] reproduction in animal feeding studies.

Another study examined the ultrastructural and immunocytochemical features of preimplanation embryos from 10 two-mo old mice fed a standard diet containing 14% GE soybean or non-GE soybean until weaning [97]. Morphological observations revealed that the embryo nuclear components were similar in the two experimental groups, but pre-mRNA maturation seemed to be less efficient in the embryos from GE-fed mice than controls. Again, this study did not provide any information on the source of the GE soy or the control, nutritional composition of the diet, and the number of female mice per group (n=5) was small. Non-adherence to standard procedures makes data interpretation difficult as it is not clear which of the multiple variables that differed between the groups were causative of the observed differences. Research published between 2002 and 2005 by researchers in Italy indicating ultrastructural changes in organs in the liver, pancreas and testes of mice fed diets supplemented with GE and non-GE soya [36,37] has likewise been criticized by independent scientific groups [98,99] regarding a lack of information concerning the source of the GE soybean, the appropriateness of the control soybean used in the diet, and the nutritional composition of the diet.

Clearly these repeated experimental design flaws in animal feeding studies evaluating GE feed are exacerbating the continued controversy associated with the safety of GE food and feed that currently divides not only the general public but frequently also the scientific community. Animal scientists have an obligation to ensure that feeding studies using GE crops are carried out according to standard protocols [5,13,65,100-102] (Table 1) to ensure data can be appropriately analyzed and unambiguously interpreted in the absence of confounding factors.

Statistical analysis and experimental design

The scientific question being addressed by feeding studies should be well-established before designing the study [79]. Designing experiments to test for intended effects is relatively straightforward. Sample size determinations are based on the size of the effect that is considered important and the required power (i.e. probability that the test will reject the null hypothesis when the alternative hypothesis is true) for a given significance level. Statistical power increases with the sample size, if all other parameters of statistical testing are held constant.

Animal feeding trials are sometimes also used to identify "unintended" effects. These are effects or results that were not expected nor considered in the experimental design and sample size calculations. If many independent tests are performed on the same sample, the probability of obtaining significant results will increase merely due to the multiple comparisons being performed. For example, when many parameters are measured it is likely just by chance one in 20 will rise to the level of statistical significance (assuming $P < 0.05$) [103]. The correct statistical methods should be used to analyze for the statistical significance of multiple comparisons. Two common methods, Bonferroni adjustment and the False Discovery Rate, are among approaches used to take multiple comparison issues into account. The false discovery rate (FDR) is a technique specifically developed for controlling the expected proportion of falsely rejected hypothesis [104]. The use of FDR or similar techniques allows this control and improves the probability of discriminating statistical differences from those generated by random chance.

It is also important to understand the biological relevance of statistically significant differences that might occur between treatment groups. The European Food Safety Authority (EFSA) clarified the difference between statistical significance and biological relevance [103]. Statistical significance is a term that has a specific and distinctive meaning when used in the context of statistical hypothesis testing. Significant does not necessarily mean "important" or "meaningful" but rather is a statistical statement on the property and information content of the observed data. Biological relevance, on the other hand, *"implies a biological effect of interest that is considered important based on expert judgment. Its use refers to an effect of interest or to the size of an effect that is considered important and biologically meaningful and which, in risk assessment, may have consequences for human health. The objective of carrying out an empirical study is usually to identify the existence of relevant biological effects at the population level using statistical tools to detect them. Therefore the identification of statistical significance is only part of the evaluation of the biological relevance"* [103]. Importantly it is stressed that the *"nature and size of biological changes or differences seen in studies that would be considered relevant should be defined before studies are initiated"* rather than be derived from a post-hoc analysis of the data. This enables the design of experiments with sufficient statistical power to be able to detect such biologically relevant effects of this size if they truly occurred.

Ignoring this distinction is a frequent criticism of studies where a statistically significant treatment effect is found in a post-hoc analysis of a data set with a small sample size and spurious conclusions regarding the

Table 1 Recommendations for the conduct of animal studies to evaluate GE crops [101]

Animals (species/categories)	Number of animals (coefficient of variation 4 to 5%)	Duration of experiments	Composition of diets	Measurements/endpoints
Poultry for meat production	10-12 pens per treatment with 9–12 birds per pen	5 wk or more	Balanced diets	Feed intake, gain, feed conversion, metabolic parameters, body composition
Poultry for egg production	12-15 replications per treatment with 3–5 layers per pen	18-40 wk of age, at least three 28-d phases	Balanced diets	Feed intake, egg production, feed conversion, egg quality
Swine	6-9 replications per treatment with 4 or more pigs per replication	Piglets (7–12 kg) 4–6 wk Growers (15–25 kg) 6–8 wk	Balanced diets	Feed intake, gain, feed conversion, metabolic parameters, carcass quality
Growing and finishing ruminants	6-10 replications per treatment with 6 or more cattle per replication	90-120 d	Balanced diets	Feed intake, grain, feed conversion, carcass data, metabolic parameters
Lactating dairy cows	12-16 cows per treatment 28 cows per treatment	Latin square 28 d periods Randomized block design	Balanced diets	Feed intake, milk performance and composition, body weight, body condition score (BCS), cell counts in milk, animal health

Table used with permission [2].

biological relevance of the finding to health are inferred. This distinction is especially relevant in the absence of knowledge regarding the normal level of biological variation that may exist between different non-GE cultivars and varieties. Many constituents in crop plants vary widely due to environmental factors (such as field location, planting, sampling time, crop management practices), and genotype and this natural variation is not typically considered to be a food safety concern. For the purposes of a safety assessment, the question is not whether a GE line has a statistically different level of some constituent from its near-isogenic nontransgenic comparator, but rather whether differences are biologically relevant to health according to expert judgment.

Feeding trials in target species

Target animal (food producing animals such as ruminants, pigs, poultry, and fish) feeding studies have not been required for regulatory approval in part because first generation GE crops have proven to be substantially equivalent to their conventional counterpart. Over the past 20 yr, the U.S. Food and Drug Administration (FDA) has found that all 148 transgenic events they have evaluated, and that includes all of the GE crops that have ever been commercialized, were substantially equivalent to their conventional counterparts [105]. These studies have spanned GE corn, soybean, cotton, canola, wheat, potato, alfalfa, rice, papaya, tomato, cabbage, pepper, raspberry and mushroom, and included traits of herbicide, drought and cold tolerance, insect and virus resistance, nutrient enhancement, and expression of protease inhibitors.

Studies with target animals conducted to date have typically been conducted to evaluate nutritional and feed equivalency of GE, rather than to evaluate safety. Flachowsky *et al.* [2] summarized the results of well over 100 studies feeding target animals (dairy cattle (12), beef cattle (14), other ruminants (10), pigs (21), broilers (48), laying hens (12), other poultry (1), others (fish, rabbits, etc.) (8)) with GE feed from various review documents [78,106]. They concluded that there is good agreement from these studies that feed from GE crops did not significantly influence feed digestibility, animal health, biologically relevant effects on animal performance, composition of animal products, or result in unintended effects (with the exception that lower mycotoxin concentrations have repeatedly been found in *Bacillus thuringiensis* toxin-expressing GE crops [107]) when compared to animals fed isogenic non-GE varieties [2].

An important consideration in target species feeding trials is the substantial costs involved in large animal feeding trials. This is especially true when contemplating long-term or mutigenerational studies on long-lived animals (Table 2). Therefore, long term studies and multigenerational experiments with target animals to date are rather rare [15]. As discussed earlier, many of these long-term studies have not adhered to standard protocols, underlying the vital need for careful consideration of experimental design given the length of time needed and expenses associated with target animal feeding studies. Increasing the number of animals, the length of the trial, and the number of generations are all associated with increased costs. High costs may prevent the public sector from conducting such studies. Long-term, multigenerational and/or target animal feeding studies should be considered and designed to address biologically-relevant questions of second generation GE crop that cannot or have not been answered using *in silico* and *in vitro* methods, or a 90-d rodent feeding study. They should be hypothesis-driven based on the novel traits and/or phenotype associated with the gene/crop combination.

There are some other practical considerations that dramatically increase the cost associated with feeding target livestock with an "as yet unapproved" GE crop. First researchers would need to obtain sufficient GE crop material and an isogenic comparator for the feeding study. Consider a 2-yr feeding study in dairy cattle involving a total of 100 animals; 50 per treatment group. Milk and meat from the cows eating the unapproved GE feed would not be able to enter the food chain and assuming a double blind study design the opportunity cost of that alone would likely be (100 cows × [$USD 5,000/year for milk × 2 yr + $800 cull cow]) in excess of $USD 1,000,000. Housing and bedding for 100 cows at $300/head/mo would be ~$USD 720,000, and then there would be the costs of sample analysis, which conservatively might add another $USD 500,000 depending upon what analytes or endpoints were examined. The cost of such a study would easily exceed $USD 2,000,000. In the absence of an unaddressed safety concern, this expense is not justified given that GE food/feed animal feeding trials of substantially equivalent GE crops that have been carried out to date have not been found to add to the

Table 2 Examples of lifespans for growing/fattening animals, in days [79]

Animal species/ categories	Conventional/more intensive	Organic/more extensive
Chickens for fattening (broilers)	3-42	56-84
Turkeys for fattening	56-168	70-112
Growing/fattening pigs	150-300	200-400
Veal calves	80-200	-
Growing/fattening bulls	300-500	400-600

Laying hens and dairy cattle are usually used for longer periods:
Laying hens: about 126–140 d for growing (pullets); about 300–360 d (one year) for the laying period.
Dairy cattle: about 22–36 mo for growing (heifers); one to ten yrs for lactation (average in Europe two to five lactations).

safety assessment, and this also avoids unnecessary animal experimentation.

It should also be noted that although comparatively few feeding trials of commercialized GE crops in target livestock are in the peer-reviewed literature, large numbers of livestock in many countries have been consuming GE feed for over a decade. For example, in 2011 alone approximately 9 billion broiler chickens, weighing over 22.5 billion kg liveweight were produced in the United States. During that year 30 million tonnes of corn and 13.6 million tonnes of soy were used as broiler and breeder poultry feed of which 88% and 94%, respectively, was likely from GE crops. Production parameters, mortality and condemnation rates for the more than 105 billion broilers that have been processed in the US since 2000 are shown in Figure 2. In 2000 approximately 25% of corn and 50% of soy grown in the US was GE and hence poultry diets have likely contained an ever increasing proportion of GE feed from 2000 to 2011. This very large field data set does not reveal overt health problems associated with the consumption of GE feed, but rather shows a continuation of industry trends that were observed prior to the introduction of GE crops (Figure 2).

Second generation crops

The second generation of GE crops, i.e. those with intentionally changed composition or output traits [102] is likely to include crops with more nutrients or less undesirable substances specifically targeted for animal feed. Other second generation target traits include plants with increased resistance to biotic and abiotic stressors such as drought and saline soils and crops that are more efficient in using limited natural resources to help address the larger challenge of improving global food security [2]. Second generation GE crops will by definition not be "substantially equivalent". Whether this represents a safety concern will depend on the trait. One study explored the use of the current safety evaluation criteria on a quality-improved GE potato and concluded that the safety of the second generation crops can be properly assessed using the existing current comparative safety assessment methodology [108]. Standard protocols outlining best practices for the conduct of animal studies to evaluate crops genetically modified for output traits have been developed [102].

Animal feeding studies may be needed to assess the bioavailability and/or digestibility of nutrients, and the efficacy of nutrient uptake from second generation GE crops [79,109]. It should be noted that target animal feeding studies to measure these parameters are not required prior to the release of new crop cultivars developed by traditional breeding, although they may be voluntarily conducted by developers to gauge animal performance on the new variety. It is difficult to scientifically justify why the process of GE should be the trigger for a target species feeding evaluation of such crops, rather than a product-based approach triggered by the novel attributes (e.g. increased oil content, decreased lignin) of the modified crop. A high oil crop produced using other plant breeding approaches (e.g. radiation mutagenesis which is known to alter gene expression patterns more than GE [13]), would be logically accompanied by the same bioavailability, animal digestibility and safety questions as a second generation GE crop with the same phenotype.

Animal preference studies

Some groups have claimed that, given a choice, animals prefer not to consume GE crops. The data to support this assertion are typically anecdotal. There are few studies in the peer-reviewed literature addressing this topic. One study evaluated beef steers grazing preferences for GE and non-GE corn residue. Sixteen steers were grazed on one pasture containing both GE and non-GE corn residue. Their grazing distribution was recorded for 50 d. There was no significant difference in the grazing preference of the animals [110]. In another study using a second generation GE potato, both mice and humans actually showed a preference for the aroma of a GE nonbrowning potato as compared to non-GE potatoes [111]. This effect was not observed when the potatoes were fresh, it was only seen 24 h after the potatoes had aged, presumably associated with the fact that the non-GE potatoes had oxidized and turned brown by that time. Several other studies have started to look at sensory analysis of second generation GE crops [112-114]. In one study, inclusion of GE tomatoes with improved antioxidant properties in the diet of cancer-susceptible mice (p53-knockout mice) significantly extended their lifespan when compared with mice fed standard diets or diets supplemented with non-GE tomatoes [115].

Fate of recombinant (rDNA) and protein from GE crops

A number of studies have been conducted to look for the presence of rDNA or the protein encoded by the rDNA construct in the milk, meat and eggs from animals fed GE feed [116-120]. To date GE DNA and expression protein products have not been detected in animal protein products derived from food animals fed GE feed. The reason that scientists are researching this topic, even though the presence of DNA and protein from conventional crops in the diet of food animal has not been considered to be problematic, is that consumers are allegedly concerned that GE DNA could alter animal health and in turn eventually pose a threat to human health [121]. The scientific merit of this perception is dubious given that animals and people eat foreign

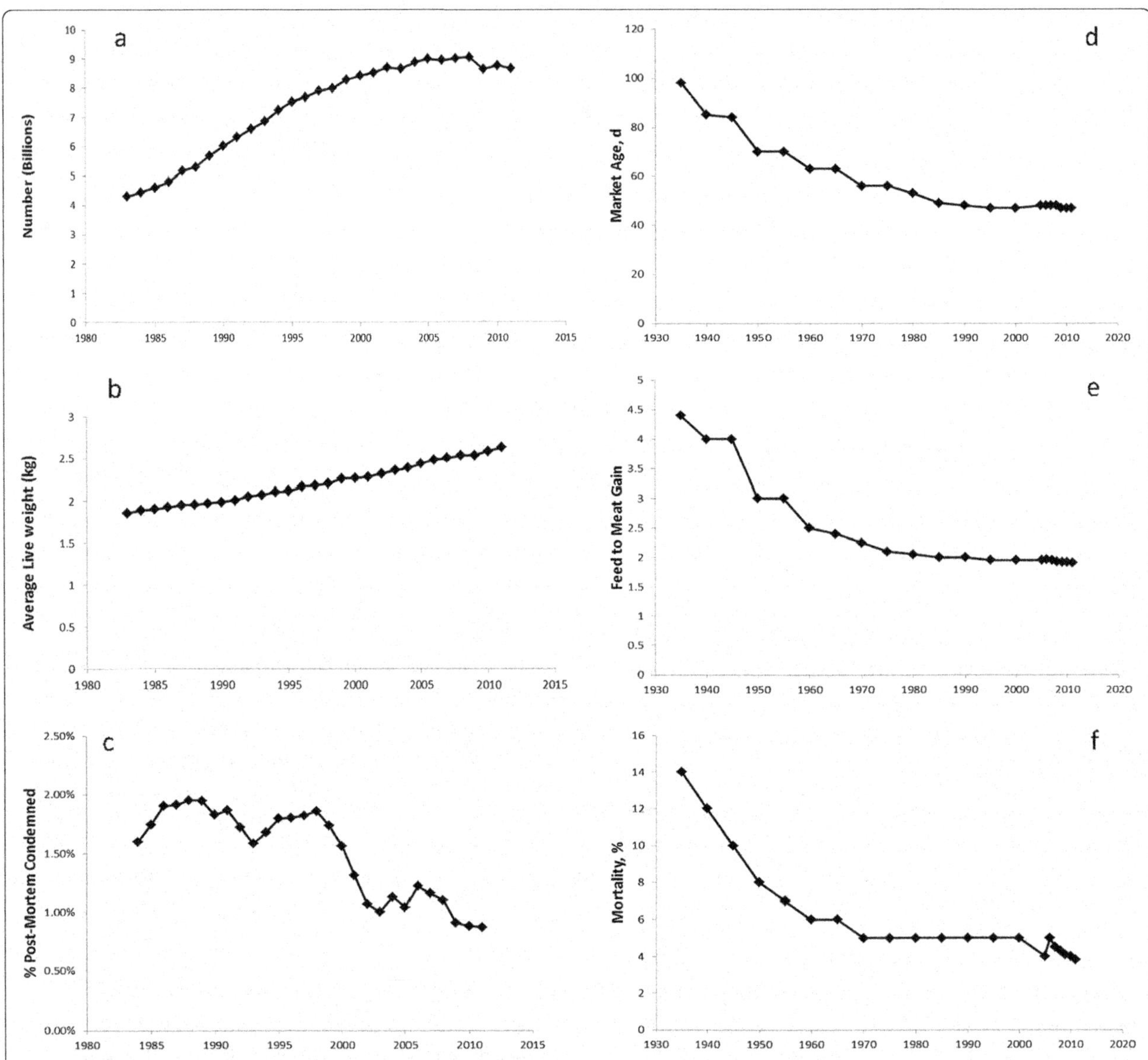

Figure 2 Summary statics of United States commercial broiler data. a) Number of chickens processed; **b**) Average weight of chickens; **c**) Percent of chickens condemned by USDA at inspection; **d**) Average d to market; **e**) Efficiency of feed utilization (kg of feed required for one kg of live weight gain; and **f**) Percent mortality. Data from USDA Economics, Statistics and Market Information System (ESMIS). (http://usda.mannlib.cornell.edu/MannUsda/viewDocumentInfo.do?documentID=1497), and the National Chicken Council, Washington DC. (http://www.nationalchickencouncil.org/about-the-industry/statistics/u-s-broiler-performance/).

DNA from various crop species every day, and DNA is generally recognized as safe whether it is derived from a GE or conventionally-bred organism.

Some studies have reported finding traces of high copy number plant nuclear and chloroplast DNA sequences in animal organs and tissues [116]. The biological importance of this finding is uncertain. To date there is no evidence that eating DNA and proteins from another species, GE or conventional, results in incorporation of food-based DNA into the genome of the consumer. No country to date has mandated the labeling of products from animals fed with GE plants [121], although voluntary market-driven approaches have resulted in some retailers offering this as a choice for their customers. It is likely to be economically if not technically impossible to use analytical procedures to determine if milk, meat or eggs are derived from animals fed with GE feed [121], and so such products will have to be sourced by supply chain management and verified by documentation.

Global adoption of GE crops and use in animal agriculture

There is a growing demand for meat and milk as the world population climbs towards 9 billion people and the income of consumers in developing countries rises, and correspondingly there is a growing demand for animal feed. Current crop yields will need to approximately double to meet the feed demands of 2050, and in the absence of newly arable land this demand will necessitate higher yielding crop varieties. Since its introduction in the mid-1990s GE technology has added an additional 110 million tonnes of soybeans and 195 million tonnes of corn, to the global production of these crops. Net level farm economic benefits resulting from GE during that 15 year period were valued at $USD 98.2 billion [122].

In 2012 about 170 million hectares of GE-plant crops (12% of total arable land) were cultivated worldwide [1]. This is a 100-fold increase from the 1.7 million hectares that were planted in 1996, making GE the fastest adopted crop technology in recent history. During the period 1996–2011 it has been estimated that the cumulative economic benefits from cost savings and added income derived from planting GE crops was $USD 49.6 billion in developing countries and $USD 48.6 billion in industrial countries [122]. Of the 17.3 million farmers who planted GE crops in 2012, 15 million were small resource-poor farmers in 20 developing countries. Approximately 14.4 million small farmers in China (4 million ha; mostly cotton, although papaya, poplar, tomato and sweet pepper have all had production approvals), and India (10.8 million ha cotton) collectively planted a record 14.8 million hectares of GE crops in 2012 [1].

Animal agriculture is highly dependent upon GE crops. Table 3 shows the importance of GE crops to the animal feed export market. This creates a problem when there are "asynchronous approvals" of GE events, where an event is fully approved for commercial use in food and feed in one country, but not in others (Figure 3). This is particularly true for trade with the European Union (EU), as it has been estimated that 98% of soybean meal and 80% of all animal feed consumed in the EU is imported, of which more than half is from GE crops imported from Brazil, the USA, and Argentina [123]. The EU imports approximately 70% of the soybean meal used in animal feed and of this 80% is GE [124]. The proportion of GE in animal feed is likely even higher in the US where 93% of soy and 88% of all corn grown were GE varieties in 2012 [1].

Table 3 Share of global crop trade accounted for by GE crops in 2011/12 (million tonnes) [125]

	Soybeans	Maize (Corn)	Cotton	Canola
Global production	238	883.5	27.0	61.6
Global trade (exports)	90.4	103.4	10.0	13.0
Share of global trade from GE producers	88.6 (98%)	70.0 (67.7%)	7.15 (71.5%)	9.9 (76%)
Estimated size of market requiring certified conventional (in countries that have import requirements)	3.0	4.4	Negligible	Negligible
Estimated share of global trade that may contain GE (i.e., not required to be segregated)	87.4	70.0	71.5	9.9
Share of global trade that may be GE	96.7%	67.7%	71.5%	76%

Notes: Estimated size of market requiring certified conventional in countries with import requirements excludes countries with markets for certified conventional for which all requirements are satisfied by domestic production (e.g. maize in the EU). Estimated size of certified conventional market for soybeans (based primarily on demand for derivatives used mostly in the food industry): EU 2.0 million tonnes bean equivalents, Japan and South Korea 1 million tonnes.

The EU does not provide for any tolerance threshold for the accidental presence of unapproved GE events that have received regulatory approval in other countries. A 0.1% "technical solution" threshold was approved for feed material authorized in a non-EU country and for which an EU authorization request for the GE event in question has been lodged with EFSA for at least 3 mo or for which the authorization has expired. The 0.1% threshold is considered to be commercially unviable [124], and as more GE crops are grown in major grain exporting countries there is a very real possibility of major trade disruptions resulting from asynchronous approvals. Livestock production accounts for 40% of the total value of agricultural production in Europe. It has been estimated that if the EU were not able to import soybean protein from outside the EU it would only be able to replace 10-20% of imports by high protein feed substitutes, and that this would result in a substantial reduction in animal protein production, exports and consumption, and a very significant increase in animal protein imports into the EU [128].

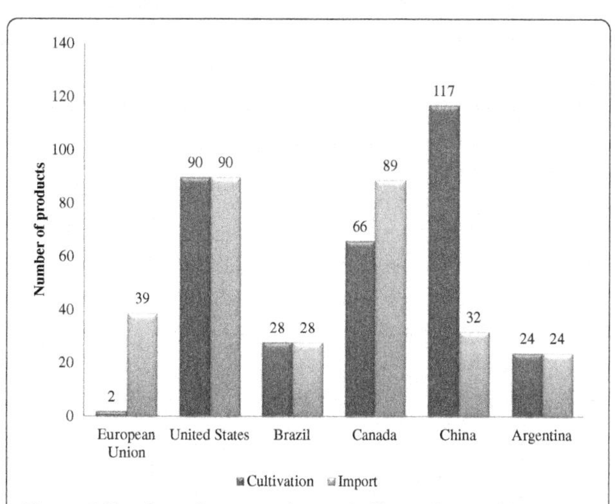

Figure 3 Number of approved genetically engineered (GE) products in the EU, USA, Brazil, Canada, China and Argentina. Data compiled from [123,126,127].

Cost:benefit analysis

In the abstract, the best approach for the regulatory evaluation of GE crops is one that allows new GE crops to be commercialized, while preventing new risks to animal and human health and the environment [129]. It is almost certain that animal agriculture globally will continue to rely on feed from GE crops. To date, commercialization of GE crop varieties has been associated with disproportionately high regulatory costs, regulatory delay, and considerable trade uncertainty. This has made their commercialization prohibitively expensive for all but the largest, multinational corporations.

Given the weight of scientific evidence on the safety of GE crops and the considerable expense involved, the decision to conduct an animal feeding study with a GE crop should be based on the need to answer a scientific question that cannot be addressed using *in silico* and *in vitro* methods. A reasonable hypothesis-driven food safety concern should be the driver for the additional expense and use of experimental animals required for such studies. The specific objectives and the rationale for choosing to perform a long-term chronic toxicity and/or carcinogenicity study should be clearly documented before conducting the experiment based on a remaining unanswered safety question following a 90-d rodent feeding trial.

Mandating long-term or target species animal feeding studies costing millions of dollars based on the process used to make a GE crop, rather than the unique traits and/or phenotype associated with the gene/crop combination is not justified based on the weight of evidence. Regulations triggered by how products are made are inconsistent with science-based risk assessment unless there is something inherently risky about the process, as compared to existing methods. A substantial body of evidence shows that GE crops are no more risky than conventionally bred crops, and mandating costly animal feeding studies in the absence of a reasonable unaddressed food safety concern associated with the novel trait and/or phenotype cannot be scientifically justified and overregulation is an indulgence that global food security can ill afford.

Moreover, the current approach to regulating GE crops does not evaluate the potential benefits that might be associated with the introduction of a GE crop. There have been substantial economic, production, and environmental benefits associated with the introduction of the first generation of GE crops [122,129-132]. All technologies are associated with both risks and benefits, and few would be adopted in the face of a risk-only analysis. In some cases GE crops may pose fewer risks than are implied by the non-GE alternative (e.g. reduced mycotoxin in *Bacillus thuringiensis* corn [107]). Perhaps as importantly, the lives saved or other benefits derived from risk assessment and management must be large enough to offset the costs and deferred potential benefits. The poorest and most vulnerable disproportionately bear the costs and impacts of excess regulation [129].

At the current time there are no international standards for assessing the potential benefits associated with the release of a new GE organism, although in many countries there are increasing calls for a risk-benefit analysis to form an integral part of GE regulatory frameworks [133]. Shifting from a risk-only regulatory focus to one that includes a risk:benefit analysis would enable a more balanced and harmonized evaluation of the likely impacts of introducing a new GE organism.

Conclusions

Hundreds of peer-reviewed animal feeding studies have repeatedly shown that GE plants can safely be used in feed, and rDNA fragments have never been detected in products (e.g. milk, meat, eggs) derived from animals that consumed GE feed. Given the 15 yr history of safe use and absence of scientific evidence to suggest GE is associated with unique risks, whole food/feed animal feeding studies on GE crops should be reserved for GE crops where the novel phenotype results in a reasonable food safety concern that remains unanswered following all other analyses. Indiscriminately requiring long-term and target animal feeding studies based on a GE process-based trigger is not scientifically justified and will have an inhibitory effect on the development and commercialization of potentially beneficial GE feed crops in the future. World-wide GE regulations have disproportionately focused only on the potential risks associated with GE technology and commercialization of GE crops has been associated with a high regulatory compliance expense which has slowed adoption, particularly in small and poor developing countries. It is time for regulatory frameworks to consider the benefits in addition to any unique risks associated with GE technology. There are many current (increased yields, reduced insecticide use, improved feed quality), and potential future benefits of GE including feed crops with enhanced nutritional characteristics and durability. Regulatory frameworks should formally evaluate the reasonable and unique risks and benefits associated with the use of both GE plants and animals in agricultural systems, and weigh them against those associated with existing systems, and the opportunity costs associated with regulatory inaction.

Abbreviations

GMO: Genetically modified organism; GE: Genetically engineered; OECD: Organisation for economic co-operation and development; GNA: *Galanthus nivalis* agglutinin; EFSA: European food safety authority; FDA: Food and drug administration; mRNA: messenger RNA (ribonucleic acid); DNA: deoxyribonucleic acid; EU: European union; rDNA: recombinant DNA.

Competing interests

The author declares no competing interests.

Author's contributions
AVE drafted and approved the final manuscript.

Author's information
AVE is an Animal Genomics and Biotechnology Extension Specialist in the Department of Animal Science at the University of California, Davis where she received a Ph.D. in Genetics. Her extension program provides research and education on the use of animal genomics and biotechnology in livestock production systems.

Acknowledgements
The author gratefully acknowledges the editorial review and assistance provided by Amy E. Young. AVE acknowledges research support from National Research Initiative Competitive Grant no. 2009-55205-05057, and Agriculture and Food Research Initiative Competitive Grant no. 2011-68004-30367 and 2013-68004-20364 from the USDA National Institute of Food and Agriculture. Preparation of this manuscript was supported by funds from the W.K. Kellogg endowment to the UC Davis Department of Animal Science.

References

1. James C: *Global Status of Commercialized Biotech/GM Crops: 2012.* 44th edition. Ithaca, NY: ISAA: ISAAA Brief; 2012.
2. Flachowsky G, Schafft H, Meyer U: **Animal feeding studies for nutritional and safety assessments of feeds from genetically modified plants: a review.** *J Verbraucherschutz Lebensmittelsicherh (J Consum Prot Food Saf)* 2012, **7:**179–194.
3. Organisation for Economic Co-operation and Development (OECD): *Safety evaluation of foods derived by modern biotechnology : concepts and principles.* Paris: Organisation for Economic Co-operation and Development; 1993.
4. Hollingworth RM, Bjeldanes LF, Bolger M, Kimber I, Meade BJ, Taylor SL, Wallace KB, Society of Toxicology ad hoc Working G: **The safety of genetically modified foods produced through biotechnology.** *Toxicol Sci* 2003, **71:**2–8.
5. European Food Safety Authority (EFSA): **Guidance for risk assessment of food and feed from genetically modified plants.** *EFSA J* 2011, **9:**2150–2187.
6. European Food Safety Authority (EFSA): **Considerations on the applicability of OECD TG 453 to whole food/feed testing.** *EFSA J* 2013, **11:**3347–3365.
7. Kalaitzandonakes N, Alston J, Bradford K: **Compliance costs for regulatory approval of new biotech crops.** *Nat Biotechnol* 2007, **25:**509–511.
8. Séralini GE, Clair E, Mesnage R, Gress S, Defarge N, Malatesta M, Hennequin D, de Vendomois JS: **Long term toxicity of a Roundup herbicide and a Roundup-tolerant genetically modified maize.** *Food Chem Toxicol* 2012, **50:**4221–4231.
9. Organisation for Economic Cooperation and Development (OECD): *Section 4 (Part 408), Health Effects: Repeated Dose 90-d Oral Toxicity Study in Rodents, Guideline for the Testing of Chemicals.* Paris, France: Organisation for Economic Co-operation and Development; 1998.
10. Cheng KC, Beaulieu J, Iquira E, Belzile FJ, Fortin MG, Stromvik MV: **Effect of transgenes on global gene expression in soybean is within the natural range of variation of conventional cultivars.** *J Agric Food Chem* 2008, **56:**3057–3067.
11. Batista R, Saibo N, Lourenço T, Oliveira MM: **Microarray analyses reveal that plant mutagenesis may induce more transcriptomic changes than transgene insertion.** *Proc Natl Acad Sci USA* 2008, **105:**3640–3645.
12. Garcia-Villalba R, Leon C, Dinelli G, Segura-Carretero A, Fernandez-Gutierrez A, Garcia-Canas V, Cifuentes A: **Comparative metabolomic study of transgenic versus conventional soybean using capillary electrophoresis-time-of-flight mass spectrometry.** *J Chromatogr, A* 2008, **1195:**164–173.
13. European Food Safety Authority (EFSA): **Safety and nutritional assessment of GM plants and derived food and feed: the role of animal feeding trials.** *Food Chem Toxicol* 2008, **46:**S2–S70.
14. Ricroch AE: **Assessment of GE food safety using '-omics' techniques and long-term animal feeding studies.** *N Biotechnol* 2012, **30:**349–354.
15. Snell C, Bernheim A, Berge JB, Kuntz M, Pascal G, Paris A, Ricroch AE: **Assessment of the health impact of GM plant diets in long-term and multigenerational animal feeding trials: a literature review.** *Food Chem Toxicol* 2012, **50:**1134–1148.
16. Zhang W, Shi F: **Do genetically modified crops affect animal reproduction? A review of the ongoing debate.** *Animal* 2011, **5:**1048–1059.
17. Buzoianu SG, Walsh MC, Gardiner GE, Rea MC, Ross RP, Lawlor PG: **Effect of feeding Bt (MON810) maize to pigs from 12 d post-weaning for 110 d on growth performance, body composition, carcass characteristics, organ weights and intestinal morphology.** *J Anim Sci* 2011, **89**(1):461. Abstract 446.
18. Buzoianu SG, Walsh MC, Gardiner GE, Rea MC, Ross RP, Lawlor PG: **Effect of feeding genetically modified Bt (MON810) maize to pigs from 12 d post-weaning for 110 d on serum and urine biochemistry.** *J Anim Sci* 2011, **89**(1):462. Abstract 447.
19. Buzoianu SG, Walsh MC, Rea MC, Cassidy JP, Ross RP, Gardiner GE, Lawlor PG: **Effect of feeding genetically modified Bt MON810 maize to approximately 40-d-old pigs for 110 d on growth and health indicators.** *Animal* 2012, **6:**1609–1619.
20. Sakamoto Y, Tada Y, Fukumori N, Tayama K, Ando H, Takahashi H, Kubo Y, Nagasawa A, Yano N, Yuzawa K, Ogata A: **A 104-week feeding study of genetically modified soybeans in F344 rats.** *Shokuhin Eiseigaku Zasshi* 2008, **49:**272–282.
21. Flachowsky G, Halle I, Aulrich K: **Long term feeding of Bt-corn–a ten-generation study with quails.** *Arch Anim Nutr* 2005, **59:**449–451.
22. Hammond B, Dudek R, Lemen J, Nemeth M: **Results of a 13 week safety assurance study with rats fed grain from glyphosate tolerant corn.** *Food Chem Toxicol* 2004, **42:**1003–1014.
23. Hammond B, Lemen J, Dudek R, Ward D, Jiang C, Nemeth M, Burns J: **Results of a 90-d safety assurance study with rats fed grain from corn rootworm-protected corn.** *Food Chem Toxicol* 2006, **44:**147–160.
24. Hammond BG, Dudek R, Lemen JK, Nemeth MA: **Results of a 90-d safety assurance study with rats fed grain from corn borer-protected corn.** *Food Chem Toxicol* 2006, **44:**1092–1099.
25. MacKenzie SA, Lamb I, Schmidt J, Deege L, Morrisey MJ, Harper M, Layton RJ, Prochaska LM, Sanders C, Locke M, *et al*: **Thirteen week feeding study with transgenic maize grain containing event DAS-01507-1 in Sprague–Dawley rats.** *Food Chem Toxicol* 2007, **45:**551–562.
26. Malley LA, Everds NE, Reynolds J, Mann PC, Lamb I, Rood T, Schmidt J, Layton RJ, Prochaska LM, Hinds M, *et al*: **Subchronic feeding study of DAS-59122-7 maize grain in Sprague–Dawley rats.** *Food Chem Toxicol* 2007, **45:**1277–1292.
27. Poulsen M, Kroghsbo S, Schrøder M, Wilcks A, Jacobsen H, Miller A, Frenzel T, Danier J, Rychlik M, Shu Q, *et al*: **A 90-d safety study in Wistar rats fed genetically modified rice expressing snowdrop lectin Galanthus nivalis (GNA).** *Food Chem Toxicol* 2007, **45:**350–363.
28. Wang Z-h, Wang Y, Cui H-r, Xia Y-w, Altosaar I, Shu Q-y: **Toxicological evaluation of transgenic rice flour with a synthetic cry1Ab gene from Bacillus thuringiensis.** *J Sci Food Agric* 2002, **82:**738–744.
29. Zhu Y, Li D, Wang F, Yin J, Jin H: **Nutritional assessment and fate of DNA of soybean meal from roundup ready or conventional soybeans using rats.** *Arch Anim Nutr* 2004, **58:**295–310.
30. Daleprane JB, Chagas MA, Vellarde GC, Ramos CF, Boaventura GT: **The impact of non- and genetically modified soybean diets in aorta wall remodeling.** *J Food Sci* 2010, **75:**T126–T131.
31. Daleprane JB, Feijo TS, Boaventura GT: **Organic and genetically modified soybean diets: consequences in growth and in hematological indicators of aged rats.** *Plant Foods Hum Nutr* 2009, **64:**1–5.
32. Sakamoto Y, Tada Y, Fukumori N, Tayama K, Ando H, Takahashi H, Kubo Y, Nagasawa A, Yano N, Yuzawa K, *et al*: **A 52-Week Feeding Study of Genetially Modified Soybeans in F344 Rats.** *Shokuhin Eiseigaku Zasshi* 2007, **48:**41–50.
33. Malatesta M, Boraldi F, Annovi G, Baldelli B, Battistelli S, Biggiogera M, Quaglino D: **A long-term study on female mice fed on a genetically modified soybean: effects on liver ageing.** *Histochem Cell Biol* 2008, **130:**967–977.
34. Malatesta M, Caporaloni C, Rossi L, Battistelli S, Rocchi M, Tonucci F, Gazzanelli G: **Ultrastructural analysis of pancreatic acinar cells from mice fed on genetically modified soybean.** *J Anat* 2002, **201:**409–415.
35. Malatesta M, Biggiogera M, Manuali E, Rocchi MB, Baldelli B, Gazzanelli G: **Fine structural analyses of pancreatic acinar cell nuclei from mice fed on genetically modified soybean.** *Eur J Histochem* 2003, **47:**385–388.
36. Malatesta M, Caporaloni C, Gavaudan S, Rocchi MB, Serafini S, Tiberi C, Gazzanelli G: **Ultrastructural morphometrical and immunocytochemical analyses of hepatocyte nuclei from mice fed on genetically modified soybean.** *Cell Struct Funct* 2002, **27:**173–180.

37. Vecchio L, Cisterna B, Malatesta M, Martin TE, Biggiogera M: **Ultrastructural analysis of testes from mice fed on genetically modified soybean.** *Eur J Histochem* 2004, **48:**448–454.
38. Sissener N, Bakke A, Gu J, Penn M, Eie E, Krogdahl A, Sanden M, Hemre G-I: **An assessment of organ and intestinal histomorphology and cellular stress response in Atlantic salmon (Salmo salar L.) fed genetically modified Roundup Ready Soy.** *Aquaculture* 2009, **298:**101–110.
39. Bakke-McKellep A, Sanden M, Danieli A, Acierno R, Hemre G-I, Maffia M, Krogdahl A: **Atlantic salmon (Salmo salar L.) parr fed genetically modified soybeans and maize: histological, digestive, metabolic, and immunological investigations.** *Res Vet Sci* 2008, **84:**395–408.
40. Aulrich K, Bohme H, Daenicke R, Halle I, Flachowsky G: **Genetically modified feeds in animal nutrition. 1st communication: Bacillus thuringiensis (Bt) corn in poultry, pig and ruminant nutrition.** *Arch Tierernahr* 2001, **54:**183–195.
41. Steinke K, Guertler P, Paul V, Wiedemann S, Ettle T, Albrecht C, Meyer HH, Spiekers H, Schwarz FJ: **Effects of long-term feeding of genetically modified corn (event MON810) on the performance of lactating dairy cows.** *J Anim Physiol Anim Nutr (Berl)* 2010, **94:**e185–e193.
42. Domon E, Takagi H, Hirose S, Sugita K, Kasahara S, Ebinuma H, Takaiwa F: **26-Week oral safety study in macaques for transgenic rice containing major human T-cell epitope peptides from Japanese cedar pollen allergens.** *J Agric Food Chem* 2009, **57:**5633–5638.
43. Scholtz N, Halle I, Danicke S, Hartmann G, Zur B, Sauerwein H: **Effects of an active immunization on the immune response of laying Japanese quail (Coturnix coturnix japonica) fed with or without genetically modified Bacillus thuringensis-maize.** *Poultry Sci* 2010, **89:**1122–1128.
44. Tyshko N, Zhminchenko V, Pashorina V, Seliaskin K, Saprykin V, Utembaeva N, Tutel'an V: **Assessment of the impact of GMO of plant origin on rat progeny development in 3 generations.** *Vopr Pitan* 2011, **80:**14–28.
45. Zhou XH, Dong Y, Wang Y, Xiao X, Xu Y, Xu B, Li X, Wei XS, Liu QQ: **A three generation study with high-lysine transgenic rice in Sprague–Dawley rats.** *Food Chem Toxicol* 2012, **50:**1902–1910.
46. Kilic A, Akay MT: **A three generation study with genetically modified Bt corn in rats: biochemical and histopathological investigation.** *Food Chem Toxicol* 2008, **46:**1164–1170.
47. Rhee GS, Cho DH, Won YH, Seok JH, Kim SS, Kwack SJ, Lee RD, Chae SY, Kim JW, Lee BM, *et al*: **Multigeneration reproductive and developmental toxicity study of bar gene inserted into genetically modified potato on rats.** *J Toxicol Environ Health A* 2005, **68:**2263–2276.
48. Daleprane JB, Pacheco J, Boaventura GT: **Evaluation of protein quality from genetically modified and organic soybean in two consecutives generations of wistar rats.** *Braz Arch Biol Technol* 2009, **52:**841–847.
49. Krzyzowska M, Wincenciak M, Winnicka A, Baranowski A, Jaszczak K, Zimny J, Niemialtowski M: **The effect of multigenerational diet containing genetically modified triticale on immune system in mice.** *Pol J Vet Sci* 2010, **13:**423–430.
50. Brake D, Thaler R, Evenson D: **Evaluation of Bt (Bacillus thuringiensis) Corn on mouse testicular development by dual parameter flow cytometry.** *J Agric Food Chem* 2004, **52:**2097–2102.
51. Brake DG, Evenson DP: **A generational study of glyphosate-tolerant soybeans on mouse fetal, postnatal, pubertal and adult testicular development.** *Food Chem Toxicol* 2004, **42:**29–36.
52. Haryu Y, Taguchi Y, Itakura E, Mikami O, Miura K, Saeki T, Nakajima Y: **Longterm biosafety assessment of a genetically modified (GM) plant: the genetically modified (GM) insect-resistant Bt11 corn does not affect the performance of multi-generations or life span of mice.** *Open Plant Sci J* 2009, **3:**49–53.
53. Baranowski A, Rosochacki S, Parada R, Jaszczak K, Zimny J, Poloszynowicz J: **The effect of diet containing genetically modified triticale on growth and transgenic DNA fate in selected tissues of mice.** *Anim Sci Paper Rep* 2006, **24:**129–142.
54. Buzoianu SG, Walsh MC, Rea MC, O'Donovan O, Gelencser E, Ujhelyi G, Szabo E, Nagy A, Ross RP, Gardiner GE, Lawlor PG: **Effects of feeding Bt maize to sows during gestation and lactation on maternal and offspring immunity and fate of transgenic material.** *PLoS One* 2012, **7:**e47851.
55. Buzoianu SG, Walsh MC, Rea MC, Cassidy JP, Ryan TP, Ross RP, Gardiner GE, Lawlor PG: **Transgenerational effects of feeding genetically modified maize to nulliparous sows and offspring on offspring growth and health.** *J Anim Sci* 2013, **91:**318–330.
56. Flachowsky G, Aulrich K, Bohme H, Halle I: **Studies on feeds from genetically modified plants (GMP) - contributions to nutritional and safety assessment. (Special Issue: Advances in feed safety.).** *Anim Feed Sci Technol* 2007, **133:**2–30.
57. Tudisco R, Mastellone V, Cutrignelli MI, Lombardi P, Bovera F, Mirabella N, Piccolo G, Calabro S, Avallone L, Infascelli F: **Fate of transgenic DNA and evaluation of metabolic effects in goats fed genetically modified soybean and in their offsprings.** *Animal* 2010, **4:**1662–1671.
58. Trabalza-Marinucci M, Brandi G, Rondini C, Avellini L, Giammarini C, Costarelli S, Acuti G, Orlandi C, Filippini G, Chiaradia E, *et al*: **A three-year longitudinal study on the effects of a diet containing genetically modified Bt176 maize on the health status and performance of sheep.** *Livest Sci* 2008, **113:**178–190.
59. Brake J, Faust M, Stein J: **Evaluation of transgenic event Bt11 hybrid corn in broiler chickens.** *Poult Sci* 2003, **82:**551–559.
60. Halle I, Aulrich K, Flachowsky G: **Four generations feeding GMO-corn to laying hens.** *Proc Soc Nutr Physiol* 2006, **15:**114.
61. Halle I, Aulrich K, Flachowsky G: **Four generations of feeding of GMO-corn to breeder quail.** *Proc Soc Nutr Physiol* 2004, **13:**124.
62. Brix AE, Nyska A, Haseman JK, Sells DM, Jokinen MP, Walker NJ: **Incidences of selected lesions in control female harlan sprague–dawley rats from two-year studies performed by the national toxicology program.** *Toxicol Pathol* 2005, **33:**477–483.
63. Arjo G, Portero M, Pinol C, Vinas J, Matias-Guiu X, Capell T, Bartholomaeus A, Parrott W, Christou P: **Plurality of opinion, scientific discourse and pseudoscience: an in depth analysis of the Seralini et al: study claiming that roundup ready corn or the herbicide roundup cause cancer in rats.** *Transgenic Res* 2013, **22:**255–267.
64. Schorsch F: **Serious inadequacies regarding the pathology data presented in the paper by Séralini et al. (2012).** *Food Chem Toxicol* 2013, **53:**465–466.
65. Food and Drug Administration (FDA): *Redbook. Guidance for industry and other stakeholders, toxicological principles for the safety assessment of food ingredients.* U.S. Department of Health and Human Services, FDA: Centre for Food Safety and Applied Nutrition; 2007.
66. Organisation for Economic Cooperation and Development (OECD): *OECD Guidelines for the Testing of Chemicals, Section 4: Health Effects Test No. 451: Carcinogenicity Studies.* Paris, France: Organisation for Economic Co-operation and Development; 2009.
67. Agence nationale de securite sanitaire de l'alimentation dleedt (ANSES): *AVIS de l'Agence nationale de securite sanitaire de l'alimantation, de l'environnement et du travail relatif a l'analyse de l'etude de Séralini et al. 2012.* 2012. http://www.anses.fr/Documents/BIOT2012sa0227.pdf.
68. Belgian Biosafety Advisory Council: *Advice of the Belgian Biosafety Advisory Council on the article by Séralini et al., 2012, on toxicity of GM maize NK603.* 2012:1–10. http://www.bio-council.be/docs/BAC_2012_0898_CONSOLIDE.pdf.
69. Brazilian National Biosafety Technical Commission (Ministry of Science Technology and Innovation National Biosafety Technical Commission): *Considered opinion.* 2012. http://www.cibiogem.gob.mx/sala-prensa/Documents/CTNBIO-Brasil-Seralini1725.pdf.
70. European Food Safety Authority (EFSA): **Review of the Seralini et al. (2012) publication on a 2-year rodent feeding study with glyphosate formulations and GM maize NK603 as published online on 19 September 2012 in food and chemical toxicology.** *EFSA J* 2012, **10:**2910.
71. Health Canada and Canadian Food Inspection Agency: *Statement on the Séralini et al. (2012) publication on a 2-year rodent feeding study with glyphosate formulations and GM maize NK603.* 2012. http://www.hc-sc.gc.ca/fn-an/gmf-agm/seralini-eng.php.
72. Nederlandse Voedsel-en Warenautoriteit (NVWA): *Wetenschappelijke beoordeling door de Nederlandse Voedsel- en Warenautoriteit van een onderzoek naar de gezondheidsrisico's voor mens en dier van Roundup-tolerante GGO-mais en Roundup herbicide aan ratten.* 2012. http://www.rijksoverheid.nl/documenten-en-publicaties/notas/2012/10/03/advies-vwa-bij-onderzoek-naar-gezondheidsgevolgen-ggo-mais-en-roundup.html.
73. Bundesinstitute fur Risikobewertung: *Feeding study in rats with genetically modified NK603 maize and with a gly-phosate containing formulation (Roundup) published by Séralini et al. (2012).* 2012. http://www.epsoweb.org/file/1095.
74. The Australian and New Zealand Food Standards Agency (FSANZ): *Response to Séralini paper on the long term toxicity of Roundup herbicide and a Roundup-tolerant genetically modified maize.* 2012. http://www.foodstandards.gov.au/consumer/gmfood/seralini/Pages/default.aspx.
75. The High Council of Biotechnologies (HCB: *Opinion of the HCD Scientific Committee on the Séralini Study (2012).* 2012. http://www.hautconseildesbiotechnologies.fr/IMG/pdf/Executive_Summary_121022.pdf.

76. Ewen SW, Pusztai A: **Effect of diets containing genetically modified potatoes expressing Galanthus nivalis lectin on rat small intestine.** *Lancet* 1999, **354:**1353–1354.
77. The Royal Society: *Review of data on possible toxicity of GM potatoes.* 1999. http://royalsociety.org/uploadedFiles/Royal_Society_Content/policy/publications/1999/10092.pdf.
78. EFSA (European Food Safety Authority): **Safety and nutritional assessment of GM plants and derived food and feed: the role of animal feeding trials.** *Food Chem Toxicol* 2008, **46**(1):S2–S70.
79. Flachowsky G, Wenk C: **The role of animal feeding trials for the nutritional and safety assessment of feeds from genetically modified plants - present stage and future challenges.** *J Anim Feed Sci* 2010, **19:**149–170.
80. Ermakova IV: *Influence of Genetically Modified-SOYA on the Birth-Weight and Survival of Rat Pups: Preliminary Study.* 2013th edition. 2005. http://www.mindfully.org/GE/2005/Modified-Soya-Rats10oct05.htm.
81. Velmirov A, Binter C, Zentek J: *Biological effects of transgenic maize NK603xMON810 fed in long term reproduction studies in mice.* Report, Forschungsberichte der Sektion IV, Band 3. Institut für Ernährung, and Forschungsinttitut für biologischen Landbau, Vienna, Austria: Bundesministerium fur Gesundheit, Familie und Jugend; 2008:105.
82. Marshall A: **GM soybeans and health safety - a controversy reexamined.** *Nat Biotech* 2007, **25:**981–987.
83. European Food Safety Authority (EFSA): *GMO Panel deliberations on the Austrian report "Biological effects of transgenic maize NK603 x MON810 fed in long term reproduction studies in mice".* 2009. http://fbae.org/2009/FBAE/website/images/pdf/biosafety/EFSA-Opinion-on-teh-Austrian-Study.pdf.
84. Dona A, Arvanitoyannis IS: **Health risks of genetically modified foods.** *Crit Rev Food Sci Nutr* 2009, **49:**164–175.
85. Delaney B, Appenzeller LM, Munley SM, Hoban D, Sykes GP, Malley LA, Sanders C: **Subchronic feeding study of high oleic acid soybeans (Event DP-305423-1) in Sprague–Dawley rats.** *Food Chem Toxicol* 2008, **46:**3808–3817.
86. Jacobs CM, Utterback PL, Parsons CM, Rice D, Smith B, Hinds M, Liebergesell M, Sauber T: **Performance of laying hens fed diets containing DAS-59122-7 maize grain compared with diets containing nontransgenic maize grain.** *Poult Sci* 2008, **87:**475–479.
87. Juberg DR, Herman RA, Thomas J, Brooks KJ, Delaney B: **Acute and repeated dose (28 d) mouse oral toxicology studies with Cry34Ab1 and Cry35Ab1 Bt proteins used in coleopteran resistant DAS-59122-7 corn.** *Regul Toxicol Pharmacol* 2009, **54:**154–163.
88. Rasmussen M, Cutler S, Wilhelms K, Scanes C: **Effects of Bt (Bacillus thuringiensis) corn on reproductive performance in adult laying hens.** *Int J Poult Sci* 2007, **6:**169–171.
89. Appenzeller LM, Malley L, Mackenzie SA, Hoban D, Delaney B: **Subchronic feeding study with genetically modified stacked trait lepidopteran and coleopteran resistant (DAS-01507-1xDAS-59122-7) maize grain in Sprague–Dawley rats.** *Food Chem Toxicol* 2009, **47:**1512–1520.
90. Appenzeller LM, Munley SM, Hoban D, Sykes GP, Malley LA, Delaney B: **Subchronic feeding study of herbicide-tolerant soybean DP-356043-5 in Sprague–Dawley rats.** *Food Chem Toxicol* 2008, **46:**2201–2213.
91. Appenzeller LM, Munley SM, Hoban D, Sykes GP, Malley LA, Delaney B: **Subchronic feeding study of grain from herbicide-tolerant maize DP-098140-6 in Sprague–Dawley rats.** *Food Chem Toxicol* 2009, **47:**2269–2280.
92. He XY, Huang KL, Li X, Qin W, Delaney B, Luo YB: **Comparison of grain from corn rootworm resistant transgenic DAS-59122-7 maize with non-transgenic maize grain in a 90-d feeding study in Sprague–Dawley rats.** *Food Chem Toxicol* 2008, **46:**1994–2002.
93. He XY, Tang MZ, Luo YB, Li X, Cao SS, Yu JJ, Delaney B, Huang KL: **A 90-d toxicology study of transgenic lysine-rich maize grain (Y642) in Sprague–Dawley rats.** *Food Chem Toxicol* 2009, **47:**425–432.
94. Healy C, Hammond B, Kirkpatrick J: **Results of a 13-week safety assurance study with rats fed grain from corn rootworm-protected, glyphosate-tolerant MON 88017 corn.** *Food Chem Toxicol* 2008, **46:**2517–2524.
95. Schroder M, Poulsen M, Wilcks A, Kroghsbo S, Miller A, Frenzel T, Danier J, Rychlik M, Emami K, Gatehouse A, *et al*: **A 90-d safety study of genetically modified rice expressing Cry1Ab protein (Bacillus thuringiensis toxin) in Wistar rats.** *Food Chem Toxicol* 2007, **45:**339–349.
96. Delaney B, Zhang J, Carlson G, Schmidt J, Stagg B, Comstock B, Babb A, Finlay C, Cressman RF, Ladics G, *et al*: **A gene-shuffled glyphosate acetyltransferase protein from Bacillus licheniformis (GAT4601) shows no evidence of allergenicity or toxicity.** *Toxicol Sci* 2008, **102:**425–432.
97. Cisterna B, Flach F, Vecchio L, Barabino SM, Battistelli S, Martin TE, Malatesta M, Biggiogera M: **Can a genetically-modified organism-containing diet influence embryo development? A preliminary study on pre-implantation mouse embryos.** *Eur J Histochem* 2008, **52:**263–267.
98. Batista R, Oliveira MM: **Facts and fiction of genetically engineered food.** *Trends Biotechnol* 2009, **27:**277–286.
99. UK Advisory Committee on Novel Foods and Processes: *ACNFP minutes: 20 July 2006.* Committee Paper for Discussion ACNFP/78/7 Advisory Committee on Novel Foods and Processes, Minutes of the Meeting Held on 20 July 2006; 2006. http://acnfp.food.gov.uk/meetings/acnfpmeet2006/acnfpjul06/acnfpminsjuly2006.
100. Nutritional and Safety Assessments of Foods and Feeds Nutritionally Improved through Biotechnology: *An Executive Summary A Task Force Report by the International Life Sciences Institute.* Washington, D.C: Comprehensive Reviews in Food Science and Food Safety; 2004:35–104.
101. International Life Sciences Institute (ILSI): *Best Practices for the Conduct of Animal Studies to Evaluate Crops Genetically Modified for Input Traits.* Washington, DC: International Life Sciences Institute; 2003:62.
102. International Life Sciences Institute (ILSI): *Best Practices for the Conduct of Animal Studies to Evaluate Crops Genetically Modified for Output Traits.* Washington, D.C: International Life Sciences Institute; 2007:202.
103. European Food Safety Authority (EFSA): **Opinion of the scientific committee/ scientific panel. Statistical significance and biological relevance.** *EFSA Journal* 2011, **9:**2372.
104. Benjamini Y, Hochberg Y: **Controlling the false discovery rate: a practical and powerful approach to multiple testing.** *J R Stat Soc Ser B Methodol* 1995, **57:**289–300.
105. Herman RA, Price WD: **Unintended compositional changes in genetically modified (gm) crops: 20 years of research.** *J Agric Food Chem* 2013: e-pub ahead of print.
106. BEETLE: *Long-term effects of genetically modified (GM) crops on health and the environment (including biodiversity).* Federal Office of Consumer Protection and Food Safety (BVL): BLaU-Umweltstudien and Genius GmbH, Berlin: Executive Summary and Main Report; 2009.
107. Hammond BG, Campbell KW, Pilcher CD, Degooyer TA, Robinson AE, McMillen BL, Spangler SM, Riordan SG, Rice LG, Richard JL: **Lower fumonisin mycotoxin levels in the grain of Bt corn grown in the United States in 2000–2002.** *J Agric Food Chem* 2004, **52:**1390–1397.
108. Llorente B, Alonso GD, Bravo-Almonacid F, Rodriguez V, Lopez MG, Carrari F, Torres HN, Flawia MM: **Safety assessment of nonbrowning potatoes: opening the discussion about the relevance of substantial equivalence on next generation biotech crops.** *Plant Biotechnol J* 2011, **9:**136–150.
109. Flachowsky G, Bohme H: **Proposals for nutritional assessments of feeds from genetically modified plants.** *J Animal and Feed Sci* 2005, **14**(Suppl 1):49–70.
110. Folmer J, Grant R, Milton C, Beck J: **Utilization of Bt corn residues by grazing beef steers and Bt corn silage and grain by growing beef cattle and lactating dairy cows.** *J Anim Sci* 2002, **80:**1352–1361.
111. Llorente B, Rodriguez V, Guillermo DA, Torres HN, Flawia MM, Bravo-Almonacid FF: **Improvement of aroma in transgenic potato as a consequence of impairing tuber browning.** *PLoS One* 2010, **5:**e14030.
112. Davidovich-Rikanati R, Sitrit Y, Tadmor Y, Iijima Y, Bilenko N, Bar E, Carmona B, Fallik E, Dudai N, Simon JE, *et al*: **Enrichment of tomato flavor by diversion of the early plastidial terpenoid pathway.** *Nat Biotech* 2007, **25:**899–901.
113. Rommens CM, Richael CM, Yan H, Navarre DA, Ye J, Krucker M, Swords K: **Engineered native pathways for high kaempferol and caffeoylquinate production in potato.** *Plant Biotechnol J* 2008, **6:**870–886.
114. Park S, Elless MP, Park J, Jenkins A, Lim W, Chambers E, Hirschi KD: **Sensory analysis of calcium-biofortified lettuce.** *Plant Biotechnol J* 2009, **7:**106–117.
115. Butelli E, Titta L, Giorgio M, Mock H-P, Matros A, Peterek S, Schijlen EGWM, Hall RD, Bovy AG, Luo J, Martin C: **Enrichment of tomato fruit with health-promoting anthocyanins by expression of select transcription factors.** *Nat Biotech* 2008, **26:**1301–1308.
116. Alexander TW, Reuter T, Aulrich K, Sharma R, Okine EK, Dixon WT, McAllister TA: **A review of the detection and fate of novel plant molecules derived from biotechnology in livestock production.** *Anim Feed Sci Technol* 2007, **133:**31–62.
117. Council for Agricultural Science and Technology (CAST): **Safety of meat, milk, and eggs from animals fed crops derived from modern biotechnology.** *Ames, Iowa: CAST* 2006, **34:**1–8.

118. Flachowsky G, Chesson A, Aulrich K: **Animal nutrition with feeds from genetically modified plants.** *Arch Anim Nutr* 2005, **59:**1–40.
119. Guertler P, Paul V, Albrecht C, Meyer HH: **Sensitive and highly specific quantitative real-time PCR and ELISA for recording a potential transfer of novel DNA and Cry1Ab protein from feed into bovine milk.** *Anal Bioanal Chem* 2009, **393:**1629–1638.
120. Guertler P, Paul V, Steinke K, Wiedemann S, Preißinger W, Albrecht C, Spiekers H, Schwarz FJ, Meyer HHD: **Long-term feeding of genetically modified corn (MON810) — Fate of cry1Ab DNA and recombinant protein during the metabolism of the dairy cow.** *Livest Sci* 2010, **131:**250–259.
121. Bertheau Y, Helbling J, Fortabat M, Makhzami S, Sotinel I, Audeon C, Kobilinsky A, Petit L, Fach P, Brunschwig P, *et al*: **Persistence of plant dna sequences in the blood of dairy cows fed with genetically modified (bt176) and conventional corn silage.** *J Agric Food Chem* 2009, **57:**509–516.
122. Brookes G, Barfoot P: **The global income and production effects of genetically modified (GM) crops 1996–2011.** *GM Crops and Food: Biotechnology in Agriculture and the Food Chain* 2013, **4:**74–83.
123. Masip G, Sabalza M, Pérez-Massot E, Banakar R, Cebrian D, Twyman RM, Capell T, Albajes R, Christou P: **Paradoxical EU agricultural policies on genetically engineered crops.** *Trends in plant science* 2013, **18:**312–324.
124. Foreign Agricultural Service (FAS): **GAIN Report: EU-27 Biotechnology Annual Report.** *USDA FAS* 2013, **Report number FR9142.** http://gain.fas.usda.gov/Recent%20GAIN%20Publications/Agricultural%20Biotechnology%20Annual_Paris_EU-27_7-12-2013.pdf.
125. Brookes G, Barfoot P: *GM crops: global socio-economic and environmental impacts 1996–2011.* UK: PG Economics Ltd; 2013. http://www.pgeconomics.co.uk/pdf/2013globalimpactstudyfinalreport.pdf.
126. Foreign Agricultural Service (FAS): **GAIN report: People's Republic of China agricultural biotechnology annual.** *USDA FAS* 2012, **Report number CH12046:**1–15. http://gain.fas.usda.gov/Recent%20GAIN%20Publications/Agricultural%20Biotechnology%20Annual_Beijing_China%20-%20Peoples%20Republic%20of_7-13-2012.pdf.
127. Foreign Agricultural Service (FAS): **GAIN report: Argentina biotechnology annual report.** *USDA FAS* 2012:1–14. http://gain.fas.usda.gov/Recent%20GAIN%20Publications/Agricultural%20Biotechnology%20Annual_Buenos%20Aires_Argentina_7-18-2012.pdf
128. Directorate-General for Agriculture and Rural Development: *Economic impact of unapproved GMOs on EU feed imports and livestock production.* 2007. http://ec.europa.eu/agriculture/envir/gmo/economic_impactGMOs_en.pdf.
129. Paarlberg R: **GMO foods and crops: Africa's choice.** *N Biotechnol* 2010, **27:**609–613.
130. Carpenter JE: **Peer-reviewed surveys indicate positive impact of commercialized GM crops.** *Nat Biotech* 2010, **28:**319–321.
131. Brookes G, Barfoot P: **Key environmental impacts of global genetically modified (GM) crop use 1996–2011.** *GM Crops Food* 2013, **4:**109–119.
132. Mannion AM, Morse S: **Biotechnology in agriculture: agronomic and environmental considerations and reflections based on 15 years of GM crops.** *Prog Phys Geogr* 2012, **36:**747–763.
133. Morris EJ: **A semi-quantitative approach to GMO risk-benefit analysis.** *Transgenic Res* 2011, **20:**1055–1071.

5

Epigenetics and transgenerational inheritance in domesticated farm animals

Amanda Feeney, Eric Nilsson and Michael K Skinner*

Abstract

Epigenetics provides a molecular mechanism of inheritance that is not solely dependent on DNA sequence and that can account for non-Mendelian inheritance patterns. Epigenetic changes underlie many normal developmental processes, and can lead to disease development as well. While epigenetic effects have been studied in well-characterized rodent models, less research has been done using agriculturally important domestic animal species. This review will present the results of current epigenetic research using farm animal models (cattle, pigs, sheep and chickens). Much of the work has focused on the epigenetic effects that environmental exposures to toxicants, nutrients and infectious agents has on either the exposed animals themselves or on their direct offspring. Only one porcine study examined epigenetic transgenerational effects; namely the effect diet micronutrients fed to male pigs has on liver DNA methylation and muscle mass in grand-offspring (F2 generation). Healthy viable offspring are very important in the farm and husbandry industry and epigenetic differences can be associated with production traits. Therefore further epigenetic research into domestic animal health and how exposure to toxicants or nutritional changes affects future generations is imperative.

Keywords: Environment, Epigenetics, Pig, Review, Transgenerational

Introduction

Mendelian genetic theories have guided much of the biological research preformed in recent history. It has long been assumed that specific phenotypes arise only from DNA sequence. However, non-Mendelian inheritance patterns challenge these theories and suggest that an alternate process might exist to account for certain mechanisms of inheritance. Epigenetics provides a molecular mechanism that can account for these non-Mendelian observations [1-3]. Epigenetics research looks into modifications and inheritance patterns that do not involve changes in the DNA sequence, but do affect genome activity and gene expression [1-4]. There are four main mechanisms by which epigenetics can alter gene expression: DNA methylation, histone modification, chromatin structure, and non-coding RNA [1,5]. Although the epigenetic processes are highly conserved among all species, the specific epigenomes are highly divergent between species. Modifications of these epigenetic processes can occur due to direct environmental exposure at critical periods in the development of the organism [1,6-8]. Clearly any generation that has direct exposure to an environmental insult may be altered in some way. Recent research shows subsequent generations that were not present at the time of the exposure can still be affected due to epigenetic transgenerational inheritance, if exposure occurred during sensitive developmental windows for the germ cells [9]. Epigenetic transgenerational inheritance is defined as germline-mediated inheritance of epigenetic information between generations, in the absence of direct environmental influences, that leads to phenotypic variation [1,9]. For example, if a pregnant animal is exposed to a toxicant during gonadal sex determination of the fetus then changes in fetal germ cell epigenetic programming may occur [8,10]. Therefore these offspring and the gametes that will form the grand-offspring are directly exposed to the toxicant, and changes seen in these F1 and F2 generations are not transgenerational [11]. However, epigenetic changes in the F3 generation (great-grand-offspring) would be considered transgenerationally inherited. In contrast, if a male or non-pregnant female adult animal is subjected to an environmental exposure, then changes seen in the F2 generation or later are considered transgenerational [11]. Changes in DNA methylation in gametes that are transmitted to subsequent

* Correspondence: skinner@wsu.edu
Center for Reproductive Biology, School of Biological Sciences, Washington State University, 99164-4236 Pullman, WA, USA

generations provide a mechanism for the inheritance of epigenetic information [12-14]. Non-coding RNA also appears to have a role in epigenetic transgenerational inheritance [15]. Much of the current research has used rodent models to demonstrate epigenetic changes after environmental insult, especially during pregnancy [8,10]. Germline epigenetic transgenerational inheritance has also been shown in plants, flies, worms, and humans [10,16-21].

Despite the amount of epigenetic and transgenerational epigenetic inheritance research being done on a multitude of mammal, insect, and plant models [8,10,16-21], a lack of research into these topics using farm animal models exists. This review will present the current epigenetic inheritance research and data using farm animal models (bovine, porcine, ovine, and gallus), Table 1. While much of the work has focused on the direct effects of environmental exposure to toxicants and nutrients, research into epigenetic transgenerational inheritance is limited. It is important that more epigenetic research be done in domesticated farm animals because of their close human relationships and potential for high pesticide exposure on farms. Pesticides have been shown to have dramatic transgenerational epigenetic effects on many animal models affecting the nervous system, reproductive and endocrine systems, and even causing cancer [9,22]. Since hybrid vigour (i.e. heterosis) has been shown to be critical in breeding of domestic animals, and epigenetics has a critical role in hybrid vigour [23], epigenetic inheritance will be important in developing optimal domestic animal breeds. Considering overpopulation issues requiring a rise in food supply, there may be more efficient ways of detecting and promoting favorable selection using epigenetics to breed for a lower instance of animal disease.

Domestic animal models

Bovine

The relationship of DNA methylation and milk production in dairy cattle has been investigated. During lactation the bovine αS1-casein gene is hypomethylated [24]. Research has characterized this gene during various physiological conditions during the lactation cycle. Vanselow *et al.* found that during lactation the (STAT)5-binding lactation enhancer, which is part of the αS1-casein encoding gene, is hypomethylated [25]. However, during *Escherichia coli* infection of the mammary gland, this region becomes methylated at three CpG dinucleotides which accompanies a shut down of αS1-casein synthesis [25]. These observations have also been shown with infection by *Streptococcus uberis* [26]. In addition, methylation of these same 3 CpG dinucleotides has been seen during non-milking periods of healthy dairy cattle when milking was ceased suddenly [27]. González-Recio *et al.* preformed a generational study to see if a mother dairy cow affected the milk production of her offspring [28]. They found that female calves born to cows that were already lactating from previous births produced between 18 and 91 kg less milk in adulthood than calves that were first-born, and that their lifespans were also shorter [28]. Because of the generational effect, researchers suggested epigenetic inheritance. However, they did not look specifically at epigenetic differences in the affected calves versus controls.

More research has been done on histone modification related to nutritional changes than on DNA methylation. Short-chain fatty acids are particularly important in ruminant digestion, and are used for cell energy production and use [29]. Butyrate, a specific short-chain fatty acid, inhibits histone deacetylases which have been shown to regulate epigenetic changes to the genome [30]. Wu, *et al.* [31] show that high doses of butyrate exposure to Madin-Darby bovine kidney epithelial cells causes cell cycle arrest, changes in gene expression, changes in nucleic acid metabolic processes, regulation of the cell cycle, and changes in DNA replication. Therefore this study claims that histone acetylation is essential for diverse cellular processes [31], but histone acetylation was not measured directly.

The influence of epigenetics on disease has been studied in many animal models such as rats, mice, and humans, but very little has been done with cattle. One bovine developmental disease called large-offspring syndrome (LOS) has been found to have epigenetic components during embryonic growth. LOS has largely been associated with reproductive technologies commonly used with cattle such as in vitro fertilization and somatic-cell nuclear transfer [32]. Symptoms usually include increases in birth weight, organ overgrowth, difficulty breathing and standing, as well as skeletal and immunological defects. There are also increased rates of fetal and neonatal deaths [33-35]. Dean *et al.* [36] has reported methylation changes in bovine embryos (morulae) between controls, *in vitro* fertilized, and somatic-cell nuclear transfer embryos, and suggests that these methylation differences may account for the different success rates and health of calves born from these reproductive technologies [36]. A number of studies have demonstrated developmental epigenetic programming in bovine germ cells [37] and bovine embryos [38], which is similar among all mammalian species. In another study focusing on innate immunity, Green *et al.* [39] looked at epigenetics and individual variation in the innate immune response of bovine dermal fibroblasts, specifically via toll receptor signaling. Exposure to de-methylating and hyper-acetylating agents led to increased expression of several cytokines as compared to controls, suggesting immune gene expression has epigenetic regulation [39].

Table 1 Environmental epigenetics and epigenetic inheritance in domestic farm animals

Environmental epigenetics and domestic farm animals		
Bovine	**Context**	**Ref.**
Mammary gland-specific hypomethylation of Hpa II sites flanking the bovine alpha S1-casein gene.	Epigenetic regulation of lactation.	[24]
DNA-remethylation around a STAT5-binding enhancer in the alphaS1-casein promoter is associated with abrupt shutdown of alphaS1-casein synthesis during acute mastitis.	Epigenetic regulation of lactation.	[25]
Transcriptome profiling of Streptococcus uberis-induced mastitis reveals fundamental differences between immune gene expression in the mammary gland and in a primary cell culture model.	Epigenetic regulation of lactation.	[26]
Conservation of methylation reprogramming in mammalian development: aberrant reprogramming in cloned embryos.	Epigenetic changes with assisted reproductive technologies	[36]
DNA methylation events associated with the suppression of milk protein gene expression during involution of the bovine mammary gland.	Epigenetic regulation of lactation.	[27]
Effect of maternal lactation during pregnancy	Epigenetic regulation of lactation.	[28]
Large offspring syndrome in cattle and sheep.	Epigenetic changes with assisted reproductive technologies	[33]
The production of unusually large offspring following embryo manipulation: Concepts and challenges.	Epigenetic changes with assisted reproductive technologies	[34]
Postnatal characteristics of calves produced by nuclear transfer cloning.	Epigenetic changes with assisted reproductive technologies	[35]
Epigenetic contribution to individual variation in response to lipopolysaccharide in bovine dermal fibroblasts.	Epigenetic role in immunity	[39]
Occurrence, absorption and metabolism of short chain fatty acids in the digestive tract of mammals.	Epigenetic changes with regulation of nutrition	[29]
Butyrate induces profound changes in gene expression related to multiple signal pathways in bovine kidney epithelial cells.	Epigenetic changes with regulation of nutrition	[30]
Transcriptome Characterization by RNA-seq Unravels the Mechanisms of Butyrate-Induced Epigenomic Regulation in Bovine Cells.	Epigenetic changes with regulation of nutrition	[31]
In vitro produced and cloned embryos: Effects on pregnancy, parturition and offspring.	Epigenetic changes with assisted reproductive technologies	[32]
Porcine		
Sulforaphane causes a major epigenetic repression of myostatin in porcine satellite cells.	Histone deacetylase inhibitor affects myostatin *in vitro*	[41]
Maternal dietary protein affects transcriptional regulation of myostatin gene distinctively at weaning and finishing stages in skeletal muscle of Meishan pigs.	Maternal diet affects offspring epigenetics	[43]
HOX10 mRNA expression and promoter DNA methylation in female pig offspring after in utero estradiol-17beta exposure.	Maternal steroid exposure affects offspring epigenetics	[48]
Investigations on transgenerational epigenetic response down the male line in F2 pigs.	Paternal diet has transgenerational epigenetic effect	[49]
Dietary Sulforaphane, a Histone Deacetylase Inhibitor for Cancer Prevention	Epigenetic changes with regulation of nutrition	[40]
Neonatal estradiol exposure alters uterine morphology and endometrial transcriptional activity in prepubertal gilts.	Steroid exposure affects epigenetics	[47]
Maternal dietary protein restriction and excess affects offspring gene expression and methylation of non-SMC subunits of condensin I in liver and skeletal muscle.	Maternal diet affects offspring epigenetics	[44]
Diet, methyl donors and DNA methylation: interactions between dietary folate, methionine and choline.	Epigenetic changes with regulation of nutrition	[45]
Ovine		
Periconceptional nutrition and the early programming of a life of obesity or adversity.	Maternal diet affects offspring epigenetics	[50]
The effect of maternal under-nutrition before muscle differentiation on the muscle fiber development of the newborn lamb.	Maternal diet affects offspring epigenetics	[51]
Effect of maternal dietary restriction during pregnancy on lamb carcass characteristics and muscle fiber composition.	Maternal diet affects offspring epigenetics	[52]

Table 1 Environmental epigenetics and epigenetic inheritance in domestic farm animals *(Continued)*

Gallus		
Comparison of the Genome-Wide DNA Methylation Profiles between Fast-Growing and Slow-Growing Broilers.	Differences in epigenetics between breeds	[55]
Insulin-like growth factor-1 receptor is regulated by microRNA-133 during skeletal myogenesis.	Epigenetic effects during muscle development	[58]
Transgenerational epigenetic effects on innate immunity in broilers: An underestimated field to be explored?	Review on role of epigenetics in innate immunity	[59]
DNMT gene expression and methylome in Marek's disease resistant and susceptible chickens prior to and following infection by MDV.	Epigenetic role in immunity	[54]
Epigenetic transgenerational inheritance study in domestic farm animals		
Porcine		
Investigations on transgenerational epigenetic response down the male line in F2 pigs.	Paternal diet has transgenerational epigenetic effect	[49]

No studies have been published showing epigenetic transgenerational inheritance in cattle.

Porcine

Swine are often used as animal models to study human disease because of the similar physiology between the two species. Because of this, much of the epigenetic porcine research involves exposure and response, with very little of the current research being transgenerational.

Epigenetic effects due to histone modification and acetylation have been studied in a porcine model both in order to increase meat production and to develop a potential treatment for muscular degenerative disease. Sulforaphane is a bioactive histone deacetylase inhibitor often found in edible vegetation like broccoli [40]. Fan *et al.* [41] treated porcine satellite cells with sulforaphane to epigenetically repress myostatin which would potentially result in more muscle growth [42]. Liu *et al.* [43] also looked at the myostatin pathway to investigate the short term and long term epigenetic changes in pigs based on maternal diet. These researchers concluded that histone modifications and changes in microRNA expression took place long term and played a part in skeletal muscle phenotype [43]. Another study looked at DNA methylation in response to altered protein and carbohydrate diets for maternal pigs during gestation [44]. Researchers found that hepatic global methylation was decreased in fetuses from protein-restricted mothers, likely caused by methionine deficiency [45]. However, skeletal muscle global methylation was not affected [44]. This study demonstrates maternal nutrition will likely have an epigenetic effect on embryonic tissue development. Epigenetic programming in the porcine germline has also been reported [46].

Research conducted by Tarletan *et al.* demonstrated that neonatal estrogen exposure in piglets can lead to epigenetic changes that affect uterine capacity and environment [47]. This leads to potentially less successful pregnancies once the piglets become adults [47]. Another environmental estrogen exposure experiment was preformed analyzing the effect on the gene HOXA10 by exposing offspring in utero to estradiol-17β. No difference in HOXA10 expression was detected in either the low dose or high dose group [48]. However, differences in HOXA10 mRNA expression were detected between pre-pubescent and post-pubescent gilts [48].

One recent transgenerational porcine study has been reported [49], Table 1. Braunschweig *et al.* preformed a three generational study to look at the effect of feeding on male epigenetic inheritance. The experimental group F0 generation males were fed a diet high in methylating micronutrients, and the resulting F2 generation had a lower fat percentage and higher shoulder muscle percentage as compared to controls. They also found significant differences in DNA methylation between the control and experimental groups, especially in the liver, which was proposed to epigenetically affect fat metabolism pathways [49].

Ovine

As shown in the bovine model and porcine model, maternal nutritional impact is a common topic in epigenetic research, and ovine studies are no exception. Zhang *et al.* [50] looked into the effects of maternal over-nutrition in sheep, both during peri-conception and during the late stages of pregnancy. They found that over-nutrition in late stages of pregnancy resulted in more visceral fat in offspring and a change in appetite that pre-disposed that lamb to over-eat in adult life. More interestingly, they also found that over-nutrition at the peri-conception period led to higher rates of visceral fat in only female ewe offspring, leading to a conclusion of sex-specific DNA methylation. They also found that when diet was restricted just before conception (maternal under-nutrition), the adrenal glands of the offspring tended to be heavier and have less methylation of the IGF2/H19 differentially methylated regions in the adrenal. Observations suggested that while a restricted peri-conception diet led to no maternal epigenetic influence on bodyweight, it did increase the stress response in these offspring [50]. Other nutritional studies have looked at muscle development in response to maternal under-nutrition during pregnancy and have shown that maternal under-nutrition causes a decrease of fast muscle fibers in early stages, but an increase in them during later stages of development [51,52]. However, these studies did not investigate epigenetic mechanisms.

No studies have been published showing epigenetic transgenerational inheritance in sheep.

Gallus

Marek's disease in chickens is a manifestation of Marek's disease virus and progresses to become a T-cell lymphoma that affects chickens and other birds. Vaccines have been developed but they are not completely successful [53]. Tian *et al.* [54] set to find out why one breeding line seemed resistant to the virus, while another was more susceptible. They found that in the virus-resistant line, DNA methylation levels in thymus cells were decreased after exposure to the virus. They also found that with pharmacological inhibition of DNA methylation *in vitro* the propagation in the infected cells was slowed. Observations suggested that DNA methylation in the host may be associated with virus resistance or susceptibility [54].

Different developmental epigenetic patterns have been studied between chicken types. One study looked at differential DNA methylation in breast muscle between slow-growing and fast-growing broiler chickens [55]. They found that between the two breeds of chickens there were 75 differentially methylated genes, including several genes belonging to the fibroblast growth factor (FGF) family. The FGF family is known for its role in many growth processes [56]. In addition, effects in the insulin growth factor receptor (IGF1R) were observed that influence skeletal muscle growth specifically [57,58].

As one review indicated, many poultry studies indicate that there may be epigenetic effects, and even transgenerational epigenetic inheritance, though very few studies actually test for DNA methylation or histone modification in their research [59].

No studies have been published showing epigenetic transgenerational inheritance in chicken.

Conclusion

While a good amount of epigenetic research has been preformed on domesticated farm animals still more needs to be done, Table 1. There is little research at all in transgenerational inheritance of these epigenetic modifications. This could be due to the fact that farm animals are more difficult and more costly to raise than other common animal research models. In addition, they have longer lifespans so transgenerational studies take more time and resources. Animal science researchers should cultivate an interest in conducting these types of experiments for a number of reasons. Healthy viable offspring are very important in the farm and husbandry industry and epigenetic differences can be associated with production traits. Recently there has been a lot of social pressure to cut down on vaccination and antibiotic use for animals raised for meat and epigenetics research may help to provide the key to lowering disease and increasing immunity. Therefore research into domestic animal health and how exposure to toxicants such as pesticides affects future generations is imperative.

Glossary

Epigenetics: Molecular factors/processes around the DNA that regulate genome activity independent of DNA sequence, and are mitotically stable.

Epigenetic: Transgenerational Inheritance: Germline-mediated inheritance of epigenetic information between generations in the *absence* of direct environmental influences, that leads to phenotypic variation.

Epimutation: Differential presence of epigenetic marks that lead to altered genome activity.

Abbreviations

F0: Generation pregnant female; F1: Generation fetus that becomes the offspring or children; F2: Generation (grandchildren); F3: Generation (great-grandchildren); LOS: Large-offspring syndrome; FGF: Fibroblast growth factor; IGF1R: Insulin growth factor receptor.

Competing interests

The authors declare that they have no competing interests.

Authors' contributions

All authors designed and wrote the study. All authors edited and approved the manuscript.

Acknowledgements

We acknowledge the assistance of Ms. Heather Johnson for assistance in preparation of the manuscript. The current address for Ms. Amanda Feeney is New York Medical College- Valhalla, New York. This research was supported by NIH grants to MKS.

References

1. Skinner MK, Manikkam M, Guerrero-Bosagna C: **Epigenetic transgenerational actions of environmental factors in disease etiology.** *Trends Endocrinol Metab* 2010, **21**:214–222.
2. Jirtle RL, Skinner MK: **Environmental epigenomics and disease susceptibility.** *Nat Rev Genet* 2007, **8**:253–262.
3. Guerrero-Bosagna C, Skinner MK: **Environmentally induced epigenetic transgenerational inheritance of phenotype and disease.** *Mol Cell Endocrinol* 2012, **354**:3–8.
4. Simmons D: **Epigenetic influence and disease.** *Nat Educ* 2008, **1**:6.
5. Egger G, Liang G, Aparicio A, Jones PA: **Epigenetics in human disease and prospects for epigenetic therapy.** *Nature* 2004, **429**:457–463.
6. Vandegehuchte MB, Janssen CR: **Epigenetics in an ecotoxicological context.** *Mutat Res Genet Toxicol Environ Mutagen* 2014, **764–765**:36–45.
7. Baccarelli A, Bollati V: **Epigenetics and environmental chemicals.** *Curr Opin Pediatr* 2009, **21**:243–251.
8. Anway MD, Cupp AS, Uzumcu M, Skinner MK: **Epigenetic transgenerational actions of endocrine disruptors and male fertility.** *Science* 2005, **308**:1466–1469.

9. Skinner MK: **Environmental epigenetic transgenerational inheritance and somatic epigenetic mitotic stability.** *Epigenetics* 2011, **6**:838–842.
10. Manikkam M, Guerrero-Bosagna C, Tracey R, Haque MM, Skinner MK: **Transgenerational actions of environmental compounds on reproductive disease and identification of epigenetic biomarkers of ancestral exposures.** *PLoS One* 2012, **7**:e31901.
11. Skinner MK: **What is an epigenetic transgenerational phenotype? F3 or F2.** *Reprod Toxicol* 2008, **25**:2–6.
12. Lees-Murdock DJ, Walsh CP: **DNA methylation reprogramming in the germ line.** *Epigenetics* 2008, **3**:5–13.
13. Reik W, Dean W, Walter J: **Epigenetic reprogramming in mammalian development.** *Science* 2001, **293**:1089–1093.
14. Smith ZD, Chan MM, Mikkelsen TS, Gu H, Gnirke A, Regev A, Meissner A: **A unique regulatory phase of DNA methylation in the early mammalian embryo.** *Nature* 2012, **484**:339–344.
15. Gapp K, Jawaid A, Sarkies P, Bohacek J, Pelczar P, Prados J, Farinelli L, Miska E, Mansuy IM: **Implication of sperm RNAs in transgenerational inheritance of the effects of early trauma in mice.** *Nat Neurosci* 2014, **17**:667–669.
16. Guerrero-Bosagna C, Settles M, Lucker B, Skinner M: **Epigenetic transgenerational actions of vinclozolin on promoter regions of the sperm epigenome.** *PLoS One* 2010, **5**:e13100.
17. Arico JK, Katz DJ, van der Vlag J, Kelly WG: **Epigenetic patterns maintained in early Caenorhabditis elegans embryos can be established by gene activity in the parental germ cells.** *PLoS Genet* 2011, **7**:e1001391.
18. Carone BR, Fauquier L, Habib N, Shea JM, Hart CE, Li R, Bock C, Li C, Gu H, Zamore PD, Meissner A, Weng Z, Hofmann HA, Friedman N, Rando OJ: **Paternally induced transgenerational environmental reprogramming of metabolic gene expression in mammals.** *Cell* 2010, **143**:1084–1096.
19. Dunn GA, Bale TL: **Maternal high-fat diet effects on third-generation female body size via the paternal lineage.** *Endocrinology* 2011, **152**:2228–2236.
20. Morgan CP, Bale TL: **Early prenatal stress epigenetically programs dysmasculinization in second-generation offspring via the paternal lineage.** *J Neurosci* 2011, **31**:11748–11755.
21. Saze H: **Transgenerational inheritance of induced changes in the epigenetic state of chromatin in plants.** *Genes Genet Syst* 2012, **87**:145–152.
22. Collotta M, Bertazzi PA, Bollati V: **Epigenetics and pesticides.** *Toxicology* 2013, **307**:35–41.
23. Groszmann M, Greaves IK, Fujimoto R, Peacock WJ, Dennis ES: **The role of epigenetics in hybrid vigour.** *Trends Genet* 2013, **29**:684–690.
24. Platenburg GJ, Vollebregt EJ, Karatzas CN, Kootwijk EP, De Boer HA, Strijker R: **Mammary gland-specific hypomethylation of Hpa II sites flanking the bovine alpha S1-casein gene.** *Transgenic Res* 1996, **5**:421–431.
25. Vanselow J, Yang W, Herrmann J, Zerbe H, Schuberth HJ, Petzl W, Tomek W, Seyfert HM: **DNA-remethylation around a STAT5-binding enhancer in the alphaS1-casein promoter is associated with abrupt shutdown of alphaS1-casein synthesis during acute mastitis.** *J Mol Endocrinol* 2006, **37**:463–477.
26. Swanson KM, Stelwagen K, Dobson J, Henderson HV, Davis SR, Farr VC, Singh K: **Transcriptome profiling of Streptococcus uberis-induced mastitis reveals fundamental differences between immune gene expression in the mammary gland and in a primary cell culture model.** *J Dairy Sci* 2009, **92**:117–129.
27. Singh K, Swanson K, Couldrey C, Seyfert H-M, Stelwagen K: **DNA methylation events associated with the suppression of milk protein gene expression during involution of the bovine mammary gland.** *Proc N Z Soc Anim Prod* 2009, **69**:57–59.
28. González-Recio O, Ugarte E, Bach A: **Trans-generational effect of maternal lactation during pregnancy: a Holstein cow model.** *PLoS One* 2012, **7**(12):e51816.
29. Bugaut M: **Occurrence, absorption and metabolism of short chain fatty acids in the digestive tract of mammals.** *Comp Biochem Physiol B* 1987, **86**:439–472.
30. Li RW, Li C: **Butyrate induces profound changes in gene expression related to multiple signal pathways in bovine kidney epithelial cells.** *BMC Genomics* 2006, **7**:234.
31. Wu S, Li RW, Li W, Li CJ: **Transcriptome characterization by RNA-seq unravels the mechanisms of butyrate-induced epigenomic regulation in bovine cells.** *PLoS One* 2012, **7**:e36940.
32. Kruip TAM, den Daas JHG: **In vitro produced and cloned embryos: effects on pregnancy, parturition and offspring.** *Theriogenology* 1997, **47**:43–52.
33. Young LE, Sinclair KD, Wilmut I: **Large offspring syndrome in cattle and sheep.** *Rev Reprod* 1998, **3**:155–163.
34. Walker SK, Hartwich KM, Seamark RF: **The production of unusually large offspring following embryo manipulation: concepts and challenges.** *Theriogenology* 1996, **45**:111–120.
35. Garry FB, Adams R, McCann JP, Odde KG: **Postnatal characteristics of calves produced by nuclear transfer cloning.** *Theriogenology* 1996, **45**:141–152.
36. Dean W, Santos F, Stojkovic M, Zakhartchenko V, Walter J, Wolf E, Reik W: **Conservation of methylation reprogramming in mammalian development: aberrant reprogramming in cloned embryos.** *Proc Natl Acad Sci U S A* 2001, **98**:13734–13738.
37. Heinzmann J, Hansmann T, Herrmann D, Wrenzycki C, Zechner U, Haaf T, Niemann H: **Epigenetic profile of developmentally important genes in bovine oocytes.** *Mol Reprod Dev* 2011, **78**:188–201.
38. Niemann H, Carnwath JW, Herrmann D, Wieczorek G, Lemme E, Lucas-Hahn A, Olek S: **DNA methylation patterns reflect epigenetic reprogramming in bovine embryos.** *Cell Reprogram* 2010, **12**:33–42.
39. Green BB, Kerr DE: **Epigenetic contribution to individual variation in response to lipopolysaccharide in bovine dermal fibroblasts.** *Vet Immunol Immunopathol* 2014, **157**:49–58.
40. Ho E, Clarke JD, Dashwood RH: **Dietary sulforaphane, a histone deacetylase inhibitor for cancer prevention.** *J Nutr* 2009, **139**:2393–2396.
41. Fan H, Zhang R, Tesfaye D, Tholen E, Looft C, Holker M, Schellander K, Cinar MU: **Sulforaphane causes a major epigenetic repression of myostatin in porcine satellite cells.** *Epigenetics* 2012, **7**:1379–1390.
42. Benny Klimek ME, Aydogdu T, Link MJ, Pons M, Koniaris LG, Zimmers TA: **Acute inhibition of myostatin-family proteins preserves skeletal muscle in mouse models of cancer cachexia.** *Biochem Biophys Res Commun* 2010, **391**:1548–1554.
43. Liu X, Wang J, Li R, Yang X, Sun Q, Albrecht E, Zhao R: **Maternal dietary protein affects transcriptional regulation of myostatin gene distinctively at weaning and finishing stages in skeletal muscle of Meishan pigs.** *Epigenetics* 2011, **6**:899–907.
44. Altmann S, Murani E, Schwerin M, Metges CC, Wimmers K, Ponsuksili S: **Maternal dietary protein restriction and excess affects offspring gene expression and methylation of non-SMC subunits of condensin I in liver and skeletal muscle.** *Epigenetics* 2012, **7**:239–252.
45. Niculescu MD, Zeisel SH: **Diet, methyl donors and DNA methylation: interactions between dietary folate, methionine and choline.** *J Nutr* 2002, **132**:2333S–2335S.
46. Hyldig SM, Croxall N, Contreras DA, Thomsen PD, Alberio R: **Epigenetic reprogramming in the porcine germ line.** *BMC Dev Biol* 2011, **11**:11.
47. Tarleton BJ, Wiley AA, Bartol FF: **Neonatal estradiol exposure alters uterine morphology and endometrial transcriptional activity in prepubertal gilts.** *Domest Anim Endocrinol* 2001, **21**:111–125.
48. Pistek VL, Furst RW, Kliem H, Bauersachs S, Meyer HH, Ulbrich SE: **HOXA10 mRNA expression and promoter DNA methylation in female pig offspring after in utero estradiol-17beta exposure.** *J Steroid Biochem Mol Biol* 2013, **138**:435–444.
49. Braunschweig M, Jagannathan V, Gutzwiller A, Bee G: **Investigations on transgenerational epigenetic response down the male line in F2 pigs.** *PLoS One* 2012, **7**:e30583.
50. Zhang S, Rattanatray L, McMillen IC, Suter CM, Morrison JL: **Periconceptional nutrition and the early programming of a life of obesity or adversity.** *Prog Biophys Mol Biol* 2011, **106**:307–314.
51. Fahey AJ, Brameld JM, Parr T, Buttery PJ: **The effect of maternal undernutrition before muscle differentiation on the muscle fiber development of the newborn lamb.** *J Anim Sci* 2005, **83**:2564–2571.
52. Daniel ZC, Brameld JM, Craigon J, Scollan ND, Buttery PJ: **Effect of maternal dietary restriction during pregnancy on lamb carcass characteristics and muscle fiber composition.** *J Anim Sci* 2007, **85**:1565–1576.
53. Davison F, Nair V: *Marek's Disease: An Evolving Problem.* London: Elsevier Science Press; 2004.
54. Tian F, Zhan F, Vanderkraats ND, Hiken JF, Edwards JR, Zhang H, Zhao K, Song J: **DNMT gene expression and methylome in Marek's disease resistant and susceptible chickens prior to and following infection by MDV.** *Epigenetics* 2013, **8**(4):431–444.
55. Hu Y, Xu H, Li Z, Zheng X, Jia X, Nie Q, Zhang X: **Comparison of the genome-wide DNA methylation profiles between fast-growing and slow-growing broilers.** *PLoS One* 2013, **8**:e56411.

56. Itoh N: **The Fgf families in humans, mice, and zebrafish: their evolutional processes and roles in development, metabolism, and disease.** *Biol Pharm Bull* 2007, **30**:1819–1825.
57. Liu JP, Baker J, Perkins AS, Robertson EJ, Efstratiadis A: **Mice carrying null mutations of the genes encoding insulin-like growth factor I (Igf-1) and type 1 IGF receptor (Igf1r).** *Cell* 1993, **75**:59–72.
58. Huang MB, Xu H, Xie SJ, Zhou H, Qu LH: **Insulin-like growth factor-1 receptor is regulated by microRNA-133 during skeletal myogenesis.** *PLoS One* 2011, **6**:e29173.
59. Berghof TV, Parmentier HK, Lammers A: **Transgenerational epigenetic effects on innate immunity in broilers: an underestimated field to be explored?** *Poult Sci* 2013, **92**:2904–2913.

A gene expression estimator of intramuscular fat percentage for use in both cattle and sheep

Bing Guo[1,2], Kritaya Kongsuwan[2,3], Paul L Greenwood[4,5], Guanghong Zhou[1], Wangang Zhang[1] and Brian P Dalrymple[2*]

Abstract

Background: The expression of genes encoding proteins involved in triacyglyceride and fatty acid synthesis and storage in cattle muscle are correlated with intramuscular fat (IMF)%. Are the same genes also correlated with IMF% in sheep muscle, and can the same set of genes be used to estimate IMF% in both species?

Results: The correlation between gene expression (microarray) and IMF% in the longissimus muscle (LM) of twenty sheep was calculated. An integrated analysis of this dataset with an equivalent cattle correlation dataset and a cattle differential expression dataset was undertaken. A total of 30 genes were identified to be strongly correlated with IMF% in both cattle and sheep. The overlap of genes was highly significant, 8 of the 13 genes in the TAG gene set and 8 of the 13 genes in the FA gene set were in the top 100 and 500 genes respectively most correlated with IMF% in sheep, *P*-value = 0. Of the 30 genes, *CIDEA*, *THRSP*, *ACSM1*, *DGAT2* and *FABP4* had the highest average rank in both species. Using the data from two small groups of Brahman cattle (control and Hormone growth promotant-treated [known to decrease IMF% in muscle]) and 22 animals in total, the utility of a direct measure and different estimators of IMF% (ultrasound and gene expression) to differentiate between the two groups were examined. Directly measured IMF% and IMF% estimated from ultrasound scanning could not discriminate between the two groups. However, using gene expression to estimate IMF% discriminated between the two groups. Increasing the number of genes used to estimate IMF% from one to five significantly increased the discrimination power; but increasing the number of genes to 15 resulted in little further improvement.

Conclusion: We have demonstrated the utility of a comparative approach to identify robust estimators of IMF% in the LM in cattle and sheep. We have also demonstrated a number of approaches (potentially applicable to much smaller groups of animals than conventional methods) to using gene expression to rank animals for IMF% within a single farm/treatment, or to estimate differences in IMF% between two farms/treatments.

Keywords: Cattle, Gene expression phenotype, IMF%, Sheep

Background

Consumers are prepared to pay more for meat with superior eating qualities [1]. Intramuscular fat (IMF), the flecks and streaks of fat within the lean sections of meat, which is also known as marbling, is associated with juiciness and flavour [2]. Recent research has shown that increased IMF% could dramatically improve the tenderness of lamb carcasses 5 days post-slaughter [3]. But compared to beef-related research (see [4,5]), few publications have focussed on the molecular mechanism of IMF deposition in sheep. In the past few years, only *FABP3 (H-FABP)*, *PPARG*, *DGAT1*, *LPL*, *ACACA*, *FASN (FAS)*, *FABP4*, *CPT1B* and *SCD* have been reported to directly influence IMF% status in sheep LM [6-9]. Thus, based on the limited information from sheep, it is hard to identify a set of genes to estimate IMF%.

In our previous studies in cattle, three gene sets, designated as the "TAG gene set" (triglyceride synthesis and storage), the "FA gene set" (fatty acid synthesis and storage) and the "PPARG gene set" (Peroxisome proliferator-activated receptor gamma), were identified based on the expression profiles of the genes in the LM across development in two crosses [4]. The expression of genes from these three gene sets, in particular the TAG gene set, was correlated with IMF deposition in cattle

* Correspondence: Brian.Dalrymple@csiro.au
[2]CSIRO Animal, Food and Health Sciences, St. Lucia QLD 4067, Australia
Full list of author information is available at the end of the article

LM [4]. The TAG gene set was used to identify the effect of HGP (hormone growth promotant) treatment, site (New South Wales [NSW] and Western Australia [WA]) and Calpain/calpastatin genotype on IMF% [4].

Cattle and sheep are evolutionarily closely related [10] and are expected to exhibit many common physiological characteristics. In this study, we hypothesised that the genes in the TAG, FA and PPARG gene sets identified in cattle could also be applied to estimate IMF% in sheep. Furthermore, based on these gene sets, we evaluated the utility of single and small sets of genes to estimate IMF% in small groups of animals from both species.

Materials and methods

Use of animals and the procedures performed in this study were approved by the Industry & Investment New South Wales (NSW) Orange Agriculture Institute Animal Ethics Committee, Commonwealth Scientific and Industrial Organisation (CSIRO) Rockhampton Animal Experimentation Ethics Committee, and the Department of Agriculture and Food, Western Australia (WA) Animal Ethics Committee.

Sheep correlation dataset

The design of the experiment has been described previously [11]. Briefly, 20 sheep were randomly assigned to five groups, four groups of treated animals received implants containing a combination of ~42 mg trenbolone acetate (TBA) and ~4.2 mg 17-βestradiol (E2), or ~50 mg TBA alone, or ~10 mg E2 alone at the start of the trial or 20 mg oxytocin delivered by Alzet osmotic pump over 30 d at the start of the experiment and again after 30 and 60 d. Following slaughter, 50 mg of LM tissue (between the 12th and 13th rib) and the strip loins (6th to 9th rib) were collected from the right sides of the carcasses for RNA preparation and meat quality analyses, respectively. IMF% was measured in duplicate on each sample by gas chromatography (GC) as previously described [11]. Gene expression was measured using the Bovine Oligo Microarray Chip (Bovine 4x44K) from Agilent Technologies (Santa Clara CA, USA and will be described in detail elsewhere [Kongsuwan et al., in preparation]). The same platform was used for the two bovine gene expression datasets described below. The Bovine Oligo Microarray platform was used as it has a larger coverage of genes than the equivalent sheep array and using the same platform simplifies data integration and analysis.

Cattle correlation dataset

Correlation between gene expression and IMF% in the LM muscle in a group of 48 intensively fed Brahman steers, including three tenderness genotypes, an environment contrast (growth at two different sites, New South Wales [NSW] and Western Australia [WA]) and with and without a hormone growth promotant (HGP) treatment, has been described previously [4]. IMF% was measured by Near Infrared Spectrophotometry (NIRS), duplicate measurements on single samples, as previously described [12]. Ultrasound estimation of IMF% was undertaken as previously described [13], values were the mean of five measurements.

Cattle DE dataset

Differential expression (DE) of genes from two cattle crosses with high and medium marbling, Wagyu cross Hereford and Piedmontese cross Hereford respectively, has been described previously [4]. The cattle in this dataset were sampled at 25 mo of age whilst at pasture.

Statistics and bioinformatics

The correlation between gene expression and IMF% was calculated using the "CORREL" function in Microsoft Excel. Student's t-test of significance was calculated using the "TTEST" function (one tailed) in Microsoft Excel.

P-values for the hypergeometric distribution for a specified number of successes in a population sample were calculated using the Excel "HYPERGEOMDIST" function.

Gene enrichment analysis was undertaken by using GOrilla network tools which uses a hypergeometric statistic to quantify functional enrichment in ranked gene lists [14]. *P*-values, and the false discovery rate (FDR) *Q*-values calculated using the Benjamini and Hochberg method [15], were provided in the results output of the GOrilla website [16].

Cluster analysis was undertaken using an expectation-maximization mixture analysis algorithm (EMMIX) [17]. All three datasets were linearly rescaled to a mean of zero and a range from −0.5 to 0.5 before analysis.

The z-score normalization was used to minimise the impact of differences in levels of expression and dynamic range of expression of genes from the combed gene expression data, individual gene expression values (log2) were normalised by dividing its difference from the mean of each measurement (across the whole set or subsets of the animals) with the relevant standard deviation using Microsoft Excel.

The random sampling and calculation of mean correlation was carried out in MATLAB software R2012a using custom scripts. Random controls with 5 genes were sampled 100,000 times by using random sampling from the rescaled cattle correlation dataset, this process was repeated 10 times. Those possessing higher correlation with IMF% in cattle in each sampling process were investigated. Then correlation of average gene expression with IMF% in these random controls in sheep was calculated.

The significance of gene rankings between groups was calculated using the Mann–Whitney Test web tool [18].

Sample size determination was then performed to estimate the minimum number of animals required to significantly differentiate two groups at a *P*-value of 0.05 and confidence interval of 95% [19], A Microsoft Excel spreadsheet "LaMorte's Power Calculator" downloaded from the web site [20] was used.

Results and discussion

Expression of genes in the TAG, FA and PPARG gene sets was correlated with IMF% in sheep

The correlation between gene expression and IMF% in sheep LM across the full set of 20 samples was calculated (Table 1). The 46 genes from the TAG, FA and PPARG gene sets and related genes identified in cattle were ranked in the cattle and sheep datasets by their correlation coefficients, and DE values in the cattle DE dataset (Table 1). 8 of the 13 genes in the previously defined TAG gene set were in the top 100 genes most correlated with IMF% in sheep, *P*-value = 0, 8 of the 13 genes in the previously defined FA gene set were in the top 500 genes most correlated with IMF% in sheep, *P*-value = 0, and 6 of the 15 genes in the previously defined PPARG gene set were in the top 1,000 genes most correlated with IMF% in sheep, *P*-value $< 10^{-10}$. Five of the 25 genes most correlated with IMF% in the whole sheep dataset and 8 of top 10 genes (Additional file 1: Table S1) in the sheep TAG, FA and PPARG gene sets correlation dataset were in the TAG gene set. This result showed the applicability of the TAG gene set in sheep, higher than the FA and PPARG gene sets, and similar to the results in cattle [4]. However, some genes ranked highly in the cattle correlation and DE datasets were ranked much lower in the sheep correlation dataset. For example, *S100G* ranked 26, 61 and 4,009 in the three datasets respectively (Table 1). Whilst such large differences in ranking may reflect species differences, it is also possible that such a large difference may be due to the use of a cattle gene expression platform for sheep mRNA. Thus the utility of some of the genes in sheep requires further investigation.

Integrated analysis

The sheep correlation dataset was generated from a small group of animals, therefore it is unavoidably noisy. Hence, integrated analysis of the cattle and sheep datasets was undertaken in order to identify robust genes for use in both species.

A cluster analysis of all the genes based on their correlation coefficients in cattle and sheep and DE values in cattle was undertaken using EMMIX [17]. Each gene was assigned a clustering parameter ranging from 0 to 1, the probability for location in the two alternative clusters. The smaller group was defined as Cluster A, the other as Cluster B (Additional file 1: Table S1). Genes with positive values in all three datasets in both clusters were selected, and then submitted to GOrilla for gene ontology enrichment analysis.

Cluster B, the larger cluster, was confirmed as background because: firstly no GO terms related to lipid metabolic process were enriched in the selected subset of genes; secondly correlation coefficients and DE values of genes in the selected subset of genes were close to 0. In Cluster A, gene ontology enrichment analysis was calculated for a number of groups of genes filtered by their clustering parameter (the probability of being a member of cluster A) (Table 2). Thirty genes were in the enriched GO term, "lipid metabolic process", with the most significant *P*-value and FDR *Q*-value in the ≥0.9 clustering parameter gene group. *P*-values and FDR *Q*-values of the groups with lower clustering parameters were progressively and dramatically reduced while only 1 or 2 new genes were added into the GO term, "lipid metabolic process" (Table 2). On the basis of this analysis, we decided to use the set of genes from ≥0.9 cut-off with the GO term "lipid metabolic process".

Of 315 genes with extreme positive values in both the cattle correlation and DE datasets identified in our previous study [4], 212 were included in the integrated analysis (96 genes were lost due to poor probe performance in the sheep dataset). The set of 30 genes identified above contained 24 of these genes (Figure 1).

Ninteen genes from the previously described TAG, FA and PPARG gene sets from cattle were included in the 30 genes. The remaining 11 genes included two categories of genes. The first category included genes with a well characterised and important role in lipid metabolic process and with relatively high correlation/DE coefficients, such as *LPL* and *G0S2* [21,22]. The second category contained genes with very low correlation/DE coefficients (close to 0) and included in the lipid metabolic process GO term, such as *ARSK*, *BDH1* and *SULT1A1* [23-25].

Of the genes reported in the literature to be important in sheep IMF deposition, *PPARG*, *LPL*, *ACACA*, *FABP4*, *FASN (FAS)* and *SCD* were included in the top 30, *FABP3 (H-FABP)*, *DGAT1*, and *CPT1B* [4,6-9] were not.

Correlation of *CIDEA* and IMF gene set(s) with IMF% in both cattle and sheep

CIDEA was the highest ranked gene based on correlation coefficients in both the cattle and sheep datasets (Table 1). Thus *CIDEA* is the best candidate for use as a single gene estimator of IMF% in both species. However a single gene may not be the best estimator of IMF%. Combinations of genes may be a better solution because they will provide multiple measurements of the lipid

Table 1 Rankings of genes in various datasets and average ranking

Genes[1]	Description	Rank			Average rank[6]	Source
		Sheep correlation[3]	Cattle correlation[4]	Cattle DE[5]		
CIDEA[2]	cell death-inducing dffa-like effector a	11	4	12	**1**	**TAG**[7]
ACSM1	acyl-CoA synthetase medium-chain family member 1	16	35	23	**3**	**TAG**
ADIPOQ	adiponectin	17	46	25	8	**TAG**
FABP4	fatty acid binding protein 4,adipocyte	24	54	19	**5**	**TAG**
PLIN1	perilipin1	25	15	49	6	**TAG**
TUSC5	tumor suppressor candidate 5	54	73	63	13	TAG
LPL	lipoprotein lipase	62	123	473	23	EMMIX A[8]
MAL2	mal, T-cell differentiation protein 2	67	33	177	16	FA
PPARG	peroxisome proliferator-activated receptor gamma	73	117	152	19	**PPARG**
DGAT2	diacylglycerol o-acyltransferase 2	74	6	30	**4**	**TAG**
AGPAT2	phosphate O-acyltransferase 2	100	21	118	14	**TAG**
G0S2	g0/g1switch 2	123	223	72	20	EMMIX A
FASN	fatty acid synthase	167	39	17	9	**FA**
THRSP	thyroid hormone responsive	187	13	1	**2**	**TAG**
ELOVL6	ELOVL fatty acid elongase 6	231	34	6	7	**FA**
TKT	transketolase	244	229	106	26	PPARG
CIDEC	cell death-inducing DFFA-like effector c	272	41	61	15	TAG
CYB5A	cytochrome b5 type A (microsomal)	307	64	483	27	PPARG
BHMT2	betaine-homocysteine S-methyltransferase 2	317	207	304	28	FA
RBP4	retinol binding protein 4, plasma	346	139	78	21	**FA**
ACSS2	acyl-CoA synthetase short-chain family member 2	440	19	43	12	**FA**
ARSK	arylsulfatase family, member K	447	948	502	42	**EMMIX A**
SCD	stearoyl-CoA desaturase	472	9	22	10	**FA**
ACACA	acetyl-CoA carboxylase alpha	484	56	63	17	**FA**
ACLY	atp citrate lyase	535	148	16	18	**PPARG**
PLS1	plastin1	659	502	6,513	49	TAG[9]
TF	transferrin	742	310	20	22	PPARG
CPT2	carnitine palmitoyltransferase 2	782	664	528	44	EMMIX A
PCK1	phosphoenolpyruvate carboxykinase 1	859	20	13	11	TAG
PTPLB	protein tyrosine phosphatase-like, member b	884	275	479	37	**PPARG**[9]
ADIG	adipogenin	937	183	36	24	TAG
PCK2	phosphoenolpyruvate carboxykinase 2 (mitochondrial)	972	92	1,629	34	FA
INTS9	integrator complex subunit 9	984	1,183	80	36	PPARG
PDE3B	phosphodiesterase 3B, cGMP-inhibited	1,426	206	1,135	40	FA
ACER3	alkaline ceramidase 3	1,519	683	969	52	**PPARG**
APOA1	apolipoprotein A-I	1,521	3,264	119	45	EMMIX A
ARSI	periplasmic arylsulfatase	1,865	114	7,806	43	EMMIX A
IDH1	isocitrate dehydrogenase 1 (NADP+), soluble	2,070	320	210	39	**PPARG**
APOE	apolipoprotein E	2,224	227	424	41	EMMIX A
GSTA1	glutathione S-transferase alpha 1	2,302	998	3	29	PPARG
SULT1A1	sulfotransferase family, cytosolic, 1A, phenol-preferring, member 1	2,341	3,667	172	53	EMMIX A
ANPEP	alanyl (membrane) aminopeptidase	2,703	2,091	87	47	FA

Table 1 Rankings of genes in various datasets and average ranking *(Continued)*

HSD17B12	hydroxysteroid (17-beta) dehydrogenase 12	2,726	74	457	35	PPARG[9]
CLU	clusterin	2,968	65	136	30	**FA**
BDH1	3-hydroxybutyrate dehydrogenase, type 1	3,240	8	1,797	31	EMMIX A
ACSS3	acyl-CoA synthetase short-chain family member 3	3,689	137	347	38	PPARG
ME1	malic enzyme 1, NADP(+)-dependent, cytosolic	3,741	69	2,457	46	EMMIX A
S100G	S100 calcium binding protein G	4,009	61	26	25	TAG
INSIG1	insulin induced gene 1	4,562	412	31	33	FA
G6PD	glucose-6-phosphate dehydrogenase	4,918	982	95	51	PPARG
GPAM	glycerol-3-phosphate acyltransferase, (mitochondrial)	5,886	1,364	185	55	TAG[9]
FBP1	fructose-1, 6-bisphosphatase 1	6,604	706	2	32	FA
QPRT	quinolinate phosphoribosyltransferase	7,565	329	137	50	PPARG
ALAD	aminolevulinate dehydratase	9,051	422	347	54	PPARG
CEBPA	CCAAT/enhancer binding protein (C/EBP), alpha	10,545	593	65	48	PPARG

[1]Full list of genes and probes is included as Additional file 1: Table S1.
[2]Genes in bold were included in the 30 gene set correlated with IMF% in cattle and sheep generated by the current study.
[3]Gene rank of the correlation coefficients in Sheep correlation dataset.
[4]Gene rank of the correlation coefficients in Cattle correlation dataset.
[5]Gene rank of DE value in Cattle DE dataset.
[6]Genes in this table were ranked from 1–55 based on their ranking in each of the cattle and the sheep correlation datasets. The average ranking across the two datasets was calculated and the genes were again ranked from 1 to 55.
[7]These genes were both included in TAG, FA and PPARG gene sets and annotated with the GO term "lipid metabolic process" and in EMMIX cluster A.
[8]Genes were annotated with the GO term "lipid metabolic process" and in EMMIX cluster A.
[9]Genes were included in each corresponding gene set in the current analysis based on the biological functions. But these genes were not the part of TAG, FA and PPARG gene sets in our previous work [4].

synthesis and storage pathway thereby reducing the measurement error. The set of 30 genes described above was used to identify a gene combination(s) able to estimate IMF% equally well in both cattle and sheep. *CIDEA*, *THRSP*, *ACSM1*, *DGAT2* and *FABP4* from the TAG gene set were selected based on their combined ranking in both the cattle and sheep correlation datasets (Table 2). Consistent with our findings, two of these five genes (*DGAT2* and *FABP4*) have also been identified to be correlated with IMF in cattle previously [26,27]. We tested the top 2, top 3, top 4 and top 5 genes above (defined as the "IMF 2–5 gene sets") to determine whether combinations of the top genes had a higher correlation with IMF% in both species than *CIDEA* alone. A simple model (average of rescaled gene expression values) was used to combine the data from the different genes.

In this study, more than 19,000 genes in cattle and more than 13,600 genes in sheep were detected by the

Table 2 Gene enrichment analysis of EMMIX Cluster A

Prob[1]	EMMIX			GOrrila			
	Genes			*P*-value	FDR *Q*-value	Genes[4]	Enrichment[5]
	Total[2]	Positive[2]	Negative[3]				
≥0.9	571	121	136	7.66E-13	8.13E-09	30	4.57
≥0.8	768	153	170	8.56E-12	9.09E-08	31	4.12
≥0.7	920	177	198	8.38E-10	2.97E-06	32	3.37
≥0.6	1109	210	237	1.79E-08	3.80E-05	33	2.93
≥0.5	1342	248	282	8.12E-08	1.08E-04	35	2.67

[1]Probability of genes to be located in cluster A.
[2]Genes with positive coordinates in all three datasets, see Additional file 1: Table S1 sheet "Genes with positive coordinates".
[3]Genes with all negative coordinates in all the three datasets.
[4]Number of genes enriched in "lipid metabolic process" GO term from genes with all positive coordinates in all the three datasets.
[5]Enrichment (N, B, n, b) is defined as follows: N - is the total number of genes; B - is the total number of genes associated with a specific GO term; n - is the number of genes in the top of the user's input list or in the target set when appropriate; b - is the number of genes in the intersection; Enrichment = (b/n) / (B/N).

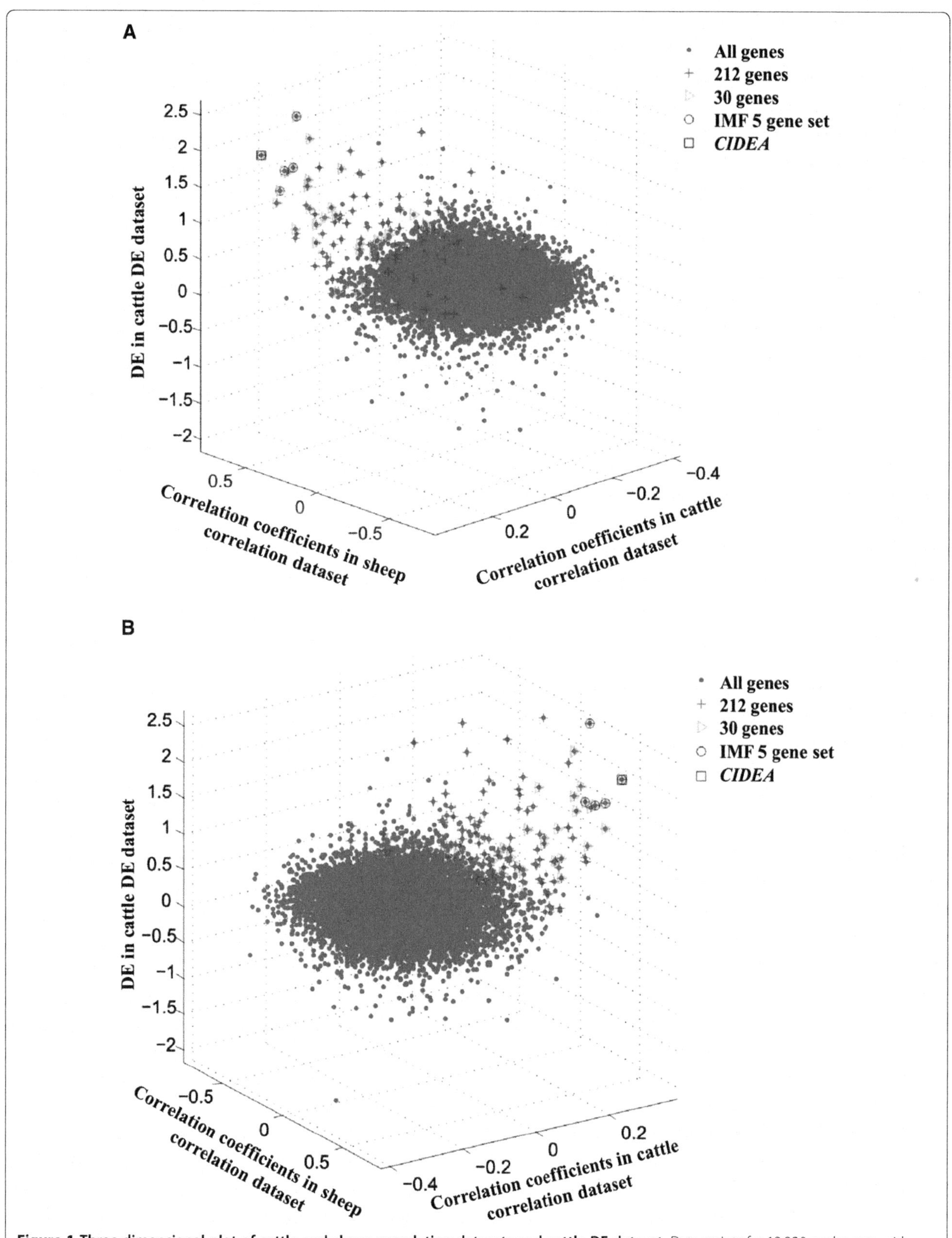

Figure 1 Three dimensional plot of cattle and sheep correlation datasets and cattle DE dataset. Data points for 13,330 probe sets with reliable data in all three datasets. The 212 genes with expression positively correlated with IMF% generated from cattle correlation and DE datasets in our previous study [4]. **A)** and **B)** are two different visual angles of this three dimensional plot.

microarray platform. Therefore, there are a very large number of possible combinations of 2–5 genes in the cattle and sheep datasets. How likely is it that a random set of 2–5 genes would be as well correlated with IMF% in both cattle and sheep as the "IMF 2–5 gene sets"? It was reasonable to test these "IMF 2–5 gene sets" based on two principles. Firstly, whether individual genes from gene combinations with higher correlation were as correlated with IMF deposition as the individual genes in the "IMF gene sets"? Secondly, whether gene combinations which exhibited a higher correlation than the "IMF 2–5 gene sets" in cattle exhibited higher correlation than the "IMF 2–5 gene sets" in sheep as well? We calculated the correlation between expression of all of the "IMF 2–5 gene sets" and IMF% and found that all the correlation coefficients were around 0.46-0.51 in both cattle and sheep (Table 3). Therefore, the IMF gene sets with 2–5 genes appeared to be equally well correlated in both cattle and sheep. For each "IMF gene set", same size randomly selected gene combinations (defined as "random controls") were sampled 100,000 times by using random sampling from the cattle correlation dataset. Those random controls which possessed higher correlation with IMF% than the corresponding "IMF gene sets" were recorded. Twenty six and 96 random controls containing sets of 2 and 3 genes respectively with higher correlation coefficients than the "IMF gene sets" were found in cattle. In contrast, very few random control gene sets with higher correlation coefficients were found containing 4 or 5 genes. Interestingly, most of the analysed random control gene sets with high correlation with IMF% contained at least one gene related to lipid metabolism from the set of 30 genes; the remaining genes were from other biological processes apparently unrelated to fat deposition. There were no random controls showing higher correlation with IMF% in sheep than the "IMF gene sets" with 2–5 genes. These results showed that the *P*-value for selecting a set of genes by chance which has as high correlation with IMF% in both cattle and sheep was $<10^{-6}$, and probably very much smaller.

As previously described [4], the 48 Brahman cattle were divided into four subgroups by experimental site (NSW and WA) and treatment (with and without HGP). In the NSW control subgroup, consistent with previous analysis [4], no significantly positive correlation between expression of any of the 5 top ranking genes and IMF% was observed (Table 3 and data not shown). In addition, in this group of animals, none of the "IMF 2–5 gene sets" showed significant correlation with IMF% (Table 3 and data not shown). This is probably due to an environmental factor such as disease or nutrition. For this reason, we repeated the analysis using the remaining three subgroups of cattle (NSW HGP-treated, WA control and WA HGP-treated), 36 animals in total. Expression of *CIDEA* was now as correlated with IMF% (0.61) as the "IMF 2–5 gene sets" (0.60-0.62) when the NSW control group was not included (Table 3).

However, the group of cattle excluding the NSW control subgroup, and the sheep may not be representative of the spectrum of animals in real production systems. Thus, we repeated the analysis on all the 48 Brahman steers and 20 sheep. In the cattle, the expression of the IMF 5 gene set showed a slightly higher correlation with IMF% than *CIDEA* alone (Table 3). In sheep, the IMF 5 gene set showed similarly high correlation with IMF% as *CIDEA* alone (Table 3).

Overall, on the basis of the correlation of gene expression with IMF%, *CIDEA* alone and the IMF 5 gene set performed very similarly in sheep and in cattle and in the subsets of cattle.

Table 3 Correlation between gene expression and IMF%

Animals	*CIDEA*	IMF 5 gene set
20 sheep	0.52	0.51
48 cattle	0.40	0.46
36 cattle[1]	0.61	0.60
NSW control group	−0.40	−0.04

[1]48 cattle excluding the NSW control group.

Relationship between gene expression and IMF% in both cattle and sheep

Is there a simple relationship between *CIDEA* and IMF 5 gene set gene expression and IMF% in both species facilitating the development of a gene-based test for sorting animals by their estimated IMF% in the LM?

The rescaled gene expression values of *CIDEA* and the IMF 5 gene set were plotted against the IMF% for cattle (Figure 2) and sheep (Figure 3). Analysing the WA control and HGP-treated animals separately demonstrated that the relationships between gene expression and IMF % were very similar (Figure 2A, C). Thus it appears that the HGP-treatment did not significantly affect the relationship between IMF% and *CIDEA*/IMF 5 gene set expression, except for the effect due to the reduced IMF% of HGP-treated animals (Figure 2C). The data from the 22 cattle (not including the NSW animals) was combined (Figure 2B, D) and similarly for the 20 sheep (Figure 3). We found similar linear relationships between IMF% with both *CIDEA* and the IMF 5 gene set (Figure 2B, D & Figure 3B) in both cattle and sheep. These results suggest that *CIDEA* and the IMF 5 gene set could be used across species and a broad range of IMF% values to estimate IMF%.

Applications of gene estimator(s) of IMF% in both cattle and sheep

We have identified robust gene signals strongly correlated with IMF% in both cattle and sheep. We have also

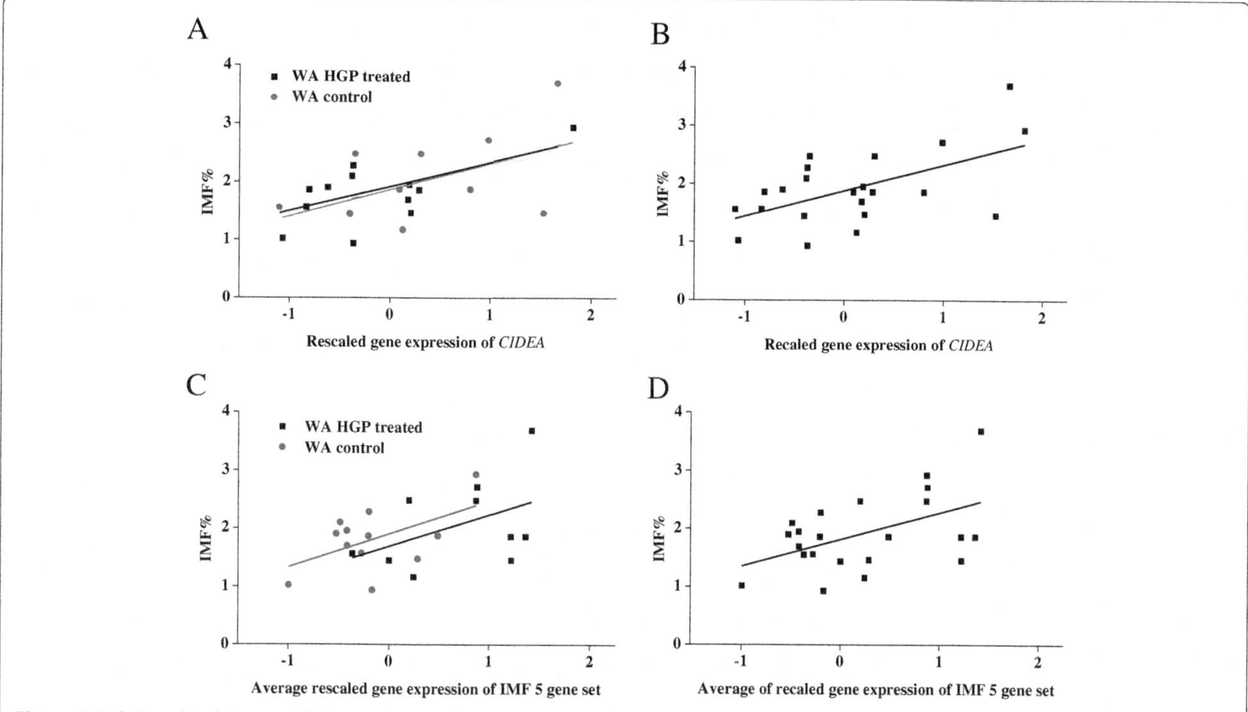

Figure 2 Relationship between IMF% and gene expression in Brahman cattle. A) IMF% and *CIDEA* expression in WA control and HGP treated animals separately. **B)** IMF% and *CIDEA* expression in WA cattle. **C)** IMF% and IMF 5 gene set expression in WA control and HGP treated animals separately. **D)** IMF% and IMF 5 gene set expression in WA cattle.

determined a simple linear relationship between IMF% and expression of the genes. How can we apply these gene expression assays in the cattle and sheep production industries to estimate IMF%? Below we discuss two example applications of the sets of genes for use in the cattle and sheep production industries: ranking animals by IMF% within a group (such as on a single farm), or comparing difference in IMF% between two groups (such as two different treatments or farms). Some methods described are only applicable within groups of animals, and the others can be applied in both situations. HGP-treatment is expected to reduce IMF% in both cattle and sheep [11,28]. This provides us with a good example to test the reliability of a gene, or gene combinations, to estimate IMF% in both species. We chose the WA HGP-treated and control groups for the example (Table 4). These animals are subsets of a larger group of 173 animals with the same treatments and from the same site [12,29]. We have shown above that at the time of sampling in this experiment HGP-treatment did not have a major effect on the relationship between IMF% and gene expression relative to the control group (Figure 2A&C). Thus the ability of the methods to differentiate between the two groups (with the HGP-treated group having a lower average measured or estimated IMF% or ranking) is a reflection of the strength of the relationship between the result of the method and

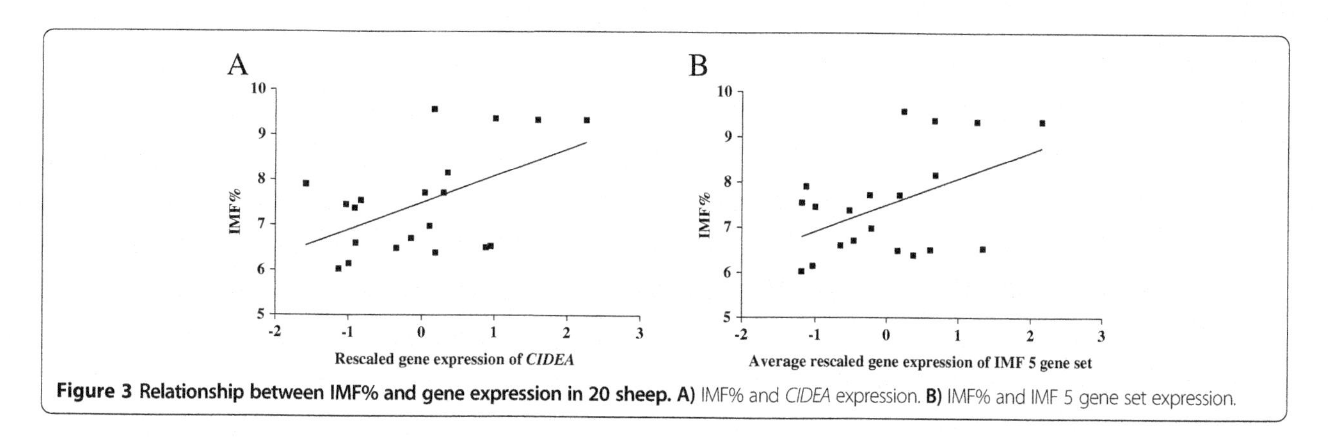

Figure 3 Relationship between IMF% and gene expression in 20 sheep. A) IMF% and *CIDEA* expression. **B)** IMF% and IMF 5 gene set expression.

Table 4 Comparison of the performance of different measures and estimators of IMF%

Method	Number of animals[1]	WA control subgroup	WA HGP subgroup	*P*-value[2]	Predicted experiment size[3]
NIRS measured IMF%	141	2.37 ± 1.00[4]	1.90 ± 0.83	0.001	198
Ultrasound estimated IMF%	173	2.66 ± 0.72[4]	2.93 ± 0.54	1.000	N/A[5]
NIRS measured IMF%	22	2.07 ± 0.77[4]	1.79 ± 0.54	0.340	294
IMF% calculated by *CIDEA* formula	22	2.30 ± 1.19[6]	1.60 ± 1.14	0.080	148
IMF% calculated by IMF 5 gene set formula	22	2.66 ± 1.04[6]	1.30 ± 0.92	0.003	26
Ranking animals using *CIDEA*	22	9.50 ± 6.62[7]	13.20 ± 6.16	0.100	156
Ranking animals using IMF 5 gene set	22	7.20 ± 5.47[8]	15.10 ± 4.99	0.0026	22
Ranking animals using TAG gene set	22	7.00 ± 4.52[8]	15.30 ± 5.48	0.0017	20
CIDEA DE	22	13.09 ± 0.44[7]	12.83 ± 0.38	0.080	130
IMF 5 gene set DE	22	0.25 ± 0.38[9]	−0.27 ± 0.31	0.0014	24
TAG gene set DE	22	0.28 ± 0.36[9]	−0.27 ± 0.33	0.00076	20

[1]Number of animals used for the analysis.
[2]For the test that average measured or estimated IMF%/gene expression/ranking in HGP-treated animals is lower than in control animals.
[3]The sample size predicted to be required to observe a significant result ($P < 0.05$) with 95% confidence intervals.
[4]Mean values, standard deviation and *P*-values are calculated using original record data.
[5]Ultrasound estimated IMF% did not detect an effect of the expected direction.
[6]Mean values, standard deviation and *P*-values are calculated using rescaled gene expression values.
[7]Mean values, standard deviation and *P*-values of ranking and DE of *CIDEA* are calculated using rescaled gene expression values.
[8]Mean values, standard deviation and *P*-values of ranking calculation of IMF 5 gene set and TAG gene set are based on Mann-Witney test.
[9]Mean values, standard deviation and *P*-values of IMF 5 gene set and TAG gene set DE are calculated using rescaled gene expression values.

the IMF% of individuals. There was no significant difference in average IMF% (NIRS measured) between the WA HGP-treated and WA control subgroups for which gene expression data was available (Table 4). As discussed previously [4], a reduction in IMF% caused by HGP-treatment would not be expected to be detectable in such small groups of animals. For the whole set of 141 WA animals (IMF% data was not available for 32 animals), HGP-treated and control animals could be significantly differentiated by directly measured IMF% (*P*-value = 0.001). But ultrasound estimated IMF% did not discriminate between the two groups, even when data from 173 animals was used (Table 4). Consistent with this IMF% estimated by ultrasound was not correlated with NIRS measured IMF%, R^2 is 0.086 (Figure 4). These results were not unexpected as ultrasound is not recommended as a method for estimating IMF% in cattle with less than 2% IMF% [13]. Half of the Brahman steers from WA had NIRS measured IMF% of less than 2%. More typical use of ultrasound in Australia to estimate IMF% is in the range 2-8% where proficient scanners achieve correlation of ultrasound estimated IMF% with measured IMF% well in excess of 0.75 [13].

To demonstrate the two applications of our findings, the regression equations based on the relationship between IMF% and expression of *CIDEA*, or the IMF 5 gene set, were used to estimate the IMF% for each animal. For both approaches there was a significant difference between the means of the estimated IMF% of the two groups ($P < 0.05$), unlike for the NIRS measured IMF% on the same 22 animals ($P = 0.34$) (Table 4). Although the correlation between *CIDEA* expression and IMF%, and the expression of the IMF 5 gene set with IMF% were very similar (Table 3). The use of more genes appears to significantly improve the accuracy of the estimation of IMF% (Table 4).

Rather than calculating an estimated IMF%, animals can be ranked on the basis of the relative gene expression values. The results showed that there was a significant difference between the average rankings of animals in the WA HGP-treated and control groups using the IMF 5 gene set (*P*-value = 0.0026) and TAG gene set (*P*-value = 0.0017), but not using *CIDEA* alone (*P*-value = 0.1) (Table 4). Again the apparent accuracy of the

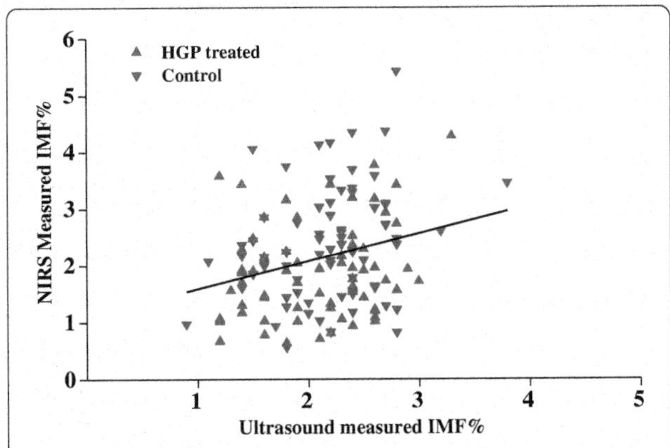

Figure 4 Correlation between ultrasound estimated IMF% and NIRS measured IMF% in Brahman cattle.

ranking method was improved significantly by the inclusion of additional genes, although increasing to 15 genes provided little additional improvement.

To compare IMF% of animals between different farms, besides the approaches above, we could also compare the DE of *CIDEA*, IMF gene set or TAG gene sets between two groups of animals. Rescaled gene expression data was used for the DE calculation of the multiple genes in the IMF 5 gene set and TAG gene set. As in the ranking method above, increasing the number of genes from one to five increased the discrimination between the two groups, but increasing the number of genes to 15 had little further effect (Table 4).

Generally speaking, in both research and production settings, the number of animals tested is the major determinant of the cost of collecting phenotypes. Methods which reduce the number of animals required to be tested and/or which can be conducted without sacrificing the animal will reduce the costs of phenotyping substantially. Using the approach of LaMorte [30], we estimated the sample size required to detect an affect at $P < 0.05$ and a confidence interval of >95% for all of the analysis methods (Table 4). Given the small size of the datasets the sample sizes for approaches using gene expression are likely to be overestimated. No reliable estimate of the sample size could be made for the use of ultrasound as the available data confirms that ultrasound is inaccurate for animals with low IMF%. To detect the effect using NIRS measured IMF% a large number of animals are required (Table 4). Using gene expression of *CIDEA* a slightly smaller sample size may be adequate. However, the use of five genes substantially reduced the predicted sample size, suggesting that around one eighth of the number of animals may be required to detect the effect of HGPs on IMF%. This improvement in performance and hence reduction in experiment size may be because the use of a multiple gene set effectively provides multiple measurements of the phenotype (deposition of TAG in lipid droplets in intramuscular adipocytes) leading to a reduced measurement error than using *CIDEA* alone, or the mean of duplicate NIRS measurements.

The estimation of IMF% using gene expression is successful probably because gene expression of the IMF 5 gene set is proportional to IMF deposition rate and in growing animals depositing IMF, and yet to reach maturity, IMF deposition rate is proportional to IMF%.

However, the improvement in performance of the gene expression over NIRS to identify the inhibitory effect of HGP-treatment may also be partly due to the following model. Within a short time after HGP treatment, intramuscular fat deposition may almost stop and as a consequence the expression of the IMF gene set genes would be predicted to be greatly reduced (Figure 5A). The concentration of circulating HGP would decrease with time (Figure 5B), so the effect of HGP on intramuscular fat deposition processes and the expression of the IMF gene set would be predicted to be gradually reduced and finally recover to the normal state described above (Figure 5A). If

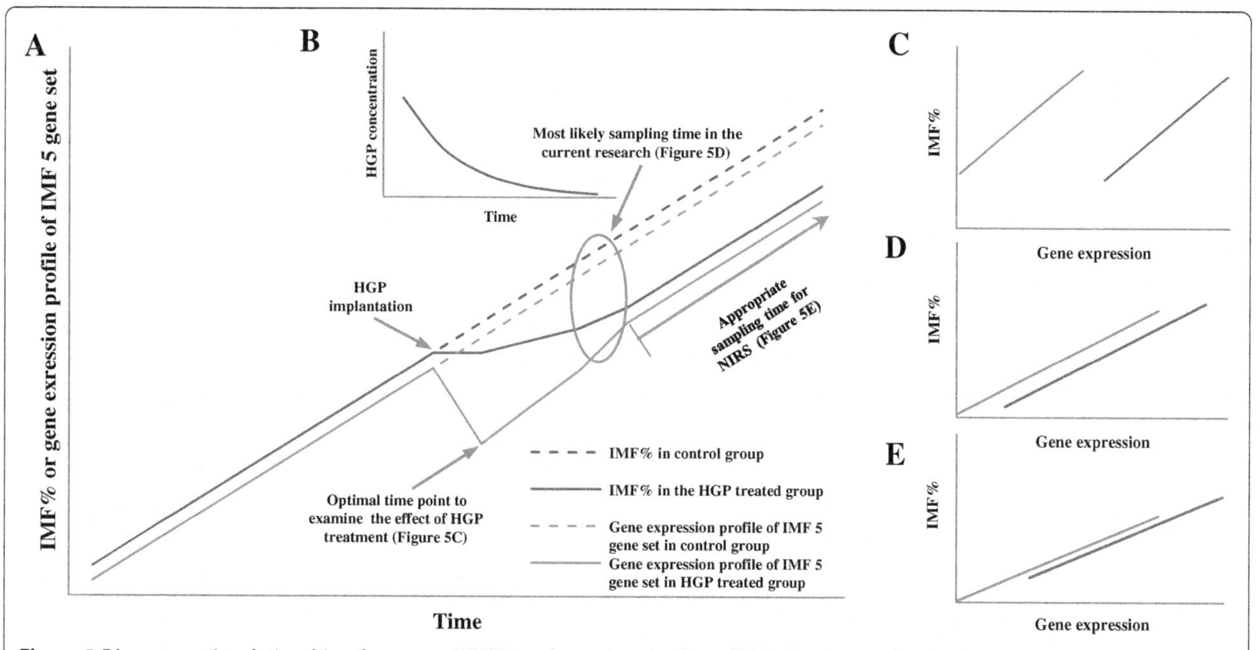

Figure 5 Diagrammatic relationship of measured IMF% and gene expression of IMF 5 gene set over time. A) Linear parallel portions of the curves for IMF% and gene expression have been separated for display purposes only. **B)** Profile of the concentration of circulating HGP after implantation. **C), D) and E)** Relationship between IMF% and gene expression of IMF 5 gene set at different time points after HGP implantation. The red solid lines represented control subgroup; blue ones represented the HGP treated subgroup.

this model is correct, then shortly after HGP treatment, there would be little difference in measured IMF%, but a large difference in gene expression, between the treated and the control animals. At this point, gene expression would have a much larger discriminating power than IMF measurement (Figure 5A, C). Subsequently, the differences between measured IMF% and estimated IMF% based on gene expression would decrease as the concentration of circulating HGP reduces. During this period, gene expression would still be predicted to have more discriminating power than the direct measurement of IMF%. Eventually, the increased power of gene expression would be reduced to that normally observed (Figure 5A, E). The exact location on this theoretical curve of the experimental animals used in this work is unknown, but may be whilst there is still some additional discriminating power due to the effect described above (Figure 5A, D). However, certainly the animals were not at the optimal discrimination point (Figures 2A, C, 5A, C). Whilst it is likely that fewer animals were required for discrimination between HGP-treated and control animals using gene expression than were required for NIRS measured IMF% due to the effect described above and the intrinsic nature of the assay, the relative contributions of the two factors is unclear.

Conclusion

By integrating data from cattle and sheep we have identified a set of 30 genes with robust correlation with IMF% in both cattle and sheep LM. Based on this gene set, we identified *CIDEA* as the gene whose expression was most correlated with IMF% in both cattle and sheep. Whilst *CIDEA* alone could be used to estimate IMF%, it is of similar utility to NIRS measured IMF%. In contrast, the non-invasive technique of ultrasound did not perform adequately on animals with low IMF%. By combining the data from 5 genes apparently improved estimates of IMF% could be calculated, with a commensurate reduction in the experiment size required to detect the impact of a treatment on IMF%. The five gene set can be used to estimate IMF% (based on the proposed relationship between the expression of the IMF 5 gene set, IMF deposition rate and IMF%) from biopsy as well as post slaughter samples, and on samples from animals with low IMF%, such as the Brahmans used in this work and younger animals of higher marbling breeds. The approach to phenotyping animals using gene expression shows promise as an alternative to current approaches for the measurement/estimation of IMF% in both cattle and sheep.

In addition, we have described a potentially generic approach to the development of robust gene expression phenotypes for other phenotypes of industry importance. The pipeline is as follows: calculate the correlation between gene expression and a phenotype (IMF% in this paper) and/or DE in two or more different groups of animals with significantly different experimental structures and phenotypic performance to generate the corresponding datasets. Then rank the genes based on the coefficients above in each dataset to primarily select a group of genes highly correlated with this phenotype and with each other across the different groups of animals. Lastly, optimise this group of genes based on their biological function to identify a gene set with appropriate size.

Additional file

Additional file 1: Table S1. Clustering parameters, correlation/DE coefficients and rankings of genes for use in cluster analysis.

Competing interests

The authors declare no competing interest regarding the content or conclusions expressed in this research.

Authors' contributions

The experimental design was mainly conceived by BPD. Samples and data were generated by KK and PLG. Data analysis was carried out by BG under the guidance of BPD. GZ, WZ and BPD resourced the project. BG and BPD were the primary authors of the paper, but all the authors contributed to, read and approved the final manuscript.

Acknowledgements

This project was partially supported by the CRC for Beef Genetic Technologies. The authors would like to thank Dr Nicholas Hudson for discussions, Dr Antonio Reverter for his help with EMMIX calculations, Mr. Zhenliang Ma at The Faculty of Engineering, Architecture & Information Technology (EAIT) of The University of Queensland Australia for his advice on programming, Dr Yutao Li for help with statistical techniques and Matthew Wolcott for access to unpublished ultrasound data.

Author details

[1]Key Laboratory of Meat Processing and Quality Control, Synergetic Innovation Center of Food Safety and Nutrition, College of Food Science and Technology, Nanjing Agriculture University, Nanjing 210095, P.R. China. [2]CSIRO Animal, Food and Health Sciences, St. Lucia QLD 4067, Australia. [3]Now at; National Institute of Animal Health, 50/2 Kasetklang, Ladyao, Bangkok 10900, Thailand. [4]CSIRO Animal, Food and Health Sciences, Armidale NSW 2350, Australia. [5]NSW Department of primary Industries, Armidale NSW 2350, Australia.

References

1. Pethick DW, Pleasants AB, Gee AM, Hopkins DL, Ross IR: **Eating quality of commercial meat cuts from Australian lambs and sheep.** *Proc New Zeal Soc An* 2006, **66**:363–367.
2. Hocquette J, Gondret F, Baéza E, Médale F, Jurie C, Pethick D: **Intramuscular fat content in meat-producing animals: development, genetic and nutritional control, and identification of putative markers.** *Animal* 2010, **4**:303–319.
3. Warner RD, Jacob RH, Edwards JEH, McDonagh M, Pearce K, Geesink G, Kearney G, Allingham P, Hopkins DL, Pethick DW: **Quality of lamb meat from the Information Nucleus Flock.** *Anim Prod Sci* 2010, **50**:1123–1134.
4. De Jager N, Hudson NJ, Reverter A, Barnard R, Cafe LM, Greenwood PL, Dalrymple BP: **Gene expression phenotypes for lipid metabolism and intramuscular fat in skeletal muscle of cattle.** *J Anim Sci* 2013, **91**:1112–1128.

5. Wang YH, Bower NI, Reverter A, Tan SH, De Jager N, Wang R, McWilliam SM, Cafe LM, Greenwood PL, Lehnert SA: **Gene expression patterns during intramuscular fat development in cattle.** *J Anim Sci* 2009, **87**:119–130.
6. Huang ZG, Xiong L, Liu ZS, Qiao Y, Liu SR, Ren HX, Xie Z, Liu GQ, Li XB: **The developmental changes and effect on IMF content of H-FABP and PPARgamma mRNA expression in sheep muscle.** *Yi Chuan Xue Bao* 2006, **33**:507–514.
7. Qiao Y, Huang Z, Li Q, Liu Z, Hao C, Shi G, Dai R, Xie Z: **Developmental changes of the FAS and HSL mRNA expression and their effects on the content of intramuscular fat in Kazak and Xinjiang sheep.** *J Genet Genomics* 2007, **34**:909–917.
8. Xu QL, Chen YL, Ma RX, Xue P: **Polymorphism of DGAT1 associated with intramuscular fat-mediated tenderness in sheep.** *J Sci Food Agric* 2009, **89**:232–237.
9. Dervishi E, Serrano C, Joy M, Serrano M, Rodellar C, Calvo JH: **The effect of feeding system in the expression of genes related with fat metabolism in semitendinous muscle in sheep.** *Meat Sci* 2011, **89**:91–97.
10. Kijas JW, Menzies M, Ingham A: **Sequence diversity and rates of molecular evolution between cattle and sheep genes.** *Anim Genet* 2006, **37**:171–174.
11. Kongsuwan K, Knox MR, Allingham PG, Pearson R, Dalrymple BP: **The effect of combination treatment with trenbolone acetate and estradiol-17beta on skeletal muscle expression and plasma concentrations of oxytocin in sheep.** *Domest Anim Endocrinol* 2012, **43**:67–73.
12. Cafe LM, McIntyre BL, Robinson DL, Geesink GH, Barendse W, Greenwood PL: **Production and processing studies on calpain-system gene markers for tenderness in Brahman cattle: 1. Growth, efficiency, temperament, and carcass characteristics.** *J Anim Sci* 2010, **88**:3047–3058.
13. Upton W, Donoghue K, Graser H, Johnston D: **Ultrasound proficiency testing.** *Proc Assoc Advmt Anim Breed Genet* 1999, **13**:341–344.
14. Eden E, Navon R, Steinfeld I, Lipson D, Yakhini Z: **GOrilla: a tool for discovery and visualization of enriched GO terms in ranked gene lists.** *BMC Bioinformatics* 2009, **10**:48.
15. Benjamini Y, Hochberg Y: **Controlling the False Discovery Rate - a Practical and Powerful Approach to Multiple Testing.** *J Roy Stat Soc B Met* 1995, **57**:289–300.
16. **GOrrila - a tool for identifying enriched GO terms.** http://cbl-gorilla.cs.technion.ac.il/.
17. McLachlan GJ, Peel D, Basford KE, Adams P: **The EMMIX software for the fitting of mixtures of normal and t-components.** *J Stat Softw* 1999, **4**:1–14.
18. **Mann–Whitney Test.** http://www.vassarstats.net/utest.html.
19. Dell RB, Holleran S, Ramakrishnan R: **Sample size determination.** *Ilar J* 2002, **42**:207–213.
20. **Lamorte Calculator.** www.uml.edu/Images/LaMorte%20calculator_tcm18-37802.xls.
21. Hung YP, Lee NY, Lin SH, Chang HC, Wu CJ, Chang CM, Chen PL, Lin HJ, Wu YH, Tsai PJ, Tsai YS, Ko WC: **Effects of PPARgamma and RBP4 gene variants on metabolic syndrome in HIV-infected patients with anti-retroviral therapy.** *PLoS One* 2012, **7**:e49102.
22. Nielsen TS, Kampmann U, Nielsen RR, Jessen N, Orskov L, Pedersen SB, Jorgensen JO, Lund S, Moller N: **Reduced mRNA and protein expression of perilipin A and G0/G1 switch gene 2 (G0S2) in human adipose tissue in poorly controlled type 2 diabetes.** *J Clin Endocrinol Metab* 2012, **97**:E1348–1352.
23. Obaya AJ: **Molecular cloning and initial characterization of three novel human sulfatases.** *Gene* 2006, **372**:110–117.
24. Martinez-Outschoorn UE, Lin Z, Whitaker-Menezes D, Howell A, Sotgia F, Lisanti MP: **Ketone body utilization drives tumor growth and metastasis.** *Cell Cycle* 2012, **11**:3964–3971.
25. McGill MR, Jaeschke H: **Metabolism and Disposition of Acetaminophen: Recent Advances in Relation to Hepatotoxicity and Diagnosis.** *Pharm Res* 2013, **30**(9):2174–2187.
26. Jeong J, Kwon E, Im S, Seo K, Baik M: **Expression of fat deposition and fat removal genes is associated with intramuscular fat content in longissimus dorsi muscle of Korean cattle steers.** *J Anim Sci* 2012, **90**:2044–2053.
27. da Costa AS, Pires VM, Fontes CM, Prates JAM: **Expression of genes controlling fat deposition in two genetically diverse beef cattle breeds fed high or low silage diets.** *BMC Vet Res* 2013, **9**:118.
28. Hunter R: **Hormonal growth promotant use in the Australian beef industry.** *Anim Prod Sci* 2010, **50**:637–659.
29. Cafe LM, McIntyre BL, Robinson DL, Geesink GH, Barendse W, Pethick DW, Thompson JM, Greenwood PL: **Production and processing studies on calpain-system gene markers for tenderness in Brahman cattle: 2. Objective meat quality.** *J Anim Sci* 2010, **88**:3059–3069.
30. **Sample size calculation in research.** http://www.uml.edu/docs/sample%20size%20calcs%20LaMorte_tcm18-37807.doc.

The liver transcriptome of two full-sibling Songliao black pigs with extreme differences in backfat thickness

Kai Xing[1], Feng Zhu[1], Liwei Zhai[1], Huijie Liu[1], Zhijun Wang[2], Zhuocheng Hou[1*] and Chuduan Wang[1*]

Abstract

Background: Fatness traits in animals are important for their growth, meat quality, reproductive performance, and immunity. The liver is the principal organ of the regulation of lipid metabolism, and this study used massive parallelized high-throughput sequencing technologies to determine the porcine liver tissue transcriptome architecture of two full-sibling Songliao black pigs harboring extremely different phenotypes of backfat thickness.

Results: The total number of reads produced for each sample was in the region of 53 million, and 8,226 novel transcripts were detected. Approximately 92 genes were differentially regulated in the liver tissue, while 31 spliced transcripts and 33 primary transcripts showed significantly differential expression between pigs with higher and lower backfat thickness. Genes that were differentially expressed were involved in the metabolism of various substances, small molecule biochemistry, and molecular transport.

Conclusions: Genes involved in the regulation of lipids could play an important role in lipid and fatty acid metabolism in the liver. These results could help us understand how liver metabolism affects the backfat thickness of pigs.

Keywords: Backfat thickness, Liver, Pig, RNA-Seq

Background

Consumer choice of pig meat is an important factor that dictates the principles of swine breeding worldwide. One aspect of this is the deposition of fat in the muscle and backfat, which is associated with growth rate, meat quality, and reproductive performance [1]. Backfat thickness is highly correlated with body fat radio, carcass cross-sectional fat area ratios, and intramuscular fat, so is a good indicator of fat deposition in pigs [2]. Therefore, the selection of backfat thickness using B-mode real-time ultrasound is a practical and economical method in pig breeding for increasing feeding efficiency, carcass value, and consumer acceptance of pork [3]. Because of the increasing problem of human obesity worldwide, it is beneficial to regulate fat deposition in pig breeding through molecular markers. Additionally, their similarity to humans in body size and other physiological/anatomical features, including their innate tendency to over consume food, means that pigs are a good animal model for studying obesity [4].

The extent of fat deposition can be determined by triacylglycerol synthesis and storage, lipid mobilization, and fatty acid oxidation [5]. The liver is one of the most important organs to regulate appetite and body weight in pigs, as well as playing a key role in regulating several metabolic processes [6]. In pigs, *de novo* cholesterol synthesis and fatty acid oxidation mainly take place in the liver [7]. Lipid hydrolysis in adipose tissue results in free fatty acid, which is combined with plasma albumin then transported to the liver for use as an energy source through oxidation [8].

RNA-Seq technology for transcriptome profiling has previously been used to explore the transcriptome of pig liver tissue. Such transcriptomes were recently compared between a full-sibling (full-sib) pair of F2 females from a White Duroc × Erhualian resource population with extreme phenotypes in growth and fat deposition [9]. Other comparisons include a backcross of two female groups (H and L) with extreme intramuscular fatty acid composition (25% Iberian ×75% Landrace) [10], and Duroc × F2 (Leicoma × German Landrace) cross

* Correspondence: zchou@cau.edu.cn; cdwang@cau.edu.cn
[1]National Engineering Laboratory for Animal Breeding and MOA Key Laboratory of Animal Genetics and Breeding, Department of Animal Genetics and Breeding, China Agricultural University, 100193 Beijing, China
Full list of author information is available at the end of the article

pigs with divergent skatole levels in backfat [11]. Although several studies have previously attempted to identify the genes and pathways involved in fatty traits in the liver, to our knowledge, they either lacked sufficient RNA-Seq liver samples or did not take into account the effect of different genetic background noise when analyzing liver regulation.

The Songliao black pig is a Chinese domestic breed with powerful stress resistance and good reproductive and fat deposition capabilities. It is therefore a good model for studying fatty deposition. In the present study, we used RNA-Seq to obtain the liver transcriptomes of two full-sib Songliao black pigs with a high variation in backfat thickness. The main aim of this study was to elucidate the genes and pathways involved in lipid metabolism in liver tissue using RNA deep sequencing technology.

Methods

Experimental design, animals, and phenotypes

The Songliao black female pig population (average age, 217 (range, 216–218) days; average live weight, 100 kg (range, 92.5-116.4 kg) was housed in consistent and standard environmental conditions with natural, uncontrolled room temperature and light. Animals were fed three times a day and had access to water *ad libitum*. Pedigree information was available for all animals. Live backfat thickness was measured on the last 3/4 rib using B-mode real-time ultrasound (HS1500, Honda, Japan). We analyzed a total of 53 individuals with full/half-sibs for backfat thickness to identify pairs with two divergent phenotypes. To minimize the noise of different genetic back grounds, full-sibs were selected as a priority.

We set out to compare transcriptome changes between two groups with a high variation in backfat thickness: pigs with higher backfat thickness (BH) and those with lower backfat thickness (BL) which had a backfat thickness 2–3 times lower than that of BH pigs. The chosen animals also had to have a similar backfat thickness within the same group (BH/BL) after adjustment for live body weight. Based on our criteria, experimental samples were made up of two pairs of pigs with extreme backfat thickness differences, both of which were full-sibs.

The chosen pigs were slaughtered according to guidelines for the ethical use and treatment of animals in experiments in China. Liver tissue was separated and stored in liquid nitrogen until analyzed. Total liver RNA was extracted using the total RNA extraction Kit (Bioteke, China) according to the manufacturer's recommendations. The quality of total RNA was assessed by the 2100 Bioanalyzer (Agilent, USA).

mRNA library construction and sequencing

mRNA was isolated from total RNA samples using oligo (DT) magnetic beads (Invitrogen, USA). Purified mRNA was first fragmented by the RNA fragmentation kit (Ambion, USA), then a one paired-end library was prepared for each sample according to the manufacturer's instructions. mRNA libraries were individually sequenced for four-samples (two from each of the BH and BL group) using the Illumina High-seq 2000 sequencing system. The libraries were sequenced using a multiplexed paired-ends protocol with 180 bp of data collected per run. The average insert size for the paired-end libraries was 180 bp. A total of four paired-end mRNA libraries were constructed individually for four liver samples.

Mapping and counting reads

Quality control and reads statistics were determined using FASTQC (http://www.bioinformatics.babraham.ac.uk/projects/fastqc/). All reads were trimmed 20-bp from the 5′ end according to the reads quality distributions. After removal of the sequencing adapt and low-complexity reads, all RNA-Seq reads were mapped on the reference pig genome (Sscrofa10.2) using TopHat v2.0.1 software [12] with default parameters. The annotation database Ensembl Genes v67 was used as a reference. Additionally, the intersect from BED Tools was used to count the number of reads mapping to exons, introns, and intergenic positions in the genome [13]. The reads count was measured using easy RNASeq software [14] to quantify the raw reads mapped on each gene.

Differential expression and novel transcript analysis

The trimmed mean of M-values (TMM) was used to normalize gene expression levels [15]. After normalization, the NOISeq package implanted in the R computation environment was used to detect differentially expressed genes (DEGs) between two groups [16]. This method infers the noise distribution from the data and performs pair wise comparisons of the samples to identify DEGs. To measure expression level changes between two conditions, NOISeq takes into consideration two statistics: M(the $\log_2$-ratio of the two conditions) and D(the absolute value of the difference between conditions). The probability thresholds were $P \geq 0.8$ and the TMM value in the lower expressed sample was ≥1. The higher the probability, the greater the change in expression between the two groups. Using a probability threshold of 0.8 means that the gene is 4 times more likely to be differentially expressed than non-differentially expressed

Table 1 The traits of backfat thickness and related fat deposition

ID	H710	H712	H906	H909
Backfat thickness of live, mm	24.9	9.4	21.7	8.8
Backfat thickness of carcass, mm	37.0	18.9	31.7	15.5
Kidney fat, kg	1.45	0.7	1.5	0.75

H710, H712 and H906, H909 are two pair full-sibs pigs. H710, H906 and H712, H909 are the BH and BL groups respectively.

Table 2 The number of reads obtained and percentages of mapped reads per sample

Sample ID[1]	Total reads number, M[2]	Mapping rate, %[3]	CDS exons, %	5'UTR, %	3'UTR, %	Intron, %
H906	53.08	84.80	72.85	1.78	10.43	6.45
H909	52.16	92.90	76.21	2.64	7.98	5.11
H710	54.88	87.86	75.85	1.81	9.77	5.01
H712	51.56	88.17	75.28	1.79	8.89	6.35

1. H710, H712 and H906, H909 are two pair full-sibs pigs. H710, H906 and H712, H909 are the BH and BL groups respectively.
2. Indicate millions of reads.
3. Use the Ensembl V68 as the reference genome annotation to classify the mapping tags into the different regions. Ratio of the tags mapping on the subregion of the gene was calculated as the tags on each region divided by the total tags on the whole genome.

[17]. Novel transcripts, differentially expressed spliced transcripts, and primary transcripts were also detected using the Cufflinks suite of software for RNA-Seq [12].

Functional enrichment analysis of differentially expressed genes

Because the pig genome is poorly annotated, pig gene IDs were converted to human gene IDs using BioMart. DEG lists were submitted to the Database for Annotation, Visualization and Integrated Discovery (DAVID) bioinformatics resource for enrichment analysis of the significant overrepresentation of GO biological processes (GO-BP), molecular function (GO-MF), cellular component (GO-CC), and KEGG-pathway category [18]. In all tests, *P*-values were calculated using Benjamini-corrected modified Fisher's exact test and ≤0.05 was taken as a threshold of significance.

To further identify the DEG interaction network in the liver, the Search Tool for the Retrieval of Interacting Genes (STRING) was used, which is based on a known protein-protein interaction database program. It generates a network of interactions from a variety of sources, including different interaction databases, text mining, genetic interactions, and shared pathway interactions [19].

Quantitative PCR and data analysis

Total RNA was extracted from the liver and converted into cDNA using the Revert Aid™ First Strand cDNA Synthesis Kit (Thermo Fisher Scientific Inc, USA) following the manufacturer's protocol. cDNA samples were analyzed with real-time reverse transcriptase (RT)-PCR using the Light Cycler® 480 Real-Time PCR System (Roche, USA). RT-PCR reactions were performed in a final volume of 20 μl with the Roche SYBR Green PCR Kit (Roche) according to the manufacturer's instructions. Pig *GAPDH* was used as an internal standard to correct the cDNA input. Triplicate RT-qPCRs were performed for each cDNA and the average Ct was used for further analysis. Relative quantification values were calculated using the $2^{-\Delta\Delta Ct}$ method.

Data availability

Complete data sets have been submitted to NCBI Sequence Read Archive (SRA) under Accession no. SRP035376, Bioproject: PRJNA234465.

Results

Analysis of RNA deep sequencing data

The backfat thickness of the pig carcass and weight of the kidney were shown to differ greatly between groups (Table 1). In general, individuals in the BH group had twice the backfat thickness compared with those in the BL group. Our experimental population is a conserved breed that has not undergone extensive selection nor hybridized with other breeds. Individuals therefore have very similar genetic backgrounds.

We obtained approximately 53 million paired-end clean reads of 90 bp for each sample, and high percentages of mapped reads ranging from 84.80 to 92.90%. The number of reads and percentages of mapped reads were similar between the two groups (Table 2). Most mapped reads were located within an exon, with percent ages ranging from 72.85 to 76.21%. Other reads mapped within the untranslated region, introns, and intergenic regions. The percentages of reads in each region are shown in Table 2.

The total number of genes expressed in the liver in the four samples ranged from 16,815 to 17,025 (Additional file 1: Table S1), with numbers of expressed genes being similar between the two groups. Correlations between biological replicate samples showed that the expressed genes were very highly reproducible, suggesting that a major fraction of the liver transcriptome is conserved between groups.

To confirm changes in transcript levels between BH and BL groups, six genes related to fatty acid synthesis or lipid metabolism were selected for RT-PCR analysis: *ACACA*, *LDHA*, *ELOVL6*, *CYP1A2*, *PDK1*, and *SCD*. We designed RT-PCR primers (Additional file 2: Table S2) for these genes, using *GAPDH* as a reference. When gene expression levels were compared, a strong correlation between RT-qPCR and RNA-Seq platforms was observed (0.72), confirming the high reproducibility of the data. For all six genes, the fold-change ratios between H and L groups were consistent with the RNA-Seq data (Figure 1).

Differentially expressed genes between H and L groups

We quantified transcript expression levels in TMM to normalize gene expression data across different samples.

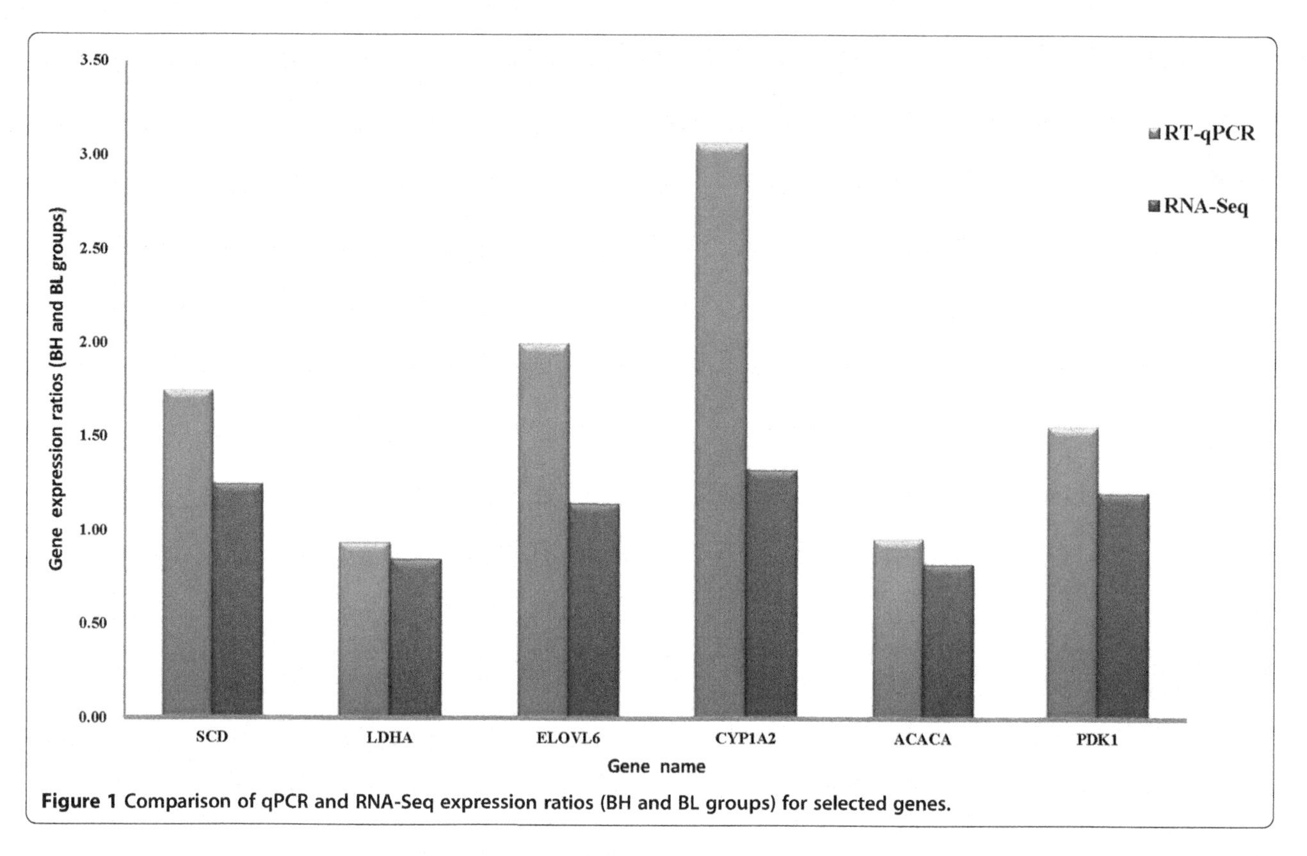

Figure 1 **Comparison of qPCR and RNA-Seq expression ratios (BH and BL groups) for selected genes.**

Differential gene expression in liver tissue was calculated from the raw reads using the NOISeq package. We treated high/low backfat thickness samples as biological replicates because they showed similar phenotypes (Table 1). This process identified 92 liver DEGs between pigs with extreme high and low backfat thickness levels (Additional file 3: Table S3) We also identified 587 and 690 genes that were only expressed in the BH or BL group, respectively (Figure 2). Comparing shared DEGs between the two different pairs of pigs, we found that DEGs of biological replicates were more homogeneous with fewer false positives. Because of the sample limitations for each pair, we only

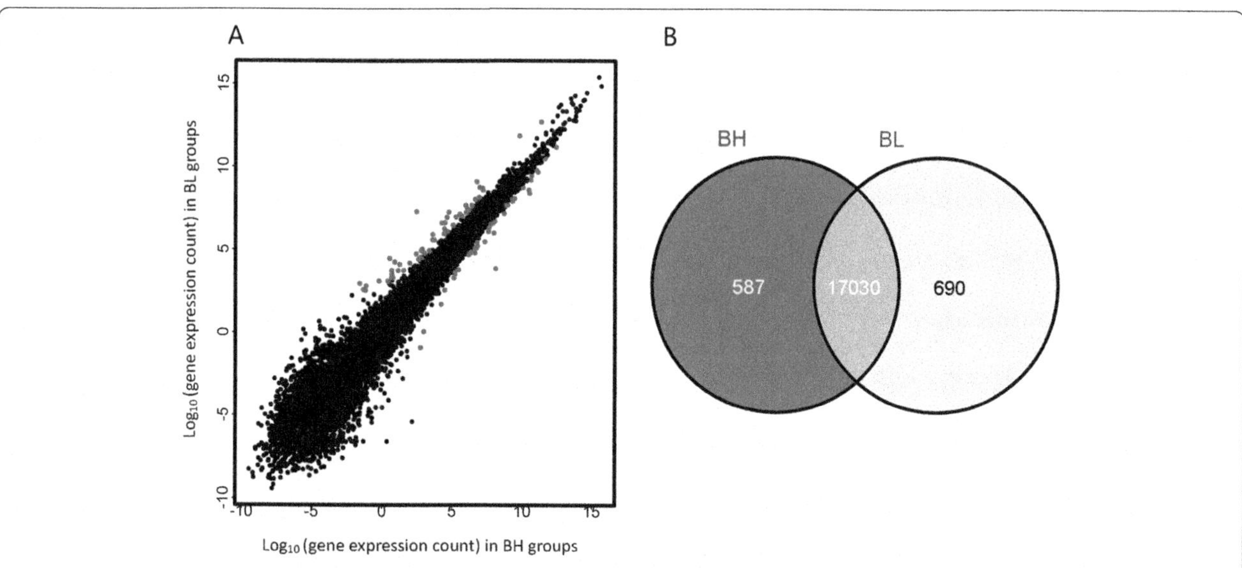

Figure 2 **Gene expression in BH and BL groups.** **A**. Red points represent genes that are significantly differentially expressed. **B**. Venn diagram showing genes only expressed in the BH group (blue circle), only expressed in the BL group (yellow circle), and common to both groups (intersection).

Table 3 Pathways enriched in up-regulated genes in the liver (BH vs. BL)

Pathway	Gene	*P* value	Benjamini
Metabolism of xenobiotics by cytochrome P450	CYP1A2, LOC100515394,LOC100526118, CYP1A1, LOC100511647, ADH1A CYP2S1	1.53E-08	4.43E-07
Drug metabolism	CYP1A2, LOC100515394, GSTM4 LOC100511647, ADH1A, CYP2A6	9.02E-07	1.31E-05
Retinol metabolism	CYP1A2, LOC100515394, CYP1A1 ADH1A, CYP2A6	1.88E-05	1.82E-04
Tryptophan metabolism	CYP1A2, CYP1A1, OGDHL	0.006753	0.047941

present functional analysis of DEGs obtained by treating the two pig pairs as two biological replicates.

Novel transcripts

We used the reference annotation based transcript algorithm implemented in Cufflinks to identify novel transcripts that were not annotated in the current Ensembl pig gene annotation database. A total of 8,226 novel transcripts were detected for FPKM ≥ 1 in the Songliao black pigs. Thirty-one spliced transcripts and 33 primary transcripts were significantly differentially expressed between the two groups (Additional file 4: Table S4).

Functional enrichment analysis of the DEGs

Of the 92 DEGs, 39 were up-regulated and 53 were down-regulated in the BH group compared with the BL group. Pig gene IDs were converted to human gene IDs, but five genes did not match with their human homologs (ENSSSCG00000008012, ENSSSCG00000016695, ENSSSCG00000030368, ENSSSCG00000010427, and ENSSSCG00000012881).

To gain an insight into the liver tissue processes that differ between the BL and BH groups of pigs, differentially up-regulated and down-regulated genes underwent separate pathway analysis and gene ontology analysis using DAVID. Human homologs were recognized for 32 of the 39 up-regulated genes, and up-regulated genes were found to be involved in the metabolism of xenobiotics, drugs, retinol, and tryptophan (Table 3). The expression of DEGs in the enriched pathways is shown in Table 4. Following GO analysis, the DEGs were shown to be related to biological processes such as amino acid biosynthesis and

Table 4 Expression levels of genes in up-regulated pathways in the liver (BH vs. BL)

Pathway	Gene symbol	BH	BL	Probability
Metabolism of xenobiotics by cytochrome P450	CYP1A2	260.26	100.93	0.83
	LOC100515394	71.57	25.53	0.83
	LOC100526118	1539.43	574.36	0.85
	CYP1A1	81.60	29.08	0.83
	LOC100511647	3211.75	1402.03	0.81
	ADH1A	2055.82	869.66	0.82
	CYP2A6	1596.90	683.59	0.82
Drug metism	CYP1A2	260.26	100.93	0.83
	LOC100515394	71.57	25.53	0.83
	GSTM4	192.04	66.50	0.85
	LOC100511647	3211.75	1402.03	0.81
	ADH1A	2055.82	869.66	0.82
	CYP2A6	1596.90	683.59	0.82
Retinol metabolism	CYP1A2	260.26	100.93	0.83
	LOC100515394	71.57	25.53	0.83
	CYP1A1	81.60	29.08	0.83
	ADH1A	2055.82	869.66	0.82
	CYP2A6	1596.90	683.59	0.82
Tryptophan metabolism	CYP1A2	81.60	29.08	0.83
	CYP1A1	81.60	29.08	0.83
	OGDHL	6.83	0.52	0.82

Probability: The higher the probability, the greater the change in expression between conditions.
BH, BL: expression level in BH or BL after TMM normalization.

metabolism, small molecule metabolism, and oxidation reduction. Forty-five of the 53 down-regulated genes had annotations in DAVID. No significant pathway was found to be associated with down-regulated genes. GO analysis showed that the biological processes enriched by DEGs are complex and relate to protein transport and enzyme activity (Figure 3).

Protein-protein interaction analysis

To gain a better understanding of the biological relationships between genes, the integral DEG list was inputted into the STRING database. Most proteins encoded by DEGs were dissociative, and two protein-protein interaction networks were identified (Figure 4). In one network, genes were associated with metabolism, detoxification, and superoxides, while genes in the other network were related to the heat stress response.

Discussion

In the present study, the percentages of mapped reads obtained per individual animal (84.80–92.90%) were higher than those in previous porcine liver transcriptome studies: 61.4–65.6% [9], 71.42–77.75% [10], and 43–84% [11]. Additionally, the percentage of reads within exons was higher than in previous studies, indicating that our results are more effective and credible.

Few studies have investigated how the liver transcriptome affects fat deposition. Our current gene expression analyses show that DEGs and pathways may play very important roles in this, and that genes involved in material metabolism and fatty acid transport are more active in the fatter groups of animals. These findings are consistent with those of previous studies. Previously, the expression levels of *CA3*, *SERPINA6*, *GATM*, *GSTA2*, and *ALAS1* were also found to be significantly different between the two groups [9]. *CA3* is associated with the internal fat rate and backfat thickness ofthe pig [20], while *SERPINA6* within the quantitative trait loci is associated with cortisol levels, fat, and muscle content, so can be considered a key regulator of obesity susceptibility [21]. These findings indicate that *CA3* and *SERPINA6* are strongly associated with fatty traits.

It is noteworthy that several DEGs identified in the present study (*FABP-1*, *LCN2*, *PLIN2*, *CYP1A1*, *CYP1A2*,

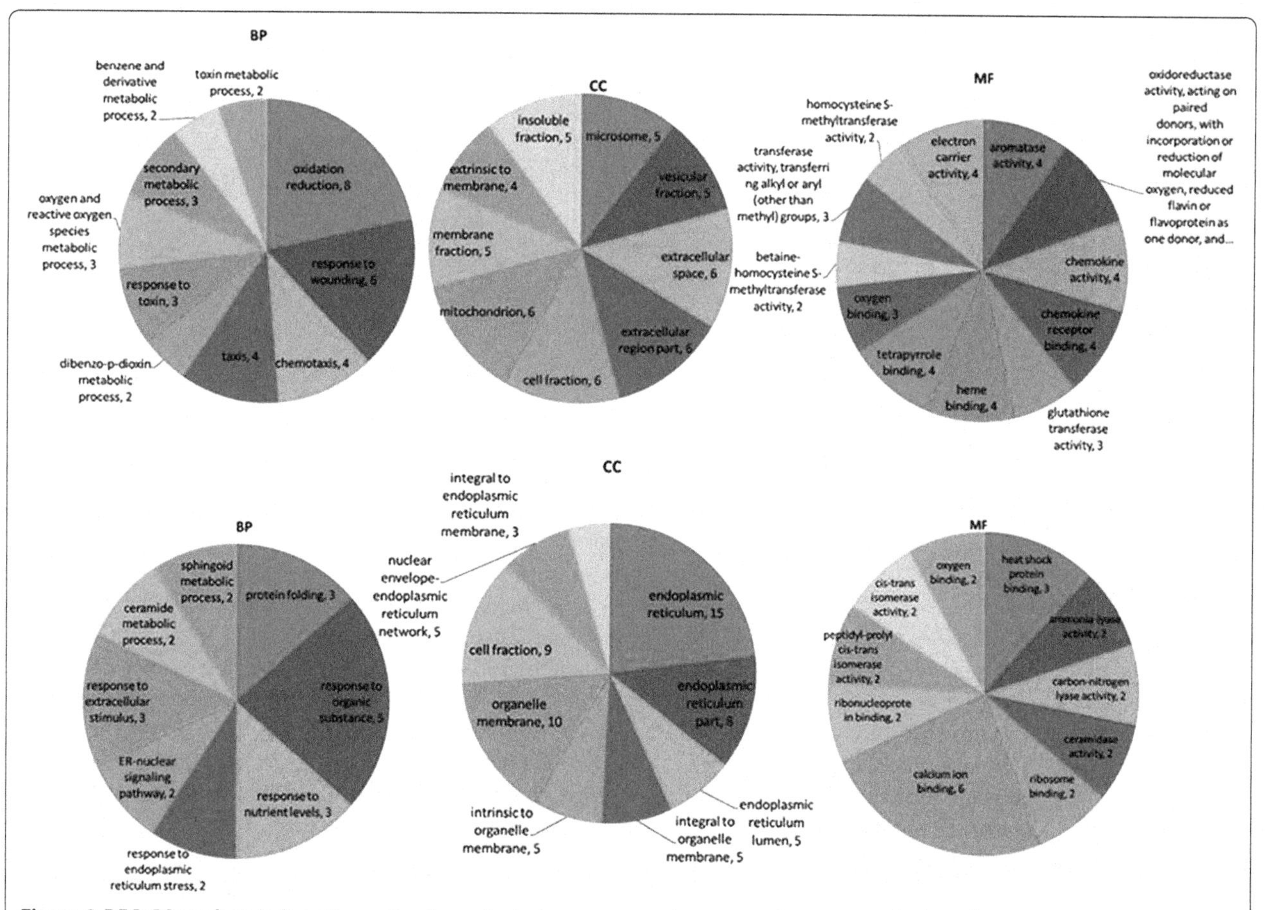

Figure 3 DEG GO analysis in liver tissue. Top three charts show gene ontology annotation processes (biological process (BP), cellular component (CC), and molecular functions (MF)) of up-regulated genes. Lower three charts show processes (BP, CC, and MF) of down-regulated genes.

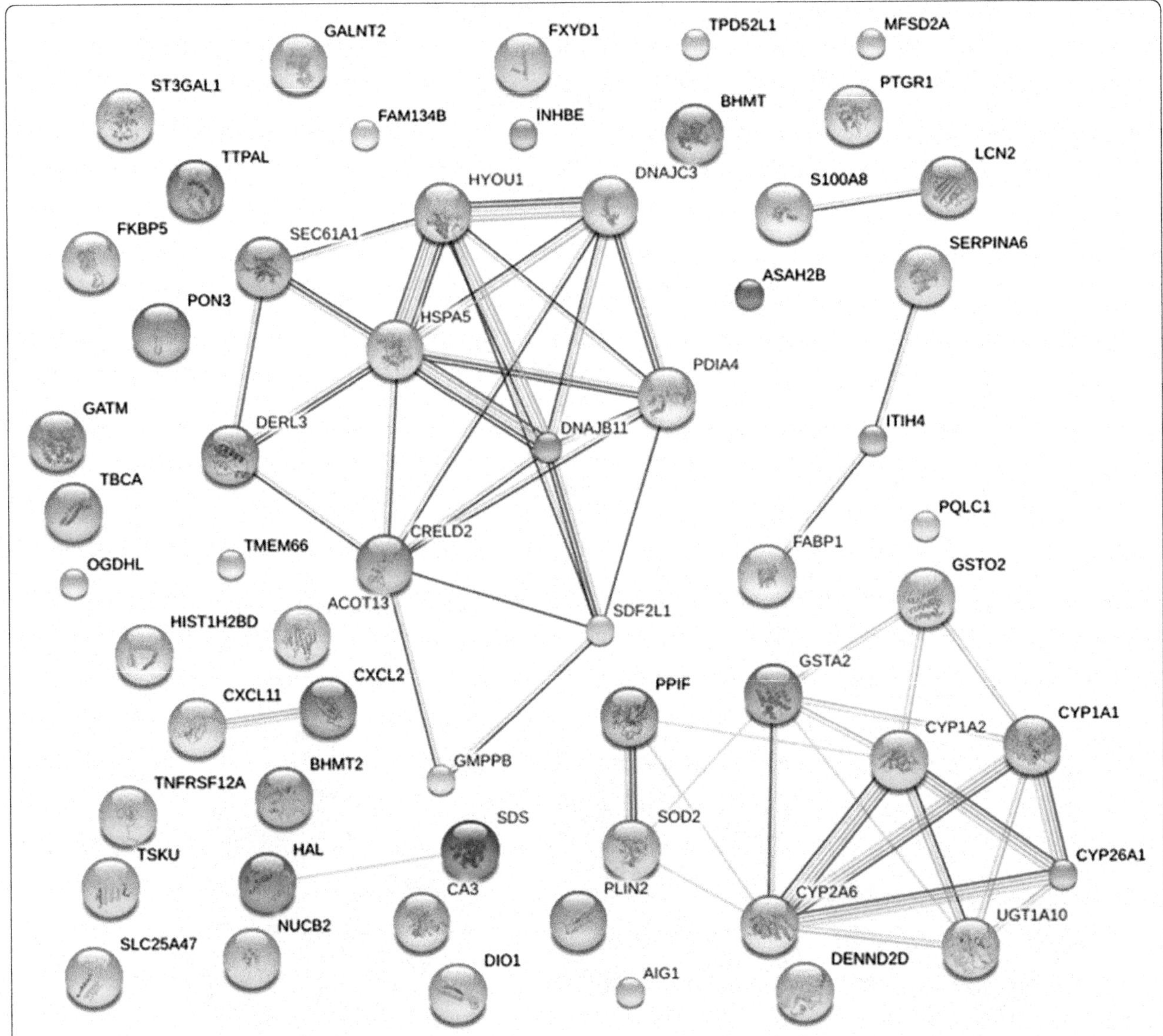

Figure 4 STRING analysis shows that DEGs are involved in known and predicted protein-protein interactions. STRING analysis of DEGs in the livers of BH and BL pigs. Network nodes represent genes shown in Additional file 2: Table S2. Lines of different color represent seven types of evidence used in predicting associations. Red line: fusion evidence; green line: neighborhood evidence; blue line: co-occurrence evidence; purple line: experimental evidence; yellow line: text mining evidence; light blue line: database evidence; black line: co-expression evidence.

CYP2A6, and *CYP26A1*) are involved in lipid metabolism. As a member of the family of fatty acid-binding proteins, *FABP-1* was expressed in all 12 tissues studied in a previous investigation, but transcript levels were more abundant in the liver and small intestine [22]. FABP-1 is involved in the transport of fatty acids to the site of β-oxidation, as well as the synthesis of triacylglycerol and phospholipids. It is also shown to interact with peroxisome proliferator-activated receptor (PPAR)α, which regulates fatty acid catabolism [23]. Compared with leaner pigs, *FABP* mRNA and protein levels were expressed at higher levels in the muscle of fatter animals [24]. In the present study, *FABP-1* was up-regulated in the liver samples of BH pigs compared with BL. Consistent with our results, *FABP-1* was also up-regulated in obese ob/ob mice compared with controls [25], suggesting that it's increased expression may enhance fatty acid transport.

The adipokine lipocalin 2 (LCN2) is a member of the lipocalin family that transports small lipophilic ligands and is highly expressed by fat cells both *in vitro* and *in vivo* [26]. Levels of *LCN2* mRNA are dramatically increased in the adipose tissue and liver of ob/ob mice, while reducing *LCN2* expression leads to a decrease in *PPARγ* expression [27]. However, we found that *LCN2* was down-regulated in the BH group compared with the BL group of pigs. Perilipin 2 (PLIN2) is a cytosolic protein that promotes the formation and stabilization of intracellular lipid droplets, which are organelles involved in the storage of lipid depots. *PLIN2* polymorphisms have been associated with carcass traits including backfat thickness in pigs [28], while increased *PLIN2* mRNA expression was detected in the skeletal muscle of pigs with higher intermuscular fat [28,29]. Our results contradict this by revealing a lower *PLIN2* mRNA expression level in the livers of pigs with higher backfat thickness. Finally, *CYP1A1, CYP1A2, CYP2A6,* and *CYP26A1* belong to the highly diverse CYP450 super family, and also showed a differential expression pattern in the current study. CYPs have been shown to play critical roles in catalyzing metabolism reactions and in the oxidation of unsaturated fatty acids [30]. Because several differences appear to exist in observed expression levels between studies, further investigation into expression differences of these genes is warranted to elucidate their role in fat deposition.

The present study identified more pathways that were up-regulated than down-regulated in the fatter group of pigs. Most of these pathways are related to the metabolism of substances such as xenobiotics, drugs, retinol, and tryptophan. The retinoid metabolism pathway has previously been shown to be up-regulated in the liver. Retinoid is crucial for most forms of life, and many studies have identified an association between retinoid and lipid metabolism [31]. Retinoids regulate metabolism by activating specific nuclear receptors, including the retinoic acid receptor and the retinoid X receptor, an obligate heterodimeric partner for other nuclear receptors including PPARs. This helps to coordinate energy balance [32] and suggests that the liver has a larger burden in fatter compared with leaner pigs.

STRING analysis in the present study indicated that genes regulating metabolism influence those related to heat stress. Heat stress was previously shown to affect lipid peroxidation, causing serious damage to membrane lipids, lipoprotein, and other lipid-containing structures [33]. In pigs, body weight is positively correlated with heat stress [34], but our current results show that mRNA levels of genes related to heat stress are down-regulated in the livers of fatter compared with leaner pigs. Future investigations into the interaction between body weight and heat stress may therefore identify novel methods to show how body weight affects healthy individuals.

Conclusions

This study undertook transcriptome analysis between two groups of Songliao black pigs with different backfat thicknesses. A total of 92 DEGs were identified between BH and BL groups. In concordance with the phenotypic differences, these genes belonged to pathways and gene networks related to lipid metabolism, regulation, and transport. Additionally, the identified DEGs related to heat stress could provide a new method of understanding and combating obesity. Our findings will be of use in understanding liver lipid regulation and in the design of new selection strategies to improve pig production.

Additional files

Additional file 1: Table S1. Gene expression count in four samples.

Additional file 2: Table S2. Primer sequences of six genes related to lipid metabolism for qRT-PCR.

Additional file 3: Table S3. Liver DEGs between BH and BL pigs with biological replicates.

Additional file 4: Table S4. Differentially expressed spliced transcripts and primary transcripts.

Abbreviations

BH: Group with higher backfat thickness; BL: Group with lower backfat thickness; DAVID: Database for annotation, visualization and integrated discovery; DEG: Differentially expressed gene; Full-sib: Full-sibling; GO-BP: Gene ontology biological processes; GO-CC: GO cellular component; GO-MF: GO molecular function; LCN2: Lipocalin 2; PLIN2: Perilipin 2; PPAR: Proliferator-activated receptor; TMM: Trimmed mean of M-values; RT: Reverse transcriptase; STRING: Search tool for the retrieval of interacting genes.

Competing interests

The authors declare that they have no competing interests.

Authors' contributions

KX carried out the experiment and drafted the manuscript. ZCH and CDW conceived the study, participated in its design and coordination, and helped draft the manuscript. FZ, HJL and ZJW helped sample, experiment and analysis of data. All authors read and approved the final manuscript.

Acknowledgment

This research was financially supported by the innovation research team for modern agricultural industry and technology in Beijing.

Author details

[1]National Engineering Laboratory for Animal Breeding and MOA Key Laboratory of Animal Genetics and Breeding, Department of Animal Genetics and Breeding, China Agricultural University, 100193 Beijing, China. [2]Tianjin Ninghe Primary Pig Breeding Farm, Ninghe 301500, Tianjin, China.

References

1. Organ R: *Key in regulating appetite and body weight.* ; 2012.
2. Suzuki K, Inomata K, Katoh K, Kadowaki H, Shibata T: **Genetic correlations among carcass cross-sectional fat area ratios, production traits, intramuscular fat, and serum leptin concentration in Duroc pigs.** *J Anim Sci* 2009, **87**:2209–2215.
3. Fontanesi L, Schiavo G, Galimberti G, Calò DG, Scotti E, Martelli PL, Buttazzoni L, Casadio R, Russo V: **A genome wide association study for backfat thickness in Italian Large White pigs highlights new regions affecting fat deposition including neuronal genes.** *BMC Genomics* 2012, **13**:583.
4. Houpt KA, Houpt TR, Pond WG: **The pig as a model for the study of obesity and of control of food intake: a review.** *Yale J Biol Med* 1979, **52**:307.
5. Reiter SS, Halsey CH, Stronach BM, Bartosh JL, Owsley WF, Bergen WG: **Lipid metabolism related gene-expression profiling in liver, skeletal muscle and adipose tissue in crossbred Duroc and Pietrain pigs.** *Comp Biochem Physiol Part D Genomics Proteomics* 2007, **2**:200–206.
6. Fam BC, Joannides CN, Andrikopoulos S: **The liver: Key in regulating appetite and body weight.** *Adipocyte* 2012, **1**:259–264.
7. Muñoz R, Estany J, Tor M, Doran O: **Hepatic lipogenic enzyme expression in pigs is affected by selection for decreased backfat thickness at constant intramuscular fat content.** *Meat Sci* 2013, **93**:746–751.
8. Nguyen P, Leray V, Diez M, Serisier S, Bloc'h JL, Siliart B, Dumon H: **Liver lipid metabolism.** *J Anim Physiol Anim Nutr* 2008, **92**:272–283.

9. Chen C, Ai H, Ren J, Li W, Li P, Qiao R, Ouyang J, Yang M, Ma J, Huang L: **A global view of porcine transcriptome in three tissues from a full-sib pair with extreme phenotypes in growth and fat deposition by paired-end RNA sequencing.** *BMC Genomics* 2011, **12**:448.
10. Ramayo-Caldas Y, Mach N, Esteve-Codina A, Corominas J, Castelló A, Ballester M, Estellé J, Ibáñez-Escriche N, Fernández AI, Pérez-Enciso M: **Liver transcriptome profile in pigs with extreme phenotypes of intramuscular fatty acid composition.** *BMC Genomics* 2012, **13**:547.
11. Gunawan A, Sahadevan S, Cinar MU, Neuhoff C, Große-Brinkhaus C, Frieden L, Tesfaye D, Tholen E, Looft C, Wondim DS: **Identification of the Novel Candidate Genes and Variants in Boar Liver Tissues with Divergent Skatole Levels Using RNA Deep Sequencing.** *PLoS One* 2013, **8**:e72298.
12. Trapnell C, Roberts A, Goff L, Pertea G, Kim D, Kelley DR, Pimentel H, Salzberg SL, Rinn JL, Pachter L: **Differential gene and transcript expression analysis of RNA-seq experiments with TopHat and Cufflinks.** *Nat Protoc* 2012, **7**:562–578.
13. Quinlan AR, Hall IM: **BEDTools: a flexible suite of utilities for comparing genomic features.** *Bioinformatics* 2010, **26**:841–842.
14. Delhomme N, Padioleau I, Furlong EE, Steinmetz LM: **easyRNASeq: a bioconductor package for processing RNA-Seq data.** *Bioinformatics* 2012, **28**:2532–2533.
15. Dillies MA, Rau A, Aubert J, Hennequet-Antier C, Jeanmougin M, Servant N, Keime C, Marot G, Castel D, Estelle J, Guernec G, Jagla B, Jouneau L, Laloë D, Le Gall C, Brigitte Schaëffer B, Le Crom S, Guedj MM, Jaffrézic F: **A comprehensive evaluation of normalization methods for Illumina high-throughput RNA sequencing data analysis.** *Brief Bioinform* 2013, **14**:671–683.
16. Tarazona S, García-Alcalde F, Dopazo J, Ferrer A, Conesa A: **Differential expression in RNA-seq: a matter of depth.** *Genome Res* 2011, **21**:2213–2223.
17. Tarazona S, Furió-Tarı P, Ferrer A, Conesa A: *NOISeq: Differential Expression in RNA-seq*; 2013.
18. Huang DW, Sherman BT, Lempicki RA: **Systematic and integrative analysis of large gene lists using DAVID bioinformatics resources.** *Nat Protoc* 2009, **4**:44–57.
19. Franceschini A, Szklarczyk D, Frankild S, Kuhn M, Simonovic M, Roth A, Lin J, Minguez P, Bork P, von Mering C: **STRING v9. 1: protein-protein interaction networks, with increased coverage and integration.** *Nucleic Acids Res* 2013, **41**:D808–D815.
20. Wu J, Zhou D, Deng C, Xiong Y, Lei M, Li F, Jiang S, Zuo B, Zheng R: **Expression pattern and polymorphism of three microsatellite markers in the porcine CA3 gene.** *Genet Sel Evol* 2008, **40**:227–239.
21. Ousova O, Guyonnet-Duperat V, Iannuccelli N, Bidanel JP, Milan D, Genet C, Llamas B, Yerle M, Gellin J, Chardon P, Emptoz-Bonneton A, Pugeat M, Mormède P, Moisan M: **Corticosteroid binding globulin: a new target for cortisol-driven obesity.** *Mol Endocrinol* 2004, **18**:1687–1696.
22. Jiang YZ, Li XW, Yang GX: **Sequence characterization, tissue-specific expression and polymorphism of the porcine (Sus scrofa) liver-type fatty acid binding protein gene.** *Yi Chuan Xue Bao* 2006, **33**:598–606.
23. Wolfrum C, Borrmann CM, Borchers T, Spener F: **Fatty acids and hypolipidemic drugs regulate peroxisome proliferator-activated receptors alpha - and gamma-mediated gene expression via liver fatty acid binding protein: a signaling path to the nucleus.** *Proc Natl Acad Sci U S A* 2001, **98**:2323–2328.
24. Zhao S, Ren L, Chen L, Zhang X, Cheng M, Li W, Zhang Y, Gao S: **Differential expression of lipid metabolism related genes in porcine muscle tissue leading to different intramuscular fat deposition.** *Lipids* 2009, **44**:1029–1037.
25. Chen R, Cao Y, Ma X, Wang Y, Zhang H, Zhang W, Zhou D: *Characterization of lipid metabolism-related genes in mouse fatty liver*; 2012.
26. Yan Q-W, Yang Q, Mody N, Graham TE, Hsu C-H, Xu Z, Houstis NE, Kahn BB, Rosen ED: **The adipokine lipocalin 2 is regulated by obesity and promotes insulin resistance.** *Diabetes* 2007, **56**:2533–2540.
27. Zhang J, Wu Y, Zhang Y, LeRoith D, Bernlohr DA, Chen X: **The role of lipocalin 2 in the regulation of inflammation in adipocytes and macrophages.** *Mol Endocrinol* 2008, **22**:1416–1426.
28. Davoli R, Gandolfi G, Braglia S, Comella M, Zambonelli P, Buttazzoni L, Russo V: **New SNP of the porcine Perilipin 2 (PLIN2) gene, association with carcass traits and expression analysis in skeletal muscle.** *Mol Biol Rep* 2011, **38**:1575–1583.
29. Gandolfi G, Mazzoni M, Zambonelli P, Lalatta-Costerbosa G, Tronca A, Russo V, Davoli R: **Perilipin 1 and perilipin 2 protein localization and gene expression study in skeletal muscles of European cross-breed pigs with different intramuscular fat contents.** *Meat Sci* 2011, **88**:631–637.
30. Lewis DF: **57 varieties: the human cytochromes P450.** *Pharmacogenomics* 2004, **5**:305–318.
31. Keller H, Dreyer C, Medin J, Mahfoudi A, Ozato K, Wahli W: **Fatty acids and retinoids control lipid metabolism through activation of peroxisome proliferator-activated receptor-retinoid X receptor heterodimers.** *Proc Natl Acad Sci U S A* 1993, **90**:2160–2164.
32. Ziouzenkova O, Plutzky J: **Retinoid metabolism and nuclear receptor responses: New insights into coordinated regulation of the PPAR-RXR complex.** *FEBS Lett* 2008, **582**:32–38.
33. Altan Ö, Pabuçcuoğlu A, Altan A, Konyalioğlu S, Bayraktar H: **Effect of heat stress on oxidative stress, lipid peroxidation and some stress parameters in broilers.** *Br Poultry Sci* 2003, **44**:545–550.
34. Ingram D, Legge K: **Effects of environmental temperature on food intake in growing pigs.** *Comp Biochem Physiol A Physiol* 1974, **48**:573–581.

8

Phenotypic and genotypic background underlying variations in fatty acid composition and sensory parameters in European bovine breeds

Natalia Sevane[1*], Hubert Levéziel[2,3], Geoffrey R Nute[4], Carlos Sañudo[5], Alessio Valentini[6], John Williams[7], Susana Dunner[1] and the GeMQual Consortium

Abstract

Background: Consuming moderate amounts of lean red meat as part of a balanced diet valuably contributes to intakes of essential nutrients. In this study, we merged phenotypic and genotypic information to characterize the variation in lipid profile and sensory parameters and to represent the diversity among 15 cattle populations. Correlations between fat content, organoleptic characteristics and lipid profiles were also investigated.

Methods: A sample of 436 largely unrelated purebred bulls belonging to 15 breeds and reared under comparable management conditions was analyzed. Phenotypic data -including fatness score, fat percentage, individual fatty acids (FA) profiles and sensory panel tests- and genotypic information from 11 polymorphisms was used.

Results: The correlation coefficients between muscle total lipid measurements and absolute vs. relative amounts of polyunsaturated FA (PUFA) were in opposite directions. Increasing carcass fat leads to an increasing amount of FAs in triglycerides, but at the same time the relative amount of PUFAs is decreasing, which is in concordance with the negative correlation obtained here between the percentage of PUFA and fat measurements, as well as the weaker correlation between total phospholipids and total lipid muscle content compared with neutral lipids. Concerning organoleptic characteristics, a negative correlation between flavour scores and the percentage of total PUFA, particularly to n-6 fraction, was found. The correlation between juiciness and texture is higher than with flavour scores. The distribution of SNPs plotted by principal components analysis (PCA) mainly reflects their known trait associations, although influenced by their specific breed allele frequencies.

Conclusions: The results presented here help to understand the phenotypic and genotypic background underlying variations in FA composition and sensory parameters between breeds. The wide range of traits and breeds studied, along with the genotypic information on polymorphisms previously associated with different lipid traits, provide a broad characterization of beef meat, which allows giving a better response to the variety of consumers' preferences. Also, the development and implementation of low-density SNP panels with predictive value for economically important traits, such as those summarized here, may be used to improve production efficiency and meat quality in the beef industry.

Keywords: Bos taurus, Beef, Fatty acids, Omega-3, Genotype assisted selection

* Correspondence: nata_sf@hotmail.com
[1]Departamento de Producción Animal, Facultad de Veterinaria, Universidad Complutense de Madrid, Madrid, Spain
Full list of author information is available at the end of the article

Background

Cattle meat provides several nutrients necessary for a balanced diet and for health preservation, especially high value proteins, minerals, B-complex vitamins and essential fatty acids (FA), and also can have an important role as a dietary source of n-3 FA and conjugated linoleic acids (CLA) [1,2]. A number of epidemiological studies have associated red meat consumption with increased disease risks [3-5], whereas other authors point out the beneficial effects of the moderate consumption of unprocessed red meat -lowers total cholesterol, LDL cholesterol and triglycerides (TG) [6,7], as well as blood pressure [8]. However, the isolation of the effects of red meat alone is difficult to accomplish [2]. Moreover, the level of intramuscular fat content and FA composition are among the main factors determining meat palatability and consumers satisfaction [9]. Muscle lipid characteristics determine meat flavour and lipid oxidation, which contributes to beef colour, and can be responsible for abnormal odours, and influences the juiciness and tenderness of meat [10]. However, meat quality traits are very complex, involve many genes and are greatly influenced by a variety of environmental factors, such as diet sex, season, age, etc. [11]. Being difficult and expensive to measure [12], they are usually not included in selection programs based on phenotypic performance, and challenge application of traditional selection methods, as well as the state-of-the-art Genomic Selection (GS) [13].

An alternative approach is to identify genes with an effect on fat composition and include these in selection objectives. Thanks to the genomic revolution of the past few years, more information and technology are available that can be used to improve meat quality. Many studies have identified QTL involved in meat quality related traits in beef cattle (e.g. [14,15]); however, the dissection of these QTL has not identified genetic variants explaining a large portion of phenotypic variance [16]. More recently, single nucleotide polymorphisms (SNP) within candidate genes have been tested for predictive value for carcass traits, and some commercial tests to genotype animals based on SNP marker panels are being proposed to breeders (see the review in [17]). The significant progress made in characterizing changes in tissue FA composition to diet, feeding system and genotype, highlights the potential for further progress to be made through genomic or marker-assisted selection in livestock and the formulation of diets to exploit the genetic potential [18]. Nevertheless, the full development of these technologies greatly depends on the precise identification of the genes and polymorphisms that have a measurable effect on muscle physiology and on meat quality, and the validation of their effects on different breeds.

In this study, we merged phenotypic and genotypic information [19-21] to characterize the variation in lipid profile and sensory parameters of *Longissimus thoracis* muscle and to represent the diversity among 15 cattle populations, reared under comparable management conditions. Correlations between fat content, organoleptic characteristics and lipid profiles were also investigated.

Methods

Animals and feed system

A total of 436 muscle samples from unrelated bulls belonging to 15 European cattle breeds were studied in the frame of the GeMQual (EC QLK5 – CT2000-0147) European project and genotyped. The breeds included specialized beef breeds, dairy breeds, and local beef breeds. The whole sample included 31 Jersey (JER), 27 South Devon (SD), 30 Aberdeen Angus (AA), and 29 Highland (HIG) from United Kingdom; 29 Holstein (HOL), 29 Danish Red (DR), and 20 Simmental (SM), from Denmark; 30 Asturiana de los Valles (AST), 31 Asturiana de la Montaña (CAS), 30 Avileña-Negra Ibérica (AVI), and 31 Pirenaica (PI) from Spain; 30 Piedmontese (PIE), and 28 Marchigiana (MAR) from Italy; and 31 Limousin (LIM), and 30 Charolais (CHA) from France.

Bulls were reared in each country in a unique location and under a uniform beef management system representative of those used in the European Union (EU) countries. Feed composition and management details are described in [22]. The diet was designed to achieve the slaughter weight of 75% of mature weight for each breed within a window of 13 to 17 mon. Animals from each breed were slaughtered the same day in either commercial or experimental abattoirs, depending on the experimental facilities of each country.

The part of this work involving live animal experimental intervention was reported by Albertí et al. [22] and followed the European standards on Care and Use of Animals (1999/575/EC). The present work was conducted on DNA and carcass samples and no Care and Use of Animals was needed.

Sampling and phenotype measurements

Carcass processing after slaughter was described by [22] and [23]. For lipid measurements, *Longissimus thoracis* muscle was excised at 24 h postmortem from the left side of the carcass between the 6th and the 13th rib and a sample was taken immediately and frozen for chemical analysis including fat concentration. The remainder was stored at 2°C ± 1°C until 48 h post-mortem. Also, samples were taken from the 48 h post-mortem section to determine total lipid content. Samples for individual FA analysis were taken from the same position on *Longissimus thoracis* from all animals and vacuum packed, frozen and transported on dry ice to University of Bristol (United Kingdom) to determine total lipid content.

Fatness score (FS) corresponds to the visual fatness cover estimated by UE standard (1 = very low, 5 = very high), and fat percentage (FP) is the proportion of subcutaneous and intermuscular fat in the rib dissection. Fat was extracted by the method of [24] separated into neutral lipid and phospholipid, methylated, separated by gas-liquid chromatography (GLC) and the individual peaks identified and quantified as described in detail by [25]. Total lipid content was taken as the sum of the neutral lipid (NL) and phospholipid (PL) fractions. Some additional phenotypes were set as are saturated FA (SFA), monounsaturated FA (MUFA), polyunsaturated FA (PUFA), n-6/n-3 ratios, polyunsaturated to saturated FA (P:S) ratios and antithrombotic potential (ATT), which is the ratio between the sum of the antithrombogenic FAs eicosatrienoic acid (20:3 n-6) and eicosapentaenoic acid (20:5 n-3), and the thrombogenic FA arachidonic acid (20:4 n-6) [(20:3 n-6 + 20:5 n-3)/20:4 n-6] [26]. Sensory panel tests assessed meat using an eight-point scale as described in [27]. The criteria assessed were flavour, texture and juiciness -the higher scores corresponding to the characteristic flavour of beef, and very tender and juicy meat, respectively. See [21] for detailed phenotype values per breed.

Marker selection

The allele frequencies of 11 polymorphisms found to be associated with different lipid traits across breeds and causing increases in traits ranging between 3.3% and 19% for one homozygous genotype compared to the other homozygous genotype (Table 1) [19,20], were used for a principal components analysis (PCA) to represent the diversity among the 15 cattle populations: calpastatin (*CAST*) g.2959G < A [20]; cofilin 1 (*CFL1*) ss77831721 [19]; EP300 interacting inhibitor of differentiation 1 (*CRI1*) ss778332128 [19]; myostatin (*GDF8*) ss77831865 [20]; insulin-like growth factor 2 receptor (*IGF2R*) ss77831885 [19]; lipoprotein lipase (*LPL*) ss65478732 [20]; matrix metalloproteinase 1 (*MMP1*) ss77831916, ss77831924 [19]; myozenin 1 (*MYOZ1*) ss77831945 [20]; phospholipid transfer protein (PLTP) ss77832104 [19]; peroxisome proliferator activated receptor γ (*PPARG*) ss62850198 [20]. The six SNPs from [19] were genotyped by Kbioscience using the proprietary Kaspar© methodology; the five SNPs from [20] were genotyped by SNP multiplex and Primer Extension amplification.

Statistical analysis

Spearman correlations were determined between fatness score, fat percentage, flavour, juiciness, texture, and different lipid profiles of *Longissimus thoracis* muscle in 15 European cattle breeds using the CORR procedure of SAS and considering the whole set of data on all animals. Allele frequency data were subjected to ANOVA, using the PROC GLM procedure of SAS and considering breed as independent variable. A PCA procedure was performed using the mean of phenotypic measurements by breed and allelic frequencies from 11 polymorphisms to determine the main traits and SNPs that explained most of the variation among the 15 cattle populations. The frequency of the allele showing a positive correlation with the trait was used (bold allele in Table 1). All these statistical analyses were carried out using the SAS statistical package v. 9.1.3 [28].

Results

Trait correlations

Table 2 shows the main correlations between FS, FP, flavour, juiciness, texture, and different lipid profiles of *Longissimus thoracis*. FS correlated positively with FP ($r = 0.62$, $P < 0.001$), and both of them with absolute amounts of lipids in muscle, including total PL and NL, SFA, MUFA, PUFA, n-3 and n-6 content, as well as with flavour score. However, both of them displayed a negative correlation to P:S ratios explained by their higher correlations with SFA content (FS $r = 0.4$, $P < 0.001$; FP $r = 0.68$, $P < 0.001$) than to PUFA (FS $r = 0.17$, $P < 0.001$; FP $r = 0.44$, $P < 0.001$), and to n-6/n-3 ratios because of their lower correlations to n-6 (FS $r = 0.13$, $P < 0.01$; FP $r = 0.37$, $P < 0.001$) compared to n-3 (FS $r = 0.28$, $P < 0.001$; FP $r = 0.56$, $P < 0.001$). The correlation between total PL and FS ($r = 0.11$, $P < 0.05$) and FP ($r = 0.44$, $P < 0.001$) is lower than between total NL and these traits (FS $r = 0.44$, $P < 0.001$; FP $r = 0.78$, $P < 0.001$). Similar correlation coefficients were also observed between PL and total lipid ($r = 0.7$, $P < 0.001$) and SFA ($r = 0.66$, $P < 0.001$), compared to NL with both total lipid and SFA ($r = 0.99$, $P < 0.001$). The correlation between SFA and MUFA ($r = 0.99$, $P < 0.001$) was higher than between SFA and PUFA ($r = 0.7$, $P < 0.001$). The proportion of 18:2 n-6 declines in muscle as fat deposition increases (correlations with FP $r = -0.77$, $P < 0.001$, total lipid $r = -0.86$, $P < 0.001$, and SFA $r = -0.88$, $P < 0.001$).

Flavour correlated positively with the organoleptic characteristics juiciness ($r = 0.21$, $P < 0.001$) and texture ($r = 0.16$, $P < 0.001$). A higher positive correlation was found between juiciness and texture ($r = 0.57$, $P < 0.001$). Finally, beef juiciness showed a small negative correlation to the amount of PL ($r = -0.13$, $P < 0.01$) and PUFA ($r = -0.1$, $P < 0.05$), particularly to n-6 content ($r = -0.13$, $P < 0.01$).

Apart from results in Table 2, it is worth highlighting the positive correlation between 18:1 trans-vaccenic FA (t18:1) and CLA cis-9,trans-11 ($r = 0.62$, $P < 0.001$).

Phenotype and genotype variation among breeds

The plot of factor pattern of the 15 cattle breeds showing the correlations to lipid traits and genotypic data from 11 polymorphisms with the two principal components is shown in Figure 1. The first two dimensions (Factor 1, 42.2%; Factor 2, 16.6%) explained 58.8% of the variation among breeds (Figure 1). When considering the different lipid traits, the first dimension was mainly influenced by total lipid measurements and flavour score, whereas on the opposite side muscle percentages of PUFA and of n-6, as well as P:S1, P:S2 and

Table 1 Allele frequencies per breed of 11 polymorphisms with effects ranging between 3.3% and 19% on different lipid traits based on results from [19,20]

Locus symbol	dbSNP[1]	Alleles[2]	Effect of the homozygous genotype for the bold allele	Frequency of bold allele														
				JER[3] (n = 30)	SD[3] (n = 27)	AA[3] (n = 30)	HIG[3] (n = 29)	HOL[3] (n = 29)	DR[3] (n = 29)	SM[3] (n = 20)	LIM[3] (n = 31)	CHA[3] (n = 31)	PIE[3] (n = 30)	MAR[3] (n = 28)	AST[3] (n = 30)	CAS[3] (n = 31)	AVI[3] (n = 30)	PI[3] (n = 31)
CAST*	g.2959G < A	**A**/G	↑ 5% FS[4]	0.77	0.83	0.88	0.94	0.69	0.66	0.78	0.68	0.65	0.70	0.64	0.73	0.84	0.85	0.65
CFL1*	ss77831721	**C**/T	↑ 8% 18:2/18:3	0.48	0.02	0.12	0.09	0.14	0.43	0.29	0.47	0.61	0.47	0.59	0.39	0.29	0.14	0.22
CRI1*	ss77832128	**G**/T	↑ 13.4% N 22:4n-6[5]	0.95	1.00	1.00	0.67	0.75	1.00	0.75	0.90	0.91	0.75	0.92	0.96	0.89	0.72	0.90
GDF8*	ss77831865	**G**/del	↑ 15% FS	1.00	0.65	1.00	1.00	1.00	1.00	1.00	1.00	1.00	1.00	1.00	0.32	0.98	1.00	0.89
IGF2R*	ss77831885	A/**G**	↑ 4.4% Flavour	0.04	0.44	0.32	0.52	0.06	0.33	0.3	0.18	0.44	0.18	0.09	0.13	0.10	0.20	0.22
LPL*	ss65478732	**T**/C	↑ N n-6[6]	0	0	0.06	0	0.1	0.05	0.06	0	0.10	0.05	0	0	0.08	0.09	0.02
MMP1*	ss77831916	**A**/G	↑ 3.3% CLA	1.00	0.95	0.79	0.76	0.82	0.85	1.00	0.97	0.95	0.95	0.74	0.85	0.98	0.91	0.91
MMP1b	ss77831924	T/**C**	↑ 14% 22:6n-3	0.50	0.37	0.55	0.44	0.50	0.50	0.50	0.50	0.48	0.47	0.54	0.47	0.50	0.52	0.50
MYOZ1*	ss77831945	C/**T**	↑ 8% 18:2/18:3	0.48	0.94	1.00	0.44	0.63	0.83	0.91	0.80	0.91	0.87	0.82	0.83	0.92	0.92	0.94
PLTP*	ss77832104	**G**/A	↑ 8% n-6/n-3	0.50	0.33	0.58	0.25	0.27	0.20	0.38	0.71	0.40	0.63	0.70	0.57	0.08	0.56	0.50
PPARG*	ss62850198	G/**A**	↑ n-3[7]	0.13	0.21	0.28	0.04	0.12	0.19	0.14	0.18	0.25	0.12	0.11	0.10	0.10	0.06	0.16

*Genes for which allele frequencies are statistically different among breeds ($P < 0.0001$).
[1]dbSNPs accession number of SNPs found to be associated with different production traits by [19,20].
[2]In bold, favourable allele, i.e. allele which improves animal performance and/or meat quality by diminishing n-6 and/or n-6:n-3 ratio, by increasing n-3, or by improving organoleptic characteristics.
[3]Complete breed names: Jersey (JER), South Devon (SD), Aberdeen Angus (AA), Highland (HIG), Holstein (HOL), Danish Red (DR), Simmental (SM), Limousin (LIM), Charolais (CHA), Piedmontese (PIE), Marchigiana (MAR), Asturiana de los Valles (AST), Asturiana de la Montaña (CAS), Avileña-Negra Ibérica (AVI), Pirenaica (PI). Between brackets: maximum n.
[4]Fatness score.
[5]N: neutral fatty acid.
[6]↑16% neutral 20:3 n-6; ↑ 19% neutral 20:4 n-6.
[7]↑ 9% 22:5 n-3; ↑ 15% 20:5 n-3; ↑ 18% 22:6 n-3.

Table 2 Correlations between fatness score, fat percentage, flavour, juiciness, texture, and different lipid profiles of *Longissimus thoracis* muscle in 15 European cattle breeds

Items	FS[1]	FP[2]	Flavour	Juiciness	Texture	Total lipid	SFA
FP	**0.62*****						
Flavour	**0.12***	**0.31*****					
Juiciness	0.07	0.01	**0.21*****				
Texture	0.06	0.01	**0.16*****	**0.57*****			
Total Lipid	**0.43*****	**0.77*****	**0.3*****	−0.04	−0.02		
SFA[3]	**0.4*****	**0.68*****	**0.3*****	−0.01	0.06	**0.95*****	
MUFA[4]	**0.41*****	**0.67*****	**0.28*****	−0.02	0.06	**0.99*****	**0.99*****
PUFA[5]	**0.17*****	**0.44*****	**0.27*****	**−0.1***	−0.02	**0.73*****	**0.7*****
% PUFA	**−0.51*****	**−0.77*****	**−0.25*****	−0.01	−0.02	**−0.9*****	**−0.92*****
% 18:2 n-6	**−0.53*****	**−0.77*****	**−0.25*****	−0.01	−0.02	**−0.86*****	**−0.88*****
Total PL	**0.11***	**0.44*****	**0.26*****	**−0.13****	−0.06	**0.7*****	**0.66*****
Total NL	**0.44*****	**0.78*****	**0.3*****	−0.03	−0.01	**0.99*****	**0.99*****
n-3 PUFA[6]	**0.28*****	**0.56*****	**0.42*****	0.07	−0.02	**0.63*****	**0.62*****
% n-3PUFA	**−0.24*****	**−0.34*****	−0.01	0.08	−0.06	**−0.62*****	**−0.62*****
n-6 PUFA[7]	**0.13****	**0.37*****	**0.21*****	**−0.13****	−0.02	**0.68*****	**0.64*****
% n-6PUFA	**−0.53*****	**−0.79*****	**−0.27*****	−0.02	−0.02	**−0.9*****	**−0.91*****
P:S1[8]	**−0.44*****	**−0.68*****	**−0.16*****	−0.01	−0.02	**−0.85*****	**−0.87*****
P:S2[9]	**−0.44*****	**−0.69*****	**−0.18*****	−0.01	−0.01	**−0.9*****	**−0.92*****
n-6/n-3	**−0.28*****	**−0.47*****	**−0.24*****	−0.08	0.04	**−0.26*****	**−0.27*****
18:2/18:3	**−0.38*****	**−0.67*****	**−0.28*****	−0.02	0.04	**−0.6*****	**−0.61*****
22:6/18:3	0.04	**−0.29*****	**−0.26*****	−0.06	−0.03	**−0.38*****	**−0.38*****
ATT[10]	**0.35*****	**0.44*****	**0.22*****	**0.13****	0.01	0.06	**0.2*****

Level of significance: ***($P < 0.001$), **($P < 0.01$), *($P < 0.05$).
[1]Fatness score: visual fatness cover estimated by UE standard.
[2]Fatness percentage: proportion of fat (subcutaneous and intermuscular) in the rib dissection.
[3]12:0 + 14:0 + 16:0 + 18:0.
[4]16:1 + t18:1 + 9c18:1+ 11c18:1 + 20:1.
[5]18:2n-6 + 18:3n-3 + 20:3n-6 + 20:4n-6 + 20:5n-3 + 22:4n-6 + 22:5n-3 + 22:6n-3.
[6]18:3n-3 + 20:5n-3 + 22:5n-3 + 22:6n-3.
[7]18:2n-6 + 20:3n-6 + 20:4n-6 + 22:4n-6.
[8](18:2n-6 + 18:3n-3) / (12:0 + 14:0 + 16:0 + 18:0).
[9](18:2n-6 + 18:3n-3 + 20:3n-6 + 20:4n-6 + 20:5n-3 + 22:4n-6 + 22:5n-3 + 22:6n-3) / (12:0 + 14:0 + 16:0 + 18:0).
[10](20:3n-6 + 20:5n-3)/20:4n-6.

18:2/18:3 ratios are plotted. The second dimension was mainly influenced by the ATT index, n-3, % n-3, and juiciness. Therefore, AA, HIG, HOL, DR and JER breeds, which displayed higher fatness [21], appeared in the positive area of the first dimension and split into two groups by the influence of the higher n-3 muscle content and flavour scores of AA and HIG breeds, and by the higher n-6 and MUFA content of HOL and DR dairy breeds. In contrast, lean breeds with high proportion of PUFAs, and high P:S and n-6 to n-3 ratios, such as PIE and AST [21], appeared at the bottom of the plot (Figure 1). Finally, SD breed stands out because of its highest ATT ratio and percentage of n-3 in muscle (Figure 1).

Concerning SNPs distribution, they plotted mainly according to their previous trait associations, although also influenced by their allele frequencies per breed (Table 1).

Discussion

All animals included in this project (GeMQual QLK5 – CT2000-0147) were fed a similar diet and reared intensively under comparable management conditions between countries. The effects of all factors other than breed (country, diet, slaughter) were controlled to minimize differences and were confounded with the breed effect. Inevitably, some variations might have occurred but special emphasis has been put to respect the diet composition in the different countries. The higher absolute n-3 PUFA muscle content

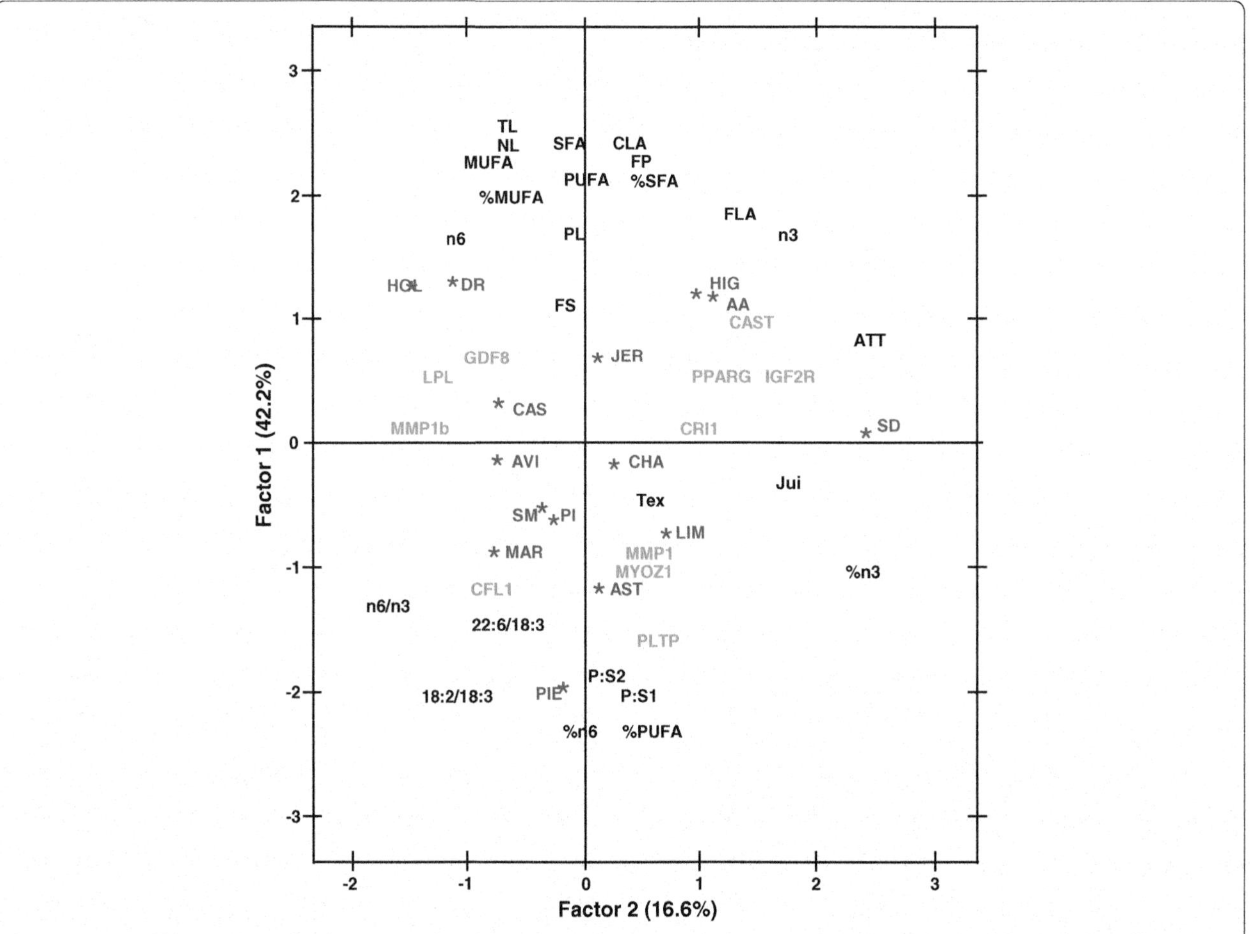

Figure 1 Plot of factor pattern for factors 1 and 2 of 15 European breeds showing the correlations to lipid traits (in black) and genotypic data (in green) from 11 polymorphisms with the two principal components. Abbreviations: TL total lipids, NL neutral lipids, PL phospholipids, FP fatness percentage, FS fatness score, Fla flavour, Tex texture, Jui juiciness, JER Jersey, SD South Devon, AA Aberdeen Angus, HIG Highland, HOL Holstein, DR Danish Red, SM Simmental, LIM Limousin, CHA Charolais, PIE Piedmontese, MAR Marchigiana, AST Asturiana de los Valles, CAS Asturiana de la Montaña, AVI Avileña-Negra Ibérica, PI Pirenaica.

found in the UK breeds, especially in the Aberdeen Angus breed, cannot be due to a grass-based diet generally used in UK [29] inexistent in this study, but rather to a specific characteristic of this fat breed.

Trait correlations

The correlation coefficients between total muscle lipid measurements (FS, FP, total lipids, SFA), and absolute vs. proportional (%) amounts of PUFA, n-6 and n-3 FAs were in opposite directions [30,31]. For example, the sum of n-3 FA showed a positive relation to FS ($r = 0.28$), FP ($r = 0.56$), total lipids ($r = 0.63$) and SFA ($r = 0.62$) for absolute amounts, but there were negative correlations between those traits and n-3 relative proportion (FS $r = -0.24$, FP $r = -0.34$, total lipids $r = -0.62$, SFA $r = -0.62$) (Table 2). In particular, the negative correlation obtained here between the percentage of 18:2 n-6 and fat measurements, as well as the weaker correlation between total PL and FS, FP, total lipid and SFA muscle content compared with NL (Table 2) [32], is in accordance with the expected proportions of PL vs. NL as animal fattens. Long chain PUFAs are mainly stored in muscle PL in cattle, which is an essential component of cell membranes and its amount remains fairly constant as the animal fattens, whereas NL increases in overall FA composition. SFA and MUFA are mainly stored in the NL fraction in triglycerides. This means increasing carcass fat leads to an increasing amount of FAs in triglycerides, but at the same time the relative amount of PUFAs is decreasing [32], which is in concordance with the negative correlation obtained here between the percentage of PUFA and fat measurements, as well as the weaker correlation between total PL and total lipid ($r = 0.7$) muscle content compared with NL ($r = 0.99$) (Table 2).

In agreement also with previous studies [33,34], we found a positive correlation between t18:1 and CLA (r =

0.62, $P < 0.001$), explained by the metabolic relationships between both FA –in ruminants the SCD enzyme forms also CLA from t18:1 in adipose tissue [29,32].

Although high levels of long chain n-3 PUFA have been described as having an impact on flavour to produce a 'grass fed' taste [35,36], and [37] found no correlation between n-6 PUFA and flavour in two beef breeds, here there was only a negative correlation to the percentage of total PUFA, particularly to n-6 fraction, whereas the percentage of n-3 in muscle did not seem to influence meat flavour in cattle not fed with a grass-based diet.

As expected, the correlation between juiciness and texture is higher than with flavour scores given that juiciness depends mainly on the meat water-binding capacity and plays a key role in meat texture [38-40], contributing to its variability [41], whereas flavour is mainly influenced by FA composition and marbling [42], as reflected by its positive correlations with all absolute fat content measurements obtained here (Table 2). Although texture and juiciness properties also are dependent on other characteristics of meat, including fat content [40], both of them showed few or no correlation with muscle fat content or FA profile (Table 2).

Phenotype and genotype variation among breeds

The distribution of breeds plotted by PCA analysis fell into three main groups (Figure 1): one group defined as having a high absolute fat content, which splits into two blocks –AA and HIG breeds on one hand, characterized by higher n-3 muscle content and flavour scores, and in the other hand HOL and DR dairy breeds, which displayed higher values of MUFA and n-6-; a second group with lower fat content and higher proportion of PUFAs, as well as PUFA vs. SFA ratios (healthier meat) PIE and AST; and a large group gathering the rest of breeds with intermediate fat content, among which it is worth highlighting SD because of its highest ATT ratio (index higher values better for health) and percentage of n-3 in muscle.

Regarding SNPs distribution, most of them plotted according to their previous trait associations and also influenced by their allele frequencies per breed (Table 1, Figure 1): *PPARG*, which influences the amount of 22:5 n-3, 20:5 n-3 and 22:6 n-3 in muscle [20], appeared near n-3 and ATT ratio -calculated as (20:3 n-6 + 20:5 n-3)/20:4 n-6- factor patterns, as well as almost equidistant to the three breeds with higher allele frequencies for the A allele –AA, CHA and SD-; *CFL1*, *PLTP* and *MYOZ1*, which were associated with n-6 to n-3 ratio [19], correlated with 18:2 n-6/18:3 n-3 and specially with n-6/n-3 ratio in Factor 1; *IGF2R*, previously linked to an increase in flavour [19], shared Factor 2 pattern with flavour and was placed almost equidistant to the three breeds with higher allele frequencies for the G allele –HIG, CHA and SD-; *CAST* was associated with an increase in FS and appeared closely related to the two breeds with higher allele frequencies for the A allele –AA and HIG, sharing also Factor 1 pattern with FS; *LPL*, associated with the increase of several neutral n-6 [20], plotted in the same Factor 2 pattern than n-6 content, but closer to and almost equidistant from the three breeds with higher allele frequencies for the T allele –HOL, AVI and CAS; and, as expected, *GDF8* SNP was placed near the trait FS [20].

Finally, there were no relationships between the two SNPs in the *MMP1* gene and the SNP in *CRI1* neither with their main trait associations –CLA, 22:6 n-3 and 22:4 n-6, respectively [19], nor with breed allele frequencies, which may be caused by the other trait associations of this SNPs with lower or unknown effects [19].

Conclusions

The wide range of traits and breeds studied, along with the genotypic information on polymorphisms previously associated with different lipid traits, provide a broad characterization of the phenotypic and genotypic background underlying variations in FA composition and sensory parameters between breeds, which allows giving a better response to the variety of consumers' preferences. Also, the development and implementation of low-density SNP panels with predictive value for economically important traits, such as those summarized here, may be used to improve production efficiency and meat quality in the beef industry as a molecular signature of GTTdelGCA CCAA for *CAST* (g.2959G < A), *CFL1* (ss77831721), *CRI1* (ss77832128), *GDF8* (ss77831865), *IGF2R* (ss778 31885), *LPL* (ss65478732), *MMP1* (ss77831916, ss77831 924), *MYOZ1* (ss77831945), *PLTP* (ss77832104), and *PPARG* (ss62850198), respectively, which would correspond to the "most favourable" haplotype.

Competing interests

The authors declare that they have no competing interests.

Authors' contributions

NS carried out the molecular genetic studies, performed the statistical analysis and drafted the manuscript. SD, HL, GN, CS, AV and JLW conceived the study, and participated in its design and coordination. SD also helped to draft the manuscript. All authors read and approved the final manuscript.

Acknowledgements

This work was supported by an EC grant QLK5 – CT2000-0147.
GeMQual Consortium: Albertí P Centro de Investigación y Tecnología Agroalimentaria, Gobierno de Aragón, 50080, Zaragoza, Spain; Amarger V. Delourme D. Levéziel H. INRA, UMR 1061, 87000 Limoges, France and Université de Limoges, UMR 1061, 87000 Limoges; Boitard S. Mangin B. INRA Chemin de Borde-Rouge-Auzeville, BP 52627, 31326 Castanet-Tolosan cedex, France;Cañón J. Checa ML. Dunner S. García D. Miranda ME. Pérez R. Dpto de Producción Animal, Facultad de Veterinaria, 28040 Madrid, Spain; Christensen M. Ertbjerg P. Department of Food Science, University of Copenhagen, 1958 Frederiksberg C., Denmark; Crisá A. Marchitelli C. Valentini A. Dipartimento di

Produzioni Animali, Università della Tuscia, via De Lellis, 01100 Viterbo, Italy; Failla S. Gigli S. CRA, Istituto Sperimentale per la Zootecnia, 00016 Monterotondo, Italy; Hocquette JF. INRA, UR1213, Unité de Recherches sur les Herbivores, Centre de Clermont-Ferrand./Theix F-63122, France; Nute G., Richardson I. Division of Farm Animal Science, University of Bristol, BS40 5DU, United Kingdom; Olleta JL., Panea B., Sañudo C. Dept de Producción Animal y Ciencia de los Alimentos, Universidad de Zaragoza, 50013, Zaragoza, Spain; Razzaq N. Roslin Institute, Roslin, Midlothian, Scotland.EH25 9PS, UK; Renand G. INRA, UR337, Station de Génétique Quantitative et Appliquée, 78352 Jouy-en-Josas cedex, France; Williams. JL. Parco Tecnologico Padano, Via Einstein, Polo Universitario, 26900 Lodi, Italy.

Author details

[1]Departamento de Producción Animal, Facultad de Veterinaria, Universidad Complutense de Madrid, Madrid, Spain. [2]INRA, UMR 1061, F-87000 Limoges, France. [3]Université de Limoges, UMR 1061, F-87000 Limoges, France. [4]Division of Farm Animal Science, University of Bristol, Bristol BS40 5DU, UK. [5]Departimento de Producción Animal y Ciencia de los Alimentos, Universidad de Zaragoza, 50013 Zaragoza, Spain. [6]Dipartimento di Produzioni Animali, Università della Tuscia, via De Lellis, 01100 Viterbo, Italy. [7]Parco Tecnologico Padano, Via Einstein, Polo Universitario, 26900 Lodi, Italy.

References

1. Givens DI, Gibbs RA: **Current intakes of EPA and DHA in European populations and the potential of animal-derived foods to increase them.** *P Nutr Soc* 2008, **67**:273–280.
2. McAfee AJ, McSorley EM, Cuskelly GJ, Moss BW, Wallace JM, Bonham MP, Fearon AM: **Red meat consumption: an overview of the risks and benefits.** *Meat Sci* 2010, **84**:1–13.
3. Cross AJ, Leitzmann MF, Gail MH, Hollenbeck AR, Schatzkin A, Sinha R: **A prospective study of red and processed meat intake in relation to cancer risk.** *PLos Med* 1973, **2007**:4.
4. World Cancer Research Fund/American Institute for Cancer Research: *Food, nutrition and the prevention of cancer: A global perspective.* Washington DC: American Institute for Cancer Research; 2007.
5. Kontogianni MD, Panagiotakos DB, Pitsavos C, Chrysohoou C, Stefanadis C: **Relationship between meat intake and the development of acute coronary syndromes: The CARDIO2000 case–control study.** *Eur J Clin Nutr* 2008, **62**:171–177.
6. Beauchesne-Rondeau E, Gascon A, Bergeron J, Jacques H: **Plasma lipids and lipoproteins in hypercholesterolaemic men fed a lipid-lowering diet containing lean beef, lean fish, or poultry.** *Am J Clin Nutr* 2003, **77**:587–593.
7. Bradlee ML, Singer MR, Moore LL: **Lean red meat consumption and lipid profiles in adolescent girls.** *J Hum Nutr Diet* 2013, **2**:292–300.
8. Hodgson J, Burke V, Beilin LJ, Puddey IB: **Partial substitution of carbohydrate intake with protein intake from lean red meat lowers blood pressure in hypertensive persons.** *Am J Clin Nutr* 2006, **83**:780–787.
9. Lee SH, Park EW, Cho YM, Kim SK, Lee JH, Jeon JT, Lee CS, Im SK, Oh SJ, Thompson JM, Yoon D: **Identification of differentially expressed genes related to intramuscular fat development in the early and late fattening stages of hanwoo steers.** *J Biochem Mol Biol* 2007, **40**:757–764.
10. Bernard C, Cassar-Malek I, Le Cunff M, Dubroeucq H, Renand G, Hocquette JF: **New indicators of beef sensory quality revealed by expression of specific genes.** *J Agr Food Chem* 2007, **55**:5229–5237.
11. Hocquette JF, Botreau R, Picard B, Jacquet A, Pethick DW, Scollan ND: **Opportunities for predicting and manipulating beef quality.** *Meat Sci* 2012, **92**:197–209.
12. Simm G, Lambe N, Bünger L, Navajas E, Roehe R: **Use of meat quality information in breeding programmes.** In *Improving the sensory and nutritional quality of fresh meat.* Edited by Kerry JP, Ledward D. UK: Woodhead Publishing Ltd; 2009:680.
13. Luan T, Woolliams JA, Lien S, Kent M, Svendsen M, Meuwissen TH: **The accuracy of Genomic Selection in Norwegian red cattle assessed by cross-validation.** *Genetics* 2009, **183**:1119–1126.
14. Casas E, Shackelford SD, Keele JW, Stone RT, Kappes SM, Koohmaraie M: **QTL affecting growth and carcass composition of cattle segregating alternate forms of myostatin.** *J Anim Sci* 2000, **78**:560–569.
15. Casas E, Shackelford SD, Keele JW, Koohmaraie M, Smith TPL, Stone RT: **Detection of quantitative trait loci for growth and carcass composition in cattle.** *J Anim Sci* 2003, **81**:2976–2983.
16. Van Eenennaam AL, Li J, Thallman RM, Quaas RL, Dikeman ME, Gill CA, Franke DE, Thomas MG: **Validation of commercial DNA tests for quantitative beef quality traits.** *J Anim Sci* 2007, **85**:891–900.
17. Ibeagha-Awemu EM, Kgwatalala P, Zhao X: **A critical analysis of production-associated DNA polymorphisms in the genes of cattle, goat, sheep, and pig.** *Mamm Genome* 2008, **19**:591–617.
18. Shingfield KJ, Bonnet M, Scollan ND: **Recent developments in altering the fatty acid composition of ruminant-derived foods.** *Animal* 2013, **7**:132–162.
19. Dunner S, Sevane N, García D, Levéziel H, Williams JL, Mangin B, Valentini A, GeMQual Consortium: **Genes involved in muscle lipid composition in 15 European Bos taurus breeds.** *Anim Genet* 2013, **44**:493–501.
20. Sevane N, Armstrong E, Cortés O, Wiener P, Pong Wong R, Dunner S, GeMQual Consortium: **Association of bovine meat quality traits with genes included in the PPARG and PPARGC1A networks.** *Meat Sci* 2013, **94**:328–335.
21. Sevane N, Cañón J, Dunner S: **GemQual Consortium: Muscle lipid composition in bulls from fifteen European breeds.** *Livestock Sci* 2014, **160**:1–11.
22. Albertí P, Panea B, Sañudo C, Olleta JL, Ripoll G, Ertbjerg P, Christensen M, Gigli S, Failla S, Concetti S, Hocquette JF, Jailler R, Rudel S, Renand G, Nute GR, Richardson RI, Williams JL: **Live weight, body size and carcass characteristics of young bulls of fifteen European breeds.** *Livest Sci* 2008, **114**:19–30.
23. Christensen M, Ertbjerg P, Failla S, Sañudo C, Richardson RI, Nute GR, Olleta JL, Panea B, Albertí P, Juárez M, Hocquette JF, Williams JL: **Relationship between collagen characteristics, lipid content and raw and cooked texture of meat from young bulls of fifteen European breeds.** *Meat Sci* 2011, **87**:61–65.
24. Folch J, Lees M, Stanley GHS: **A simple method for the isolation and purification of lipids from animal tissues.** *J Biol Chem* 1957, **226**:497–509.
25. Scollan ND, Choi NJ, Kurt E, Fisher AV, Enser M, Wood JD: **Manipulating the fatty acid composition of muscle and adipose tissue in beef cattle.** *Br J Nutr* 2001, **85**:115–124.
26. Ulbricht TLV, Southgate DAT: **Coronary heart disease: seven dietary factors.** *Lancet* 1991, **338**:985–992.
27. Wood JD, Nute GR, Fursey GAJ, Cuthbertson A: **The effect of cooking conditions on the eating quality of pork.** *Meat Sci* 1995, **40**:127–135.
28. SAS Institute Inc: **Statistical Analysis with SAS/STAT® Software V9.1.3.** Cary, NC, USA: SAS Institute Inc; 2009.
29. Scollan ND, Hocquette JF, Nuernberg K, Dannenberger D, Richardson RI, Maloney A: **Innovations in beef production systems that enhance the nutritional and health value of beef lipids and their relationship with meat quality.** *Meat Sci* 2006, **74**:17–33.
30. Dinh TTN: **Lipid and Cholesterol Composition of the Longissimus Muscle from Angus, Brahman, and Romosinuano.** In *Master of Science Thesis in Animal and Food Science.* Graduate Faculty of Texas Tech University: Dinh TTN; 2007.
31. Hoehne A, Nuernberg G, Kuehn C, Nuernberg K: **Relationships between intramuscular fat content, selected carcass traits, and fatty acid profile in bulls using a F2-population.** *Meat Sci* 2012, **90**:629–635.
32. Wood JD, Enser M, Fisher AV, Nute GR, Sheard PR, Richardson RI, Hughes SI, Whittington FM: **Fat deposition, fatty acid composition and meat quality: A review.** *Meat Sci* 2008, **78**:343–358.
33. Enser M, Scollan N, Choi N, Kurt E, Hallett K, Wood J: **Effect of dietary lipid on the content of CLA in beef cattle.** *Anim Sci* 1999, **69**:143–146.
34. Lawless F, Stanton C, L'Escop P, Devery R, Dillon P, Murphy JJ: **Influence of breed on bovine milk cis-9, trans-11-conjugated linoleic acid.** *Livest Prod Sci* 1999, **62**:43–49.
35. Wood JD, Richardson RI, Nute GR, Fisher AV, Campo MM, Kasapidou E, Sheard PR, Enser M: **Effects of fatty acids on meat quality: a review.** *Meat Sci* 2004, **66**:21–32.
36. Warren HE, Scollan ND, Nute GR, Hughes SI, Wood JD, Richardson RI: **Effects of breed and a concentrate or grass silage diet on beef quality in cattle of 3 ages. II: Meat stability and flavour.** *Meat Sci* 2008, **78**:270–278.

37. Costa P, Lemos JP, Lopes PA, Alfaia CM, Costa AS, Bessa RJ, Prates JA: **Effect of low- and high-forage diets on meat quality and fatty acid composition of Alentejana and Barrosã beef breeds.** *Animal* 2012, **6**:1187–1197.
38. Harries JM, Rhodes DN, Chrystall BB: **Meat texture: I. Subjective assessment of the texture of cooked beef.** *J Texture Stud* 1972, **3**:101–114.
39. Dransfield E, Francombe MA, Whelehan OP: **Relationships between sensory attributes in cooked meat.** *J Texture Stud* 1984, **15**:33–48.
40. Dransfield E, Nute G, Roberts T, Boccard R, Touraille C, Buchter L, Casteels M, Cosentino E, Hood D, Joseph R: **Beef quality assessed at European research centres.** *Meat Sci* 1984, **10**:1–20.
41. Juarez M, Aldai N, Lopez-Campos O, Dugan MER, Uttaro B, Aalhus JL: **Beef Texture and Juiciness.** In *Handbook of Meat and Meat Processing.* Edited by Hui YH. Boca Raton: CRC press; 2011:177–206.
42. Melton SL, Amiri M, Davis GW, Backus WR: **Flavor and chemical characteristics of ground beef from grass-, forage-, grain- and grain-finished steers.** *J Anim Sci* 1982, **55**:77–87.

Identification and quantitative mRNA analysis of a novel splice variant of *GPIHBP1* in dairy cattle

Jie Yang, Xuan Liu, Qin Zhang and Li Jiang*

Abstract

Background: Identification of functional genes affecting milk production traits is very crucial for improving breeding efficiency in dairy cattle. Many potential candidate genes have been identified through our previous genome wide association study (GWAS). Of these, *GPIHBP1* is an important novel candidate gene for milk production traits. However, the mRNA structure of the bovine *GPIHBP1* gene is not fully determined up to now.

Results: In this study, we identified a novel alternatively splice transcript variant (X5) which leads to a 31 bp insertion in exon 3 and also confirmed the other four existed transcripts (X1, X2, X3 and X4) of the bovine *GPIHBP1* gene. We showed that transcript X5 with a 31 bp insertion and transcript X1 with an 8 bp deletion might have tremendous effect on the protein function and structure of *GPIHBP1*, respectively. With semi-quantitative PCR and quantitative real-time RT-PCR, we found that the mRNA expression of *GPIHBP1*, *GPIHBP1*-X1 and *GPIHBP1*-X5 in mammary gland of lactating cows were much higher than that in other tissues.

Conclusions: Our study reports a novel alternative splicing of *GPIHBP1* in bovine for the first time and provide useful information for the further functional analyses of *GPIHBP1* in dairy cattle.

Keywords: Alternative splice variant, Cattle, Expression pattern, *GPIHBP1*

Background

Our previous genome-wide association study (GWAS) in Chinese Holstein population revealed Glycosylphospha tidylinositol-anchored HDL binding protein1 (*GPIHBP1*) is a potential candidate functional gene for milk production traits [1]. A SNP which is located 1,295 bp upstream from the translation initiation site of *GPIHBP1* gene showed strong association with milk yield trait, protein yield and fat percentage with *P* values 1.02E-10, 1.55E-07 and 6.30E-20, respectively. To confirm the association between the *GPIHBP1* gene and milk production traits, we selected a SNP within 5′UTR of *GPIHBP1* in another Chinese Holstein population for further association study. This SNP also showed very significant association with milk yield trait, fat percent trait and protein yield trait (unpublished data). Therefore, *GPIHBP1* was considered as a novel promising candidate functional gene in dairy cattle.

The GPIHBP1 protein is a glycosylphosphatidylinositol (GPI)-anchored protein of the lymphocyte antigen 6 family. It contains an N-terminal signal peptide, an acidic domain, a lymphocyte antigen 6 (Ly6) domain, and a hydrophobic carboxyl-terminal motif [2]. In the endoplasmic reticulum, the signal peptide is removed and the carboxyl-terminal hydrophobic sequence is replaced by a GPI-anchor [3]. Thus, the acidic domain and Ly6 motif are of great importance for mature GPIHBP1. Recent studies showed that they play an important role in the capacity of GPIHBP1 to bind lipoprotein lipase (LPL) [4]. It has been demonstrated that some mutations, such as C65Y, C89F and Q115P, in the most highly conserved portion of the Ly6 domain lead to the abolishment of GPIHBP1 to bind LPL [5-7], and a mutation in the C-terminal hydrophobic domain, G175R, markedly reduces the ability of GPIHBP1 to reach the cell membrane and bind LPL [7].

GPIHBP1 is responsible for actively transporting LPL across endothelial cells [8]. Once inside capillaries, LPL hydrolyzes the triglycerides in plasma lipoproteins and provides the lipids from blood for production of milk lipids [9,10]. Thus, *GPIHBP1* plays a critical role in the lipolytic processing of triglyceride-rich lipoproteins. Rios

* Correspondence: jiangli198374@163.com
National Engineering Laboratory for Animal Breeding; Key Laboratory of Animal Genetics, Breeding and Reproduction, Ministry of Agriculture of China; College of Animal Science and Technology, China Agricultural University, Beijing 100193, China

et al. [11] found that in human a deletion of 17.5 Kb containing the entire *GPIHBP1* gene resulted in extremely high plasma triglyceride and cholesterol level. Beigneux *et al.* [12] reported that glycosylation of Asn-76 within the Ly6 domain of the mouse GPIHBP1 was critical for its appearance on the cell surface. Beigneux *et al.* [12] showed the *GPIHBP1*-knockout (*GPIHBP1-/-*) mice displayed severe hypertriglyceridemia, with a plasma triglyceride level of 1,000-6,000 mg/dL at 7-10 week of age. It was reported that *GPIHBP1* was highly expressed in heart and adipose tissue in mice [12,13] and its tissue expression pattern was similar to that of *LPL* [13]. Recent studies showed that *GPIHBP1* was the key element for transport and localization of *LPL* [8,14,15] and might serve as a platform for lipolysis on endothelial cells [3,16].

Up to now, the genomic organization of *GPIHBP1* remains undetermined yet. The mRNA structure of the bovine *GPIHBP1* gene has been keeping on changing in the NCBI database in the most recent years. In the present study, we investigated a new splice variant of bovine *GPIHBP1*. In order to layout the groundwork for its biological function validation in dairy cattle, we also performed quantitative analysis of the mRNA expression patterns of *GPIHBP1* and its novel splice variant in different tissues. We aimed to establish which splice variant is predominantly expressed in bovine tissues.

Methods

Animals and tissue sample collection

Three Chinese Holstein cows which were in the same period of lactation were selected from Beijing Sanyuan Dairy Farm Center. All of them were fed in a consistent environmental condition. Eight tissues samples (heart, liver, lung, kidney, mammary gland, ovary, uterus and muscle) from each cow were collected within 30 min after slaughter and stored at liquid nitrogen. The whole procedure for collection of the tissue samples of all animals was carried out in strict accordance with the protocol approved by the Animal Welfare Committee of China Agricultural University (Permit number: DK996).

RNA extraction and reverse transcription

The total RNA was extracted from the eight tissues of the three cows by using Trizol reagent (Invitrogen, CA, USA). The quantity and quality of RNA were measured via an ND-2000 spectrophotometer (Thermo, USA). Reverse transcription (RT) was carried out in a solution of 20 μL, containing 12 μL Mix (0.5 μL Primer (50 μmol/L) oligio (dt), 0.5 μL Random primer, 1 μL dNTPs (10 mmol/L), 5 μg total RNA and ddH_2O up to 12 μL), 4 μL 5 × First-Strand buffer, 2 μL 0.1 mol/L dTT, 1 μL RNaseout (40U/μL), and 1 μL SuperScrip III RT (200U/μL) (Life, USA). The Mix was heated at 65°C for 5 min and then incubated on ice for at least 1 min. Tubes containing all contents were incubated at 25°C for 5 min, 50°C for 60 min and 70°C for 15 min. To ensure the quality of the first strand cDNA, 1 μL of cDNA was used in a PCR reaction to amplify the glyceraldehyde phosphate dehydrogenase (*GAPDH*) gene.

Polymerase chain reaction and clone sequencing

PCR reactions were performed to amplify the coding regions of *GPIHBP1*. The primers (Primer1, Primer2 and Primer3, Additional file 1: Table S1) were designed using the Primer 3 web-tool (http://frodo.wi.mit.edu/primer3/) and the Oligo 6.0 software. For each amplicon, 1 μL of cDNA (1,000 ng/μL), 2.0 μL of 10× PCR buffer, 250 μmol/L of each dNTP, 0.5 units of HotstarTaq polymerase (Takara Biotechnology, Tokyo, Japan), and 0.5 μmol/L of primer (Life Technologies) were used in a total 20 μL reaction. The reaction was denatured for 10 min at 95°C, then 35 cycles of 94°C for 30 s, special annealing temperature for 30 s and 72°C for 30 s, and a final extension of 72°C for ten min. The products were electrophoresed on 2% agarose gels and stained with ethidium bromide.

The purified double-stranded DNA (Omega, USA) was cloned in pMD18-T (Takara Biotechnology, Tokyo, Japan). The products of the ligation reactions were transformed into competent cells. Twenty colonies per sample were selected randomly for sequencing. With the DNAMAN 7.0 software, we performed multiple sequences alignment analysis.

Predicted structures of the GPIHBP1 protein

The T coffee website tool (http://tcoffee.vital-it.ch/apps/tcoffee/do:regular) was used to align amino acid sequences of the bovine and human GPIHBP1 proteins. We predicted the open reading frame of the bovine *GPIHBP1* transcript X5 using ORF Finder (http://www.ncbi.nlm.nih.gov/projects/gorf/). Secondary structures of the GPIHBP1 proteins were predicted using the PSIPRED v3.3 website tool (http://bioinf.cs.ucl.ac.uk/psipred/). SignalP 4.1 (http://www.cbs.dtu.dk/services/SignalP/) [17] was used to predict the presence and location of the signal peptide of GPIHBP1. Big-PI Predictor (http://mendel.imp.ac.at/gpi/gpi_server.html) [18] was utilized to predict GPI anchor sites in protein sequence. The human CD59 (membrane-bound glycoprotein) gene which also has the UPAR/Ly6 domain [19], was used as the reference for predicted bovine GPIHBP1 tertiary structures using the SWISS MODEL method (http://swissmodel.expasy.org/) [20]. The reported human CD59 (membrane-bound glycoprotein) served as the reference for predicted bovine GPIHBP1 tertiary structures using SWISS MODEL method (http://swissmodel.expasy.org/) [20].

Real time RT-PCR

Real-time PCR (RT-PCR) was performed on the eight tissues of three cows. TaqMan Real-time PCR assays were performed using 7500Fast (Life, USA). The PCR amplification mix consisted of 2 μL 10× PCR Buffer, 1.2 μL Mg^{2+} (50 mmol/L), 0.5 μL dNTPs (10 mmol/L), 0.5 μL of each primer (10 μmol/L, Additional file 1: Table S2), 0.2 μL Taqman probe (GPI-Probe, X1-Probe and X5-Probe, 10 μmol/L, Additional file 1: Table S2), 1 μL cDNA, 0.2 μL Taq polymerase and 13.9 μL ddH_2O in a final volume of 20 μL. The reaction was performed with the conditions as follows: an initial 2 min hold at 95°C, 50 cycles of 95°C for 10 s, 60°C for 30 s. The assays were carried out in triplicate and the average C_T values were obtained to calculate gene expression level. In addition, parallel assays using the same cDNA were carried out using the primers (Additional file 1: Table S2) and probe (*GPADH*-Probe, Additional file 1: Table S2) to the housekeeping gene *GAPDH*. The relative mRNA expression levels of *GPIHBP1* and two alternative splice variants were normalized to the *GAPDH* gene by the $2^{-\Delta\Delta CT}$ method [21].

Results

Identification of a novel mRNA spliced variant of *GPIHBP1*

Three primers (Additional file 1: Table S1) were used to amplify the coding region of *GPIHBP1* in samples of mammary gland. After PCR amplification with primer 1 and primer 3, we observed three (Figure 1A) and one (Figure 1C) PCR bands, respectively. These 4 bands correspond to the expected fragment size (Additional file 1: Table S1) derived from the bovine *GPIHBP1* sequence in the NCBI database. With primer 2, two bands with fragment lengths of 352 bp and 344 bp, respectively, would be expected according to bovine *GPIHBP1* sequence. However, the PCR products showed the two bands were almost merged into one band (Figure 1B), since there is only 8 bp difference in size between the two fragments. Interestingly, we observed an additional band of 383 bp which was present in all samples (Figure 1B, only two samples are shown). To verify the results of primer 2, we purified the PCR products and cloned them in Pmd18-T (Takara Biotechnology, Tokyo, Japan). We then randomly selected twenty colonies to sequence. It turned out that 17 of them were of 352 bp length, one of them showed a deletion of 8 bp (5′-GGCCGCAG-3′, Chr14:2550837-Chr14:2550830) in exon 3 and was of length of 344 bp, and two of them showed an insertion of 31 bp (5′-TGGAGGTTTACAGGTGTCCCTGCGCGGCCAG-3′, Chr14:2551602-Chr14:2551572) in exon 3 and were of length of 383 bp. Thus, both the two expected fragments and the novel fragment were confirmed.

Currently, four transcript variants (X1, X2, X3 and X4) of *GPIHBP1* are presented in NCBI. Transcript X1, which leads to colonies with a deletion of 8 bp nucleotides, was reported very recently. However, the colonies with an insertion of 31 bp nucleotides suggest that there may exist a novel transcript variant in the bovine *GPIHBP1* gene. This novel transcript variant was named transcript variant X5

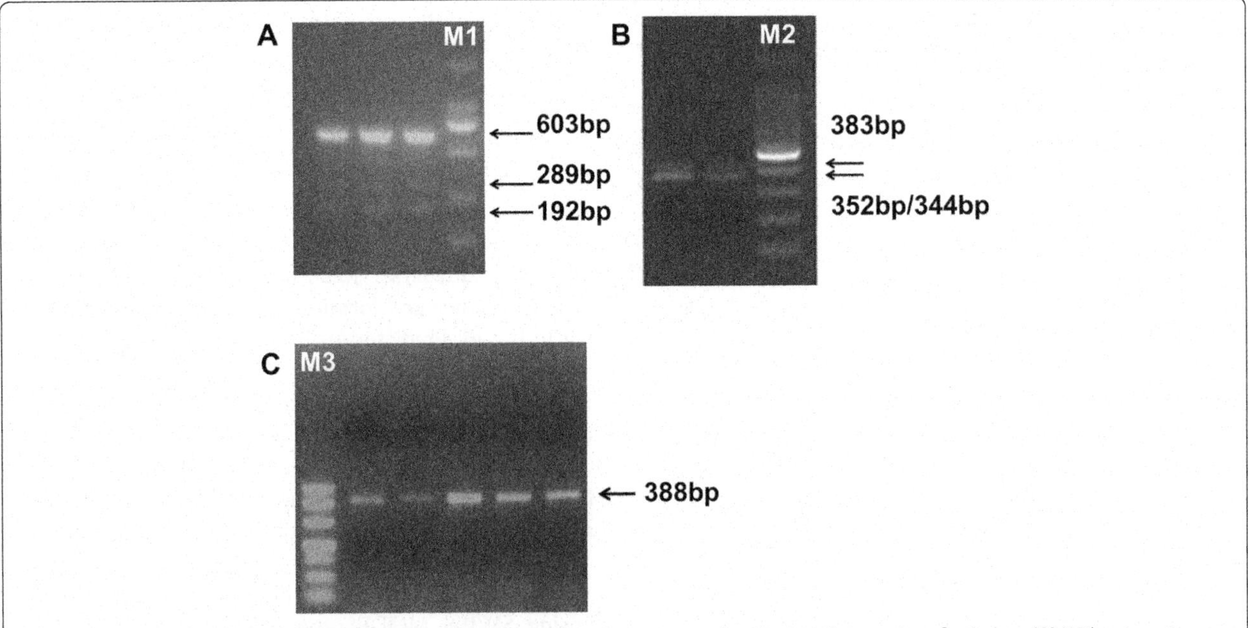

Figure 1 PCR analysis of the *GPIHBP1* coding region in samples of mammary gland using three pairs of primers. (A) With primer 1, transcript X2 (603 bp), transcript X3 (289 bp) and transcript X4 (192 bp) were observed as expected (according to the mRNA sequence of the bovine *GPIHBP1* gene in NCBI database). M1: DL2000, 2,000 bp, 1,000 bp, 750 bp, 500 bp, 250 bp, 100 bp. **(B)** With primer 2, in addition to the expected PCR band (352 bp and 344 bp), a band of 383 bp was also observed in all samples. M2: 100 bp DNA Ladder from 1,500–100 bp. **(C)** With primer 3, only the expected band (388 bp) was observed in all samples. M3: DL500, 500 bp, 400 bp, 300 bp, 200 bp, 150 bp, 100 bp and 50 bp.

[GenBank accession number: KJ502292]. To further confirm that the 8 bp deletion and the 31 bp insertion were not a cloning artifact, we designed two pairs of primers (Additional file 1: Table S3) specifically for the 8 bp deletion and the 31 bp insertion sequences, respectively, and performed PCR amplification. As a result, we obtained a fragment of 207 bp for transcript X1 and a fragment of 101 bp for transcript X5. The existence of transcript variants X1 and X5 were thus confirmed again. The structures of different splice forms of *GPIHBP1* were shown in Figure 2. And there were notable differences in 5′ untranslated region (UTR) of different *GPIHBP1* transcripts. Out of five different splice forms, three (X2,X3 and X4) have the same translation initiation site.

Characteristics of the *GPIHBP1* splice variants

Transcript variants X2, X3 and X4 have the same open reading frame (ORF) and encode a 171-amino acid protein that was named bovine GPIHBP1 P2. In contrast, the transcript variant X1 contains a different ORF and encodes a 142-amino acid protein, which named bovine GPIHBP1 P1. However, the ORF of transcript X5 was still not known clearly up to now. The ORF Finder software (http://www.ncbi.nlm.nih.gov/gorf/gorf.html) was used to predict all possible ORF of transcript variants X5. As a result, five potential ORF were predicted which had initiation codon and termination codon. The amino acid sequences corresponding to the five ORF were obtained using the DNAMAN 7.0 software and named bovine GPIHBP1 P5.1, P5.2, P5.3, P5.4, and P5.5, respectively (Additional file 1: Table S4).

Predicted structures of the GPIHBP1 protein

The predicted secondary structures of the bovine GPIHBP1 amino acid sequences were compared with that of the human GPIHBP1 protein. The α-Helix and β-sheet structures of bovine GPIHBP1 P1 and P2 were similar to the human GPIHBP1 protein secondary structure (Figure 3A). By using SignalP 4.1 [17], we found that only human GPIHBP1, bovine GPIHBP1 P2 and bovine GPIHBP1 P5.5 had N-terminal signal peptide which contained a predicted helical structure (Figure 3A, B). The GPI-modification sites of the human GPIHBP1 protein, bovine GPIHBP1 P1 and bovine GPIHBP1 P2 were also predicted ($P < 0.01$) using Big-PI Predictor [18,22] (Additional file 1: Table S5). We found that bovine GPIHBP1 P1 and P2 and the human GPIHBP1 protein had similar position for alternative GPI-modification site. The predicted tertiary structures of the bovine GPIHBP1 P1, bovine GPIHBP1 P2 and human GPIHBP1 sequences are shown in Figure 4. It can be seen that these tertiary structures were similar to the reported UPAR-LY6 domain of the human CD59 protein [19]. Their modeling ranges of residues were 62-138aa, 28-104aa and 58-133aa, respectively, which contained four or five β-sheet structures. However, due to the low alignment quality between target and specified template, the tertiary structures of the predicted bovine GPIHBP1 P5.1, P5.2, P5.3, P5.4, and P5.5 could not be constructed.

Tissue mRNA expression pattern of three splice variants of *GPIHBP1*

It can be seen from Figure 2 that the difference in 5′ untranslated region (UTR) of *GPIHBP1* transcripts X2, X3 and X4 did not affect the structure of protein. In contrast, the 8 bp deletion of transcript X1 and 31 bp insertion of transcript X5 had tremendous effect on the structure and function of protein. Thus, semi-quantitative PCR and TaqMan Real-time PCR were employed with specific primers for the 8 bp deletion of transcript X1, 31 bp insertion of

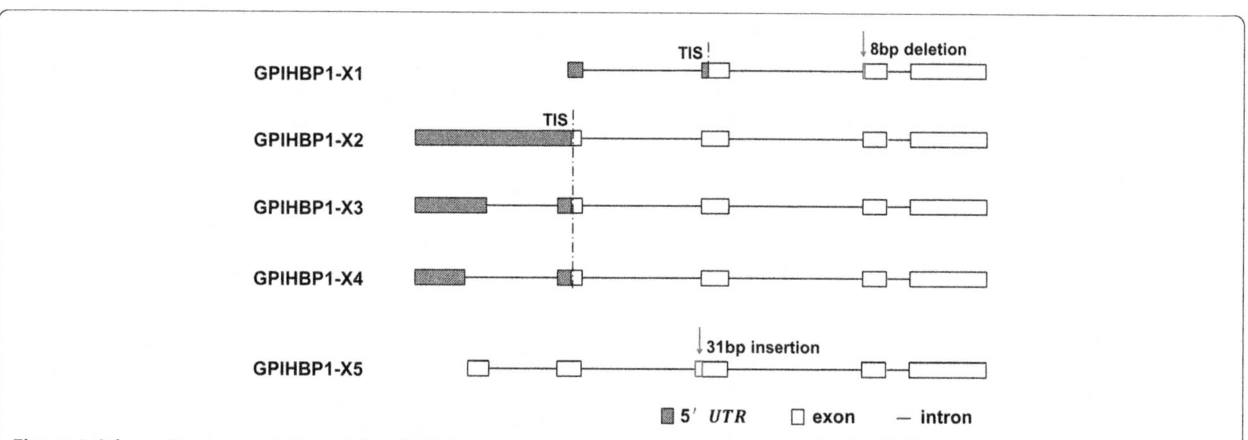

Figure 2 Schematic representation of the *GPIHBP1* alternative splice variants. Transcripts X1 (XM_002692563.2), X2 (XM_005215283.1), X3 (XM_005215284.1) and X4 (XM_005215285.1) were obtained from NCBI. Transcript X5 was observed by RNA-seq (unpublished data). Transcript X1 contained a 8 bp deletion in exon 3. Transcript X5 contained a 31 bp insertion in exon 3. Although transcript X2, X3 and X4 had different 5′ untranslated region, they had the same translation initiation site (TIS).

```
A Human-GPIHBP1    MKA--LGAVLLALLLCGRPGRGQTQQEEEEEDEDHGPDDYDEEDEDEVEE 48
  Bovine-P1        MTTRTLGAKATTM-------------------------------------R 14
  Bovine-P2        MKA--LAAVLLALLWCRLQGRGRAQ-EDEDDDPDAGREGYDDEDE----E 43

  Human-GPIHBP1    EETNRLPGGRSRVLLRCYTCKSLPRDERCNLTQNCSHGQTCTTLIAHGNT 98
  Bovine-P1        TRRKRRPACLRAAGTQCYTCQSLHKGESCEQVQSCVLPGTCKAIVSSWNT 64
  Bovine-P2        EEEAGVPAGSRDSGPQCYTCQSLHKGESCEQVQSCVLPGTCKAIVSSWNT 93

  Human-GPIHBP1    ESGLLTTHSTWCTDSCQPITKTVEGTQVTMTCCQSSLCNVPPWQSSRVQD 148
  Bovine-P1        ESGPQTTYSGWCADTCQAISRTVDGSLTTISCCQSSLCNTPPWQ-----D 109
  Bovine-P2        ESGPQTTYSGWCADTCQAISRTVDGSLTTISCCQSSLCNTPPWQ-----D 138

  Human-GPIHBP1    PTGKGAGGPRGSSETVGAALLLNLLAGLGAMGARRP 184
  Bovine-P1        PQGRGAGGPQGSPATVAATVLLSLLASLQEMG---L 142
  Bovine-P2        PQGRGAGGPQGSPATVAATVLLSLLASLQEMG---L 171

B Bovine-p5.1      MVCRHMSSHQQDGGRISDDHILLPVQPVQHPTLARPPGEGGRRPPGEPCDCGRHR
                   PAQPPGQPSGDGALSGALPPRAQHGLHPGWPALCPLSPVPSPRPWRINWKHL

  Bovine-p5.2      MVWGRGQETADRVLASQGAGHAGREEGAPHSEPHLLKAGQEAEQDGGGHSRRAP
                   GAACPPPLGVLPGWGVAQAGLAAGYGRQRSVHRPADGLTCVCTPSRVGGLGT

  Bovine-p5.3      MCLHTIPSRWSGDLTQCSRRRRWLCMSRGARSSAPARSSPPCAGTGTCSTAALSP
                   CCPQARRPPLPPRPHRRSLRAQRPGRHPRPPAPALSLAAQGHL

  Bovine-p5.4      MTTRTLGAKATTMRTRRKRRPACLRAAGTQGRSATRASPCTRGRAASRCRAACS
                   PGHAKPSSPPGTLSQVPRPPTRDGVQTHVKPSAGRWTD

  Bovine-p5.5      MKALAAVLLALLWCRLQVEVYRCPCAARERAGAGGRG
```

Figure 3 Predicted secondary structures of bovine and human GPIHBP1. Residues involved in the N-signal peptide formation are indicated by underline, α-helices are in red, β-sheets are in green, GPI-modification site is in blue, alternative GPI-modification site is in purple. **(A)** Amino acid sequence alignments of bovine and human GPIHBP1. The predicted secondary structures for bovine-p1 (XP_002692609.2) and bovine-p2 (XP_005215340.1, XP_005215341.1 and XP_005215342.1) are similar for several regions with that of human GPIHBP1. **(B)** The predicted secondary structures for bovine-p5.1, bovine-p5.2, bovine-p5.3, bovine-p5.4 and bovine-p5.5 of transcript X5.

transcript X5 and overall GPIHBP1 transcripts in eight tissues of three cows.

TaqMan Real-time PCR analysis was conducted to further identify the tissue mRNA expression pattern of bovine *GPIHBP1*. After normalization with the corresponding mRNA expression level of the housekeeping gene *GAPDH*, analysis of variance (ANOVA) and multiple comparisons were conducted with R software. We found that mRNA expression level of *GPIHBP1*, *GPIHBP1*-X1 and *GPIHBP1*-X5 all were significantly different among eight tissues ($P < 0.05$), in which the mRNA expression levels were significantly higher in mammary gland than other tissues ($P < 0.05$). And all of *GPIHBP1*-X1, *GPIHBP1*-X5 and overall *GPIHBP1* had much lower expression level in liver, kidney and muscle (Figure 5).

Discussion

Tissue-specific mRNA expression patterns are important for revealing functional candidate genes associated with milk production traits [23]. The specifically high expression

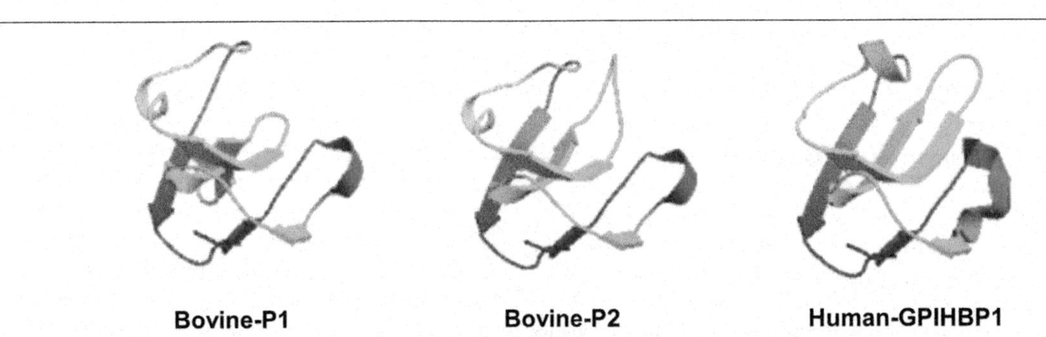

Figure 4 Predicted tertiary structures of bovine and human GPIHBP1. The reported human CD59 was used as the reference to obtain predicted GPIHBP1 tertiary structures by the SWISS MODEL method. The rainbow color code describes the tertiary structures from the N-termini (blue) to C-termini (red) for GPIHBP1 UPAR/Ly6 domains. Arrows indicate the directions for β-sheets.

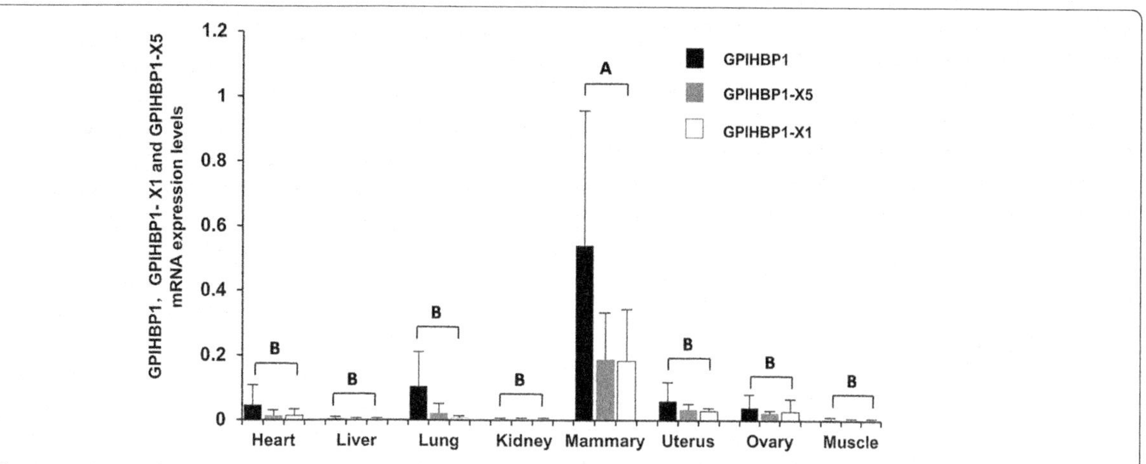

Figure 5 The mRNA expression patterns of *GPIHBP1*, *GPIHBP1*-X1 and *GPIHBP1*-X5 revealed by RT-PCR. The histograms represent the mRNA expression level of *GPIHBP1*, *GPIHBP1*-X1 and *GPIHBP1*-X5 in eight tissues of three cows. mRNA expression levels in mammary gland of *GPIHBP1*, *GPIHBP1*-X1 and *GPIHBP1*-X5 all were highest among 8 tissues. The different capital letters indicated significant differences in the expression among eight tissues at $P < 0.01$.

of *GPIHBP1* in mammary gland suggests that it may play an important role in milk production traits or mammogenesis. It has been reported that *GPIHBP1* was highly expressed in mammary fat and heart tissues in mice [12,13]. Previous studies [24-26] showed that the lipoprotein lipase-mediated processing of lipoproteins within mammary gland is important for providing the lipid nutrients to produce milk fat. And the lipoprotein lipase (*LPL*) expression pattern in bovine mammary gland at different stages of lactation was quite similar to the lactation curve, which suggest that *LPL* is important for maintenance of milk synthesis [27]. Meanwhile, some studies on hyperlipidemia showed that *GPIHBP1* served as the transporter and the platform for the lipoprotein lipase-mediated lipolysis processing [28]. Therefore, *GPIHBP1* is essential for LPL to realize its biological function and play an important role in the process of producing milk fat and maintenance of milk synthesis.

Commonly, alternative splicing may change the structure of transcripts of a gene and the protein encoded by the gene, leading to profound functional alternation. It has been demonstrated that alternative splicing could affect the binding properties, intracellular localization, enzymatic activity, protein stability and post-translational modifications of a large number of proteins [29]. The effects of alternative splicing range from complete loss of function or gain of a new function to very subtle modulations that are difficult to detect [30]. Changes in alternative splicing of a gene can modulate its mRNA expression levels by subjecting mRNAs to nonsense-mediated decay (NMD) and alter the structure of protein [30]. Alternative splicing is regulated by splicing codes, including exonic splicing enhancers (ESEs), exonic splicing silencers (ESSs), intronic splicing enhancers (ISEs) and intronic splicing silencers (ISSs) [31]. Tissue-specific mRNA expression pattern could be associated with absence or presence of splicing codes in various tissues.

In this study, we identified that there were five transcripts (X1, X2, X3, X4 and X5) of the bovine *GPIHBP1* gene. The proteins of transcripts X2, X3 and X4 have the classical structure of the GPIHBP1 protein consisting of the N-terminal signal peptide, UPAR-Ly6 domain and C-terminal GPI-Modification Site. The protein P1 encoded by transcript X1 has the UPAR-Ly6 domain and the C-terminal GPI-Modification Site, but it lacks the signal peptide and acidic domain. It is not clear if this protein would be ever produced and secreted because of lacking the signal peptide. However, even if it is at all secreted as normal, it is also a non-functional GPIHBP1 because it lacks acidic domain, which makes it unable to bind to LPL [32]. The splicing resulting in the transcript X5 has a tremendous effect on the protein structure. The predicted secondary structures of bovine GPIHBP1 P5.1, P5.2, P5.3, P5.4 and P5.5 are quite different from that of bovine GPIHBP1 P1, P2 and human GPIHBP1 (Figure 3B). They do not have the UPAR-LY6 domain, which is considered as a very important functional region of GPIHBP1 [32]. Therefore, this novel splicing variant may regulate the transcript abundance of *GPIHBP1* in mammary gland of dairy cattle by nonsense-mediated decay and thus affect milk production traits indirectly.

Conclusions

This study is the first report on alternative splicing of bovine *GPIHBP1* gene. We identified a novel alternatively spliced transcript variant of *GPIHBP1* gene (*GPIHBP1*-X5) with 31 bp insertion in the exon and also confirmed other four existed transcripts (X1, X2, X3 and X4) of the *GPIHBP1* in Chinese Holstein cow. And we found that the 8 bp deletion of transcript X1 and 31 bp insertion of transcript X5 have tremendous effect on protein function and structure, respectively. Based on the results of Taq-Man RT-PCR, we found that *GPIHBP1*-X1, *GPIHBP1*-X5 and *GPIHBP1* expressed in higher level in mammary gland than in other tissues of lactating dairy cow. In conclusions, our findings provided more information for the further functional analyses of *GPIHBP1* in dairy cattle.

Additional file

Additional file 1: Table S1. Sequences of forward (F) and reverse (R) primers for PCR amplification of the coding region of GPIHBP1. **Table S2.** Sequences of forward (F) and reverse (R) primers and probes used for TaqMan real time PCR. **Table S3.** Specific primers for confirming the existence of 8 bp deletion and 31 bp insertion. **Table S4.** The amino acid sequences of five potential open reading frames of GPIHBP1 X5. **Table S5.** Predicted GPI-modification sites for amino acid sequence of GPIHBP1 splice variants.

Abbreviations

GPIHBP1: Glycosylphosphatidylinositol-anchored HDL binding protein1; LPL: Lipoprotein lipase; UTR: Untranslated region; ORF: Open reading frame; NMD: Nonsense-mediated decay; ESEs: Exonic splicing enhancers; ESSs: Exonic splicing silencers; ISEs: Intronic splicing enhancers; ISSs: Intronic splicing silencers.

Competing interests

All authors have reviewed and approved this final version of the manuscript. There is no competing interest in our submission.

Authors' contributions

JY carried out the experiment and drafted the manuscript, LJ and QZ participated in its design and coordination, and helped draft the manuscript, XL designed part of primers. All authors read and approved the final manuscript.

Acknowledgements

This work was financially supported by the National Natural Science Foundations of China [31201772]; Chinese Universities Scientific Fund [2014JD021]; the 948 Program of the Ministry of Agriculture of China [2011-G2A(3)]; the National High Technology Research and Development Program of China [863 Program 2011AA100302] and the Program for Changjiang Scholars and Innovative Research Team in University (IRT1191).

References

1. Jiang L, Liu J, Sun D, Ma P, Ding X, Yu Y, Zhang Q: **Genome wide association studies for milk production traits in Chinese Holstein population.** *PLoS One* 2010, **5**(10):e13661.
2. Beigneux AP, Davies BS, Bensadoun A, Fong LG, Young SG: **GPIHBP1, a GPI-anchored protein required for the lipolytic processing of triglyceride-rich lipoproteins.** *J Lipid Res* 2009, **50**(Suppl):S57–S62.
3. Young SG, Davies BS, Fong LG, Gin P, Weinstein MM, Bensadoun A, Beigneux AP: **GPIHBP1: an endothelial cell molecule important for the lipolytic processing of chylomicrons.** *Curr Opin Lipidol* 2007, **18**(4):389–396.
4. Gin P, Yin L, Davies BS, Weinstein MM, Ryan RO, Bensadoun A, Fong LG, Young SG, Beigneux AP: **The acidic domain of GPIHBP1 is important for the binding of lipoprotein lipase and chylomicrons.** *J Biol Chem* 2008, **283**(43):29554–29562.
5. Franssen R, Young SG, Peelman F, Hertecant J, Sierts JA, Schimmel AW, Bensadoun A, Kastelein JJ, Fong LG, Dallinga-Thie GM, Beigneux AP: **Chylomicronemia with low postheparin lipoprotein lipase levels in the setting of GPIHBP1 defects.** *Circ Cardiovasc Genet* 2010, **3**(2):169–178.
6. Beigneux AP, Franssen R, Bensadoun A, Gin P, Melford K, Peter J, Walzem RL, Weinstein MM, Davies BS, Kuivenhoven JA, Kastelein JJ, Fong LG, Dallinga-Thie GM, Young SG: **Chylomicronemia with a mutant GPIHBP1 (Q115P) that cannot bind lipoprotein lipase.** *Arterioscler Thromb Vasc Biol* 2009, **29**(6):956–962.
7. Charriere S, Peretti N, Bernard S, Di Filippo M, Sassolas A, Merlin M, Delay M, Debard C, Lefai E, Lachaux A, Moulin P, Marcais C: **GPIHBP1 C89F neomutation and hydrophobic C-terminal domain G175R mutation in two pedigrees with severe hyperchylomicronemia.** *J Clin Endocrinol Metab* 2011, **96**(10):E1675–E1679.
8. Davies BS, Beigneux AP, Barnes RH II, Tu Y, Gin P, Weinstein MM, Nobumori C, Nyren R, Goldberg I, Olivecrona G, Bensadoun A, Young SG, Fong LG: **GPIHBP1 is responsible for the entry of lipoprotein lipase into capillaries.** *Cell Metab* 2010, **12**(1):42–52.
9. Dallinga-Thie GM, Franssen R, Mooij HL, Visser ME, Hassing HC, Peelman F, Kastelein JJ, Peterfy M, Nieuwdorp M: **The metabolism of triglyceride-rich lipoproteins revisited: new players, new insight.** *Atherosclerosis* 2010, **211**(1):1–8.
10. Beigneux AP, Davies BS, Tat S, Chen J, Gin P, Voss CV, Weinstein MM, Bensadoun A, Pullinger CR, Fong LG, Young SG: **Assessing the role of the glycosylphosphatidylinositol-anchored high density lipoprotein-binding protein 1 (GPIHBP1) three-finger domain in binding lipoprotein lipase.** *J Biol Chem* 2011, **286**(22):19735–19743.
11. Rios JJ, Shastry S, Jasso J, Hauser N, Garg A, Bensadoun A, Cohen JC, Hobbs HH: **Deletion of GPIHBP1 causing severe chylomicronemia.** *J Inherit Metab Dis* 2012, **35**(3):531–540.
12. Beigneux AP, Weinstein MM, Davies BS, Gin P, Bensadoun A, Fong LG, Young SG: **GPIHBP1 and lipolysis: an update.** *Curr Opin Lipidol* 2009, **20**(3):211–216.
13. Beigneux AP, Davies BS, Gin P, Weinstein MM, Farber E, Qiao X, Peale F, Bunting S, Walzem RL, Wong JS, Blaner WS, Ding Z, Melford K, Wongsiriroj N, Shu X, de Sauvage F, Ryan R, Fong LG, Bensadoun A, Young SG: **Glycosylphosphatidylinositol-anchored high-density lipoprotein-binding protein 1 plays a critical role in the lipolytic processing of chylomicrons.** *Cell Metab* 2007, **5**(4):279–291.
14. Davies BS, Goulbourne CN, Barnes RH II, Turlo KA, Gin P, Vaughan S, Vaux DJ, Bensadoun A, Beigneux AP, Fong LG, Young SG: **Assessing mechanisms of GPIHBP1 and lipoprotein lipase movement across endothelial cells.** *J Lipid Res* 2012, **53**(12):2690–2697.
15. Olivecrona G, Ehrenborg E, Semb H, Makoveichuk E, Lindberg A, Hayden MR, Gin P, Davies BS, Weinstein MM, Fong LG, Beigneux AP, Young SG, Olivecrona T, Hernell O: **Mutation of conserved cysteines in the Ly6 domain of GPIHBP1 in familial chylomicronemia.** *J Lipid Res* 2010, **51**(6):1535–1545.
16. Young SG, Zechner R: **Biochemistry and pathophysiology of intravascular and intracellular lipolysis.** *Genes Dev* 2013, **27**(5):459–484.
17. Petersen TN, Brunak S, Von Heijne G, Nielsen H: **SignalP 4.0: discriminating signal peptides from transmembrane regions.** *Nat Methods* 2011, **8**(10):785–786.
18. Eisenhaber B, Bork P, Eisenhaber F: **Prediction of potential GPI-modification sites in proprotein sequences.** *J Mol Biol* 1999, **292**(3):741–758.
19. Leath KJ, Johnson S, Roversi P, Hughes TR, Smith RA, Mackenzie L, Morgan BP, Lea SM: **High-resolution structures of bacterially expressed soluble human CD59.** *Acta Crystallogr Sect F: Struct Biol Cryst Commun* 2007, **63**(Pt 8):648–652.
20. Arnold K, Bordoli L, Kopp J, Schwede T: **The SWISS-MODEL workspace: a web-based environment for protein structure homology modelling.** *Bioinformatics* 2006, **22**(2):195–201.
21. Livak KJ, Schmittgen TD: **Analysis of relative gene expression data using real-time quantitative PCR and the 2(-Delta Delta C(T)) Method.** *Methods* 2001, **25**(4):402–408.
22. Sunyaev SR, Eisenhaber F, Rodchenkov IV, Eisenhaber B, Tumanyan VG, Kuznetsov EN: **PSIC: profile extraction from sequence alignments with position-specific counts of independent observations.** *Protein Eng* 1999, **12**(5):387–394.
23. Weikard R, Goldammer T, Brunner RM, Kuehn C: **Tissue-specific mRNA expression patterns reveal a coordinated metabolic response associated with genetic selection for milk production in cows.** *Physiol Genomics* 2012, **44**(14):728–739.
24. Rudolph MC, McManaman JL, Phang T, Russell T, Kominsky DJ, Serkova NJ, Stein T, Anderson SM, Neville MC: **Metabolic regulation in the lactating mammary gland: a lipid synthesizing machine.** *Physiol Genomics* 2007, **28**(3):323–336.

25. Jensen DR, Gavigan S, Sawicki V, Witsell DL, Eckel RH, Neville MC: **Regulation of lipoprotein lipase activity and mRNA in the mammary gland of the lactating mouse.** *Biochem J* 1994, **298**(Pt 2):321–327.
26. Fisher EA: **GPIHBP1: lipoprotein lipase's ticket to ride.** *Cell Metab* 2010, **12**(1):1–2.
27. Bionaz M, Loor JJ: **Gene networks driving bovine milk fat synthesis during the lactation cycle.** *BMC Genomics* 2008, **9**:366.
28. Adeyo O, Goulbourne CN, Bensadoun A, Beigneux AP, Fong LG, Young SG: **Glycosylphosphatidylinositol-anchored high-density lipoprotein-binding protein 1 and the intravascular processing of triglyceride-rich lipoproteins.** *J Intern Med* 2012, **272**(6):528–540.
29. Kelemen O, Convertini P, Zhang Z, Wen Y, Shen M, Falaleeva M, Stamm S: **Function of alternative splicing.** *Gene* 2013, **514**(1):1–30.
30. Stamm S, Ben-Ari S, Rafalska I, Tang Y, Zhang Z, Toiber D, Thanaraj TA, Soreq H: **Function of alternative splicing.** *Gene* 2005, **344**:1–20.
31. Wang Z, Burge CB: **Splicing regulation: from a parts list of regulatory elements to an integrated splicing code.** *RNA* 2008, **14**(5):802–813.
32. Beigneux AP, Gin P, Davies BS, Weinstein MM, Bensadoun A, Fong LG, Young SG: **Highly conserved cysteines within the Ly6 domain of GPIHBP1 are crucial for the binding of lipoprotein lipase.** *J Biol Chem* 2009, **284**(44):30240–30247.

A direct real-time polymerase chain reaction assay for rapid high-throughput detection of highly pathogenic North American porcine reproductive and respiratory syndrome virus in China without RNA purification

Kang Kang[1,2†], Keli Yang[3†], Jiasheng Zhong[2], Yongxiang Tian[3], Limin Zhang[2], Jianxin Zhai[4], Li Zhang[4], Changxu Song[5], Christine Yuan Gou[6], Jun Luo[1*] and Deming Gou[2*]

Abstract

Background: Porcine reproductive and respiratory syndrome virus (PRRSV), and particularly its highly pathogenic genotype (HP-PRRSV), have caused massive economic losses to the global swine industry.

Results: To rapidly identify HP-PRRSV, we developed a direct real-time reverse transcription polymerase chain reaction method (dRT-PCR) that could detect the virus from serum specimen without the need of RNA purification. Our dRT-PCR assay can be completed in 1.5 h from when a sample is received to obtaining a result. Additionally, the sensitivity of dRT-PCR matched that of conventional reverse transcription PCR (cRT-PCR) that used purified RNA. The lowest detection limit of HP-PRRSV was 6.3 $TCID_{50}$ using dRT-PCR. We applied dRT-PCR assay to 144 field samples and the results showed strong consistency with those obtained by cRT-PCR. Moreover, the dRT-PCR method was able to tolerate 5-20% (v/v) serum.

Conclusions: Our dRT-PCR assay allows for easier, faster, more cost-effective and higher throughput detection of HP-PRRSV compared with cRT-PCR methods. To the best of our knowledge, this is the first report to describe a real-time RT-PCR assay capable of detecting PRRSV in crude serum samples without the requirement for purifying RNA. We believe our approach has a great potential for application to other RNA viruses.

Keywords: Highly pathogenic, Porcine reproductive and respiratory syndrome virus, Real-time RT-PCR

Background

Porcine reproductive and respiratory syndrome virus (PRRSV) is a pathogen responsible for significant economic losses to the global swine industry in recent years. The virus causes reproductive failure in pregnant sows and respiratory tract illness in young pigs [1,2]. PRRSV is a single-stranded 15-kb RNA virus of the family *Arteriviridae* [3]. It can be subdivided into types 1 (European genotype) [2] and 2 (Northern American genotype) [4,5]. In 2006, a high fever syndrome in swine due to a highly pathogenic type 2 PRRSV (HP-PRRSV) broke out in China [6]. HP-PRRSV infected more than 2 million pigs in 2006 with an average fatality rate of 25%.

The current methods for detecting PRRSV include virus isolation [7], immunoperoxidase monolayer assay (IPMA) [8], and enzyme-linked immunosorbent assay (ELISA) [9,10]. However, these methods are not specific enough to discriminate between HP-PRRSV and classical strains (C-PRRSV). HP-PRRSV has a discontinuous deletion of 30 amino acids in its non-structural protein 2 (NSP2) [6]. Discrimination of HP-PRRSV from C-PRRSV can be achieved by reverse transcription polymerase chain

* Correspondence: luojun1@yahoo.com; dmgou@hotmail.com
†Equal contributors
1College of Animal Science and Technology, Northwest A&F University, Yangling, Shaanxi 712100, China
2College of Life Sciences, Shenzhen Key Laboratory of Microbial Genetic Engineering, Shenzhen University, Shenzhen 518060, China
Full list of author information is available at the end of the article

reaction (RT-PCR) assays targeting differences in the *nsp2* gene [11-15]. However, current PCR methods rely on purification of RNA from a sample, which is time-consuming; if such a step was eliminated, the speed of viral detection could be greatly enhanced.

We sought to develop a direct real-time RT-PCR (dRT-PCR) assay for HP-PRRSV detection in crude samples without subjecting them to RNA extraction and purification procedures. We believe that such a method would greatly simplify and accelerate high throughput viral analysis, along with reducing associated costs and chances of cross contamination.

Materials and methods

Viruses, cells and clinical specimens

Classical PRRSV strain CH-1a (GenBank:AY032626) was kindly provided by Dr. Hanzhong Wang (Wuhan Institute of Virology, Chinese Academy of Sciences, Wuhan, China). The highly pathogenic type 2 PRRSV strain 07HBEZ was isolated in 2007 (GenBank:FJ495082.2). Other viruses, including classical swine fever virus (CSFV), pseudorabies virus (PRV), porcine circovirus type 2 (PCV2), porcine parvovirus (PPV) and rotavirus (RV) were stored in our laboratory and used to confirm the specificity of dRT-PCR assay we developed. MARC-145 cells were cultured and maintained in Dulbeco's modified Eagle's medium (DMEM) supplemented with 10% newborn calf serum (Gibco) at 37°C, 5% CO_2. We collected 144 porcine serum samples from more than 10 pig farms across Hubei Province, China from July to September 2012. Briefly, blood samples were taken from the anterior vena cava and placed in 5-mL centrifuge tubes lacking any coagulant or anticoagulant. Samples were centrifuged at $1,000 \times g$ for 10 min and the serum was stored at −80°C until required.

Sample collection of porcine sera complied with the regulation of the Ministry of Agriculture of China. Our study was carried out in strict accordance with the recommendations in the Guide for the Care and Use of Laboratory Animals of the Ministry of Agriculture of China. Our study was approved by the Committee on the Ethics of Animal Experiments of Shenzhen University, China (Permit Number: 0156-05/13).

Primers and probe

HP-PRRSV has a discontinuous deletion of 30 amino acids in the *nsp2* gene. Based on the alignment of *nsp2* sequences published in GenBank, primers were designed using the Primer Premier 5.0. A Taqman probe spanning the flanking sequence of the deleted region of the *nsp2* gene was designed. The primer and probe sequences used in our study were summarized in Table 1.

Direct real-time RT-PCR

We used the Direct One-Step S/P qRT-PCR Taqprobe Kit (VitaNavi Technology, Manchester, USA) to conduct our dRT-PCR assay. This kit contained an S/P RT-PCR enzyme mix comprising reverse transcriptase and highly resistant Taq DNA polymerase, and a 2× S/P RT-PCR master mix containing a PCR enhancer cocktail (PEC)). A volume of each sample (1–5 μL of serum per 25 μL reaction) was directly added to the master reagents that were supplemented with the designed primers and probe. Viral RNA release, reverse transcription (RT) and PCR were all carried out in the same reaction tube. Reactions were optimized by altering the concentration of primers (0.2 ~ 0.4 μmol/L), probe (0.2 ~ 0.4 μmol/L), and enzyme mix (0.75-1.25 μL per 25 μL reaction); and by examining RNA release/RT temperature (55 ~ 60°C), annealing/extension temperature (60 ~ 65°C), annealing/extension time (30 ~ 120 s) and the number of PCR cycles (35 ~ 45 cycles). As a positive control, we used purified RNA from a 100 μL sample of HP-PRRSV-containing serum. This sample had been subjected to RNA extraction using the TaKaRa MiniBEST Viral RNA/DNA Extraction Kit Ver. 4.0 (TaKaRa, Dalian, China) according to the manufacturer's instructions. The RNA was eluted into 100 μL of RNase-free water, restoring the original specimen volume. PRRSV-negative serum was used as a negative control. Reactions were performed in triplicate and carried out on a StepOne Plus thermocycler (Applied Biosystems). The threshold was set at 0.01 and the threshold cycle (Ct) value was analyzed. Amplification products were confirmed by electrophoresis and DNA sequencing.

Conventional RT-PCR assay

We used conventional RT-PCR (cRT-PCR) assay employing purified RNA sample as previously described [15].

Sensitivity of dRT-PCR assay

HP-PRRSV-infected cell supernatants with viral titers of $10^{4.8}$ $TCID_{50}$/μL was 10-fold serially diluted with PRRSV-negative serum. A 1 μL aliquot of each diluted serum sample was assayed by dRT-PCR in a 25 μL reaction

Table 1 Primer and probe sequences used in our study

Name	Sequence 5′→3′	Location (GenBank no.)	Product
dHP-PRRSV-F	GGGTCGGCACCAGTTCC	1555-1571 (FJ495082.2)	73 bp
dHP-PRRSV-R	AATCCAGAGGCTCATCCTGGT	1607-1627 (FJ495082.2)	
dHP-PRRSV-pro	FAM-CACCGCGTATAACTGTGACAACAACGC-BHQ1	1574-1600 (FJ495082.2)	

volume. Each reaction was performed in triplicate. The correlation between the titer of HP-PRRSV and Ct value was analyzed.

Specificity of dRT-PCR assay

The specificity of the dRT-PCR assay was identified by analyzing seven different viruses (HP-PRRSV, C-PRRSV, CSFV, PRV, PCV2, PPV and RV).

Serum tolerance of dRT-PCR assay

A 1 μL of HP-PRRSV-positive serum was diluted with PRRSV-negative serum to acquire different concentrations of 5, 10, 15, 20, and 25% (v/v) serum in a 25 μL reaction, respectively. Then, all concentrations of sera and a 1 μL of HP-PRRSV RNA were analyzed in parallel in dRT-PCR assays. A 100 μL of HP-PRRSV-positive serum sample was subjected to RNA extraction. The RNA was eluted into 100 μL of RNase-free water, restoring the original specimen volume. Ct values were analyzed and PCR products were evaluated by electrophoresis. Tolerance test was also performed by detecting the same amount of purified RNA added with no serum or with negative serum in various concentrations (5-25%, v/v).

Statistical analyses

Statistical analyses were performed with GraphPad Prism 5. Correlation coefficient (R^2) was calculated by linear regression analysis. Repeatability and reproducibility of the dRT-PCR assay was determined by analyzing the mean values and standard deviations of Ct values.

Results

dRT-PCR assay

Optimized dRT-PCR assay (25 μL) contained 12.5 μL 2× S/P RT-PCR master mix, 1 μL S/P RT-PCR enzyme mix, 0.4 μmol/L forward primer, 0.4 μmol/L reverse primer, 0.2 μmol/L probe, template (1–5 μL) and water. Reactions were conducted at 55°C for 30 min (for viral RNA release and RT), then 95°C for 5 min, followed by 40 cycles of 94°C for 20 s and 60°C for 40 s.

Sensitivity of dRT-PCR assay

The sensitivity of dRT-PCR was compared to that of conventional RT-PCR (cRT-PCR) using purified RNA. When using 1 μL HP-PRRSV serum sample, the average Ct value produced by dRT-PCR was 25.91, while that for cRT-PCR was 26.93 (Figure 1), indicating the sensitivity of dRT-PCR was even better than that of cRT-PCR. We further tested the sensitivity using HP-PRRSV-infected cell supernatants of known viral titer. The correlation coefficient (R^2) for virus input and Ct values was 0.996 when viral titer ranged from $10^{4.8}$ $TCID_{50}$/μL to $10^{0.8}$ $TCID_{50}$/μL (Figure 2). The lowest detection limit of HP-PRRSV was 6.3 $TCID_{50}$.

Specificity and accuracy of dRT-PCR assay

The Taqman probe we designed only worked when the 30-amino acid coding sequence of *nsp2* gene was deleted, allowing for specific discrimination of HP-PRRSV from C-PRRSV and other viruses (CSFV, PRV, PCV2, PPV and RV; Additional file 1). To further evaluate the accuracy of dRT-PCR assay, we analyzed a profusion of field serum samples. These samples had been previously evaluated using a cRT-PCR method developed by Yang [15]. A double-blind analysis was performed for the dRT-PCR assay. Of the 144 samples assayed, 94 (65.3%) were HP-PRRSV positive while 50 (34.7%) were negative, showing strong consistency with that obtained by cRT-PCR method (Table 2).

Repeatability and reproducibility of dRT-PCR

To test the repeatability and reproducibility of the dRT-PCR, three serum samples with differing viral titers were selected. Each sample was assayed six times, and these experiments were repeated three times on different days; with a total of 18 replicates available for analysis (Figure 3 and Additional file 2). The standard deviations of Ct values ranged from 0.13 to 0.26, indicating a good repeatability and reproducibility of the method.

Inhibitor tolerance of dRT-PCR assay

To determine to what extent this method could endure inhibitors, we performed dRT-PCR assay using a wide range of serum concentrations (5-25%, v/v). Reaction mixture adding with purified RNA was transparent after the reaction was completed while the reaction mixtures using crude serum samples were no longer transparent at the end of the reactions (Figure 4A). They became increasingly turbid as the serum concentration increased. Interestingly, the dRT-PCR assay remained functional when the serum concentration was in the range 5–20% as there were amplification plots (Ct value < 35), and the target PCR band could be observed in electrophoresis (Figure 4B and C). The most robust signals were detected when the serum concentration was between 5% and 10%.

We also tested the tolerance of dRT-PCR in a way that used the same amount of purified RNA added with no serum or with negative serum in various concentrations (5-25%, v/v). The results showed that significant signals were obtained among 5-10% serum although the reactions were gradually inhibited as serum concentrations increased (Figure 4D). It also suggested that notable RNA loss might have occured during RNA purification process whereas dRT-PCR assay could effectively avoid this problem.

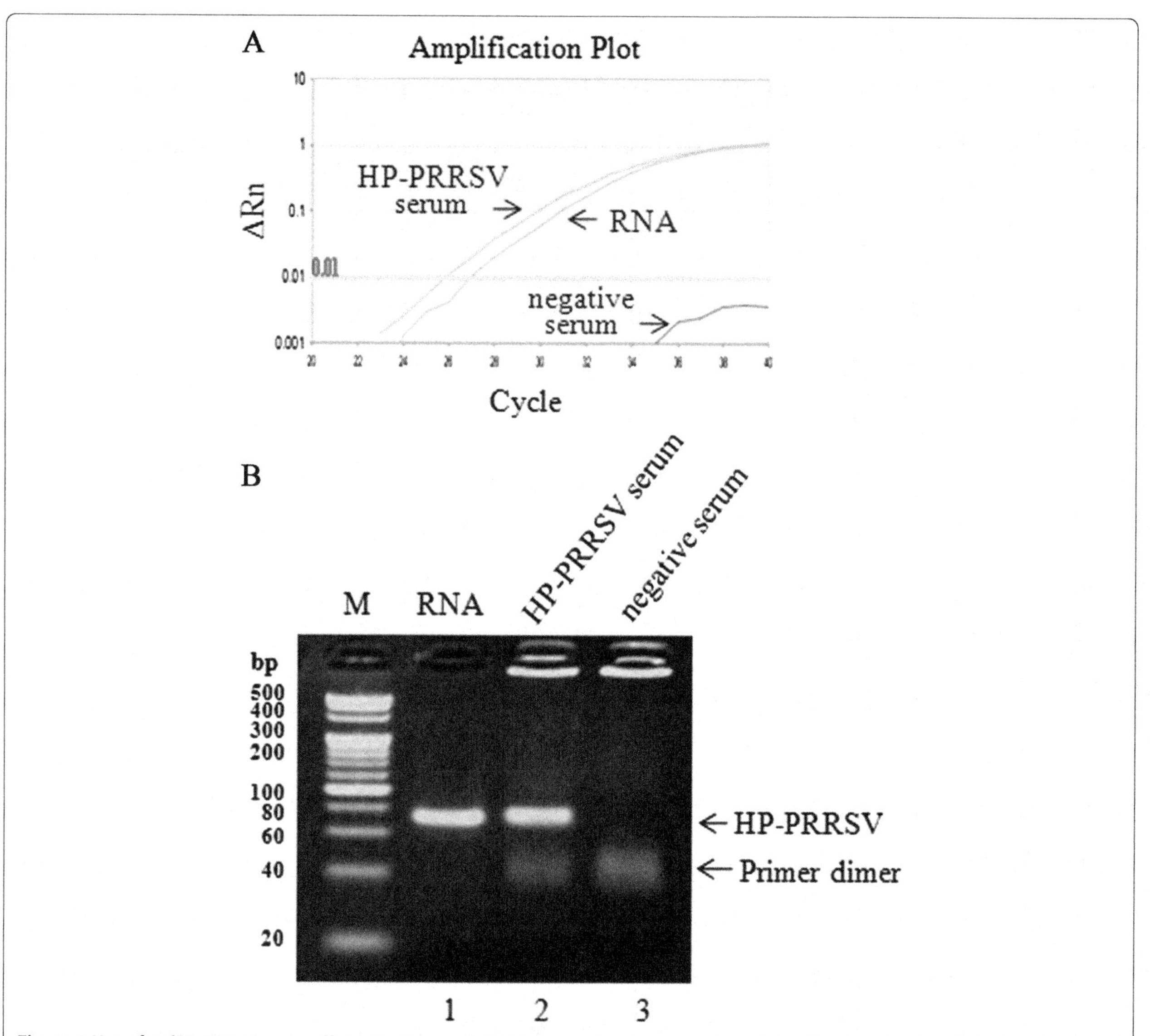

Figure 1 Use of a dRT-PCR assay to efficiently detect HP-PRRSV in crude serum samples. (A) Amplification plots for dRT-PCR and cRT-PCR assays. Threshold was set at 0.01. **(B)** Amplicons from the dRT-PCR assays and controls were electrophoresed on 3.5% (w/v) agarose gels. The arrows indicate the 73-bp target bands for HP-PRRSV. A 1 μL of purified HP-PRRSV RNA (Lane 1), 1 μL of HP-PRRSV-containing serum (Lane 2) and 1 μL of PRRSV-negative serum (Lane 3) were added to a 25 μL dRT-PCR mixture, respectively. All assays were performed in triplicate.

Discussion

Development of a rapid and high-throughput RT-PCR method for monitoring HP-PRRSV infections could be significant for quickly sensing and responding to epidemics. One strategy to shorten detection time would be to eliminate the time-consuming RNA purification process. However, abundant PCR inhibitors exist in crude biological samples, such as IgG, hemoglobin, hemin and lactoferrin in blood or serum/plasma [16,17], and humic acid in soil [18,19]. The presence of these inhibitors suppresses the activity of DNA polymerase, necessitating a nucleic acid purification prior to conducting any type of PCR assay.

In 2009, Barnes reported an N-terminal 278-aa truncated Taq DNA polymerase Klentaq1 that could amplify genomic DNA in the presence of 5-10% whole blood [20]. This seemed unlikely given that the full-length Taq DNA polymerase enzyme can be inhibited with 0.1 ~ 1% blood; however, this finding inspired researchers to screen for more enzyme mutants with higher resistance. Further studies showed that amino acids relating to high resistance were located at codons 706–708. Subsequently, all 19 possible amino-acid variants of Taq DNA polymerase at codon 708 were tested. Two mutant alternatives, KT 10 (E708K) and Taq 22 (E708Q) were revealed to exhibit higher tolerance and remained functional even in 20% blood, and when humic acid (0.2–0.4 μg/mL) and lactoferrin (2.6–5.2 mmol/L) were present.

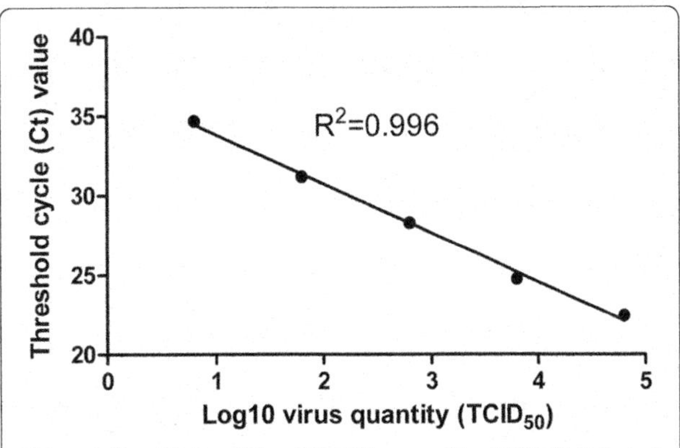

Figure 2 Sensitivity of the dRT-PCR assay. The HP-PRRSV-infected cell supernatant was 10-fold serially diluted with PRRSV-negative serum. Viral titers ranged from $10^{4.8}$ $TCID_{50}$/μL to $10^{0.8}$ $TCID_{50}$/μL. Each diluted sample was assayed in triplicate.

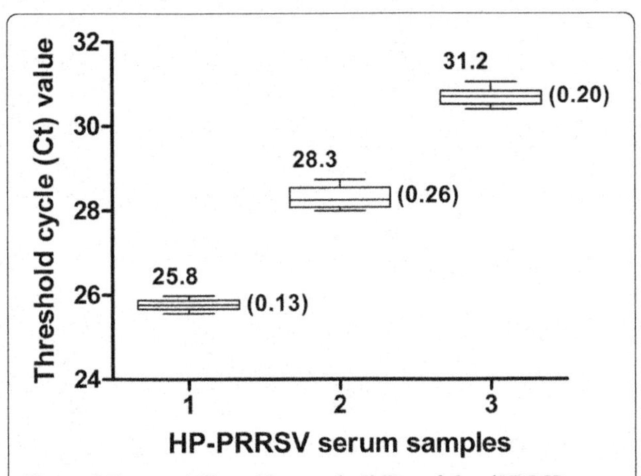

Figure 3 Repeatability and reproducibility of the dRT-PCR assay. Three field serum samples with different virus titers were selected and assayed. Mean values and standard deviations of Ct values are depicted above and beside the box plots, respectively.

Betaine is a commonly used PCR enhancer capable of reducing melting temperature and inhibiting the formation of secondary structure in templates [21,22]. L-carnitine and D-(+)-trehalose function in similar ways to betaine [23]. The nonionic detergent NP-40 can enhance DNA sequencing by lowering the number of non-specific bands [24]. A PEC of L-carnitine, D-(+)-trehalose and NP-40 can help in the amplification of GC-rich templates and boost reactions rich with inhibitors [23]. This PEC is included in the 2× S/P RT-PCR master mix we used for our dRT-PCR assay to overcome the various inhibitors in serum.

Real-time fluorescence PCR using crude blood or serum/plasma samples is more challenging than PCR methods that use purified nucleic acids as the fluorescent signal is quenched by compounds such as hemoglobin and heme. There were studies reporting direct RT-PCR assays without RNA purification for the detection of Norovirus and bovine viral diarrhoea virus [25,26]. Nonetheless, none of them were real-time PCR assay. African Chikungunya virus could be detected by TaqMan RT-PCR assay without RNA extraction with less than 1% serum [27]. However, our dRT-PCR assay remains functional in the presence of rather high concentrations of serum because of the previously described tolerant Taq DNA polymerase and PEC. To the best of our knowledge, this is the first successful real-time RT-PCR assay for HP-PRRSV using a crude specimen where nucleic acid extraction and purification has not been conducted.

Another factor ensuring success in our dRT-PCR assay is the efficient release of viral RNA and the subsequent RT reaction. In conventional RT-PCR, chemicals such as TRIzol help denature the capsid protein and release the viral RNA. A high temperature also facilitates release of viral RNA. As an example, pretreatment at 85°C for 1 min was used to release noroviruses RNA [25]; Treatment at 95°C for 4 min was used to liberate of bovine viral diarrhoea virus RNA [26]. However, pretreatment at high temperatures will result in degradation of RNA. In our experiment, we adopted a simple one-step procedure for RNA release and RT. Treating samples at 55°C for 30 min was efficient for both RNA release and subsequent RT for HP-PRRSV. The detergent NP-40 also promotes denaturation of viral capsid protein. Other compounds in the PEC also improve the resistance of reverse transcriptase to inhibitors in serum [23].

As shown in Figures 1 and 4, the sensitivity of the dRT-PCR assay was comparable to, and in some cases, exceeded that of conventional RT-PCR assay. These results suggested that remarkable RNA loss might frequently occurr during RNA purification step in conventional RT-PCR assay, while dRT-PCR assay could effectively conquer this problem.

Given that real-time RT-PCR can be performed in 96- and 384-well formats currently, thousands of field samples can be assayed within a single day when the need of RNA purification is eliminated, thus our dRT-PCR assay providing a high-throughput detection for HP-PRRSV. It is of great significance during an epidemic outbreak.

Table 2 Application of dRT-PCR and cRT-PCR assays on field serum samples

Method	No. of specismen	HP-PRRSV positive	HP-PRRSV negative	Positive rate, %	Correlation, %
dRT-PCR	144	94	50	65.3	100
cRT-PCR[a]	144[b]	94	50	65.3	100

[a]The cRT-PCR data were provided by Yang [15].
[b]These samples had also been detected using immunochromatochemistry by Yang [15].

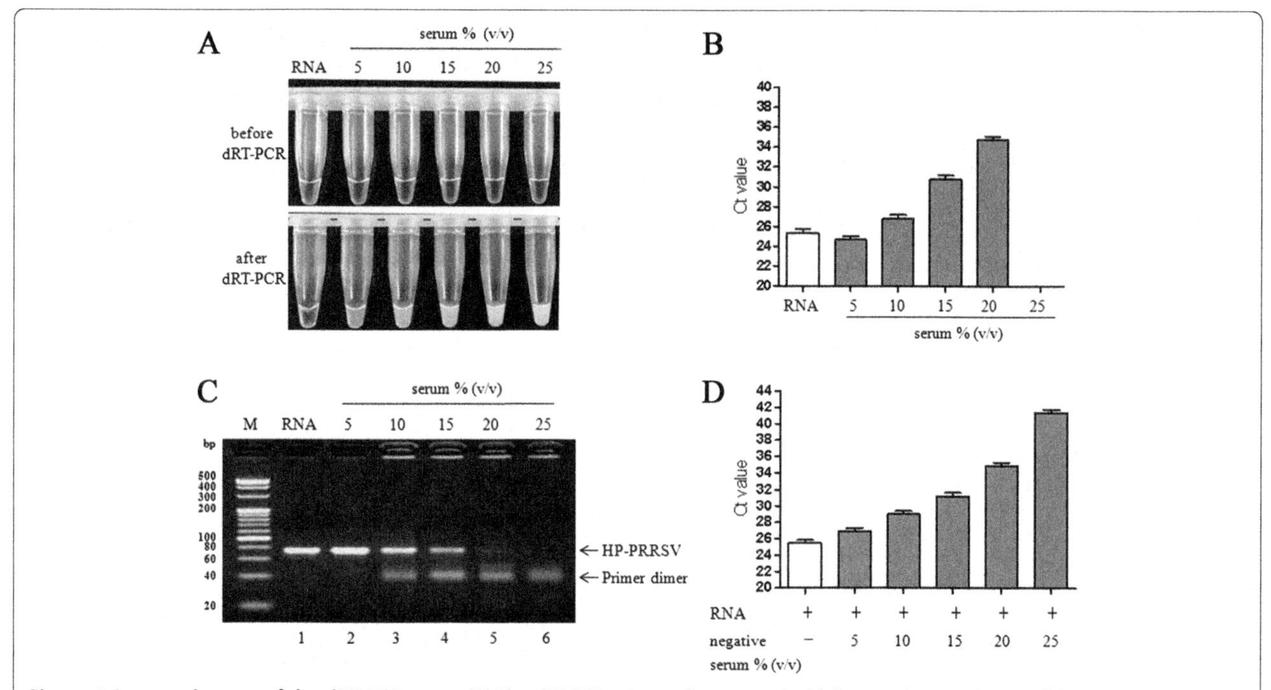

Figure 4 Serum tolerance of the dRT-PCR assay. (A) The dRT-PCR mixtures became turbid following the completion of thermal cycling. **(B)** Ct values of dRT-PCR assay where different concentration of HP-PRRSV serum and purified RNA were used. **(C)** Amplicons of dRT-PCR assays were electrophoresed on 3.5% (w/v) agarose gel. Lane 1: 1 μL of purified HP-PRRSV RNA derived from 1 μL of HP-PRRSV-positive serum; Lane 2-6: 1 μL of HP-PRRSV-positive serum diluted with negative serum to acquire different serum concentrations of 5, 10, 15, 20, and 25% (v/v) in a 25 μL reaction, respectively. The arrow indicates the 73-bp target bands for HP-PRRSV. **(D)** Ct values of dRT-PCR assays detecting the same amount of purified RNA supplemented with no serum or with negative serum in various concentrations (5-25%, v/v).

Conclusions

We developed a direct real-time RT-PCR assay to determine the presence of HP-PRRSV in crude serum sample. Our assay allowed for more rapid and higher throughput detection of the virus. This approach has a great potential for application to other RNA viruses.

Additional files

Additional file 1: Specificity of the dRT-PCR assay. Primers and probe based on the 30 amino-acid deletion in the HP-PRRSV *nsp2* gene were used in dRT-PCR assays for seven different viruses. Except for HP-PRRSV, there was no significant amplification signal for six other viruses (C-PRRSV, CSFV, PRV, PCV2, PPV and RV), indicating high specificity of the dRT-PCR.

Additional file 2: Ct values for HP-PRRSV in repeatability and reproducibility assays using our developed dRT-PCR method.

Competing interests

The authors declare that they have no competing interests.

Authors' contributions

DG and JL conceived and designed the experiments. KK, KLY, JSZ and LMZ performed the experiments. YXT, JXZ and LZ analyzed the data. KK wrote the manuscript. CXS and CYG revised the manuscript. All authors read and approved the final manuscript.

Acknowledgements

This study was supported in part by the National Basic Research Program of China (973 Program, 2012CB124701); National Natural Science Foundation of China No. 81170047, 81370151 (to DG); Shenzhen overseas high-level talents innovation program No.YFZZ20111009 (to DG); Shenzhen Nanshan Core Technology Program No. KC2013JSJS0020A; Shenzhen Municipal Basic Research Program No. JCYJ20130329120507746 (to KK); and Postdoctoral Science Foundation of China No. 2013 M542203 (to KK). Hubei Province Research and Development Project No. 2011BBB080 (to KY); Project supported by the Key Natural Science Foundation of Hubei Province, China No. 2012FFA067 (to YT) and the Opening Subject of Hubei Key Laboratory of Animal Embryo and Molecular Breeding No.2012ZD156 (to KY). The funders had no role in study design, data collection and analysis, decision to publish, or preparation of the manuscript.

Author details

[1]College of Animal Science and Technology, Northwest A&F University, Yangling, Shaanxi 712100, China. [2]College of Life Sciences, Shenzhen Key Laboratory of Microbial Genetic Engineering, Shenzhen University, Shenzhen 518060, China. [3]Hubei Key Laboratory of Animal Embryo and Molecular Breeding, Institute of Animal Husbandry and Veterinary, Hubei Academy of Agricultural Sciences, Wuhan 430064, China. [4]Shenzhen Ao Dong Inspection and Testing Technology Co,. Ltd, Shenzhen 518000, China. [5]Veterinary Medicine Institute, Guangdong Academy of Agricultural Sciences, Guangzhou 510640, China. [6]Northwestern University, Evanston, IL, USA.

References

1. Rossow KD: **Porcine reproductive and respiratory syndrome.** *Vet Pathol* 1998, **35:**1–20.
2. Albina E: **Epidemiology of porcine reproductive and respiratory syndrome (PRRS): an overview.** *Vet Microbiol* 1997, **55:**309–316.
3. Cavanagh D: **Nidovirales: a new order comprising Coronaviridae and Arteriviridae.** *Arch Virol* 1997, **142:**629–633.
4. Elazhary Y, Weber J, Bikour H, Morin M, Girard C: **'Mystery swine disease' in Canada.** *Vet Rec* 1991, **129:**495–496.

5. Wensvoort G, Terpstra C, Pol JM, ter Laak EA, Bloemraad M, de Kluyver EP, Kragten C, van Buiten L, den Besten A, Wagenaar F: **Mystery swine disease in The Netherlands: the isolation of Lelystad virus.** *Vet Q* 1991, **13**:121–130.
6. Tian K, Yu X, Zhao T, Feng Y, Cao Z, Wang C, Hu Y, Chen X, Hu D, Tian X, Liu D, Zhang S, Deng X, Ding Y, Yang L, Zhang Y, Xiao H, Qiao M, Wang B, Hou L, Wang X, Yang X, Kang L, Sun M, Jin P, Wang S, Kitamura Y, Yan J, Gao GF: **Emergence of fatal PRRSV variants: unparalleled outbreaks of atypical PRRS in China and molecular dissection of the unique hallmark.** *PLoS One* 2007, **2**:e526.
7. Mengeling WL, Lager KM, Wesley RD, Clouser DF, Vorwald AC, Roof MB: **Diagnostic implications of concurrent inoculation with attenuated and virulent strains of porcine reproductive and respiratory syndrome virus.** *Am J Vet Res* 1999, **60**:119–122.
8. Nodelijk G, Wensvoort G, Kroese B, van Leengoed L, Colijn E, Verheijden J: **Comparison of a commercial ELISA and an immunoperoxidase monolayer assay to detect antibodies directed against porcine respiratory and reproductive syndrome virus.** *Vet Microbiol* 1996, **49**:285–295.
9. Albina E, Leforban Y, Baron T, Plana Duran JP, Vannier P: **An enzyme linked immunosorbent assay (ELISA) for the detection of antibodies to the porcine reproductive and respiratory syndrome (PRRS) virus.** *Ann Rech Vet* 1992, **23**:167–176.
10. Dea S, Wilson L, Therrien D, Cornaglia E: **Detection of antibodies to the nucleocapsid protein of PRRS virus by a competitive ELISA.** *Adv Exp Med Biol* 2001, **494**:401–405.
11. Chai Z, Ma W, Fu F, Lang Y, Wang W, Tong G, Liu Q, Cai X, Li X: **A SYBR Green-based real-time RT-PCR assay for simple and rapid detection and differentiation of highly pathogenic and classical type 2 porcine reproductive and respiratory syndrome virus circulating in China.** *Arch Virol* 2013, **158**:407–415.
12. Wernike K, Hoffmann B, Dauber M, Lange E, Schirrmeier H, Beer M: **Detection and typing of highly pathogenic porcine reproductive and respiratory syndrome virus by multiplex real-time rt-PCR.** *PLoS One* 2012, **7**:e38251.
13. Chen NH, Chen XZ, Hu DM, Yu XL, Wang LL, Han W, Wu JJ, Cao Z, Wang CB, Zhang Q, Wang BY, Tian KG: **Rapid differential detection of classical and highly pathogenic North American Porcine Reproductive and Respiratory Syndrome virus in China by a duplex real-time RT-PCR.** *J Virol Methods* 2009, **161**:192–198.
14. Xiao XL, Wu H, Yu YG, Cheng BZ, Yang XQ, Chen G, Liu DM, Li XF: **Rapid detection of a highly virulent Chinese-type isolate of Porcine Reproductive and Respiratory Syndrome Virus by real-time reverse transcriptase PCR.** *J Virol Methods* 2008, **149**:49–55.
15. Yang K, Li Y, Duan Z, Guo R, Liu Z, Zhou D, Yuan F, Tian Y: **A one-step RT-PCR assay to detect and discriminate porcine reproductive and respiratory syndrome viruses in clinical specimens.** *Gene* 2013, **531**:199–204.
16. Queipo-Ortuno MI, De Dios Colmenero J, Macias M, Bravo MJ, Morata P: **Preparation of bacterial DNA template by boiling and effect of immunoglobulin G as an inhibitor in real-time PCR for serum samples from patients with brucellosis.** *Clin Vaccine Immunol* 2008, **15**:293–296.
17. Al-Soud WA, Radstrom P: **Purification and characterization of PCR-inhibitory components in blood cells.** *J Clin Microbiol* 2001, **39**:485–493.
18. Tsai YL, Olson BH: **Rapid method for separation of bacterial DNA from humic substances in sediments for polymerase chain reaction.** *Appl Environ Microbiol* 1992, **58**:2292–2295.
19. Watson RJ, Blackwell B: **Purification and characterization of a common soil component which inhibits the polymerase chain reaction.** *Can J Microbiol* 2000, **46**:633–642.
20. Kermekchiev MB, Kirilova LI, Vail EE, Barnes WM: **Mutants of Taq DNA polymerase resistant to PCR inhibitors allow DNA amplification from whole blood and crude soil samples.** *Nucleic Acids Res* 2009, **37**:e40.
21. Henke W, Herdel K, Jung K, Schnorr D, Loening SA: **Betaine improves the PCR amplification of GC-rich DNA sequences.** *Nucleic Acids Res* 1997, **25**:3957–3958.
22. Musso M, Bocciardi R, Parodi S, Ravazzolo R, Ceccherini I: **Betaine, dimethyl sulfoxide, and 7-deaza-dGTP, a powerful mixture for amplification of GC-rich DNA sequences.** *J Mol Diagn* 2006, **8**:544–550.
23. Zhang Z, Kermekchiev MB, Barnes WM: **Direct DNA amplification from crude clinical samples using a PCR enhancer cocktail and novel mutants of Taq.** *J Mol Diagn* 2010, **12**:152–161.
24. Petry H, Bachmann B, Luke W, Hunsmann G: **PCR sequencing with the aid of detergents.** *Methods Mol Biol* 1996, **65**:105–109.
25. Nishimura N, Nakayama H, Yoshizumi S, Miyoshi M, Tonoike H, Shirasaki Y, Kojima K, Ishida S: **Detection of noroviruses in fecal specimens by direct RT-PCR without RNA purification.** *J Virol Methods* 2010, **163**:282–286.
26. Bachofen C, Willoughby K, Zadoks R, Burr P, Mellor D, Russell GC: **Direct RT-PCR from serum enables fast and cost-effective phylogenetic analysis of bovine viral diarrhoea virus.** *J Virol Methods* 2013, **190**:1–3.
27. Pastorino B, Bessaud M, Grandadam M, Murri S, Tolou HJ, Peyrefitte CN: **Development of a TaqMan® RT-PCR assay without RNA extraction step for the detection and quantification of African Chikungunya viruses.** *J Virol Methods* 2005, **124**:65–71.

11

Pigs fed camelina meal increase hepatic gene expression of cytochrome 8b1, aldehyde dehydrogenase, and thiosulfate transferase

William Jon Meadus[1*], Pascale Duff[1], Tanya McDonald[2] and William R Caine[3]

Abstract

Camelina sativa is an oil seed crop which can be grown on marginal lands. Camelina seed oil is rich in omega-3 fatty acids (>35%) and γ-tocopherol but is also high in erucic acid and glucosinolates. Camelina meal, is the by-product after the oil has been extracted. Camelina meal was fed to 28 d old weaned pigs at 3.7% and 7.4% until age 56 d. The camelina meal supplements in the soy based diets, improved feed efficiency but also significantly increased the liver weights. Gene expression analyses of the livers, using intra-species microarrays, identified increased expression of phase 1 and phase 2 drug metabolism enzymes. The porcine versions of the enzymes were confirmed by real time PCR. Cytochrome 8b1 (CYP8B1), aldehyde dehydrogenase 2 (Aldh2), and thiosulfate transferase (TST) were all significantly stimulated. Collectively, these genes implicate the camelina glucosinolate metabolite, methyl-sulfinyldecyl isothiocyanate, as the main xeniobiotic, causing increased hepatic metabolism and increased liver weight.

Keywords: Camelina meal, Gene expression, Glucosinolates, Pig liver

Background

Camelina sativa, a member of the family *Brassicaceae* is related to rapeseed [1]. It has commercial value as an oil seed crop for biofuels and biolubricants and can be grown on marginal lands [2]. Camelina seed has an oil content of over 40% (dry weight) and this oil is high in omega-3 fatty acids, gamma tocopherol [3] but also in the monounsaturated omega-9 fatty acid, erucic acid (C22:1 ω-9) [4]. Typical seed crushers will extract the oil content from the seed down to 4% (dry weight) leaving the meal. The Camelina meal still has a problem because, after the oil has been extracted, it can have a total glucosinolate content of ~ 24 μmol/g [5]. In canola meal, when total glucosinolate content is higher than 15 μmol/g of feed, it will reduce feed intake and growth in finishing pigs [6,7].

Glucosinolates are considered bitter to humans [8]. Chickens [9], fish (trout) [10] and pigs [4] will initially show reduced feed intakes depending on the dosage. The maximum recommended dose of glucosinolates for monogastric animals, such as swine, is approximately 2 umoles/g of feed. The glucosinolates are metabolized by endogenous plant enzymes called myrosinase or β-thioglucosidases from gut bacteria, into biologically active compounds, isothiocyanates, indoles and nitriles [8]. Higher doses of the glucosinolate metabolite, thiocyanate, can affect the transport of iodine to thyroid [11]. Glucosinolate metabolic products are mainly associated with the induction of Phase I and Phase 2 biotransformation enzymes [12]. Phase 1 enzymes catalyse a variety of hydrolytic, oxidative and reductive reactions, including the cytochrome P450s involved in metabolizing xenobiotics and toxins [13]. Phase 2 enzymes, such as glutathione S-transferase and UDP-glucuronyl transferase, form conjugation products with xenobiotics and are readily excreted [14]. When feeding xenobiotics, liver is typically the most responsive tissue for phase 1 and 2 expression. However, the phase 1 and 2 enzymes are expressed almost ubiquitously throughout the body, in multiple species, including humans.

Glucosinoates metabolites from camelina, rapeseed and canola differ slightly in the type and quantity. Glucosinolates metabolites from rapeseed and Canadian canola (Brassica napsis) are predominately progoitrin (2-hydroxy-3-butenyl) and gluconapin (3-butenyl) structure [15].

* Correspondence: jon.meadus@agr.gc.ca
[1]AAFC-Lacombe, 6000 C&E Trail, Lacombe, AB, Canada T4L 1 W1
Full list of author information is available at the end of the article

Glucosinolates from camelina are predominantly glucocamelina, which is metabolized into 10-methylsulphinyldecyl isothiocyanate [16]. The structure of 10-methylsulfinyldecyl is closely analogous to sulforaphane which is common to cruciferous vegetables such as broccoli, mustard and cabbage and may be protective against cancer and cardiovascular disease [17]. The present study was undertaken as a preliminary investigation of pigs receiving camelina meal as alternative feed source and as an animal model to assess the potential health benefits.

Methods

Animals and feed

The feeding trial was performed in Lacombe Research center piggery, under the supervision of trained Agriculture and Agri-Food staff that monitored the animals in accordance with guidelines of the Canadian Council on Animal Care. The pig breed in the study was a Large White X Duroc (Hypor Inc, Regina, SK). The camelina meal was provided by Canpresso (Midland, SK, Canada). The HPLC analysis of camelina for glucocamelinin, 10-methylsulfinyldecyl, and 11-methyl-sulfinyldecyl content was made using method AK 1-92 of the American Oil Chemist Society (AOCS) by Bioprofile Testing Labs (St. Paul, MN, USA) (Table 1). The molecular weight of glucocamelinin $C_{18}H_{35}O_{10}S_3N$ is estimated to be 521.65 g/mol; therefore the glucocamelinin content of 22.84 μmol/kg meal, is equivalent to 12.36 g/kg of camelina meal. The feeding trial was composed of three groups fed either, the control (CON) diet, the 3.7% camelina meal supplemented LOW diet, or the 7.4% camelina meal supplemented HIGH diet, for 28 d (Table 2). The soy concentrate (Cargill, Elk River, MN) and corn starch based grower diets were balanced with canola oil to be iso-caloric but not iso-nitrogenous. Compositions of the diets were calculated to meet the Nutrient requirements of swine [18]. The HIGH diet had slightly more estimated crude protein 144.0 g/kg than the control diet at 141.3 g/kg (Table 2). The piglets were started on standard weaner diet [18] prepared by Wetaskiwin Co-op Feeds (Wetaskiwin, Alberta) until the feeding trial test diets, which were started, 2-wk post weaning, at age of 28 d, on 27 barrows, weighing 12.7 ±1.73 kg. The piglets were held individually in metabolic crates penned on concrete floor with slatted sections. The feed was provided *ad libitum* and they had full access to drinking water for the 28 d trial. Their weights and feed were monitored daily; until they reached 56 d. Pigs were euthanized in accordance to CCAC guidelines [19]. Organ tissues were removed *post mortem*, weighed (Table 3) and stored at -20°C.

Table 1 Chemical analysis of camelina meal's crude fat, crude protein and glucosinolate content

Item[1], g/kg	
Dry matter	928
Crude protein	363
Crude fat	143
Calcium	2.1
Phosphorus	9.6
Glucosinolates, μmol/g	23.79
Glucocamelinin	22.84
10-methyl-sulfinyldecyl	0.95

[1]Dry matter basis.

Table 2 Composition of the experimental cornstarch soy-concentrate-based diets without (Control) or with Low (37 g/kg) or High (74 g/kg) levels of added camelina meal

Diet	Control	Low camelina	High camelina
Ingredients[1], g/kg			
Corn starch	639.1	634.1	629.2
Soy concentrate	200.0	183.0	166.0
Camelina meal	-	37.0	74.0
Canola oil	50.0	44.7	39.4
Solka-floc (cellulose)	60.0	50.3	40.6
Dicalcium phosphate (21%)	14.0	14.0	14.0
Sodium chloride	2.7	2.7	2.7
Magnesium oxide (56%)	2.3	2.3	2.3
Sodium bicarbonate	2.0	2.0	2.0
Calcium carbonate	12.0	12.0	12.0
Trace mineral mix	1.0	1.0	1.0
Choline chloride (60%)	1.0	1.0	1.0
Selenium (1,000 mg/kg)	0.5	0.5	0.5
ADE vitamin mix[2]	0.7	0.7	0.7
Vitamin E (1,000 IG/g)	0.2	0.2	0.2
Lysine hydrochloride	5.5	5.5	5.5
DL-Methionine	3.0	3.0	3.0
L-Threonine	3.0	3.0	3.0
Chromic oxide	3.0	3.0	3.0
Calculated contents, g/kg[1]			
ME, kcal/kg[3]	3,754	3,764	3,775
Protein[4]	141.5	144.9	146.3
Dry matter (DM)	932	926	926
Total glucosinolates, μmol/kg	-	0. 88	1.76

[1]As-fed basis.
[2]ADE vitamin mix: vitamin A (50,000 IU), D3 (5,000 IU), E(50 IU) /g.
[3]Metabolizable Energy; (ME; kcal kg^{-1}) calculated based on values, as follows: 3985, cornstarch; 8410, canola and camelina oil; 3250, soy concentrate; 2717, camelina meal.
[4]Protein calculated based on values, as follows: 65% soy concentrate + 36.3% camelina meal + lysine + methionine + threonine.

Table 3 Daily feed intake, average daily gain, feed conversion efficiency and organ tissue weights of pigs after a 28 d trial of cornstarch soy-concentrate-based diets without (Control) or with Low (3.7 g/kg) or High (7.4 g/kg) levels of added camelina meal

Diet	Control	Low camelina	High camelina	SEM[1]	*P*
Item					
Start weight (kg)	13.3	12.5	12.3	0.7	0.39
Final weight (kg)	17.13	17.05	17.05	0.8	0.99
Average daily gain (g/d)	185.1	236.8	233.8	17.7	0.10
Average feed intake (g/d)	605.9	559.2	518.1	18.1	0.06
Feed conversion (feed/gain)	3.72[a]	2.37[b]	2.32[b]	0.3	0.04
Organ tissue, g					
Liver	346.4[b]	418.4[a]	421.7[a]	19.0	0.02
Heart	82.8	90.2	89.0	3.0	0.21
Spleen	35.4	40.1	43.3	2.7	0.14
Thyroid	19.4	17.7	15.9	2.8	0.48

[1]Standard error of the mean ($n = 27$).
[a,b]Values in the same row with different superscripts differ ($P < 0.05$).

RNA microarray analysis and quantitative PCR

Total pig liver RNA was examined for gene expression changes by microarray analysis using, the Rat Drug metabolism: phase 1 array (PARN-068) and the Human Drug Metabolism: phase 2 array (PAHS-069) (SABioscience /Qiagen, Mississauga, ON). Briefly, Liver samples were collected from the pigs within 10 min post slaughter and stored at -80°C in RNAlater (Qiagen). Total RNA was extracted from the livers (100 mg) by homogenizing in 5 moles/L guanidium isothiocyanate and then binding on silica columns for DNase treatment and washing before collecting the RNA in water according to the manufacturer's methods of the Aurum Total RNA fatty and fibrous tissue kit (BioRad, Mississauga, Ontario). The quality and quantity of the RNA was assessed spectrophotometrically at 260 nm, 280 nm and 320 nm. The RNA quality was also checked for good, intact, 28S and 18S rRNA bands by visualization on a 1.0% agarose gel electrophoresis stained with 1 ug/mL ethidium bromide

For microarray analysis, total RNA was collected from the control fed pigs and the 7.4% camelina fed pigs and pooled within the three treatments and was converted to cDNA with a RT^2 First Strand Kit (Qiagen Canada, Mississauga, ON). The pooled cDNA was tested by microarray analysis, on the Rat drug metabolism: phase 1 array (PARN-068) and the Human Drug Metabolism: phase 2 array (PAHS-069) (SABioscience /Qiagen, Mississauga, ON) using the RT^2 SYBR Green polymerase chain reaction (PCR) Master mix according to manufacturer's recommendations on a Mx3000P QPCR real time PCR machine with SYBR gene detection (Agilent Technologies Canada Inc., Mississauga, ON). PCR conditions were hot start at 95°C for 10 min followed by 95°C for 15 sec, 55°C for 15 s, and 60°C for 60 s, for 40 cycles. Quantitative gene analysis was based on comparative $2^{-\Delta\Delta}$Ct method using the RT^2 profiler PCR data analysis program 3.5. (SABiosciences/ Qiagen, Mississauga, ON).

The genes identified by the arrays were used in the search for their closest porcine equivalent in the National Center for Biotechnology Information (NCBI) U.S. National Library of Medicine, GenBank [20] using the Basic Local Alignment Search Tool (Blast) version 2.2.27 program. Primers sequences were selected, using the Primer 3 program v.0.4.0, to amplify the porcine version of the gene transcript [21]. Porcine transcripts were confirmed by sequencing the PCR products on a CEQ8000 machine using a GenomeLab™ DTCS Quick Start dideoxy sequencing kit as per manufacturer's instructions (Beckman Coulter, Mississauga, ON, Canada).

Total RNA was collected from the animals (n = 27) in the three diets and was used to prepare cDNA. RNA (2ug) was combined with, 0.5 ug oligo-dT, 200 mmoles/L dNTPs and preheated at 65°C for 2 min to denature secondary structures. The mixture was then cooled rapidly to +20°C and then 10 uL 5X RT Buffer, 10 mmoles/L DTT and 200 U MMLV Reverse Transcriptase (Sigma-Aldrich, Oakville, ON, Canada) was added for a total volume of 50 uL. The RT mix was incubated at 37°C for 90 min. then stopped by heating at 95°C for 5 min. The cDNA stock was stored at – 20°C. The yield of cDNA was measured according to the PCR signal generated from the internal standard housekeeping gene β-actin amplified from 0.1 μL of the cDNA solution. The volume of each cDNA sample was adjusted to give approximately the same exponential phase PCR signal strength for β-actin after 20 cycles [22]. The primers for the porcine version of the microarray selected genes are shown in (Table 4). The cDNA [100 ng] was used in a RT^2 SYBR Green master mix with 10 umoles/L of each primer run on Mx300P QPCR machine at 95°C/10 min for hot start followed by 40 cycles of 95°C/30 s, 56°C/30 s and 72°C/60 s. Relative gene analysis was based on comparative $2^{-\Delta\Delta Ct}$ method [23]. The reference genes were averaged between internal housekeeping genes, GADPH and β–actin.

Statistical analysis

Animals' responses to the diets were analyzed using ANOVA followed by Duncan's test for differences in the group means. The gene analysis data was calculated using the comparative $2^{-\Delta\Delta Ct}$ method and significance between treatment groups was determined using the Student's t-test. Statistical significance was accepted at $P < 0.05$ and trends were indicated at $P < 0.1$. All data were run on SAS version 9 [24].

Results and discussion

Diet

The estimated chemical composition of the camelina meal supplemented diets for crude protein and crude fat are outlined (Tables 1 and 2). The amount of crude protein in the meal was estimated to be ~363 g/kg and the amount of crude fat was 143 g/kg, which is high at ~14%, considering the oil was extracted by cold pressure [6]. The concentration of glucosinolates was measured to be ~23.70 μmol/g in the camelina meal using a chromatography method [25]. This content is significantly higher than the average value of 7.8 μmol glucosinolate /g of meal from modern conventional varieties of canola [15]. However, the estimated final concentration of glucocamelinin in the LOW and HIGH supplemented diets was well within the safe region at 0.82 μmol/kg and 1.63 μmol/kg of feed (Table 2).

Camelina contains a high concentration of the antioxidant, gamma tocopherol 1,000 μg/g, which is substantially higher, relative to other oils seed crops such as, canola and flax [3]. Gamma tocopherol is the most commonly found form of vitamin E in plants [26]. It provides antioxidant protection to the oils and gives a nutty aroma to camelina oil. For humans however, alpha-tocopherol is the preferred form and the only isomer selected by the tocopherol transfer protein (α-TTP) and tocopherol associated protein (α-TAP) in the liver [27]. Gamma tocopherol is absorbed but quickly excreted in the urine.

In addition to glucosinolates, Camelina also contains the monounsaturated omega-9 fatty acid, erucic acid. Erucic acid (C20:1 w-9) is a large component of mustard oil and can account to approximately 5% of camelina seed and is suspected to reduce the palatability of feed for monogastric swine [28]. The allowable amount of erucic acid in canola seed in Canada is set at 2% [29]. Erucic acid content is a large component of mustard oil and its consumption by human is still controversial. It implicated in causing thrombocytopenia but it is also part of Lorenzo's oil [5] used to the treat adrenoleukodystrophy [6]. The crude fat in meal of camelina after crushing and processing was 14.3% and this fat is expected to contain less than 0.715 g erucic acid/kg of meal. The level of erucic acid in the LOW diet was expected to be 0.026 g/kg feed. In the HIGH diet, the erucic acid content was estimated to be 0.053 g/kg of feed (~ 0.005%), which is below the level expected for any detectable physiological effect [30].

Animals and weights

Mean initial weights of the pigs were 12.7 ±1.73 kg for the start of trial and at the end of the 28 day trial; the pigs weighed an average of 17.1 ±2.12 kg at 56 d (Table 3). The animals on the camelina meal supplemented diets had a significant improvement in their feed efficiency. This may have been due to an increase in the available lysine protein and metabolizable energy available from the camelina meal supplement; although, the soy concentrate, corn starch based control diet, met NRC [18] recommended lysine levels estimated at ~ 14.6 g/kg feed for 20 kg pigs (Table 2). It may have been also due to a reduced feed intake in the LOW and HIGH diets (Table 3). The pigs did have some problems adjusting to the camelina meal diets. The pigs on HIGH diet did have an initial aversion to the meal, with reduced average daily feed intakes of 518.1 ±70.4 g/d, as compared to the CON diet intake at 605.9 ±111.3 g/d but this was only shown as a trend $P < 0.1$. There was a significant increase in the liver weights

Table 4 Forward and reverse *porcine* primer sequences and accession numbers of genes used in quantitative RT-PCR analysis

Gene	Forward primer	Reverse primer	GenBank accession #	cDNA size, bp
CYP8b1	5′-aagtgggccggctccagtgt-3′	5′-gcccgagccccatggcatag-3′	NM_214426.1	625
Aldh2	5′-gcatcggcatgttgcgccct-3′	5′-ggtaggtccggtcccgctca-3′	NM_001044611.2	374
TST	5′-cgggctcaagggcggtacct-3′	5′-tttgcccacggggcatggac-3′	XM_001926303	435
Gst*mu*5	5′-tcgcccgcaagcacaacatgt-3′	5′-acaagcagtgcaagtccgcct-3′	AK233626.1	453
β-actin	5′-acatcaaggagaagctgtgc-3′	5′-ttggcgtagaggtccttgc-3′	AY550069	256

from the pigs fed the HIGH diet. The liver weights at the end of the trial was 346.4 g/pig for the CON group, 418.4 g/pig for MED group and 427.7 g/pig for HIGH fed group (Table 3). This indicates extra hepatic activity caused by increasing the camelina meal. The pigs fed the LOW and HIGH camelina meal diet also indicated a reduction in their thyroid weights (Table 3). This may be due to a disruption of iodine absorption and thyroid activity caused by glucoinosinolates in the camelina meal and will have to investigated further; possibly in a longer (>28 d) trial in which the iodine level in the pigs are measured.

Hepatic gene expression

The total RNA from the pigs livers fed the High level of camelina meal was compared with animal fed the CON diets and was tested on two available microarrays representing 168 genes involved in drug metabolism (Figure 1). The Rat Drug Metabolism Phase I Enzymes RT^2 Profiler PCR Array PARN – 068A contains 84 genes involved in phase I drug metabolism. Phase I drug metabolism enzymes make compounds more hydrophilic. This array represents genes involved in Phase I drug metabolism reactions including oxidation, reduction, hydrolysis, cyclization, decyclization and members of the Cytochrome P450 enzyme family. The Human Drug Metabolism Phase II Enzymes RT^2 Profiler™ PCR Array PAHS-069Z, contains 84 genes involved in the enzymatic processes of drug biotransformation. Phase II drug metabolism enzymes catalyze the conjugation of lipophilic compounds with hydrophilic functional groups or moieties to form water-soluble conjugates that can be cleared from cells and from the body. This array represents genes encoding the phase II drug metabolism enzymes catalyzing such reactions as glutathione conjugation, glucuronidation, sulfation, methylation, amino acid conjugation, epoxidation, and esterification. The pig RNA transcripts were expected to be only moderately homologous with humans and rats; therefore, an annealing temperature reduction of 55°C for 15 s was made in the PCR program to overcome some differences between primer sequences.

The rat Drug Metabolism Phase 1 microarray identified the cytochrome P450 -8b1 (Cyp8b1) and the aldehyde dehydrogenase 2 (Aldh2) gene transcription as being stimulated in the pig livers by camelina meal feeding. The human drug metabolism phase 2 array identified glutathione S-transferase mu 5 (GSTM5) and thiosulfate transferase (TST), as being significantly up-regulated > 4-fold (Figure 1) relative to the control livers, as determined by the RT^2 profiler PCR data analysis program 3.5.

A comparison of the gene sequences identified in the rat array and porcine (Sus scrofa) version of the genes was 87% between the rat Aldh2 GenBank# NM_032416 and

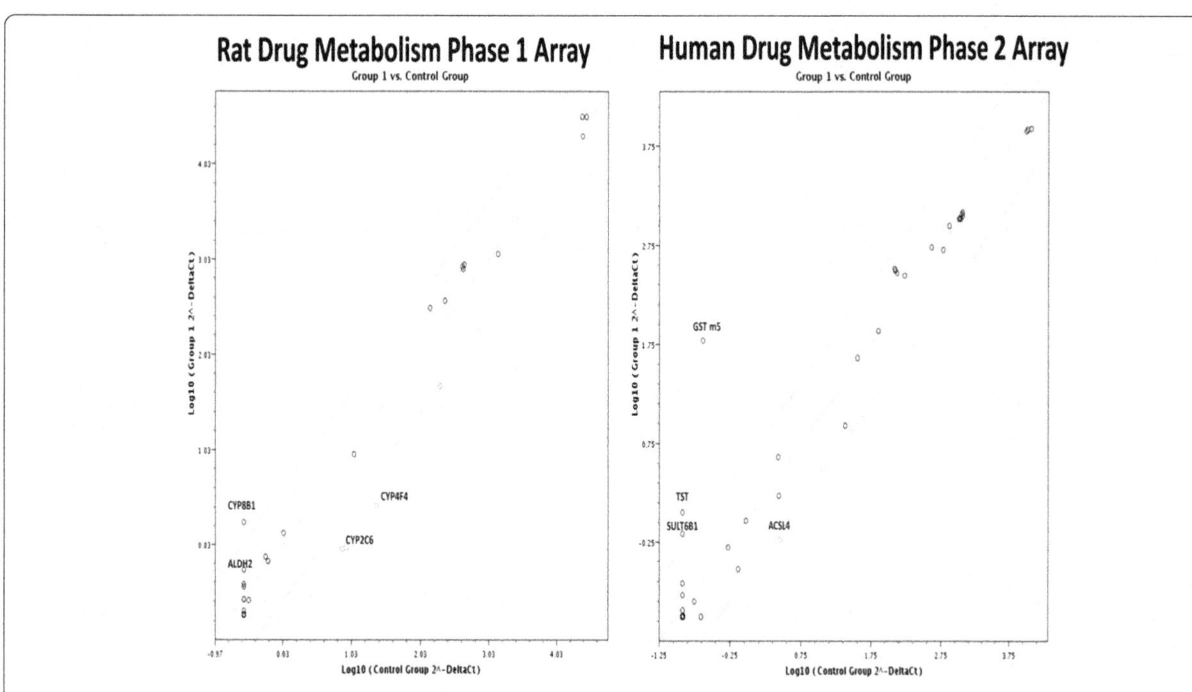

Figure 1 Total liver RNA examined for gene expression changes by microarray analysis using, the *Rat* drug metabolism: phase 1 array (PARN-068) and the *Human* Drug Metabolism: phase 2 array (PAHS-069). The genes that were significantly up or down regulated in the microarrays are labelled. The RNA was from a pooled sample of pigs fed the control verses the High 7.4% camelina diet as outlined in the materials and methods.

the porcine GenBank# NM_001044611.2 and 80% between the rat Cyp8b1 GenBank# NM_031241 and the porcine GenBank# NM_214426.1. Comparison of the genes identified in the human array with porcine (Sus scrofa) mRNA was 86% between the human GSTm5 and the porcine GSTm2 GenBank# AK233626.1 and 89% between the human TST GenBank# NM_003312 and the porcine TST GenBank# XM001926303.2. New porcine specific primers were created based on the aforementioned porcine sequences (Table 4).

The genes investigated directly for mRNA expression by quantitative real-time PCR using porcine specific primers on the livers of the fed pigs LOW and HIGH amounts of camelina meal are shown in Figure 2. Cyp8b1 mRNA was up-regulated approximately 80-fold in the liver tissue of pigs supplemented with High camelina meal. The transcripts for the TST and Aldh2 were increased approximately 1.8 and 3.2 fold but these were only weakly significant ($P < 0.1$). The Gstm5 transcript was not stimulated by adding camelina to their diet.

Cytochrome P450 8B1 (Cyp8b1) is primary a microsomal sterol hydroxylase involved in bile acid formation [31]. Porcine Cyp8b1 will catalyse the hydroxylation of cholic acid into hyocholic acid. This activity is developmentally dependent giving higher expression in the fetal than the adult pig liver. The basic structure of cholic acid is cholesterol, which is quite different from the metabolites of camelina meal. Glucosinolates of camelina are mainly glucocamelina, which is a methyl-sulfinyldecyl isothiocyanates . The transcriptional activation Cyp8b1 is probably the same mechanism as caused by broccoli. Metabolites of broccoli glucosinolate contain sulforaphane which is a methyl-sulfinylbutane isothiocyanate. Sulforaphane activates phase 2 drug enzymes by the nuclear factor, E2 p-45-related factor 2 (Nrf2) transcription factor [32]. The Nrf2 factor binds the antioxidant response element (ARE) to activate transcription of the respective genes [33]. The same Nrf2 transcription factors have been shown to activate the other phase 2 enzymes induced by camelina meal including the Aldh2, TST and GstM1 [34]. Future trials should examine if camelina thiocyanate and its derivatives are actually inducing the Nrf2 transcription factor and the genes which respond to the ARE, such as the phase 1 enzyme Cyp1A1.

Conclusions

Camelina sativa has high erucic acid and high glucosinolate content but it can be grown on marginal land and has good oil yield. The continued industrial use of camelina oil as a biofuel may provide a cheap source of protein as meal for the pig industry but more study into the nutritional aspects is needed. Recent information on the biological action of glucosinolates and erucic acid has suggested that these are not as detrimental as originally thought [4]. Rats are relatively slow at digesting erucic acid and due to its accumulation in the heart , thought to be cardiotoxic [35] but recent studies have shown that the animals will adjust to the feed and increase erucic acid metabolism within their peroxisomes [36]. Glucoinsolates were thought to be only anti-nutritional but now the research evidence has demonstrated that glucosinolates may act as a chemo-protective, anti-cancer agent [12]. This trial on camelina meal feeding in pigs, indicates a potential anti-carcinogen benefit, through the stimulation hepatic expression of phase 1 and 2 xenobiotic detoxifying enzymes.

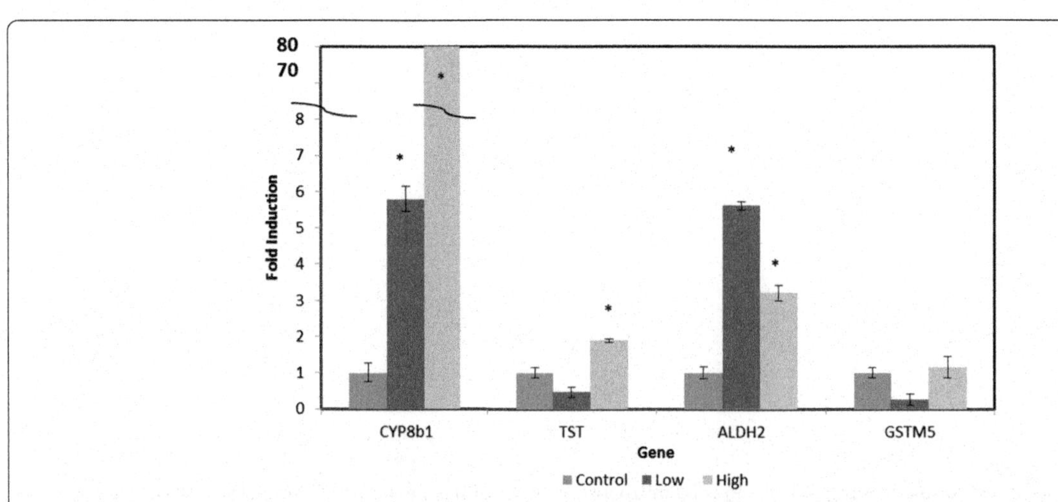

Figure 2 The effect of camelina meal on the pig liver gene expression level of Cyp8B1, TST, Aldh2, and Gstm5. Transcripts were determined by real-time PCR with *porcine* specific primers. Pig diets were supplemented with either the Control (0%), Low (3.7%) or High (7.4%) levels of camelina meal for 24d. Error bars represent the standard error of the mean. * indicates $P > 0.05$ relative to the Control using Student's *t-test*.

Abbreviations
AldH2: Aldehyde dehydrogenase; CYP8b1: Cytochrome P450 8b1; Gstm5: GlutathioneS- transferase mu 5; PCR: Polymerase chain reaction; TST: Thiosulfate transferase.

Competing interests
The authors declare that they have no competing interests.

Authors' contributions
WJM performed the analysis for gene expression and supported the findings with statistical analysis and was the main author of the manuscript. PD carried out the real-time PCR trials. TM was involved in providing the camelina meal. WC performed the animal trials including the design of the diets. All authors read and approved the final manuscript.

Acknowledgements
This work was supported by Agriculture and Agri-food Canada. The authors would also like to thank the staff as the Lacombe piggery for their assistance.

Author details
[1]AAFC-Lacombe, 6000 C&E Trail, Lacombe, AB, Canada T4L 1 W1. [2]School of Innovation, Olds College, 4500-50th St., Olds, AB, Canada T4H 1R6. [3]Caine Research Consulting, Box #1124, Nisku, AB, Canada T9E 8A8.

References
1. Gehringer A, Friedt W, Luhs W, Snowdon RJ: **Genetic mapping of agronomic traits in false flax (Camelina sativa subsp. sativa).** *Genome* 2006, **49:**1555–1563.
2. Gesch RW, Archer DW: **Double-cropping with winter camelina in the northern Corn Belt to produce fuel and food.** *Ind Crop Prod* 2013, **44:**718–725.
3. Zubr J, Matthaus B: **Effects of growth conditions on fatty acids and tocopherols in < i > Camelina sativa</i > oil.** *Industrial crops and products* 2002, **15:**155–162.
4. Ni Eidhin D, Burke J, Lynch B, O'Beirne D: **Effects of dietary supplementation with camelina oil on porcine blood lipids.** *J Food Sci* 2003, **68:**671–679.
5. Schuster A, Friedt W: **Glucosinolate content and composition as parameters of quality of < i > Camelina</i > seed.** *Industrial crops and products* 1998, **7:**297–302.
6. Bell JM: **Nutrients and toxicants in rapeseed meal: a review.** *J Anim Sci* 1984, **58:**996–1010.
7. Caine WR, Aalhus JL, Dugan MER, Lien KA, Larsen IL, Costello F, McAllister TA, Stanford K, Sharma R: **Growth performance, carcass characteristics and pork quality of pigs fed diets containing meal from conventional or glyphosate-tolerant canola.** *Can J Anim Sci* 2007, **87:**517–526.
8. Mithen RF, Dekker M, Verkerk R, Rabot S, Johnson IT: **The nutritional significance, biosynthesis and bioavailability of glucosinolates in human foods.** *J Sci Food Agric* 2000, **80:**967–984.
9. Pekel AY, Patterson PH, Hulet RM, Acar N, Cravener TL, Dowler DB, Hunter JM: **Dietary camelina meal versus flaxseed with and without supplemental copper for broiler chickens: live performance and processing yield.** *Poult Sci* 2009, **88:**2392–2398.
10. Morais S, Edvardsen RB, Tocher DR, Bell JG: **Transcriptomic analyses of intestinal gene expression of juvenile Atlantic cod (Gadus morhua) fed diets with Camelina oil as replacement for fish oil.** *Comp Biochem Physiol B Biochem Mol Biol* 2012, **161:**283–293.
11. Schone F, Jahreis G, Lange R, Seffner W, Groppel B, Hennig A, Ludke H: **Effect of varying glucosinolate and iodine intake via rapeseed meal diets on serum thyroid hormone level and total iodine in the thyroid in growing pigs.** *Endocrinologia experimentalis* 1990, **24:**415–427.
12. Zhang Y, Talalay P, Cho CG, Posner GH: **A major inducer of anticarcinogenic protective enzymes from broccoli: isolation and elucidation of structure.** *Proc Natl Acad Sci U S A* 1992, **89:**2399–2403.
13. Gross-Steinmeyer K, Stapleton PL, Tracy JH, Bammler TK, Strom SC, Eaton DL: **Sulforaphane- and phenethyl isothiocyanate-induced inhibition of aflatoxin B1-mediated genotoxicity in human hepatocytes: role of GSTM1 genotype and CYP3A4 gene expression.** *Toxicol Sci* 2010, **116:**422–432.
14. Nho CW, Jeffery E: **The synergistic upregulation of phase II detoxification enzymes by glucosinolate breakdown products in cruciferous vegetables.** *Toxicol Appl Pharmacol* 2001, **174:**146–152.
15. Shahidi F, Gabon JE: **Individual glucosinolates in six canola varieties.** *J Food Qual* 1989, **11:**421–431.
16. Kjaer AG R, Jensen RB: **Isothiocyanates XXI. 10-methylsuphinyldecyl isothiocyanate, a new mustard oil present as a glucoside (glucocamelinin) in Camelina species.** *Acta Chem Scand* 1956, **10:**1614–1619.
17. Anwar-Mohamed A, El-Kadi AO: **Sulforaphane induces CYP1A1 mRNA, protein, and catalytic activity levels via an AhR-dependent pathway in murine hepatoma Hepa 1c1c7 and human HepG2 cells.** *Cancer Lett* 2009, **275:**93–101.
18. NRC: *Nutrient Requiements of Swine.* 10th edition. Washington DC: National Academy of Sciences; 1998.
19. CCAC: *CCAC Guidelines on: The Care and Use of Farm Animals in Research, Teaching, and Testing.* 8th edition. Ottawa, ON: Canadian Council on Animal Care; 2009.
20. Benson DA, Karsch-Mizrachi I, Clark K, Lipman DJ, Ostell J, Sayers EW: **GenBank.** *Nucleic Acids Res* 2012, **40:**D48–D53.
21. Rozen SHJS (Ed): *Primer3 on the WWW for General Users and for Biologist Programmers.* Totowa, NJ, USA: Humana Press; 2000.
22. Meadus WJ, Duff P, Rolland D, Aalhus JL, Uttaro B, Dugan MER: **Feeding docosahexaenoic acid to pigs reduces blood triglycerides and induces gene expression for fat oxidation.** *Can J Anim Sci* 2011, **91:**601–612.
23. Schmittgen TD, Livak KJ: **Analyzing real-time PCR data by the comparative C(T) method.** *Nature protocols* 2008, **3:**1101–1108.
24. SAS (Ed): *SAS user's guide: Stastics. SAS for Windows, Version 9.1.* Cary, NC, USA: SAS Institute Inc; 2003.
25. AOCS/Ak-1-92: *Determination of Glucosinolate Content in Rapeseed and Canola by HPLC method Ak 1-92,* Official methods and recommended practices of the American Oil Chemists' Society 6th ed 2nd print Champain, Ill; 2012:8.
26. Wagner KH, Kamal-Eldin A, Elmadfa I: **Gamma-tocopherol–an underestimated vitamin?** *Ann Nutr Metab* 2004, **48:**169–188.
27. Doring F, Rimbach G, Lodge JK: **In silico search for single nucleotide polymorphisms in genes important in vitamin E homeostasis.** *IUBMB life* 2004, **56:**615–620.
28. Habeanu M, Hebean V, Taranu I, Ropota M, Lefter N, Marin D: **Dietary ecologic camelina oil - a beneficial source of n-3 PUFA in muscle tissue and health status in finishing pigs.** *Romanian Biotechnological Letters* 2011, **16:**6564–6571.
29. CFIA-DD96-07: *Determination of Environmental Safety of Monsanto Canada Inc.'s Roundup® Herbicide-Tolerant Brassica napus Canola Line GT200,* Canadian Food Inspection Agency, vol. DD96-07 supplement; 2011.
30. Borg K: **Physiopathological effects of rapeseed oil: a review.** *Acta Med Scand Suppl* 1975, **585:**5–13.
31. Lundell K, Wikvall K: **Gene structure of pig sterol 12alpha-hydroxylase (CYP8B1) and expression in fetal liver: comparison with expression of taurochenodeoxycholic acid 6alpha-hydroxylase (CYP4A21).** *Biochimica et biophysica acta* 2003, **1634:**86–96.
32. Thimmulappa RK, Mai KH, Srisuma S, Kensler TW, Yamamoto M, Biswal S: **Identification of Nrf2-regulated genes induced by the chemopreventive agent sulforaphane by oligonucleotide microarray.** *Cancer Res* 2002, **62:**5196–5203.
33. Huang J, Tabbi-Anneni I, Gunda V, Wang L: **Transcription factor Nrf2 regulates SHP and lipogenic gene expression in hepatic lipid metabolism.** *Am J Physiol Gastrointest Liver Physiol* 2010, **299:**G1211–G1221.
34. Qin S, Chen J, Tanigawa S, Hou DX: **Microarray and pathway analysis highlight Nrf2/ARE-mediated expression profiling by polyphenolic myricetin.** *Mol Nutr Food Res* 2013, **57:**435–446.
35. Murphy CC, Murphy EJ, Golovko MY: **Erucic acid is differentially taken up and metabolized in rat liver and heart.** *Lipids* 2008, **43:**391–400.
36. de Wildt DJ, Speijers GJ: **Influence of dietary rapeseed oil and erucic acid upon myocardial performance and hemodynamics in rats.** *Toxicol Appl Pharmacol* 1984, **74:**99–108.

12

Transcriptome responses to heat stress in hypothalamus of a meat-type chicken

Hongyan Sun[1], Runshen Jiang[2], Shengyou Xu[2], Zebin Zhang[1], Guiyun Xu[1], Jiangxia Zheng[1] and Lujiang Qu[1*]

Abstract

Background: Heat stress has resulted in great losses in poultry production. To address this issue, we systematically analyzed chicken hypothalamus transcriptome responses to thermal stress using a 44 k chicken Agilent microarray,

Methods: Hypothalamus samples were collected from a control group reared at 25°C, a heat-stress group treated at 34°C for 24 h, and a temperature-recovery group reared at 25°C for 24 h following a heat-stress treatment. We compared the expression profiles between each pair of the three groups using microarray data.

Results: A total of 1,967 probe sets were found to be differentially expressed in the three comparisons with $P < 0.05$ and a fold change (FC) higher than 1.5, and the genes were mainly involved in self-regulation and compensation required to maintain homeostasis. Consistent expression results were found for 11 selected genes by quantitative real-time PCR. Thirty-eight interesting differential expression genes were found from GO term annotation and those genes were related to meat quality, growth, and crucial enzymes. Using these genes for genetic network analysis, we obtained three genetic networks. Moreover, the transcripts of heat-shock protein, including Hsp 40 and Hsp 90, were significantly altered in response to thermal stress.

Conclusions: This study provides a broader understanding of molecular mechanisms underlying stress response in chickens and discovery of novel genes that are regulated in a specific thermal-stress manner.

Keywords: Chicken, Heat shock protein, Heat stress, Hypothalamus, Microarray, Transcription

Introduction

With increasing importance of global warming issues, heat stress, also known as hyperthermia, could play a vital factor in affecting animal performance. Although chickens can use a variety of physiological mechanisms to regulate their body temperatures, heat stress could adversely affect chickens, especially broilers, in matters such as high mortality and consistent decrease in egg weight, shell thickness, rate of egg production, growth rate, meat quality, body weight [1,2], and immune capability as well as blood biochemical indicators [3,4]. St. Pierre et al. [5], reported that heat stress caused economic losses of between $1.69 billion and $2.36 billion per year in livestock industries in United States, with $51.8 million in the broiler sector alone.

During the past few decades, researchers have paid increased attention to physiological, biochemical, and immune capability changes of heat-stressed chickens. Deyhim and Teeter [6] found physiological responses to heat stress involve changes in respiration rate and blood pH, plasma concentration of ions, cardiovascular functions and hormonal effects. Heat stress also increased glucose and circulatory cortisol levels and decreased blood protein level. Khajavi et al. [7], reported that occurrence of $CD4^+$ and $CD8^+$ lymphocytes and antibody production decreased under heat stress.

Recently, transcriptome comparison has become a popular methodology for studying heat stress because it can elucidate the heat stress mechanism based on genetic level. Li et al. [8], using broiler breast tissue, investigated 110 differentially expressed genes during chronic cyclic heat stress, including 4 new genes that were related to heat stress. Wang et al. [9], reported 169 up-regulated and 140 down-regulated genes responded to

* Correspondence: quluj@163.com
[1]Department of Animal Genetics and Breeding, National Engineering Laboratory for Animal Breeding, College of Animal Science and Technology, China Agricultural University, Beijing 100193, China
Full list of author information is available at the end of the article

acute heat stress; most of the differential expression genes were related to transport, signal, and metabolism.

Previous transcriptome research studies have chosen liver, breast, and testis as materials [8-10] to decipher heat stress, but no one has used the hypothalamus as a subject for studying the mechanism of heat stress and genetic networks. The hypothalamus, linking the nervous system to the endocrine system via the pituitary gland, plays the role of a central regulator in temperature regulation in poultry. It is well known the preoptic area (POA) and anterior hypothalamic area (AH) of the hypothalamus is the thermoregulatory center [11] and that hypothalamic areas such as the paraventricular hypothalamic nucleus (PVN), supraoptic nucleus (SO) [12-14], and lateral hypothalamic (LH) are activated under heat stress [15,16].

In addition, although RNA sequencing is currently a popular technology for studying transcriptome, it is very costly and results in less mature analysis than a microarray. The objective of this study is thus to use a microarray to detect the versatile gene expression profile in the hypothalamus, to determine the genes associated with heat stress, and to gain insight into the potential mechanisms underlying chickens' response to heat stress. Characterizing the hypothalamus transcriptome of heat-stressed broilers will help clarify the effects of heat stress on broiler nerve centre and signal transduction. This information will also provide a platform for future investigation into genetic networks relevant to commercial broilers' response to heat stress.

Materials and methods

Experimental animals and sampling

All experimental protocols were approved by the Committee for the Care and Use of Experimental Animal at Anhui Agricultural University. The purpose of this study was to find how broiler hypothalamus responds to heat stress in influencng growth rate and meat quality at the genetic level. Moreover, considering the particular economic value of hens, our priority was to use hen broilers as research subjects. Because broilers don't have sweat glands, it is difficult for them to shed heat in hot weather (30-35°C). At a temperature of 34°C with 10-50% relative humidity, 50-80% of the heat loss was by evaporative heat shed through panting (http://www.heatstress.info/HeatStressExplained/BodyHeatregulationinPoultry.aspx). In this study, we used a full-sib and half-sib chicken population to reduce the impact of genetic background variation. Thus, twelve 56-day-old full-sib and half-sib commercial female broilers were randomly divided into three groups, including a control group, a heat-treated group, and a temperature-recovery group with four replicates in each group. All the birds were initially reared in an artificial climate chamber (relative humidity of 40-60%) with free access to food and water at 25°C. At the age of 56 days, both the heat-treated (HS) group and the temperature-recovery (TR) group were subjected to a temperature of 34°C for 24 h. After heat stress, hypothalamuses of four HS hens were dissected from their brains. The other chickens as a control group (CL) were further kept at a normal temperature of 25°C for 24 h. After exposure to heat, the hypothalamuses of CL and TR groups were collected. The hypothalamus collection procedure was as follows: (1) After decapitation, carefully remove the brain; then bisect it by making a section perpendicular to the cortical surface at the caudal end of the hemisphere. (2) Before identifying the hypothalamus, carefully remove the meninges and then, using tweezers, carefully pull the meninges layers away from the brain. (3) When the hypothalamus has been found, carefully remove the whole specimen. All the tissues were preserved overnight in RNAfixer (BioTeke Co. Ltd, Beijing, China) at 4°C and transferred to −80°C until RNA isolation. The hypothalamuses from the three groups were then subjected to gene expression analysis.

Total RNA isolation, labeling, and microarray hybridization

Total RNA was isolated from the hypothalamus using TRIZOL Reagent according to manufacturer's instructions (Life Technologies, CA, US). RNA quantity and purity of each sample were detected using a NanoDrop ND-1000 spectrophotometer (Thermo Fisher Scientific Inc, Delaware, US). The RNA integrity of each sample was determined using an Agilent 2100 Bioanalyzer. The processed RNA had a 28S/18S > 1.5 and the RNA Integrity Numbers (RINs) for the samples were all higher than 7.6, indicating that the quality of the samples was suitable for microarray analysis. Four biological replicates were used in each group and cyanine 3-labeled nucleotide was conducted to prevent dye bias during sample labeling. A 2 μg isolated total RNA from each tissue sample was reverse-transcribed into cDNA, then into cRNA incorporating cyanine 3-labelled nucleotide. The labeled RNA was purified using QIAGEN RNeasy® Mini Kit (Qiagen, CA, US). A total of 875 ng Cy3 labeled cRNA was hybridized to 44 K chicken Agilent arrays (ID: 026441). Equal amounts of cRNA were hybridized using a Gene Expression Hybridization Kit (5188–5242, Agilent, USA). Arrays were performed at 60°C for 17 h in Agilent microarray hybridization chambers. The hybridized slides were washed using a Gene Expression Wash Buffer Kit (Agilent technologies, Santa Clara, CA, US) before stabilization and drying solutions were applied (Agilent Technologies, CA, US). Arrays were scanned using an Agilent Microarray Scanner (Agilent Technologies, Santa Clara, CA, US) with a scan resolution of 5 μmol/L, PMT 100%, 10%. The data were then compiled with Feature Extraction software 10.7 (Agilent Technologies, Santa Clara, CA, US).

Microarray data analysis

Array normalization and error detection analysis was carried out using a Quantile algorithm supplied with Gene Spring Software 11.0 (Agilent Technologies, CA, US). The "filter on flags feature" with software algorithms determined whether spots were "present", "marginal", or "absent" and values representing poor intensity and low dependability were removed from the raw data. The "absent" spots represented signals not positive, significant or above background level. A "present" spot was one for which the output was uniform, saturated, and significantly above background level. Spots were deemed "marginal" when the signals satisfied the main requirements but were outliers relative to the typical values for the other genes. We used only the items categorized as "present" and "marginal" for further analysis and compared the differential expression transcripts between each pair of the three groups. i.e., HS compared to CL (HS vs. CL), HS compared to TR (HS vs. TR), or TR compared to CL (TR vs. CL).

Principal-component analysis (PCA) and heat-map analysis were performed by SAS software from SHANGHAI BIOTECHNOLOGY CORPORATION eBioService (http://www.ebioservice.com). The differentially-expressed (DE) genes in each comparison were selected using SAS online software (http://www.ebioservice.com) and a local R package. With respect to the number of expression genes in the hypothalamus, we set the cutoff as $P < 0.05$, FDR < 0.15 and fold change (FC) > 1.5 to obtain more DE genes. Although this FDR cutoff is much higher than the usual FDR < 0.05, there are many good published papers using a variety of different FDR cutoff values according to reported experiments [17,18]. The significant DE gene clusters from heat-map results were used to do pathway analysis using Ingenuity Pathway Analysis (IPA) (Ingenuity® Systems, www.ingenuity.com). IPA is a very useful online tool for analyzing biological and molecular networks; it transforms gene network into relevant pathways using the genes' functional annotation and known molecular interactions [19-21].

Quantitative real time PCR (qRT-PCR)

Eleven genes were quantitatively determined using real-time PCR with the same 12 samples tested by the microarray to confirm gene-expression patterns observed using the microarray. PCR-specific primers for target genes were designed with Primer Express software (Version 3.0, Applied Biosystems, US) and were synthesized by Shanghai Sangong Biological Engineering Technology & Services Co, Ltd. (Shanghai, China). All the primers were sequenced and worked well. β-actin was used as the internal control for the target genes. Real-time PCR primer information is shown in Table 1. The reverse transcription reaction system was conducted in a total volume of 25 μL, 1 μg of total RNA, 10 mmol/L oligo T primer and murine leukemia virus (MLV) reverse transcriptase (Promega Biotech Co., Ltd, Beijing, China). The PCR system was quantified using the ABI 7300 system (Applied Biosystems, US) in a final volume of 15 μL with 1 μL of cDNA, 200 nmol/L of each primer and a 1 × PCR mix (Power SYBR® Green PCR Master Mix, Applied Biosystems, US). The optimum thermal cycles were performed as follows: denaturation at 95°C for 10 min, then 40 cycles of 95°C for 15 s, 60°C for 1 min, 95°C for 15 s, 60°C for 30 s, and 95°C for 15 s. Every sample was supplied in duplicate and the average critical threshold cycle (Ct) was used for calculating the relative quantification by fold-change and statistical significance. At the same time, to ensure that a single PCR product was amplified in each reaction, dissociation curve analysis at the end of amplification and gel electrophoresis was performed after the real-time PCR. The resulting data for each gene were calculated by the expression $2^{-\Delta\Delta\Delta Ct}$, where the $-\Delta\Delta Ct$ is the sum of :[$Ct_{gene} - Ct_{\beta\text{-actin}}$] (treatment) – [$Ct_{gene} - Ct_{\beta\text{-actin}}$] (control) for the relative gene expression level among the three comparison groups. All analyses were subjected to a Student's t-test to confirm statistically-significant gene expression.

Results

Effects of heat stress on gene expression in different comparison groups

A comparative analysis was performed by comparing gene expression profiling between each pair of the three groups CL, HS, and TR. Generally, more DE genes were present in the comparison of HS vs. CL and HS vs. TR compared to that of TR vs. CL. In summary, there were 1,239, 846, and 9 probe sets DE in HS vs. CL, HS vs. TR, and TR vs. CL, respectively. These results showed that the TR group is more similar to the CL group. Moreover, 638 of 1,239 probe sets, 493 of 846 probe sets and 5 of 9 probe sets are up-regulated in HS vs. CL, HS vs. TR, and TR vs. CL, respectively (Figure 1). The FC range was 1.5 to 25 in HS vs. CL, 1.5 to 36 in HS vs. TR, and 1.5 to 16 in TR vs. CL. In addition, the pattern of the total 1,743 differential expression probe sets is shown in Figure 2. There were 890, 499, and 3 probe sets uniquely DE in HS vs. CL, HS vs. TR, and TR vs. CL, respectively. Contrasts HS vs. CL and HS vs. TR shared 345 DE genes (Figure 2). All the microarray data have been deposited in the GEO repository (GSE37400, NCBI tracking system #16562670).

Heat map and Principal component analysis

Principal component analysis (PCA) was used to classify the broilers based on their gene expression patterns in

Table 1 Primer sequences for qRT-PCRs

Gene symbol	Accession no.	Forward (5′-3′)	Reverse (5′-3′)	Product size, bp
DNAJC13	XM_418787	TGAAAGGAGCAGGTTTGGTGAT	GTGTGCAAGTGACGAGGTAAGG	115
HSPCB	NM_206959	CAGGCGCAGACATCTCCAT	TGACGACTCCCAGGCATACTG	125
HMOX1	NM_205344	TCCCTCCACGAGTTCAAGCT	CGGAAAATAAACAGGAGCATAGACA	114
TH	NM_204805	GAGAGCCATGCTGAACCTGTT	GGGCTTTCGGCTGAGTCTT	130
EDN3	AB235921	TTTCGGTGCTCTTGTTTGGAT	CGGTCCTTCTCTGTTGTTTTCAC	110
GCH1	NM_205223	CAGGAACGCCTTACCAAACAA	CTGGTTGCCGTTTTACTGTTCA	140
GPR23	CR353562	TCAGATCGGGACAAACAAAGAGA	CCGCACGAGAGCATACAAGA	115
SDC1	XM_419972	TGTAGTGACAGAGGAGCCAGTTG	TGAAGCACCGAAGGGAAGTG	135
PDK4	CR387492	AGTTCCATCAGAAAAGCCCAGAT	TCCTTGTGCCATTGTAGGGACTA	110
ITGA8	NM_205288	TAGCCGTGGCAGAACCTTACA	CCATAGCCGGGACCTGATTT	140
SFRP1	NM_204553	CCAGTTCCCACAGGATTATGTCT	CCTCCGACTTCATCTCGTTGTC	116
β-actin	NM_205518	GAGAAATTGTGCGTGACATCA	CCTGAACCTCTCATTGCCA	132

one-way analysis of variance (ANOVA) with $P < 0.05$ and FC > 1.5 among three groups (12 samples). The HS group had a transcription profile different from those of the CL and TR groups, proving that heat stress reprograms gene expression in broiler hypothalamus and has a unique expression pattern. Using PCA to observe the clustering of samples, we found that gene expression of HS group showed the largest variation, with the four samples distributing over a wide range. One sample of HS treatment was located in TR treatment, which may be because that bird responded to heat stress very quickly and had its own method for protecting itself. The four CL samples were highly homogenous, while the four TR chickens exhibited proximal patterns distributed similarly to the CL samples (Figure 3). Moreover, a heat map was also developed to investigate the expression pattern and gene clusters (Figure 4). The heat map result is consistent with the PCA result in terms of sample expression pattern. The gene clusters were also subjected to further analysis similar to the gene network analysis (IPA).

Functional categories of differentially expressed genes

Gene ontology was used to evaluate the function of DE genes in three comparisons (Figures 5 and 6). All the DE genes in HS vs. CL, HS vs. TR, and TR vs. CL were performed by gene ontology (GO) terms through the DAVID platform. The biological process (BP) components were presented as functional clusters. The significantly-enriched GO categories were selected only when $P < 0.05$ at the fifth level. In summary, up-regulated genes in HS vs. CL and HS vs. TR were mainly enriched in regulation of cell morphogenesis involved in differentiation, neurogenesis, cellular component morphogenesis, neuron development, neuron projection morphogenesis, and transmembrane transport. GO items such as ion transport and cognition were associated only with down-regulation in TR vs. CL. The down-regulated genes between HS vs. CL and TR vs.

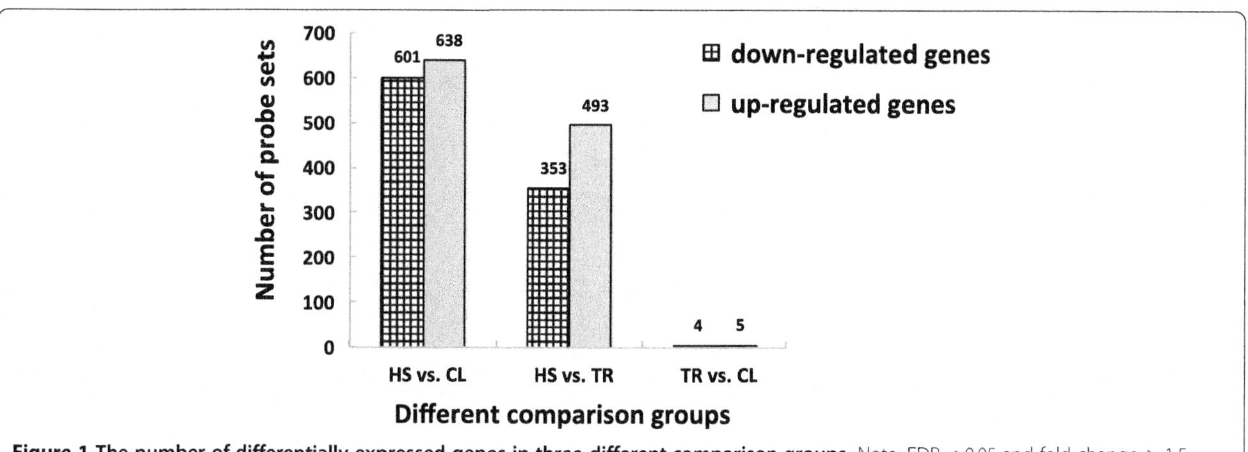

Figure 1 The number of differentially expressed genes in three different comparison groups. Note: FDR < 0.05 and fold change ≥ 1.5; HS: heat stress group; CL: control group; TR: temperature recovery group.

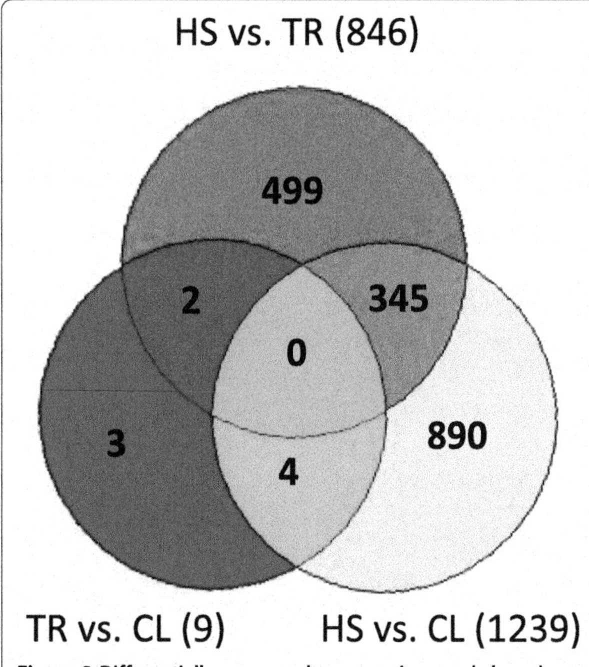

Figure 2 Differentially expressed genes unique and shared between different contrast groups. Note: FDR < 0.05 and fold change ≥ 1.5; HS: heat stress group; CL: control group; TR: temperature recovery group.

CL showed muscle-organ development, striated muscle-tissue development, and cardiac muscle-tissue development; muscle-tissue development was enriched, indicating that muscle development is significantly inhibited during heat stress. In HS vs. CL, the up-regulated genes were mostly concentrated in regulation of gene expression, regulation of macromolecule biosynthetic process, regulation of nucleobase, nucleoside, nucleotide and nucleic acid metabolic processes, regulation of transcription, and DNA metabolic processes. The regulation of protein amino acid phosphorylation and lymph vessel development were only changed in HS vs. TR. Also, among the many functional GO annotation categories, a number of genes were involved that might relate to or affect meat quality, growth factor, and enzyme (Additional file 1: Table S1).

Based on the fact that genes might be related to meat quality, growth factor, and enzyme from the GO annotation, a functional network of these DE genes and their interrelationships were analyzed using Ingenuity Pathway Analysis (IPA); three genetic networks (P-value = 0.0001) were created from those DE genes: "Cardiovascular disease, lipid metabolism, molecular transport" (*ACTC1, BMP3, CETP, DBI, FABP7, EMP1, FIGF, LIPG, MYH11, PDK4, PTGS2, SDC1*, and TH), "Cell-to-cell signalling and interaction, molecular transport, small molecule

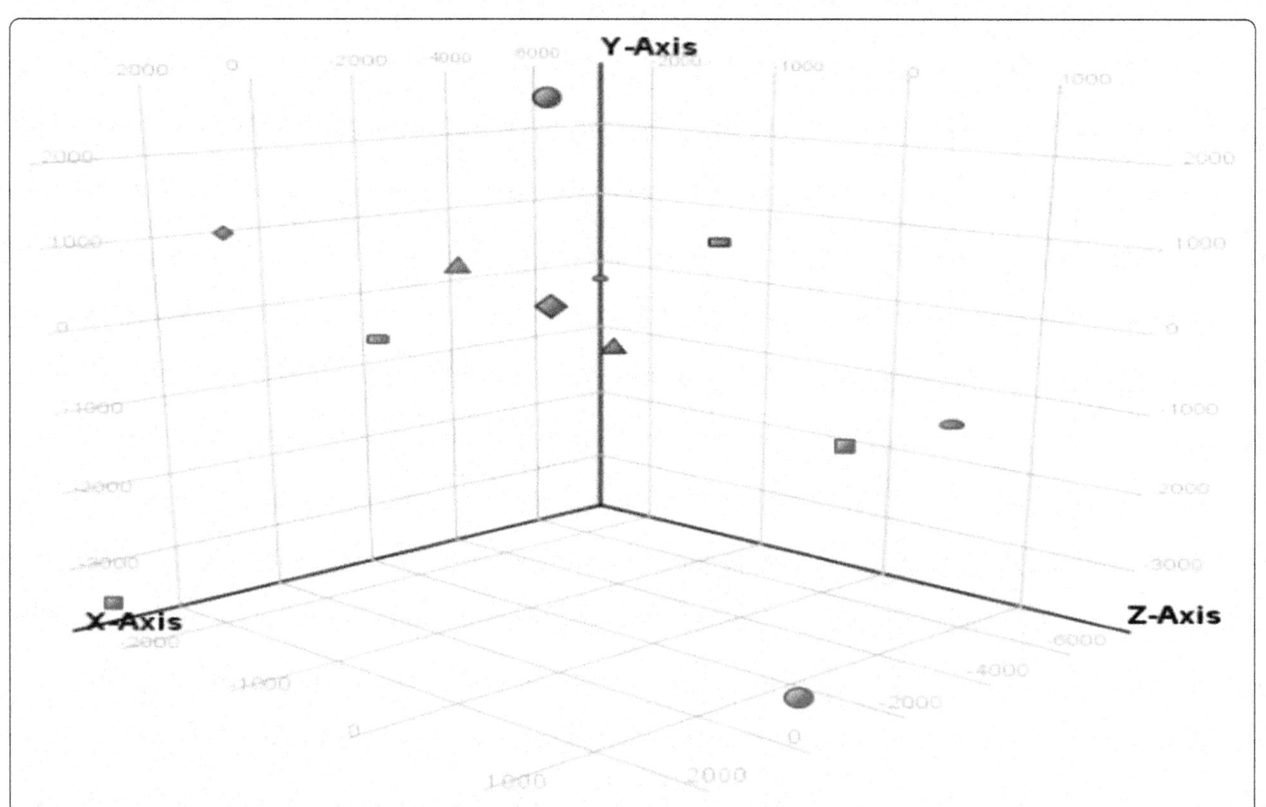

Figure 3 Principal component analysis. Note: red color is control group; blue color is heat stress group; brown color is temperature recovery group.

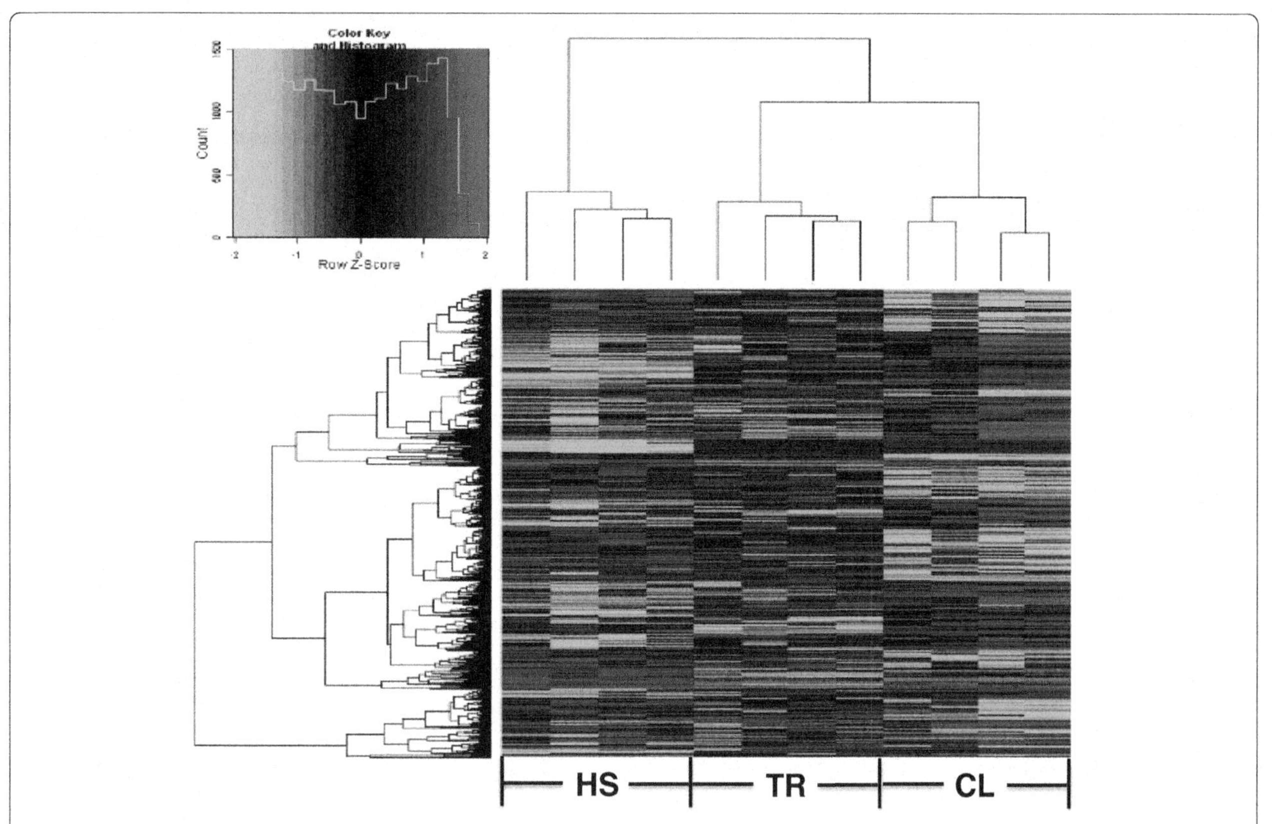

Figure 4 Heat map. The X-axis is sample expression pattern in different treatment group. The first four samples are heat stress (HS) treatment. The middle four samples are temperature recovery (TR) treatment after heat stress. The last four samples are control group. The Y-axis is the gene clusters across HS, TR, and CL treatment.

biochemistry" (*ACTC1, AOAH, DRD2, GCH1, GDPD5, JPH1, LMX1A, LOC430178,* and *UCP3*), and "Lipid metabolism, molecular transport, small molecule biochemistry" (*CROT, CUBN, DRD2, LPAR4, MYH7, PLA1A,* and *TH*) (Figures 7, 8 and 9).

Validation of quantitative real time PCR results

We detected 11 DE genes using qRT-PCR to verify our microarray analysis. These eleven genes were distributed in three comparison groups. *DNAJC13* (HSP40), *HSPCB* (HSP90), *TH, GCH1, GPR23* were DE in both HS vs. CL and HS vs. TR. *ITGA8* transcripts significantly differed in both HS vs. TR and TR vs. CL. *HMOX1, EDN3, SDC1, PDK4* were uniquely changed in HS vs. CL and SFRP was altered in TR vs. CL. Among the 11 tested genes, two genes (*DNAJC13, HSPCB*) were included from the heat-shock protein family; four DE genes (*TH, HMOX1, GCH1, GPR23*) were related to crucial enzymes under heat stress; three DE genes (*SDC1, EDN3, SFRP*) were involved in growth, and the other two DE genes were randomly selected from our DE genes data set. The qRT-PCR results showed that the eleven selected genes presented the same expression pattern as those in the microarray although the fold change differed somewhat for the two methods. We thus have confidence in the reliability of the microarray results (Table 2).

Discussion

Much literature has shown that meat quality and growth can be influenced by environmental heat stress [22-25], but few studies explain this phenomenon at the gene level for broilers, especially with a focus on the hypothalamus. In this study we found 24 DE genes related to meat quality, 8 DE genes in the growth category, and 6 crucial DE enzyme genes associated with acute heat stress and based on heat-map gene clusters and GO term annotation. In addition, these meat quality and growth related DE genes, together with the crucial enzyme genes and other DE genes were subjected to genetic network analysis through IPA. We found 3 networks: Cardiovascular Disease, Lipid Metabolism, Molecular Transport; Cell-to-Cell Signaling and Interaction, Molecular Transport, Small Molecule Biochemistry; and Lipid Metabolism, Molecular Transport, Small Molecule Biochemistry.

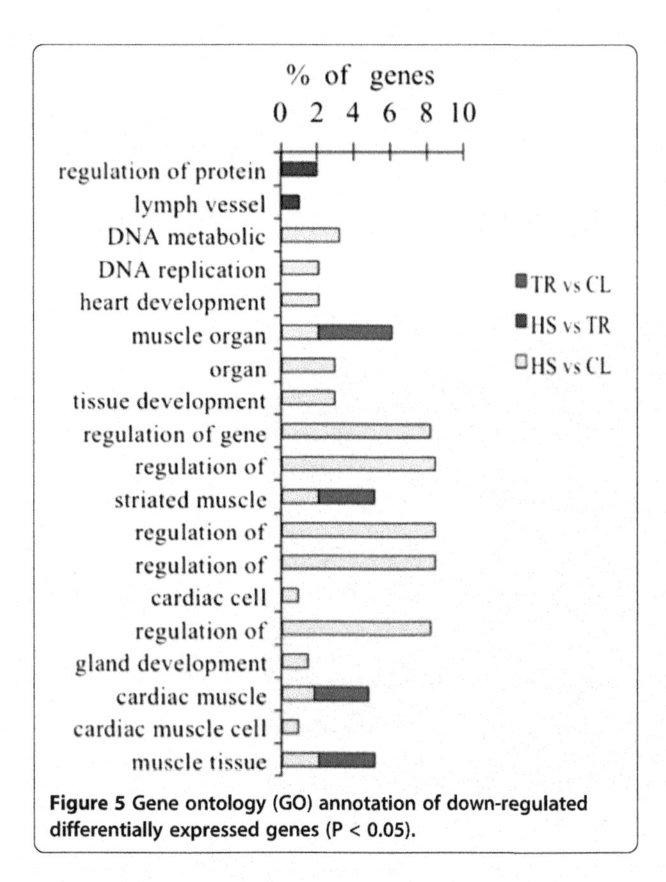

Figure 5 Gene ontology (GO) annotation of down-regulated differentially expressed genes (P < 0.05).

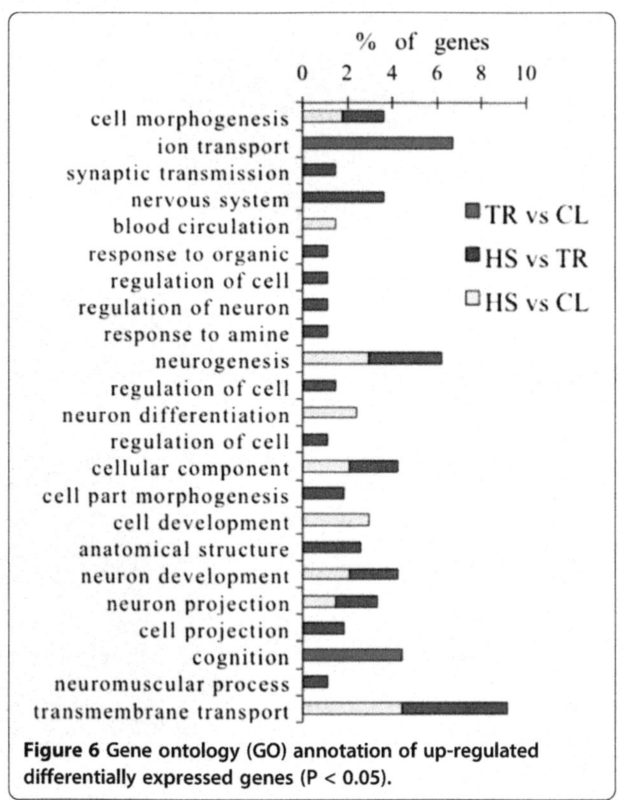

Figure 6 Gene ontology (GO) annotation of up-regulated differentially expressed genes (P < 0.05).

Gene network analysis

Cardiovascular Disease, Lipid Metabolism, Molecular Transport. The hub in the "Cardiovascular Disease, Lipid Metabolism, Molecular Transport" network is prostaglandin G/H synthase 2 precursor (*PTGS2*), also known as *COX2*, an enzyme that can catalyse the biosynthesis of prostaglandins (PGs) [26]. *PTGS2* directly interacts in this network with heme oxygenase (decycling) 1 (*HMOX1* or *HO1*), platelet-derived growth factor beta (*PDGFB*), insulin (Ins), vascular endothelial growth factor (*VEGF*), signal transducer and activator of transcription 5B (*STAT5B*), and so on. *HMOX1*, *PDGFB*, Ins, *VEGF*, and *STAT5B* are key nodes in the "Cardiovascular Disease, Lipid Metabolism, Molecular Transport" network and also interact with other DE genes.

Arnaud et al. [27], reported that *PTGS2* had capability to participate in heat stress-induced myocardial preconditioning *in vivo*. *HMOX1* also has an important role in cellular homeostasis because of its pro- and antioxidative function [28,29]. Bloomer et al. [30], observed that *HMOX1* was sensitive to temperature and its expression was increased immediately after heat stress. *PTGS2* can regulate *HMOX1* expression and this expression is up-regulated in cardiac fibroblasts when the broilers were under stress [31]. In addition, *HMOX1* can be regulated by c-fos induced growth factor (*FIGF*), a member of the PDGF/VEGF family,that can associate with mitogenic and morphogenic activity on fibroblast cells [32,33]. *PDGFB* can be affected by *PTGS2*, *HMOX1*, actin, alpha, cardiac muscle 1 (*ACTC1*), myosin, heavy chain 11, smooth muscle (*MYH11*), bone morphogenetic protein 3 (osteogenic) (*BMP3*). Signoroni et al. [34], demonstrated that *PTGS2* can modulate constitutive interaction of a PDGF receptor. *ACTC1* plays a crucial role in early muscle development and fetal development [35,36], and reduced *ACTC1* expression had an essential role in congenital heart disease and atrial septa defect [37,38]. *ACTC1* also is involved in "Cell-to-Cell Signaling and Interaction, Molecular Transport, Small Molecule Biochemistry" and "Lipid Metabolism, Molecular Transport, Small Molecule Biochemistry". Many research studies [39,40] have indicated that gene *ACTC1* contributes to muscle development and meat quality. *MYH11* is a unique gene of smooth muscle cells. Landreh et al. [41], suggested that the cultured PDGF receptor α- positive peritubular cell (PTC) can increase *MYH11* (PTC-specific gene) expression. PDGF receptorα- positive PTC is a steroidogenic member of the stem Leydig cell (SLCs). BMP3 has been identified as playing a crucial role in both skeletal development and bone homeostasis [42].

In this study, we found that *PTGS2*, *FIGF*, *ACTC1*, *MYH11*, and *BMP3* were up-regulated under heat stress and their fold changed from 1.51 to 15.14 while the expression of *HMOX1* and *PDGFB* decreased during heat

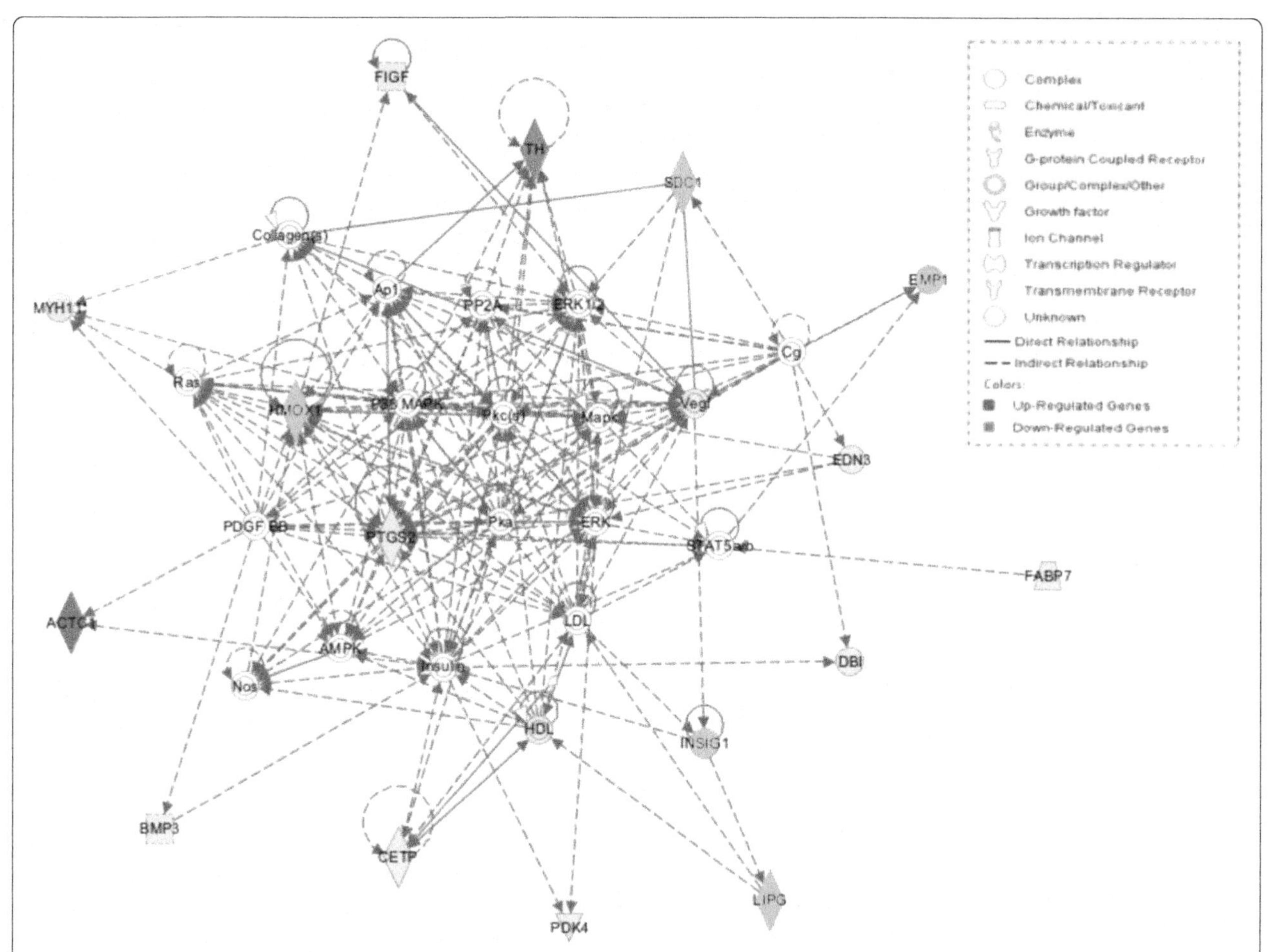

Figure 7 Cardiovascular disease, lipid metabolism, molecular transport. Pathway analysis of gene functions in broiler hypothalamus transcriptome in response to acute heat stress. Red color shows up-regulation and green color shows down-regulation (IPA). The intensity of green and red molecule colors indicates the degree of down or upregulation, respectively. White molecules are not differential expression, but are included to illustrate association with significantly up-regulated and down-regulated genes.

stress and their fold changes were −1.64 and −1.80, respectively. All these DE genes are in HS vs. CL or HS vs. TR. However, except for *HMOX1*, those DE genes have not been reported as relating to heat stress. This might be relatively few researchers have studied meat quality and growth under heat stress at the gene level. Thus, based on this study, we can infer that heat stress significantly influences broiler body temperature, meat quality, sperm quality, growth, and skeletal quality through up-regulation or down-regulation of unique genes in the hypothalamus. Furthermore, the changes of expression of muscle-related genes indicate that muscle development is inhibited to a certain degree in the short term after heat treatment.

Heat stress also can change the hormone level. Tyrosine hydroxylase (*TH*), the precursor of catecholamines (the sympathetic nervous system) (SNS) neurotransmitters, is a rate-limiting enzyme for catecholamine biosynthesis and biomarker enzyme for adrenaline expression [16,43]. In this study, *TH* was a key node and involved in "Cardiovascular Disease, Lipid Metabolism, Molecular Transport" and "Lipid Metabolism, Molecular Transport, Small Molecule Biochemistry" networks. It can catalyse tyrosine into dopa that can be transformed into catecholamine by the dopamine enzyme. *TH* can also regulate the receptor of dopa (*DRD2*) amd dopa could also be formed into noradrenaline and adrenaline through a special enzyme. Garcia et al. [44], reported that *TH* expression was decreased under chronic heat stress. However, in our acute heat-stress study, *TH* and *DRD2* were both significantly expressed with 14.91 and 1.63 FC, respectively. These genes are DE expressed in both HS vs. CL and HS vs. TR. Our current findings reveal that host-enhanced SNS activity leads to an up-regulated catecholamine metabolic pathway and increases noradrenaline and adrenaline to reinforce a body's resistance to heat stress.

There are also still a number of interested DE genes in the three gene networks but for reasons of space not

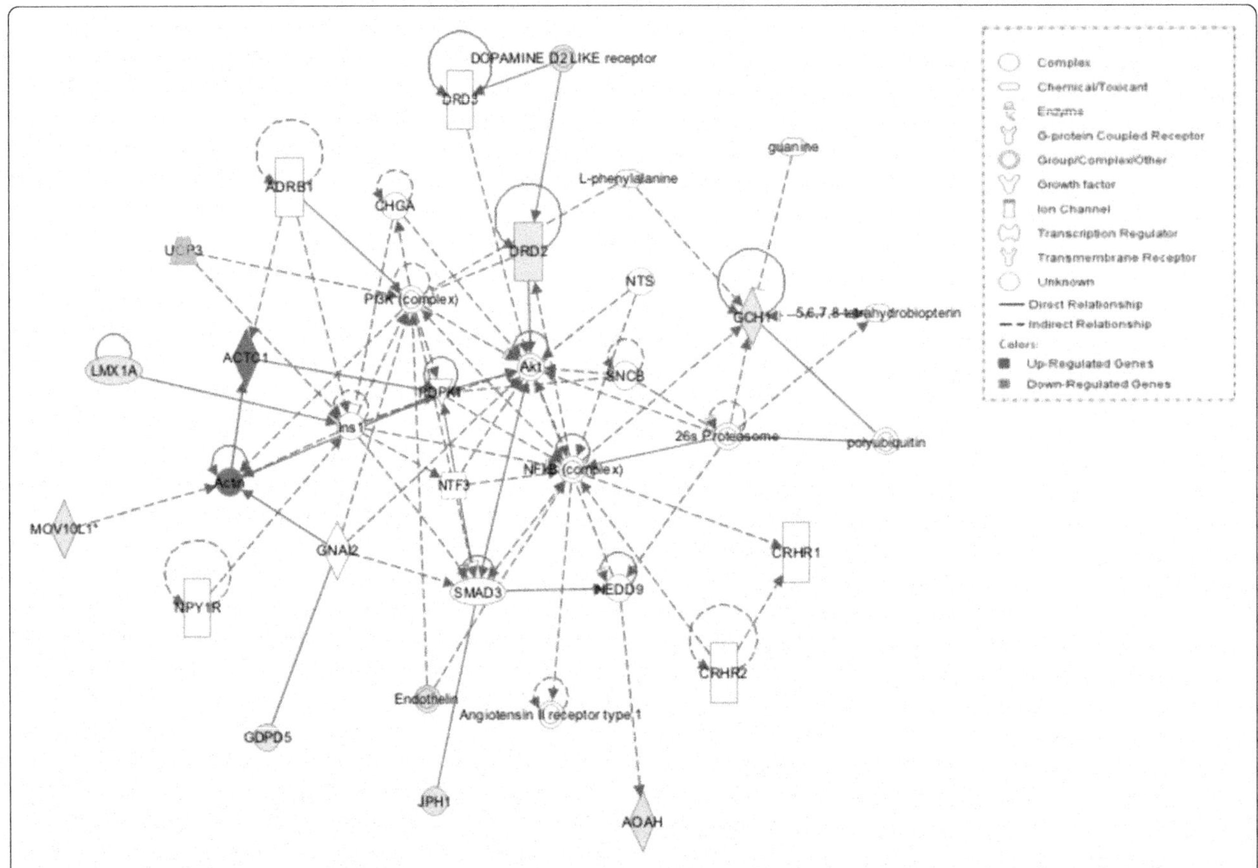

Figure 8 Cell-cell signaling and interaction, molecular transport, small molecular biochemistry. Pathway analysis of gene functions in broiler hypothalamus transcriptome in response to acute heat stress. Red color shows up-regulation and green color shows down-regulation (IPA). The intensity of green and red molecule colors indicates the degree of down or upregulation, respectively. White molecules are not differential expression, but are included to illustrate association with significantly up-regulated and down-regulated genes.

discussed here; they included *SDC1, EMP1, EDN3, DBI, INSIG1, HDL, CETP, PDK4* in "Cardiovascular Disease, Lipid Metabolism, Molecular Transport", *GCH1, LMX1A, MOV10L1, GDPD5, JPH, AOAH* in "Cell-to-Cell Signaling and Interaction, Molecular Transport, Small Molecule Biochemistry", *CUBN, MYH7, CROT, LPAP4, PLA1A* in "Lipid Metabolism, Molecular Transport, Small Molecule Biochemistry". All these DE genes were related to meat quality, growth, and enzyme; for example, *MYH7* is related to drip loss and cooking loss [45,46]. More details are given in a supplemental file. Moreover, some of those DE genes have been identified as being associated with meat quality under heat stress by other researchers. For example, Yu et al. [47], reported that myogenic factor 6 (*MYF6*) and fatty acid binding protein 4, adipocyte (*FABP4*) were involved in meat quality under heat stress.

Heat shock protein

During heat stress, a class of heat shock protein family could also be significantly changed. When a host encounters heat stress, the expression of highly-conserved heat-shock protein (HSP) will be either increased or decreased in various tissues and cells to protect the body from excessive damage [48]. HSP are ubiquitous in organisms ranging from from bacteria to mammals throughout the whole biosphere. They can help the body adapt to adverse environments by serving as molecular chaperone, antioxidant, synergistic immune, and anti-apoptotic [49]. HSP70 can also play a significant role in heat stress, but we didn't find that it was DE in the current study. This might be because of the distribution of HSP70 in terms of different expression in diverse tissues or organs, perhaps especially enriching heart, liver, kidney, and spleen but not gathering in hypothalamus. In addition, HSP70 usually changes within eighteen hours, while after eighteen hours its expression may be back to normal [50]. In this study we found that the expression of *HSPCB* (HSP90) and *DNAJC13* (HSP40) underwent significant changes. *HSPCB* families can play a key role in DNA replication and transcription, protein folding and mutation, protein transport, proteolysis, cell signaling, and immune response [51,52]. *HSPCB* is also necessary in hormone receptor and tyrosine protein kinase [53,54]. *DNAJC13* can interact with HSP70 in

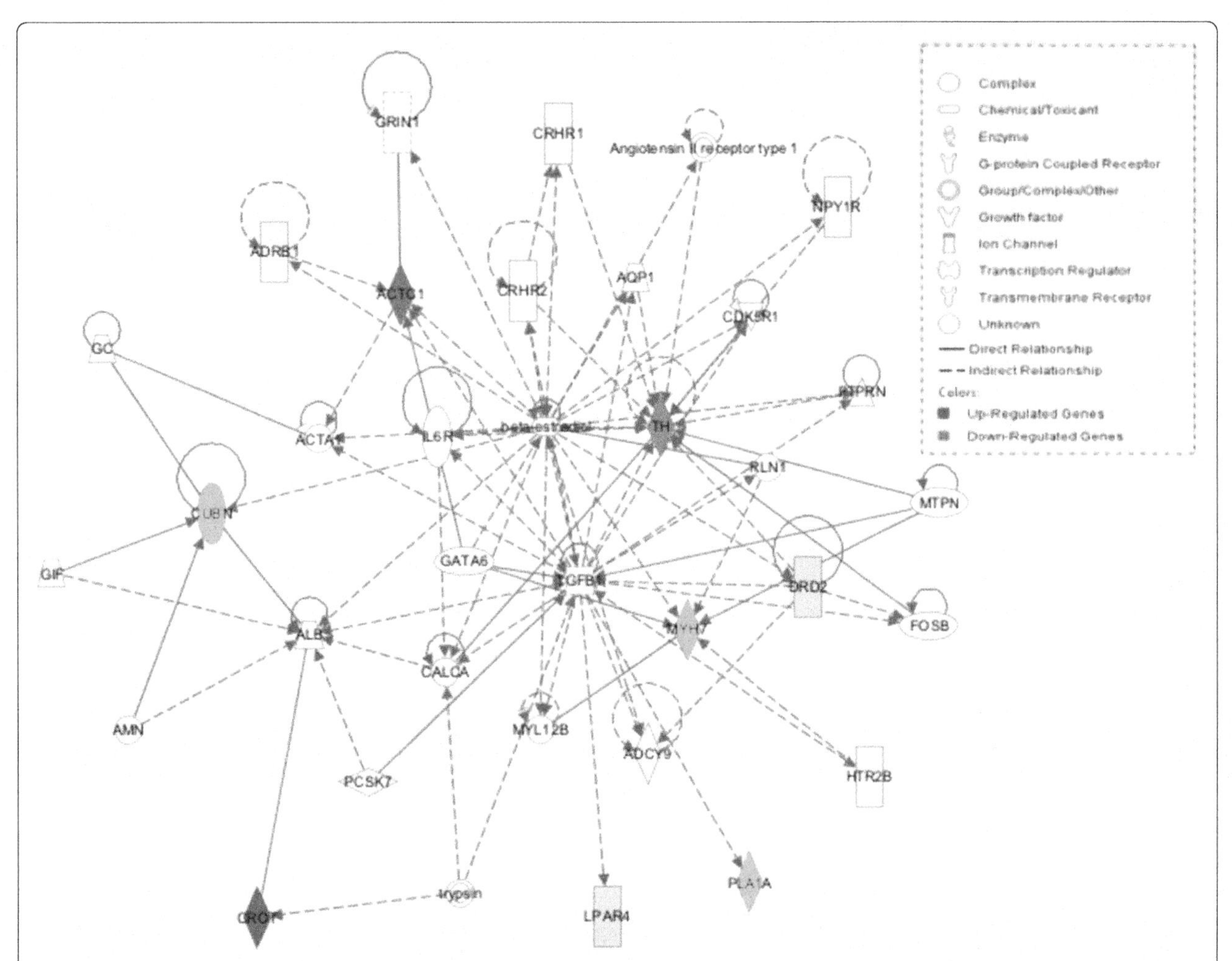

Figure 9 Lipid metabolism, molecular transport, small molecule biochemistry. Pathway analysis of gene functions in broiler hypothalamus transcriptome in response to acute heat stress. Red color shows up-regulation and green color shows down-regulation (IPA). The intensity of green and red molecule colors indicates the degree of down or upregulation, respectively. White molecules are not differential expression, but are included to illustrate association with significantly up-regulated and down-regulated genes.

Table 2 Fold change of 11 selected genes in each comparison group by microarray and qRT-PCR

Gene symbol	HS vs CL fold-change		HS vs TR fold-change		TR vs CL fold-change	
	Microarray	qRT-PCR	Microarray	qRT-PCR	Microarray	qRT-PCR
HSPCB	−1.61*	−1.94*	−1.53*	−1.52*	--	--
DNAJC13	1.90**	1.74**	1.95**	1.53*	--	--
TH	11.76*	12.30*	18.86*	12.70**	--	--
GCH1	2.64**	2.51*	2.42*	1.54*	--	--
HMOX1	−1.63*	−1.94*	--	--	--	−1.85*
EDN3	1.82*	2.12*	--	--	--	--
GPR23	1.59**	2.79*	--	--	--	--
SDC1	−1.61*	−1.55*	--	--	--	--
PDK4	2.48*	1.67**	--	--	--	--
ITGA8	--	--	−3.38**	−1.46*	−2.51*	−1.12
SFRP1	--	--	--	--	1.57**	1.55*

*Means significant difference between the treatments ($P < 0.05$). **means very significant difference between the treatments ($P < 0.01$). "--" means no significant difference between the treatments or the fold change is less than 1.5.

renaturation of heat-denatured proteins [55] and *DNAJC13* can also bind directly to HSP70 by recruiting specific substrates to stimulate both its intrinsically weak ATPase activity and its interaction with polypeptide substrates [56,57]. *HSPCB* and *DNAJC13* are also DE in HS vs. CL and HS vs. TR. The expression of *DNAJC13* was up-regulated while the expression of *HSPCB* was down-regulated in the heat-stress process, indicating that heat stress can exert great harm to the body and the body can actively produce various mechanisms to resist heat stress.

Conclusions

Transcriptome of the chicken hypothalamus could be altered extensively when chickens are exposed to heat stress. We performed global gene-expression analysis of hypothalamus from chickens subjected to heat stress, temperature recovery, and normal temperature. We used bird hypothalamus as material and used a microarray to detect the versatile transcript changes respond to broiler heat stress. In hypothalamus, the TR group and the CL group behaved similarly to the HS group. Various enzymes are changed following the level of hormone under heat stress. At the same time, a number of DE genes related to meat quality and growth were significantly DE. The expression of heat shock protein families also presented certain changes, especially in *DNAJC13* and *HSPCB*. The results of this study provides novel insight into the effects of heat stress on meat quality, growth, and essential enzyme for hormone pathway by using hypothalamus transcriptome analysis; it offers a platform for future investigations into genetic networks for studying broiler heat-stress response.

Additional file

Additional file 1: Table S1. List of some differentially expressed (DE) genes in muscle, growth and enzyme.

Competing interests

The authors declare that they have no competing interests.

Authors' contributions

HS, RJ and LQ conceived of the study, carried out the experiments and drafted the manuscript. SX assisted with the animal management and the interpretation of its results. ZZ, GX and JZ participated in the study's design and coordination. All authors read and approved the final manuscript.

Acknowledgements

This work was financially supported by Beijing innovation team attached to poultry industry technology system (CARS-PSTP).

Author details

[1]Department of Animal Genetics and Breeding, National Engineering Laboratory for Animal Breeding, College of Animal Science and Technology, China Agricultural University, Beijing 100193, China. [2]College of Animal Science and Technology, Anhui Agricultural University, Hefei 230036, China.

References

1. Muiruri HK, Harrison PC. Effect of roost temperature on performance of chickens in hot ambient environments. Poult Sci. 1991;70:2253–8.
2. Wolfenson D, Bachrach D, Maman M, Graber Y, Rozenboim I. Evaporative cooling of ventral regions of the skin in heat-stressed laying hens. Poult Sci. 2001;80:958–64.
3. Mahmoud KZ, Beck MM, Scheideler SE. Acute high environmental temperature and Calcium estrogen relationships in the hen. Poult Sci. 1996;75:1555–62.
4. Borges SA, Fischer da Silva AV, Majorka A, Hooge DM, Cummings KR. Physiological responses of broiler chickens to heat stress and dietary electrolyte balance (sodium plus potassium minus chloride, milliequivalents per kilogram). Poult Sci. 2004;83:1551–8.
5. St-Pierre NR, Cobanov B, Schnitkey G. Economic losses from heat stress by U.S. livestock industries. J Dairy Sci. 2003;86:E52–77.
6. Deyhim F, Teeter RG. Sodium and potassium chloride drinking water supplementation effects on acid-base balance and plasma corticosterone in broilers reared in thermoneutral and heat-distressed environment. Poult Sci. 1991;70:2551–3.
7. Khajavi M, Rahimi S, Hassan ZM, Kamali MA, Mousavi T. Effect of feed restriction early in life on humoral and cellular immunity of two commercial broiler strains under heat stress conditions. Br Poult Sci. 2003;44:490–7.
8. Li C, Wang X, Wang G, Li N, Wu C. Expression analysis of global gene response to chronic heat exposure in broiler chickens (Gallus gallus) reveals new reactive genes. Poult Sci. 2011;90:1028–36.
9. Wang SH, Cheng CY, Tang PC, Chen CF, Chen HH, Lee YP, et al. Differential gene expressions in testes of L2 strain Taiwan country chicken in response to acute heat stress. Theriogenology. 2013;79:374–82.
10. Lin H, Decuypere E, Buyse J. Acute heat stress induces oxidative stress in broiler chickens. Comp Biochem Physiol A Mol Integr Physiol. 2006;144:11–7.
11. Kiyohara T, Miyata S, Nakamura T, Shido O, Nakashima T, Shibata M. Differences in Fos expression in the rat brains between cold and warm ambient exposures. Brain Res Bull. 1995;38:193–201.
12. McKitrick DJ. Expression of Fos in the hypothalamus of rats exposed to warm and cold temperature. Brain Res Bull. 2000;53:307–15.
13. Patronas P, Horowitz M, Simon E, Gerstberger R. Differential stimulation of c-fos expression in hypothalamic nuclei of the rat brain during short-term heat acclimation and mild dehydration. Brain Res. 1998;798:127–39.
14. Kanosue K, Zhang YH, Yanase-Fujiwara M, Hosono T. Hypothalamic network for thermoregulatory shivering. Am J Physiol. 1994;267:R275–82.
15. Storey JD, Tibshirani R. Statistical significance for genomewide studies. Proc Natl Acad Sci. 2003;100:9440–5.
16. Smeets WJAJ, Reiner A. Phylogeny and Development of Catecholamine Sytems in the CNS of Vertebrates. Cambridge, England: University Press; 1994.
17. Do DD, Strathe AB, Ostersen T, Pant SD, Kadarmideen HN. Genome-wide association and pathway analysis of feed efficiency in pigs reveal candidate genes and pathways for residual feed intake. Front Genet. 2014;5:1–10.
18. Uthe JJ, SMD Bearson, L Qu, JC Dekkers, D Nettleton, YR Torres, et al. Integrating comparative expression profiling data and association of SNPs with *Salmonella* shedding for improved food safety and porcine disease resistance. Animal Genetics. 2010. doi:10.1111/j.1365-2052.2010.02171.x
19. Li CJ, Li RW, Wang YH, Elsasser TH. Pathway analysis identifies perturbation of genetic networks induced by butyrate in a bovine kidney epithelial cell line. Funct Integr Genomics. 2007;7:193–205.
20. Calvano SE, Xiao W, Richards DR, Felciano RM, Baker HV, Cho RJ, et al. A network-based analysis of systemic inflammation in humans. Nature. 2005;437:1032–7.
21. Lagoa CE, Bartels J, Baratt A, Tseng G, Clermont G, Fink MP, et al. The role of initial trauma in the host's response to injury and hemorrhage: insights from a correlation of mathematical simulations and hepatic transcriptomic analysis. Shock. 2006;26:592–600.
22. Zhang ZY, Jia GQ, Zuo JJ, Zhang Y, Lei J, Ren L, et al. Effects of constant and cyclic heat stress on muscle metabolism and meat quality of broiler breast fillet and thigh meat. Poult Sci. 2012;91:2931–7.
23. Lu Q, Wen J, Zhang H. Effect of chronic heat exposure on fat deposition and meat quality in two genetic types of chicken. Poult Sci. 2007;86:1059–64.
24. Wang RR, Pan XJ, Peng ZQ. Effects of heat exposure on muscle oxidation and protein functionalities of pectoralis majors in broilers. Poult Sci. 2009;88:1078–84.

25. Yalçin S, Ozkan S, Türkmut L, Siegel PB. Siegel Responses to heat stress in commercial and local broiler stocks. 2. Developmental stability of bilateral traits. Br Poult Sci. 2001;42:153–60.
26. Arnaud C, Joyeux-Faure M, Godin-Ribuot D, Ribuot C. COX-2: an *in vivo* evidence of its participation in heat stress-induced myocardial preconditioning. Cardiovasc Res. 2003;58:582–8.
27. Gong P, Cederbaum AI, Nieto N. Heme oxygenase-1 protects HepG2 cells against cytochrome P450 2E1-dependent toxicity. Free Radic Biol Med. 2004;36:307–18.
28. Vane JR, Bakhle YS, Botting RM. Cyclooxygenases 1 and 2. Annu. Rev. Pharmacol. Toxicol. 1998;38:97–120.
29. Hori R, Kashiba M, Toma T, Yachie A, Goda N, Makino N, et al. Gene transfection of H25A mutant heme oxygenase-1 protects cells against hydroperoxide-induced cytotoxicity. J Biol Chem. 2002;277:10712–8.
30. Bloomer SA, Zhang HJ, Brown KE, Kregel KC. Differential regulation of hepatic heme oxygenase-1 protein with aging and heat stress. J Gerontol A Biol. 2009;64:419–25.
31. Takeda K, Lin J, Okubo S, Akazawa-Kudoh S, Kajinami K, Kanemitsu S, et al. Transient glucose deprivation causes upregulation of heme oxygenase-1 and cyclooxygenase-2 expression in cardiac fibroblasts. J Mol Cell Cardiol. 2004;36:821–30.
32. Chen S, An J, Lian L, Qu L, Zheng J, Xu G, et al. Polymorphisms in AKT3, FIGF, PRKAG3, and TGF-β genes are associated with myofiber characteristics in chickens. Poult Sci. 2013;92:325–30.
33. Rocchigiani M, Lestingi M, Luddi A, Orlandini M, Franco B, Rossi E, et al. Human FIGF: cloning, gene structure, and mapping to chromosome Xp22.1 between the PIGA and the GRPR genes. Genomics. 1998;47:207–16.
34. Signoroni S, Frattini M, Negri T, Pastore E, Tamborini E, Casieri P, et al. Cyclooxygenase-2 and platelet-derived growth factor receptors as potential targets in treating aggressive fibromatosis. Clin Cancer Res. 2007;13:5034–40.
35. Gunning P, O'Neill G, Hardeman E. Tropomyosin-based regula- tion of the actin cytoskeleton in time and space. Physiol Rev. 2008;88:1–35.
36. McNally E, Dellefave L. Sarcomere mutations in cardiogenesis 29. and ventricular noncompaction. Trends Cardiovasc Med. 2009;19:17–21.
37. Jiang HK, Qiu GR, Li LJ, Xin N, Sun KL. Reduced ACTC1 expression might play a role the onset of congenital heart disease by inducing cardiomyocyte apoptosis. Circ J. 2010;74:2410–8.
38. Matsson H, Eason J, Bookwalter CS, Klar J, Gustavsson P, Sunnegårdh J, et al. Alpha-cardiac actin mutations produce atrial septal defects. Hum Mol Genet. 2008;17:256–65.
39. Nam J, Lee DG, Kwon J, Choi CW, Park SH, Kwon SO, et al. Comparative proteome analysis of porcine *longissimus dorsi* on the basis of pH 24 of post-mortem muscle. J AGR SCI. 2012;4:48–56.
40. Wu T, ZH Zhang, ZQ Yuan, LJ Lo, J Chen, YZ Wang, et al. Distinctive genes determine different intramuscular fat and muscle fiber ratios of the *longissimus dorsi* muscles in jinhua and landrace pigs. PLoS ONE. 8: e53181. doi:10.1371/journal.pone.0053181
41. Landreh L, Stukenborg JB, Söder O, Svechnikov K. Phenotype and steroidogenic potential of PDGFRα-positive rat neonatal peritubular cells. Mol Cell Endocrinol. 2013;372:96–104.
42. Gamer LW, Ho V, Cox K, Rosen V. Expression and function of BMP3 during chick limb development. Dev Dyn. 2008;237:1691–8.
43. Ernsberger U, Patzke H, Tissier-Seta JP, Reh T, Goridis C, Rohrer H. The expression of tyrosine hydroxylase and the transcription factors cPhox-2 and Cash-1: evidence for distinct inductive steps in the differentiation of chick sympathetic precursor cells. Mech Dev. 1995;52:125–36.
44. Garcia C, Schmitt P, D'Aléo P, Bittel J, Curé M, Pujol JF. Regional specificity of the long-term variation of tyrosine hydroxylase protein in rat catecholaminergic cell groupsafter chronic heat exposure. J Neurochem. 1994;62:1172–81.
45. Thi Phuong Loan H. Analysis of functional candidate genes related to ubiquitination process for meat quality in commercial pigs. Bonn: University-und Landesbibliothek Bonn; 2014.
46. Neuhoff C. Transcriptomics and proteomics analysis to identify molecular mechanisms associated with meat quality traits. Bonn: Univ., Diss; 2014.
47. Yu K, Shu G, Yuan F, Zhu X, Gao P, Wang S, et al. Fatty acid and transcriptome profiling of longissimus dorsi muscles between pig breeds differing in meat quality. Int J Biol Sci. 2013;9:108–18.
48. Burdon RH. Heat shock and the heat shock proteins. Biochem J. 1986;240:313–24.
49. Benjamin IJ, McMillan DR. Stress (heat shock) proteins: molecular chaperones in cardiovascular biology and disease. Circ Res. 1998;83:117–32.
50. Wang S, Xie W, Rylander MN, Tucker PW, Aggarwal S, Diller KR. HSP70 kinetics study by continuous observation of HSP-GFP fusion protein expression on a perfusion heating stage. Biotechnol Bioeng. 2008;99:146–54.
51. Ellis RJ, Hartl FU. Principles of protein folding in the cellular environment. Curr Opin Struct Biol. 1999;9:102–10.
52. Jolly C, Morimoto RI. Role of the heat shock response and molecular chaperones in oncogenesis and cell death. J Natl Cancer Inst. 2000;92:1564–72.
53. Buchner J. Supervising the fold: functional principles of molecular chaperones. FASEB J. 1996;10:10–9.
54. Caplan AJ. Hsp90's secrets unfold: new insights from structural and functional studies. Trends Cell Biol. 1999;9:262–8.
55. Minami Y, Hohfeld J, Ohtsuka K, Hartl FU. Regulation of the heat-shock protein 70 reaction cycle by the mammalian DnaJ homolog, Hsp40. J Biol Chem. 1996;271:19617–24.
56. Szabo A, Langer T, Schroder H, Flanagan J, Bukau B, Hartl FU. The ATP hydrolysis-dependent reaction cycle of the Escherichia coli Hsp70 system DnaK, DnaJ, and GrpE. Proc Natl Acad Sci. 1994;91:10345–9.
57. Hartl F. Molecular chaperones in cellular folding. Nature. 1996;381:571–9.

13

Impact of source tissue and *ex vivo* expansion on the characterization of goat mesenchymal stem cells

Nuradilla Mohamad-Fauzi[1,3], Pablo J Ross[1], Elizabeth A Maga[1] and James D Murray[1,2*]

Abstract

Background: There is considerable interest in using goats as models for genetically engineering dairy animals and also for using stem cells as therapeutics for bone and cartilage repair. Mesenchymal stem cells (MSCs) have been isolated and characterized from various species, but are poorly characterized in goats.

Results: Goat MSCs isolated from bone marrow (BM-MSCs) and adipose tissue (ASCs) have the ability to undergo osteogenic, adipogenic and chondrogenic differentiation. Cytochemical staining and gene expression analysis show that ASCs have a greater capacity for adipogenic differentiation compared to BM-MSCs and fibroblasts. Different methods of inducing adipogenesis also affect the extent and profile of adipogenic differentiation in MSCs. Goat fibroblasts were not capable of osteogenesis, hence distinguishing them from the MSCs. Goat MSCs and fibroblasts express CD90, CD105, CD73 but not CD45, and exhibit cytoplasmic localization of OCT4 protein. Goat MSCs can be stably transfected by Nucleofection, but, as evidenced by colony-forming efficiency (CFE), yield significantly different levels of progenitor cells that are robust enough to proliferate into colonies of integrants following G418 selection. BM-MSCs expanded over increasing passages *in vitro* maintained karyotypic stability up to 20 passages in culture, exhibited an increase in adipogenic differentiation and CFE, but showed altered morphology and amenability to genetic modification by selection.

Conclusions: Our findings provide characterization information on goat MSCs, and show that there can be significant differences between MSCs isolated from different tissues and from within the same tissue. Fibroblasts do not exhibit trilineage differentiation potential at the same capacity as MSCs, making it a more reliable method for distinguishing MSCs from fibroblasts, compared to cell surface marker expression.

Keywords: Adipose, Bone marrow, Characterization, Differentiation, Goat, Mesenchymal stem cells

Background

Mesenchymal stem cells (MSCs), also known as multipotent stromal cells, are one of the most studied adult stem cells for their ease of culture *ex vivo* and their multipotentiality, as well as supportive functions *in vivo*. Believed to reside in virtually all post-natal tissues [1], MSCs have been isolated and characterized from a variety of tissue types, most commonly from bone marrow [2,3] and adipose tissue [4]. Even though MSCs isolated from different tissues appear similar in morphology and their ability to differentiate into osteogenic, chondrogenic and adipogenic lineages [5], they can display phenotypical differences, including in differentiation potential [6-9] and cell surface proteins [10,11]. Bone marrow-derived MSCs (BM-MSCs) have a higher capacity to differentiate into osteogenic and chondrogenic lineages [6,7], whereas adipose-derived MSCs (ASCs) are better at differentiating into adipocytes [8,9]. The influence of source tissue, along with variations due to species, donor, and culture methods, have challenged the consistency of reports in the literature and complicated the understanding of MSC biology.

Goats are widely used as large animal models for bone tissue engineering as they have knee joints that are

* Correspondence: jdmurray@ucdavis.edu
[1]Department of Animal Science, University of California, Davis, California 95616, USA
[2]Department of Population Health and Reproduction, University of California, Davis, California 95616, USA
Full list of author information is available at the end of the article

similar to humans [12]. Due to the osteogenic and chondrogenic potential of MSCs, goat MSCs are utilized in tissue engineering applications for bone and cartilage regeneration, such as repairing segmental bone defects [13,14] and restoring articular cartilage [15,16]. Some studies also employed genetic engineering of goat MSCs for the purpose of cell-based therapy, such as generating GFP-positive MSCs for tracking engraftment of transplanted cells [17], enhancing differentiation potential of transplanted MSCs via expression of transgenes [18] and expressing factors to facilitate the tissue generation process [19]. Reports on the characterization of goat MSCs are relatively more recent and amount to little compared to MSCs isolated from other more commonly used species such as the human, mouse and pig. Many of the reports utilizing goat MSCs, mostly from bone marrow, for tissue engineering lack characterization information. There are relatively few reports on goat MSCs isolated from other tissues, such as adipose tissue [20,21] and umbilical cord [22,23]. Little information exists on comparisons between goat MSCs isolated from different source tissues, and on comparisons with fibroblasts, which are morphologically similar to and share characteristics with MSCs [24,25].

Adipogenic differentiation *in vitro* is easy to characterize, as distinct morphological changes that occur are easily visualized and molecular markers of adipogenesis are well-described. Exploring adipogenic differentiation of MSCs in culture may provide a window into understanding adipogenesis, especially upstream in the pathway where lineage commitment occurs. This cannot be studied in pre-adipocyte cell lines such as 3T3-1 cells, which are already lineage-committed. Additionally, studying adipogenic differentiation in MSCs may have implications for meat animals such as goats and cows [26,27], as intramuscular adipocyte differentiation is initiated by MSCs [28]. As goat MSCs are researched extensively for applications in bone and cartilage tissue regeneration, measuring adipogenesis may provide valuable information to inform the selection of MSC cultures for these applications.

For use in experiments and clinical applications, MSCs are needed in amounts that are larger than the starting population isolated from a sample and must be expanded *in vitro*. With increasing population doublings, studies have observed an increase in doubling time [29], increase in cellular senescence [30], changes in cell surface marker profile [31], and loss in multipotentiality [32,33]. This suggests that *ex vivo* culture conditions for MSCs outside of their niches *in vivo* is not sufficient for maintaining MSC characteristics over long-term expansion. To date, changes in MSC characteristics due to long-term *ex vivo* culture have not yet been characterized in MSCs isolated from goats.

In this study, we report characterization of three lines of putative MSCs isolated from bone marrow and adipose tissue of neonatal kid goats. We provide a comparison between MSC lines isolated from the same tissue type as well as from different source tissues. Osteogenic, chondrogenic and adipogenic differentiation, as well as the expression of cell surface markers, were investigated. Adipogenic differentiation capacity was measured by both the extent of Oil Red O staining and mRNA expression of genes involved in adipogenesis. These characteristics also were compared to fibroblasts isolated from goat ear tissue. Using MSCs, we also assessed colony-forming efficiency, expression of pluripotency markers and transfection efficiency as well as integration of an introduced plasmid construct. BM-MSCs were also expanded up to 20 passages, and examined for their adipogenic differentiation, colony-forming efficiency, cell surface marker expression and potential for genetic modification.

Methods

Isolation and establishment of cell lines

Bone marrow and adipose tissue samples were collected from two male neonatal kid goats (9003 and 9004). MSCs were isolated using methods described in Monaco et al. [8]. Three lines were established: one bone marrow-derived line from individual 9004 (9004 BM-MSC), as well as one bone marrow- and one adipose-derived line from individual 9003 (9003 BM-MSC and 9003 ASC, respectively).

Ear fibroblasts (1014 EF) were isolated from a juvenile (2–3 months old) male goat from the UC Davis herd. A biopsy of the ear was taken and stored in PBS until it was processed. The outer skin was removed with a scalpel, and the remaining tissue was diced into approximately 3 mm × 3 mm pieces, which were plated in a 35-mm dish. Fibroblasts that migrated out of the tissue were subsequently trypsinized and expanded. 1014 EF was used as a control cell line for subsequent differentiation experiments and cell surface marker analysis.

Unless otherwise noted, cells were cultured in expansion medium: high glucose DMEM (Gibco Life Technologies 12100–046) with 10% fetal bovine serum (FBS, JR Scientific), in 5% CO_2 at 37°C. Cells were expanded by passaging at 1:3 ratio at each passage, and were cryopreserved in 75% DMEM, 10% DMSO and 15% FBS, to be thawed at corresponding passages for subsequent experiments. Passage 5 (P5) cells were used for all experiments comparing different MSC lines with fibroblasts. 9004 BM-MSC at passage 10 (P10), passage 15 (P15) and passage 20 (P20) cells were used for all experiments comparing MSCs of different passages. MSCs and fibroblasts were counted with a hemocytometer and seeded the same way within each experiment. The term 'MSCs' in this report will be used to refer to both BM-MSCs and ASCs.

Colony-forming unit assay

MSCs were counted and 150 cells were seeded in a 10-cm dish. The cells were then cultured in 10 mL expansion medium for 10 d, without physical disturbance or medium changes. At Day 10, cells were rinsed with phosphate-buffered saline (PBS), fixed with 4% paraformaldehyde for 30 min, and stained with 0.5% crystal violet solution for 10 min to visualize colonies. The number of colonies per dish was determined under light microscopy, with colonies of 50 cells or more counted as positive. Colony-forming efficiency (CFE) was then expressed as the percentage of colonies over the number of cells plated at Day 0. This assay was done for each cell line in triplicates (P5 MSCs) or 4 replicates (P10-20 MSCs), and the experiment was repeated once more. Data from all replicates were combined to get the average percent GFP-positive cells for each cell line.

Karyotype analysis

To harvest mitotic chromosomes for karyotype analysis, MSCs were cultured to 60-70% confluency and colcemid was added to the culture medium at 0.07 μg/mL. After incubation for 60 min at 37°C, culture medium was discarded and the cells were trypsinized and centrifuged. The cell pellet was resuspended in 0.075 mol/L KCl hypotonic salt solution and incubated at 37°C for 20 min. Then, Carnoy's fixative (3 absolute methanol:1 glacial acetic acid) was added to the suspension, after which cells were centrifuged and the supernatant discarded. The cell pellet was resuspended in Carnoy's fixative and incubated at room temperature for 15 min, centrifuged again and the fixative wash procedure repeated twice. Finally, the cell pellet was resuspended in fixative and the cell suspension dropped from a Pasteur pipette onto cold, wet slides. Slides were air dried overnight, stained with 3% Giemsa solution and air dried. Metaphase spreads were visualized under light microscopy at 100X magnification, and images were taken using an attached camera. Chromosomes were then counted using ImageJ (National Institutes of Health) to determine total chromosome number and presence or absence of chromosomal abnormalities. Non-overlapping chromosome spreads were picked for counting that were roughly circular, with chromosomes spread sufficiently to enable counting. For each cell line, at least 20 spreads were counted.

Osteogenic differentiation assay

MSCs and fibroblasts were seeded at a density of 10,000 cells/cm^2 in 6-well plates and cultured in expansion medium to approximately 80% confluency in 2 d, then the medium was replaced by osteogenic differentiation medium: DMEM, 10% FBS, 100 nmol/L dexamethasone (Sigma), 0.05 mmol/L ascorbic acid-2-phosphate (Sigma) and 10 mmol/L β-glycerophosphate (Sigma). Cells were cultured in differentiation medium for 21 d, with medium changes every 3 d. Control cells were cultured in expansion medium in the same 6-well plate. After 21 d, cells were rinsed with PBS twice, fixed with 4% paraformaldehyde for 30 min, rinsed three times with distilled water, covered with 2% Alizarin Red S solution and incubated at room temperature for 10 min. Cells were then rinsed with distilled water and visualized under light and phase microscopy.

Chondrogenic differentiation assay

Chondrogenic differentiation was carried out using the micromass method described by Zuk et al. [4]. MSCs and fibroblasts were trypsinized and resuspended in expansion medium to a concentration of 1×10^7 cells/mL. Then, 10 μL drops of approximately 1×10^5 cells of the cell suspension was carefully pipetted into the wells of 6-well plates. The cells were incubated for 2 h to allow for attachment, and then the wells were carefully filled with differentiation medium (or expansion medium for control cells) without disturbing the cell clumps. Chondrogenic differentiation medium consisted of DMEM, 10% FBS, 0.1 μmol/L dexamethasone, 50 μg/mL ascorbic acid-2-phosphate, 1X Insulin-Transferrin-Selenium (Life Technologies), and 10 ng/mL TGF-β3 (Life Technologies), which was added fresh each time. Cells were cultured for 21 d, with medium changes every 3 d. Cells were rinsed with PBS and fixed with 4% paraformaldehyde at room temperature for 15 min, after which they were incubated in 0.1 N HCl (pH 1) for 5 min, then stained with 1% Alcian Blue in 0.1 N HCl overnight. Cells were rinsed twice with 0.1 N HCl for 5 min with shaking to remove non-specific staining, after which they were air-dried and visualized under light microscopy.

Adipogenic differentiation assay

Two methods of inducing adipogenic differentiation were investigated. In the first method (designated "Adipo I"), MSCs and fibroblasts were plated in 6-well plates at a density of 10,000 cells/cm^2 and cultured to 80% confluency. Then, expansion medium was replaced with adipogenic induction medium: DMEM, 10% FBS, 1.0 μmol/L dexamethasone, 0.5 mmol/L 3-isobutyl-1-methylxanthine (Sigma), 200 μmol/L indomethacin (Sigma) and 10 μmol/L insulin (Akron Biotech). Cells were cultured for 21 d with medium changes every 3 d.

In the second method (designated "Adipo II"), adipogenesis was induced using the method described by Pittenger et al. [34], with modifications. MSCs and fibroblasts were plated in 6-well plates at a density of 10,000 cells/cm^2 and cultured to full confluency. Then, expansion medium was replaced with adipogenic induction medium (as described above) for 3 d, after which the medium was replaced with adipogenic maintenance medium – DMEM,

10% FBS, and 10 μmol/L insulin – for 1 d, followed by a return to adipogenic induction medium. After three cycles of induction, cells where maintained in adipogenic maintenance medium for another 7 d.

In both methods, control cells were cultured in expansion medium instead of induction medium for the length of the experiments. At the end of the assays, cells were rinsed with PBS twice and fixed with 4% paraformaldehyde for 60 min. Cells were then rinsed with distilled water and incubated in 60% isopropanol for 2–5 min. Finally, cells were covered with Oil Red O solution for 5 min, rinsed, and covered with hematoxylin as a counterstain for 1 min. Stained cells were visualized under light microscopy. To quantify adipogenic differentiation capacity, images of cells were captured to represent at least 4,000 cells per replicate of each cell line. Lipid-positive cells that stained with Oil Red O were counted using ImageJ (National Institutes of Health). When lipid-positive cells were too sparse, each well of the plate was divided into sectors, and positive cells were counted by hand using the microscope. Adipocyte count is expressed as the percentage of lipid-positive cells over the total number of nuclei counted. This assay was done in triplicates and the experiment repeated once more.

RNA extraction and cDNA synthesis

Prior to RNA extraction, cells were harvested, centrifuged and flash frozen as pellets. For measuring the expression of cell surface markers, cells were harvested at 70-80% confluency. For measuring expression of adipogenic genes, cells were harvested at the end of the differentiation assay. Total RNA was extracted using TRIzol Reagent (Life Technologies) according to the manufacturer's instructions. RNA quality was confirmed via agarose gel electrophoresis, and 1 μg of total RNA was subsequently DNAse-treated (Thermo Scientific) and used for first-strand cDNA synthesis using Oligo(dT) primers and RevertAid First Strand cDNA Synthesis Kit (Thermo Scientific) according to the manufacturer's instructions.

RT-PCR

Published primer sequences were used for amplifying *GAPDH, OCT4, SOX2* [35] and *NANOG* [36]. Primer sequences for amplifying *THY1* (CD90), *NT5E* (CD73), *ENG* (CD105), *PTPRC* (CD45) and *CD34* (CD34) were designed based on caprine or bovine sequences using Primer3 and spanned exon junctions or included introns, to control for possible genomic DNA contamination [Table 1]. Cycling parameters for amplifying *GAPDH, OCT4, SOX2* and *NANOG* were acquired from the papers referenced, whereas cycling parameters used for amplifying the other genes were as follows: initial denaturation at 94°C for 2 min, followed by 32 cycles of denaturation at 94°C for 30 s, annealing at respective temperatures for 30 s, 72°C for 40 s, and then a final extension at 72°C for 5 min.

Quantitative RT-PCR

Primers for amplifying *GAPDH* [37], CD90, CD105, CD73, *PPARG* [38] and *FABP4* [39] are listed in Table 2. Primer sets that were not obtained from references were designed based on caprine or bovine sequences using Primer3 and spanned exon junctions or included introns. mRNA expression of these genes was quantified using Fast SYBR Green Master Mix (Life Technologies) on the ABI 7500 Fast thermocycler (Applied Biosystems), using the following thermocycling parameters: 95°C for 20 s, followed by 40 cycles of 95°C for 3 s and 60°C for 30 s, followed by a melt curve stage of 95°C for 15 s, 60°C for 1 minute, 95°C for 15 s and 60°C for 15 s. Relative gene expression was obtained by normalizing with GAPDH expression as the internal control and using the $2^{-\Delta\Delta CT}$ method [40,41], calculating differences in mRNA expression as fold changes relative to expression in 1014 EF for comparisons between cell lines, or to 9004 BM-MSC at P5 for comparisons between passages.

Immunofluorescence

Cells of each cell line were plated into a 4-well plate. At 50-70% confluency after 2 d, the cells were rinsed with PBS twice and fixed with 4% paraformaldehyde for 10 min at room temperature. After rinsing with PBS three times, the cells were blocked with PBS-Tween (PBST) and 3% normal donkey serum (NDS) for 30 min at room temperature. Cells were probed with a goat polyclonal primary antibody against human OCT4 (Santa Cruz Biotechnology) (1:300) in PBST and 1% NDS overnight at 4°C with shaking. After that, the cells were rinsed with PBST and 1% NDS three times for 10 min each, and then probed with a donkey anti-goat secondary antibody (1:500), conjugated with Alexafluor 488, in PBST and 1% NDS for 1 h at room temperature with shaking. Then, the cells were rinsed in PBST three times for 10 min each. In the last rinse, Hoescht stain was added at 10 μg/mL to stain the nuclei. The wells were kept wet in PBS and the cells were visualized under fluorescent microscopy. Bovine blastocysts were used as a positive control. Blastocysts were rinsed, blocked and probed as described above, then stored in PBST until visualization. For visualization, stained blastocysts were mounted on a glass slide in Prolong Gold Anti-Fade reagent (Life Technologies).

Transfection efficiency of circular GFP (transient)

For transfection experiments, plasmid pEGFP-N1 (Clontech) was used. This 4.7 kb plasmid contains the enhanced green fluorescent protein gene (EGFP) controlled by a human cytomegalovirus (CMV) promoter and the neomycin resistance gene (Neo^R) controlled by a simian

Table 1 Primer sequences used in RT-PCR

Gene	Primer sequences (5′ → 3′)	Product size, bp	Anneal T, °C	Source
GAPDH	F: TCATGACCACAGTCCATGCCATCACT	253	60	[35]
	R: GATGTCATCATATTTGGCAGGTTTCTCC			
OCT4	F: AGGTGTTCAGCCAAACGACTATCTG	192	55	[35]
	R: TCGGTTCTCGATACTTGTCCGCTT			
NANOG	F: GCCGAGGAATAGCAATGG	444	53	F: AY786437.2
	R: TACAAATCTTCAGGCTGTATGTTG			R: [36]
SOX2	F: TGCAGTACAACTCCATGACCAGCT	281	55	[35]
	R: GTAGTGCTGGGACATGTGAAGTCTG			
CD90 (*THY1*)	F: GCACCATGAACCCTACCATC	241	55	NM_001034765.1
	R: TTGGTTCGGGAGCTGTATTC			
CD73 (*NT5E*)	F: CTGAGACACCCGGATGAGAT	160	57	NM_174129.3
	R: ACTGGACCAGGTCAAAGGTG			
CD105 (*ENG*)	F: AGATGCCAACATCACACAGC	129	60	NM_001076397.1
	R: TCCAGACGAAGGAAGATGCT			
CD45 (*PTPRC*)	F: GGGAGGAGGGAAAGCAAACC	146	60	XM_005691061.1
	R: GCAGCTCTTCCCCATTCCAG			
CD34 (*CD34*)	F: AGGTGTGCTCCTTGCTCCT	134	55	NM_174009
	R: GCCCATCTCTCTCAGGTCAG			

The sequences of primers used are listed with the genes amplified, PCR product sizes (bp), annealing temperatures and sources. Primers were acquired from the indicated references, or designed using Primer3 based on the GenBank accessions provided.

virus-40 (SV40) promoter. Plasmid DNA was isolated using PerfectPrep EndoFree Plasmid Maxi Kit (5 Prime) according to manufacturer's instructions, and then concentrated using a vacuum centrifuge until DNA concentration was at least 1 μg/μL for use in transfection.

MSCs were counted with a hemocytometer, and 500,000 cells were aliquoted for transfection with 10 μg plasmid DNA. Transfection was performed via the Nucleofector II system (Lonza, Amaxa, Germany), using the Human MSC Nucleofector kit. The bone marrow-derived cell lines (9004 BM-MSC and 9003 BM-MSC) were Nucleofected using program G-017, whereas the adipose-derived cell line (9003 ASC) were Nucleofected using program C-020. After transfection, cells were plated in a T-25 flask for culture in

Table 2 Primer sequences used in quantitative RT-PCR

Gene	Primer sequences (5′ → 3′)	Product size, bp	Anneal T, °C	Source
GAPDH	F: TTGTGATGGGCGTGAACC	127	60	[37]
	R: CCCTCCACGATGCCAAA			
PPARG	F: AAAGCGTCAGGGTTCCACTA	201	60	[38]
	R: CCCGAACCTGATGGCGTTAT			
FABP4	F: TGAGATGTCCTTCAAATTGGG	101	60	[39]
	R: CTTGTACCAGAGCACCTTCATC			
CD90 (*THY1*)	F: GAATCCCACCGTCTCCAATA	190	60	XM_005689553.1
	R: CTTGTGGCTTCCTGTGTCCT			
CD73 (*NT5E*)	F: CTGAGACACCCGGATGAGAT	160	57	NM_174129.3
	R: ACTGGACCAGGTCAAAGGTG			
CD105 (*ENG*)	F: GCAAAGCATCTTCCTTCGTC	56	60	NM_001076397.1
	R: GATTGCAGGAGAACGGTGAG			

The sequences of primers used are listed with the genes amplified, PCR product sizes (bp), annealing temperatures and sources. Primers were acquired from the indicated references, or designed using Primer3 based on the GenBank accessions provided.

expansion medium. Fluorescent cells were counted at 24 h post-transfection using a fluorescent microscope. Transfection efficiency was expressed as the percentage of green cells over total cells counted. This assay was done for each cell line in triplicates (P5 MSCs) or 4 replicates (P10-20 MSCs), and the experiment was repeated once more. Data from all replicates were combined to get the average percent GFP-positive cells for each cell line.

Integration of Neo^R construct

For these transfection experiments, pEGFP-N1 was isolated using PerfectPrep EndoFree Plasmid Maxi Kit (5 Prime) according to manufacturer's instructions, and then linearized by ApaLI digestion and purified by ethanol precipitation. Plasmid DNA was resuspended in UltraPure H_2O (Gibco Life Technologies) such that DNA concentration was at least 1 μg/μL for use in transfection. MSCs were counted with a hemocytometer, and 500,000 cells were aliquoted for transfection with 10 μg linearized plasmid DNA as described in the previous section. After transfection, cells were plated in a T-25 flask for culture in expansion medium. At 48 h post-transfection, expansion medium was aspirated and replaced with selective medium: DMEM, 10% FBS and 800 μg/mL G418 (Gibco Life Technologies). Bone marrow-derived MSCs (9004 BM-MSC and 9003 BM-MSC) were maintained in selection for 12 d, whereas adipose-derived MSCs (9003 ASC) required a longer selection window of 16 d. At the end of selection, cells were rinsed with PBS and fixed with 4% paraformaldehyde for 30 min. Then, cells were stained with 0.5% crystal violet solution for 10 min to visualize neomycin-resistant colonies. Only discrete colonies that measured ≥ 2 mm in diameter were counted. Integration of linear pEGFP-N1 was measured by taking the number of integrant colonies per total cells transfected. This assay was performed for each cell line in 4 replicates and the experiment repeated once more. Data from all 8 replicates were combined to get the average number of integrant colonies for each cell line.

Statistical analysis

Using the software SAS 9.3 (SAS Institute), data was analyzed using one-way ANOVA and statistically grouped using Tukey's multiple means comparison test. Data was blocked by experiment, or the blocks removed when the block effect was not significant. P values are reported for the ANOVA, while statistical differences with a P value of less than 0.05 was considered as significant for the Tukey analysis. Data is represented as means ± SEM.

Results

Colony-forming efficiency (CFE)

All three MSC lines were able to form colonies after low-density plating (150 cells/55 cm^2). As MSCs are heterogeneous, colonies formed were a mixture of dark, dense colonies of smaller, proliferative cells [Additional file 1: Figure S1A] and colonies of larger, flatter cells that were lighter in color [Additional file 1: Figure S1B]. At P5, there were significant differences in CFE between the cell lines ($P = 0.0008$). 9004 BM-MSC exhibited a CFE of 61.0% ± 2.2, which was significantly higher ($P < 0.05$) than those of 9003 ASC (48.0 ± 4.2) and 9003 BM-MSC (40.7 ± 1.9) [Figure 1A].

To investigate the effect of passaging on CFE and other MSC characteristics, 9004 BM-MSC was expanded to passages 10, 15 and 20. At higher passages, 9004 BM-MSC retained colony-forming ability with statistical differences observed in CFE between passages ($P = 0.03$). The highest CFE was observed in P20 MSCs, in which 53.0% ± 2.1 of cells plated formed colonies. P15 MSCs exhibited a CFE of 47.4% ± 3.4, while P10 MSCs yielded a CFE of 43.8% ± 2.7, which was significantly lower compared to P20 MSCs ($P < 0.05$) [Figure 1B].

Karyotype analysis

All three MSC lines demonstrated chromosomal stability *in vitro*, at least up to 5 passages in culture. For each cell line, the majority of chromosome spreads showed a normal ploidy level of 2N = 60, with no sign of gross chromosomal abnormalities observed. The incomplete spreads were usually characterized by the loss of 1 to 3 chromosomes, which is consistent with chromosome loss caused by the spreading technique. Examination of complete metaphase spreads isolated from 9004 BM-MSC at P10, P15 and P20 revealed a complete set of 60 chromosomes with no sign of gross chromosomal abnormalities.

Osteogenic differentiation

All three MSC lines showed capacity for undergoing induced osteogenic differentiation [Figure 2]. Mineralized calcium deposits stained orange-red with Alizarin Red S. Control MSCs and the fibroblast line did not exhibit calcium deposition, showing inability to undergo induced osteogenic differentiation. Differences in the extent of calcium mineralization between the MSC lines can be observed by visual comparison. The bone-marrow derived MSCs lines appear to have higher capacity for osteogenic differentiation, as shown by the greater extent of staining in 9004 BM-MSC [Figure 2A] and 9003 BM-MSC [Figure 2C], compared to the adipose-derived 9003 ASC [Figure 2E].

Chondrogenic differentiation

MSCs and ear fibroblasts were able to undergo chondrogenesis. Differentiated cells underwent morphological changes, migrating to each other to form ridges and clumps, whereas control cells remained in monolayer [Figure 3]. These ridges and clumps of cells stained

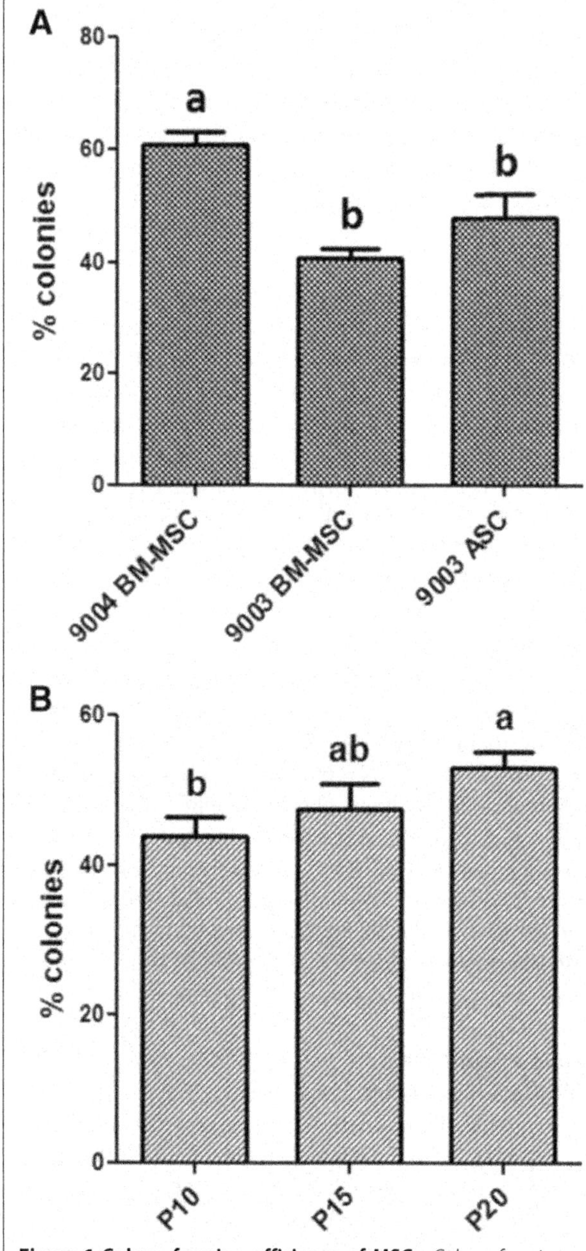

Figure 1 Colony-forming efficiency of MSCs. Colony-forming efficiency (CFE) of P5 MSCs from different sources **(A)** and 9004 BM-MSC cultured to passages 10, 15 and 20 **(B)**. CFE is expressed as percentage of colonies formed per total cells plated and presented as means (± SEM). Statistically significant groups by Tukey's test are indicated by letters ($P < 0.05$).

positive with Alcian Blue, which stains sulfated glycosaminoglycans prevalent in cartilage. Undifferentiated monolayers showed little to no staining. Fewer but larger micromasses were observed in the BM-MSCs, whereas the ASCs yielded numerous but smaller micromasses. In 1014 ear fibroblasts, there were large micromasses as well as small ones akin to those found in 9003 ASC.

Adipogenic differentiation: cytochemical staining of adipocytes

Both differentiation methods demonstrated the capacity of MSCs and ear fibroblasts to undergo adipogenic differentiation [Figure 4]. Control cells that were cultured in expansion medium did not undergo adipogenesis and did not stain positive with Oil Red O [Figure 4I-L]. However, there were visual differences in the morphology of adipocytes obtained from each differentiation method. Cells that were differentiated with the Adipo I method accumulated many small lipid droplets in the cytoplasm [Figure 4A-D], whereas there were visually fewer but bigger lipid droplets in cells differentiated using the Adipo II method [Figure 4E-H].

Adipo I yielded greater percentages of lipid-positive cells, with significant differences between the cell lines ($P < 0.0001$) [Figure 5A]. 9003 ASC exhibited the highest percentage of lipid-positive cells (8.9% ± 1.0), significantly higher than the percentage observed in 9004 BM-MSC (1.1% ± 0.08, $P < 0.05$). 9003 BM-MSC and 1014 EF yielded the lowest percentage of lipid-positive cells (0.038% ± 0.001 and 0.036% ± 0.02, respectively), significantly lower than 9004 BM-MSC ($P < 0.05$). In Adipo II, the number of lipid-positive cells was much lower, but statistical differences were still found among the cell lines ($P < 0.0001$) [Figure 5C]. 9003 ASC still yielded the highest percentage of lipid positive cells (0.40% ± 0.08), significantly higher than 1014 EF (0.092% ± 0.02, $P < 0.05$). 9004 BM-MSC only produced 0.051% ± 0.005 lipid-filled cells, while 9003 BM-MSC induced with the Adipo II method did not yield any lipid-positive cells.

Upon visual observations throughout the 21-d assay period of Adipo I, it appeared that the numbers of adipocytes in 9004 BM-MSC and 9003 BM-MSC peaked at the midpoint of the assay, after which lipid-filled cells were lost. As the experiment was not planned to be a time-course, samples were not taken at this time point. Nevertheless, the BM-MSCs alone were differentiated with the Adipo I method for 10 d in a separate experiment and the results confirmed the earlier observations [Additional file 1: Figures S2 and S3]. As for 9003 ASC and 1014 EF, the number of adipocytes steadily increased throughout the 21-d period. This phenomenon was not observed in BM-MSCs differentiated in the Adipo II method, in which adipocyte numbers steadily increased without loss.

Adipogenic differentiation: expression of adipogenesis marker genes

The differences in percentage of lipid-positive cells correlated with the differences seen in the mRNA expression of *PPARG* and *FABP4*. In Adipo I, there were significant differences in *PPARG* and *FABP4* expression among the cell lines ($P < 0.0001$ for both) [Figure 5B]. 9003 ASC showed the highest expression of *PPARG* at

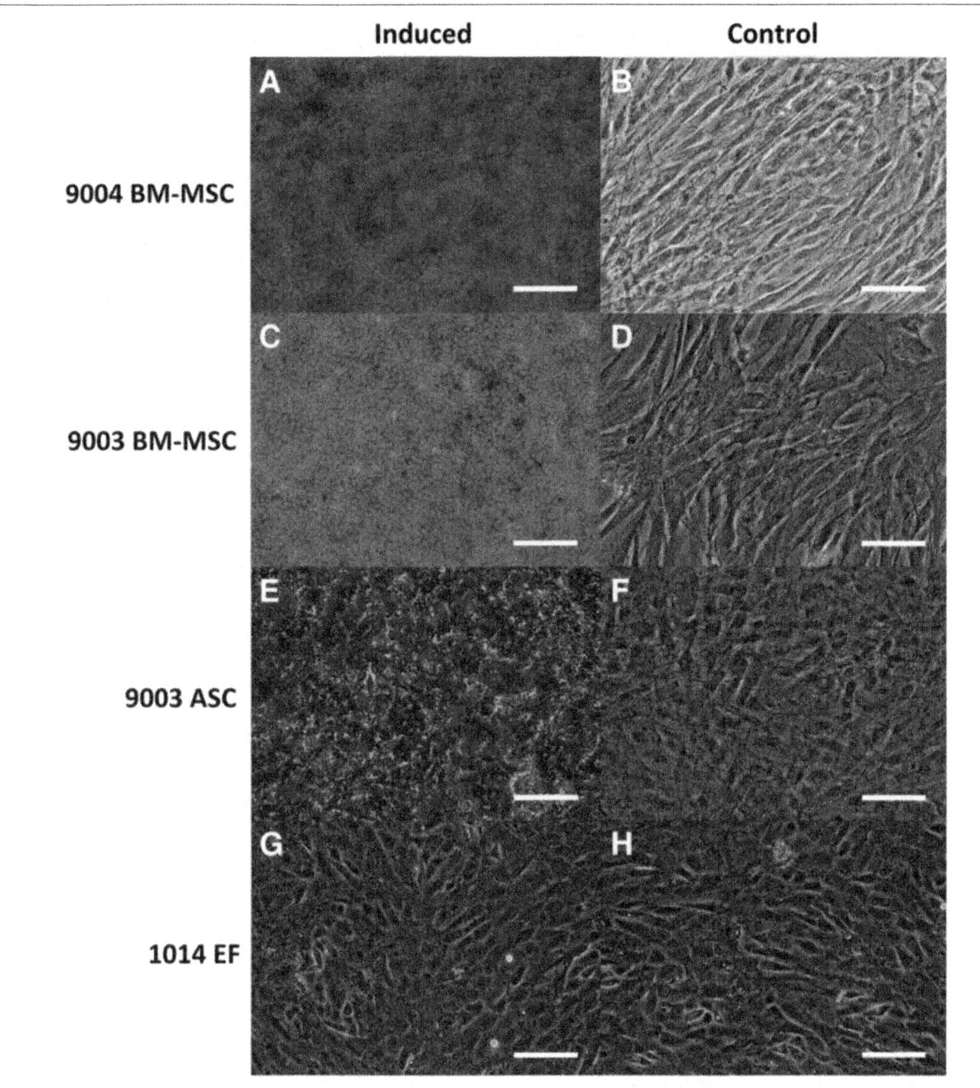

Figure 2 Osteogenic differentiation of MSCs and fibroblasts. P5 MSCs from 9004 BM-MSC **(A, B)**, 9003 BM-MSC **(C, D)**, 9003 ASC **(E, F)** and 1014 EF **(G, H)** were cultured in osteogenic differentiation or control medium, respectively and stained with Alizarin Red S, which stains calcium deposits red. Calcium mineralization in 9003 ASC **(E)** appeared less extensive than that of the BM-MSCs **(A and C)**. 1014 EF failed to undergo osteogenesis, as evident from the absence of staining. **(G, H)**. Representative images are shown at 200X magnification. Scale bars represent 100 μm.

286.4 ± 9.3-fold of expression in the control cell line 1014 EF, significantly higher than the other cell lines ($P < 0.05$). Expression of *PPARG* in 9004 BM-MSC and 9003 BM-MSC was higher than 1014 EF as well, but the difference was not significant. The expression of *FABP4* in 9003 ASC was also significantly higher than the other cell lines, at 120.7 ± 38.9-fold higher than the control ($P < 0.05$). The expression in 9004 BM-MSC was only 1.6 ± 0.6-fold higher than the control, whereas 9003 BM-MSC expressed fewer *FABP4* transcripts at 0.08 ± 0.01-fold of the control.

MSCs differentiated with the Adipo II method showed lower differential expression of these genes relative to the fibroblasts. Significant differences between the cell lines were observed for *PPARG* and *FABP4* ($P < 0.0001$ for both), and the highest expression of both genes was again seen in 9003 ASC [Figure 5D]. *PPARG* expression was significantly higher in 9003 ASC (28.2 ± 5.8-fold, $P < 0.05$) compared to the other cell lines, and consistently higher but not significantly so in the BM-MSCs relative to the fibroblasts, with 9004 BM-MSC showing 4.6 ± 1.2-fold and 9003 BM-MSC showing 2.4 ± 0.3-fold higher expression relative to 1014 EF. *FABP4* expression again was concordant with observations in adipocyte counts: 9003 ASC had significantly higher expression at 4.1 ± 0.4-fold above the control cell line 1014 EF ($P < 0.05$), while *FABP4* expression was lower in 9004 BM-MSC, at 0.42 ± 0.07-fold of the control. 9003 BM-MSC exhibited

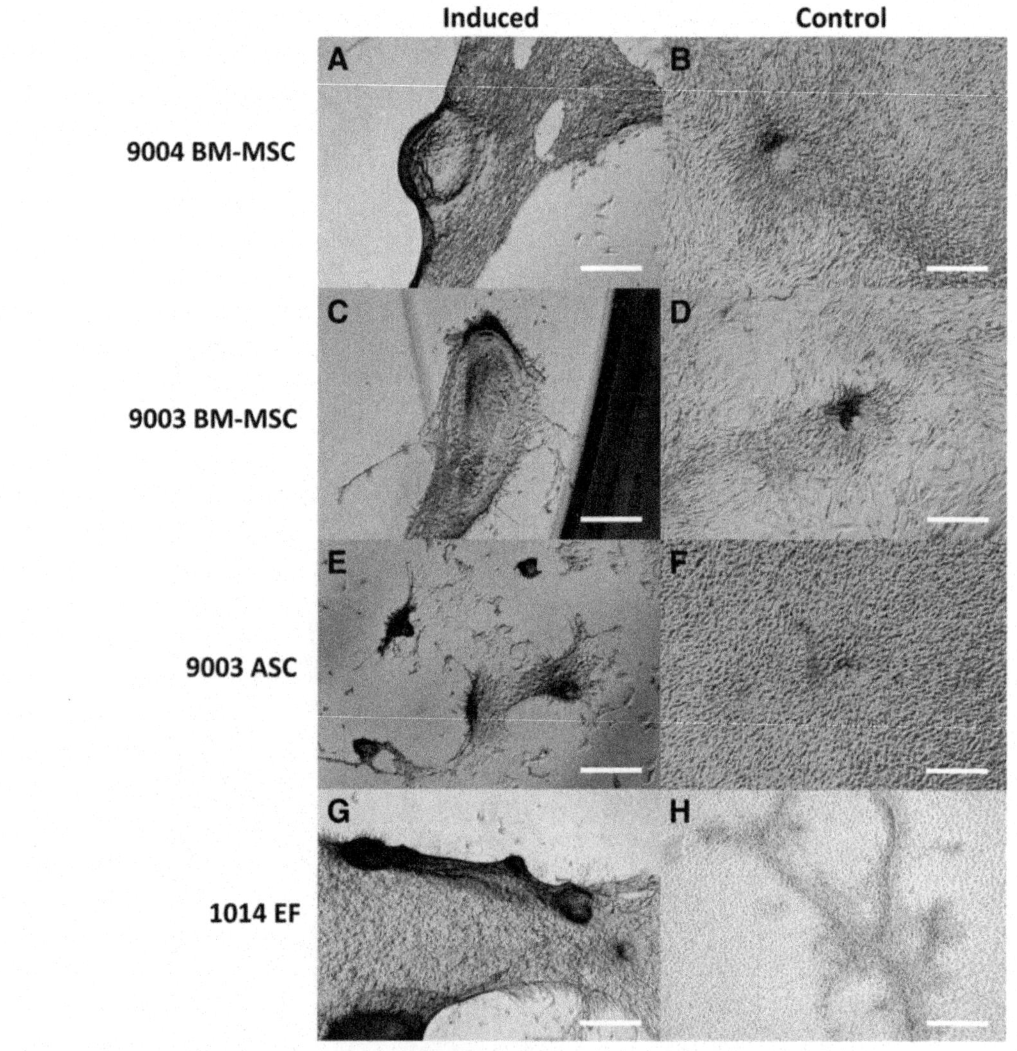

Figure 3 Chondrogenic differentiation of MSCs and fibroblasts. P5 MSCs from 9004 BM-MSC **(A, B)**, 9003 BM-MSC **(C, D)**, 9003 ASC **(E, F)** and 1014 EF **(G, H)** were cultured in chondrogenic differentiation or control medium, respectively and stained with Alcian Blue, which stains cartilage blue. Cellular condensation, as well as ridge and micromass formations that stain positive were observed in 9004 BM-MSC, 9003 BM-MSC, 9003 ASC and 1014 EF cultured in chondrogenic medium (**A**, **C**, **E**, **G** respectively). Some staining was observed in cells cultured in control medium, but cells generally remained in monolayer (**B**, **D**, **F**, **H** respectively). Representative images are shown in phase contrast at 40X magnification. Scale bars represent 500 μm.

the lowest expression of these genes, even yielding significantly lower relative expression of *FABP4* than that of the fibroblasts ($P < 0.05$).

Adipogenic differentiation during long-term culture

The number of lipid-positive cells increased with increasing passages, and was significantly different between passages ($P < 0.0001$) [Figure 6A]. After 10 d of differentiation, 56.9% ± 2.9 of P20 MSCs stained positive for Oil Red O, which was significantly higher than the 45.9% ± 3.7 that stained positive in P15 MSCs ($P < 0.05$). P10 MSCs yielded 13.1% ± 0.7 lipid-positive cells, significantly lower than P15 and P20 MSCs ($P < 0.05$). The same trend was also observed in gene expression data, which is represented as relative to control, non-induced MSC cultures at the corresponding passage [Figure 6B]. Significant differences in *PPARG* and *FABP4* expression were observed as passage increased ($P = 0.0006$ and $P < 0.0001$, respectively). *PPARG* expression in P20 MSCs was the highest, at 3.4 ± 0.5-fold higher than the level in control non-induced P20 cells, followed by a 3.0 ± 0.3-fold increase in P15 MSCs. *PPARG* expression in P10 MSCs was significantly lower, at 1.4 ± 0.3-fold increase relative to control cultures ($P < 0.05$). In the case of *FABP4* expression, we observed the same pattern of significance as compared to the percentage of adipocytes counted. P20 MSCs expressed

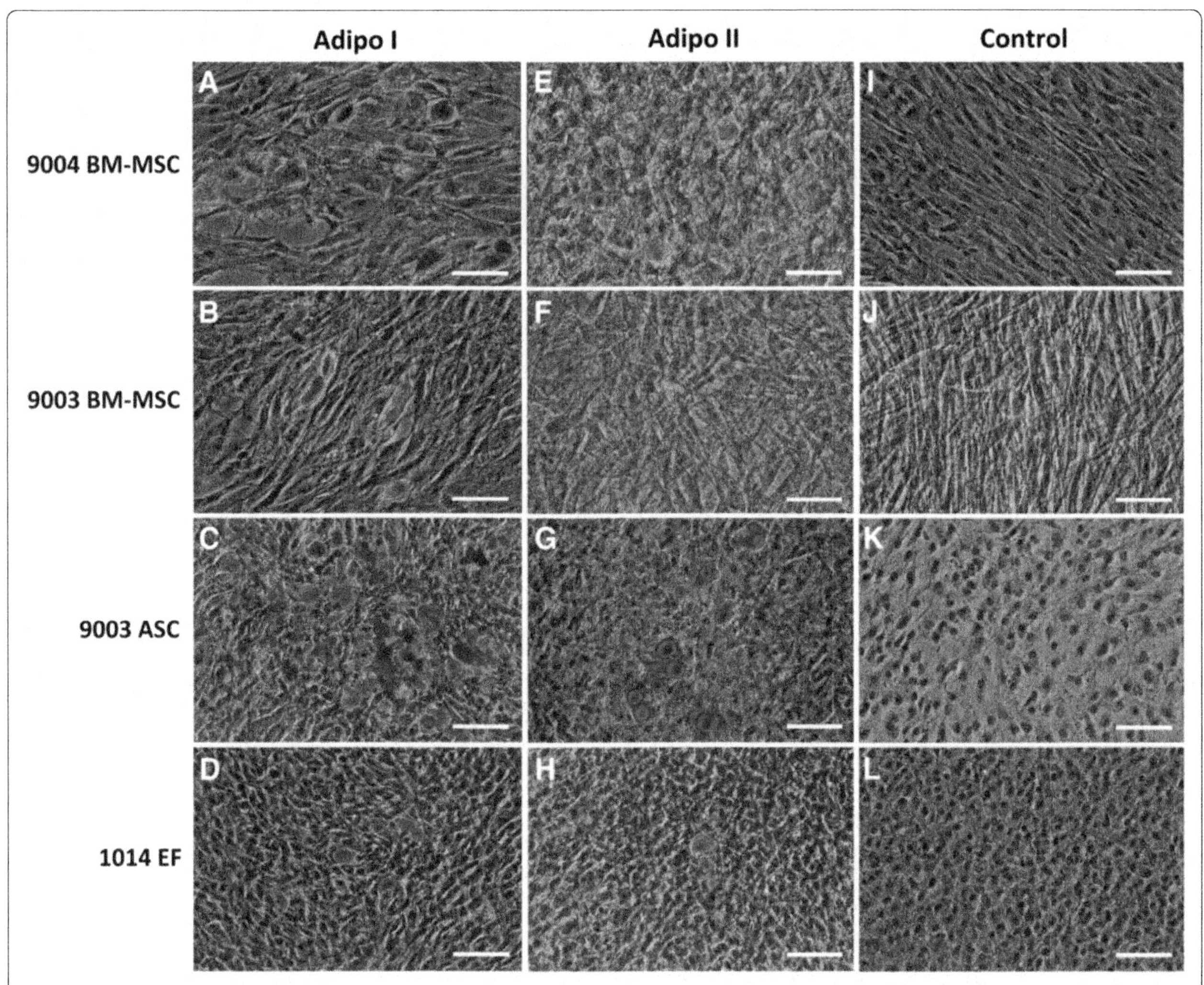

Figure 4 Oil Red O staining of MSCs and fibroblasts differentiated using two methods of adipogenic induction. P5 9004 BM-MSC **(A, E, I)**, 9003 BM-MSC **(B, F, J)**, 9003 ASC **(C, G, K)** and 1014 EF **(D, H, L)** were differentiated by the Adipo I method **(A-D)**, Adipo II method **(E-H)**, or cultured in expansion medium **(I-L)**. Differentiated adipocytes accumulated lipid droplets in the cytoplasm that stain red with Oil Red O. Cells cultured in control medium **(I-L)** and 9003 BM-MSC induced by Adipo II **(F)** did not yield lipid-filled adipocytes. Representative images are shown in phase contrast at 200X magnification. Scale bars represent 100 μm.

FABP4 at 5047.4 ± 434.8-fold higher relative to non-induced cells, significantly higher than expression in P15 cells at 3451.3 ± 320.0-fold ($P < 0.05$). *FABP4* expression was at 774.9 ± 136.0-fold relative to control cells in P10 cells, which was significantly lower compared to P15 and P20 cells ($P < 0.05$).

Upon further examination of the Oil Red O-stained cells, it was apparent that P15 and P20 MSCs contained more polygonal cells, compared to the largely spindle-shaped cells in P10 MSCs (data not shown). Control cells incubated in expansion medium did not yield lipid-positive cells (data not shown).

Expression of cell surface marker genes

RT-PCR analysis revealed that all three MSC lines and 1014 EF expressed CD90, CD105 and CD73 [Figure 7], which are markers that have been defined as positive cell surface markers for human MSCs [5]. Expression of the hematopoietic marker CD45 was not detected in any of the cell lines, whereas CD34 was detected only in 9003 ASC [Figure 7].

Quantitative RT-PCR analysis showed significant differences in CD90 expression ($P = 0.0123$), in that it was lower in MSCs compared to 1014 EF [Figure 8A]. CD90 expression in 9004 BM-MSC was 0.93 ± 0.1-fold of the expression in 1014 EF, whereas 9003-BM MSC and 9003 ASC showed significantly lower expression, with 0.54 ± 0.06-fold and 0.51 ± 0.02-fold expression compared to 1014 EF ($P < 0.05$). On the other hand, CD105 expression was overall higher in MSCs compared to 1014 EF ($P = 0.0004$), with 9003 BM-MSC showing significantly higher expression at 18.3 ± 3.2-fold compared to 1014

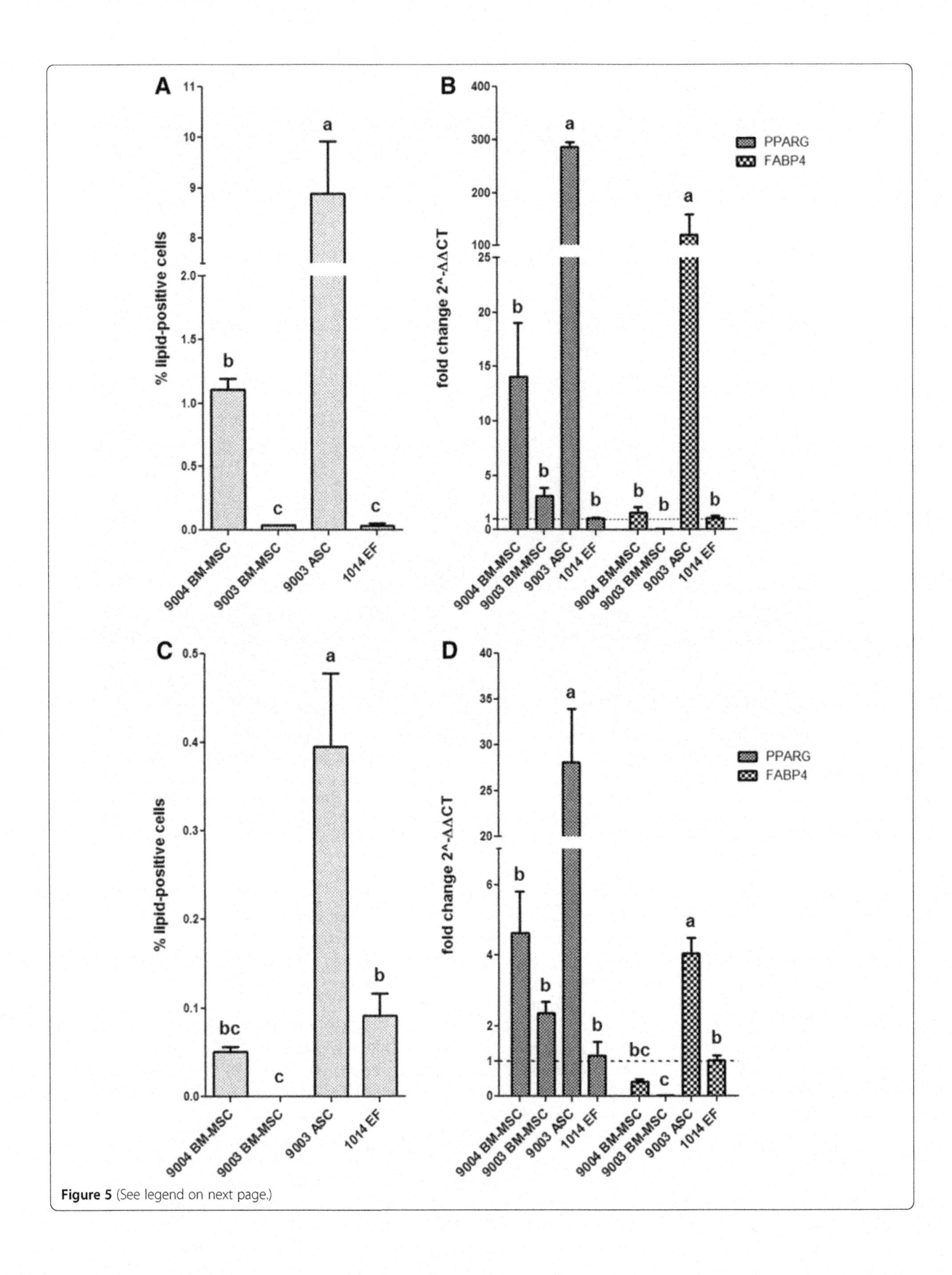

Figure 5 (See legend on next page.)

(See figure on previous page.)
Figure 5 Adipogenic differentiation capacity of P5 MSCs and fibroblasts. The percentage of lipid-positive cells over total cells counted (± SEM) is shown for each cell line differentiated with Adipo I **(A)** and Adipo II **(C)**. Quantitative RT-PCR analysis of PPARG and FABP4 expression was performed on each cell line differentiated with Adipo I **(B)** and Adipo II **(D)**, and data is presented as fold change (± SEM) in expression relative to expression levels in 1014 EF (fold change ~1, indicated by the dotted line). Statistically significant groups by Tukey's multiple means comparison test are indicated by letters ($P < 0.05$ for **A-D**).

EF ($P < 0.05$). CD105 expression was 6.6 ± 0.2-fold and 3.8 ± 0.9-fold higher in 9004 BM-MSC and 9003 ASC, respectively, but the difference was not significant. CD73 expression was not significantly different in MSCs compared to 1014 EF ($P = 0.2645$).

To determine whether the expression of cell surface markers change during *ex vivo* expansion, quantitative PCR was performed on RNA isolated from 9004 BM-MSCs at P10, P15 and P20 to examine the gene expression of CD90, CD105 and CD73. Relative differences in mRNA expression were calculated using expression levels in P5 cells as the baseline. The expression of CD90 and CD73 were found to not change with passaging ($P = 0.15$ and $P = 0.83$ respectively) [Figure 8B]. However, the expression of CD105 increased significantly between P5 and P10 cells ($P = 0.0084$), with expression at 3.6 ± 0.2-fold in P10 MSCs relative to P5 MSCs ($P < 0.05$). CD105 expression appeared to plateau sometime after passage 10, showing no difference between P10, P15 and P20 MSCs.

Expression of pluripotency genes and immunofluorescent detection of OCT4

RT-PCR showed that low expression of *OCT4* was detected in all three MSC lines as well as 1014 EF [Figure 7]. *SOX2* was also detected in the MSCs and fibroblasts, though expression in 9003 ASC appeared to be much lower. *NANOG* was faintly detected only in 9003 ASC and 1014 EF, and was either very lowly expressed or not expressed at all in 9004 BM-MSC and 9003 BM-MSC.

Immunofluorescent staining showed that all the MSC lines and the 1014 EF stained positive for OCT4 protein. Fluorescence was clearly restricted to the cytoplasm, while the nuclei remained dark and unstained [Figure 9]. Staining was also uneven – some cells fluoresced stronger than others, although it was not clear if this was a result of uneven OCT4 expression or due to uneven staining. Considering the faint bands detected from RT-PCR analysis, it is more likely that OCT4 expression in the cells was uneven. There seemed to be no difference in fluorescence intensity among the cell lines. In the bovine blastocysts, fluorescence was restricted to only the nuclei [Figure 9], consistent with known distribution of OCT4A protein in blastocysts [42,43] and suggesting that the antibody used bound to both the OCT4A and OCT4B isoforms.

Transfection and integration of introduced DNA

No significant difference was found in the percentage of GFP-positive cells among the MSC lines at P5 ($P = 0.6049$). Transfection efficiency was 45.3% ± 5.5 in 9004 BM-MSC,

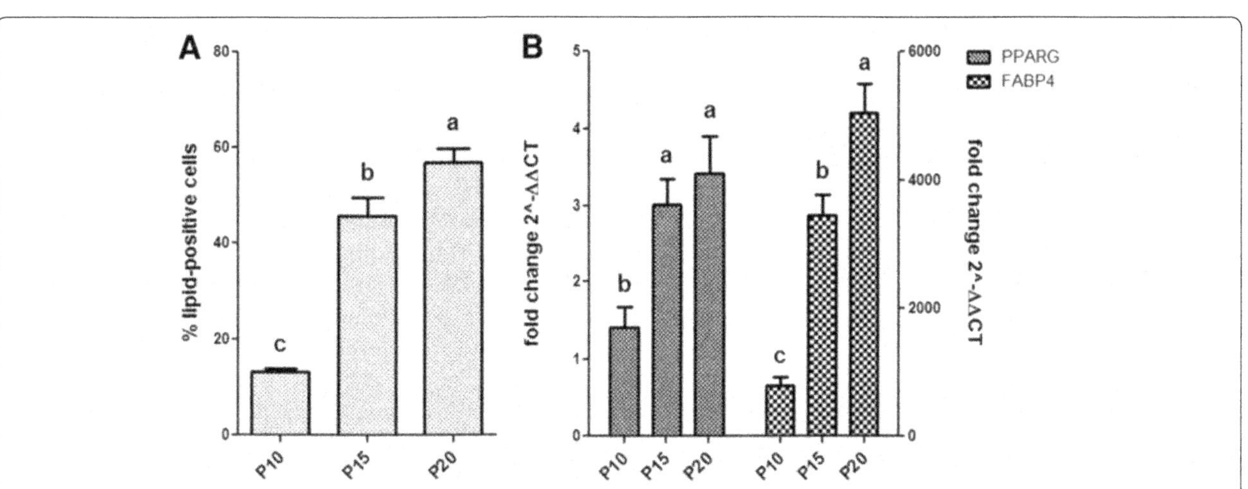

Figure 6 Adipogenic differentiation capacity of BM-MSCs at higher passages. The percentage of lipid-positive cells over total cells counted (± SEM) for 9004 BM-MSC at passages 10, 15 and 20 **(A)** and quantitative RT-PCR analysis of PPARG and FABP4 expression for each passage presented as fold change (± SEM) in expression relative to control cells at corresponding passages **(B)**. Statistically significant differences are indicated in letters according to analysis by Tukey's multiple means comparison ($P < 0.05$).

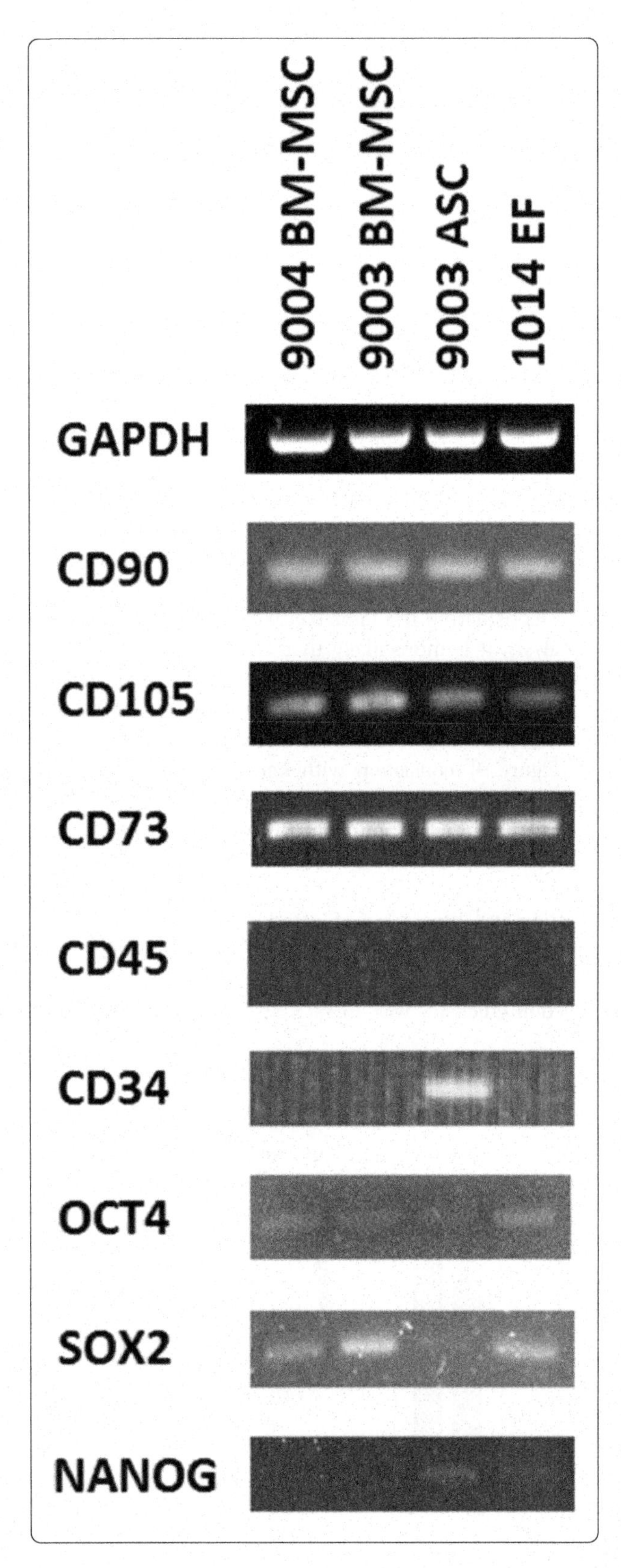

Figure 7 RT-PCR analysis of gene expression in MSCs and fibroblasts. RT-PCR was performed using primers that amplify the cell surface markers CD90, CD105, CD73, CD45 and CD34, and the pluripotency markers *OCT4*, *SOX2* and *NANOG*. Primers amplifying *GAPDH* were used as a positive control for the cDNA. PCR products were visualized in 2% agarose.

50.4% ± 5.4 in 9003 BM-MSC and 51.7% ± 2.7 in 9003 ASC. Following transfection of linearized pEGFP-N1 and G418 selection in MSCs, surviving colonies of G418-resistant cells were counted. Significant differences in integration per 500,000 cells transfected were found ($P < 0.0001$). The number of integrant colonies was significantly different between each MSC line, with 9004 BM-MSC yielding the highest number of integrants (35.2 ± 3.3), followed by 9003 ASC (25.5 ± 5.8) and 9003 BM-MSC (12.8 ± 1.1) ($P < 0.05$) [Figure 10A].

To investigate whether long-term expansion of MSCs *in vitro* affect their potential for genetic modification, the transfection and integration of 9004 BM-MSCs at higher passages were examined. No significant difference was observed in the percentage of GFP-positive cells among MSCs at different passages ($P = 0.62$). Transfection efficiency was 24.4% ± 1.6 in P10 MSCs, 22.4% ± 1.5 in P15 MSCs and 23.4% ± 1.0 in P20 MSCs. Following transfection of linearized pEGFP-N1 and G418 selection in MSCs, a decreasing trend in integrant colony number was observed as passage number increased ($P = 0.035$). The number of integrants was the highest in P10 MSCs (15.6 ± 1.4), and progressively decreased in number in P15 MSCs (15.3 ± 1.0) and P20 MSCs (11.5 ± 1.0), with P20 having significantly fewer integrants than P10 ($P < 0.05$) [Figure 10B].

Discussion

Our results indicate that the 9004 BM-MSC, 9003 BM-MSC and 9003 ASC are true mesenchymal stem cells, as evidenced by their ability to undergo trilineage differentiation [5]. 1014 EF was able to undergo adipogenic and chondrogenic differentiation, but not osteogenesis, hence distinguishing fibroblasts from MSCs. Fibroblasts can be isolated from multiple tissues; as a result, differentiation potentials reported for fibroblasts have been variable, such as being only capable of chondrogenesis [25,44], exhibiting trilineage differentiation at a lower capacity [25,45], or not at all [3]. In this study, goat ear fibroblasts lacked osteogenic capacity. In all cases, fibroblasts do not exhibit trilineage differentiation potential at the same capacity as MSCs, making it a more reliable method for distinguishing MSCs from fibroblasts, compared to cell surface marker expression.

Based on our results, adipogenic differentiation capacity in MSCs appears to be impacted by source tissue.

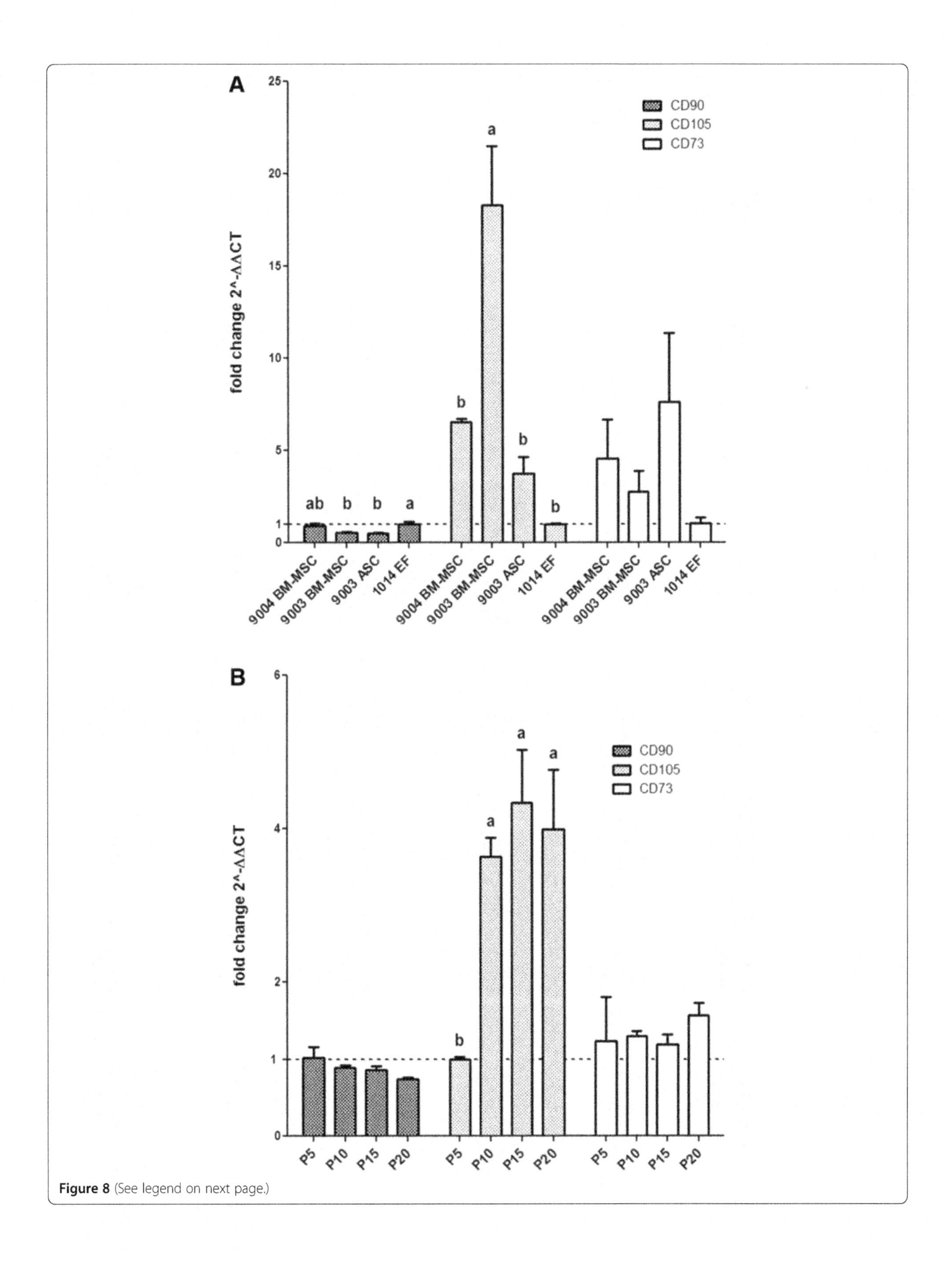

Figure 8 (See legend on next page.)

(See figure on previous page.)
Figure 8 Cell surface marker expression in MSCs and fibroblasts. Quantitative RT-PCR was used to measure the expression of the cell surface markers CD90, CD105 and CD73 in P5 MSCs **(A)** and in 9004 BM-MSCs at passages 10, 15 and 20 **(B)**. Expression levels are presented as fold change (± SEM) relative to 1014 EF **(A)** or expression levels in P5 9004 BM-MSC **(B)** (fold change ~1, indicated by the dotted line). Statistically significant groups by Tukey's multiple means comparison test are indicated by letters ($P < 0.05$). Significant differences between the cell lines were observed in CD90 and CD105 expression. Significant difference between the different passages was observed in CD105, but not in CD90 and CD73.

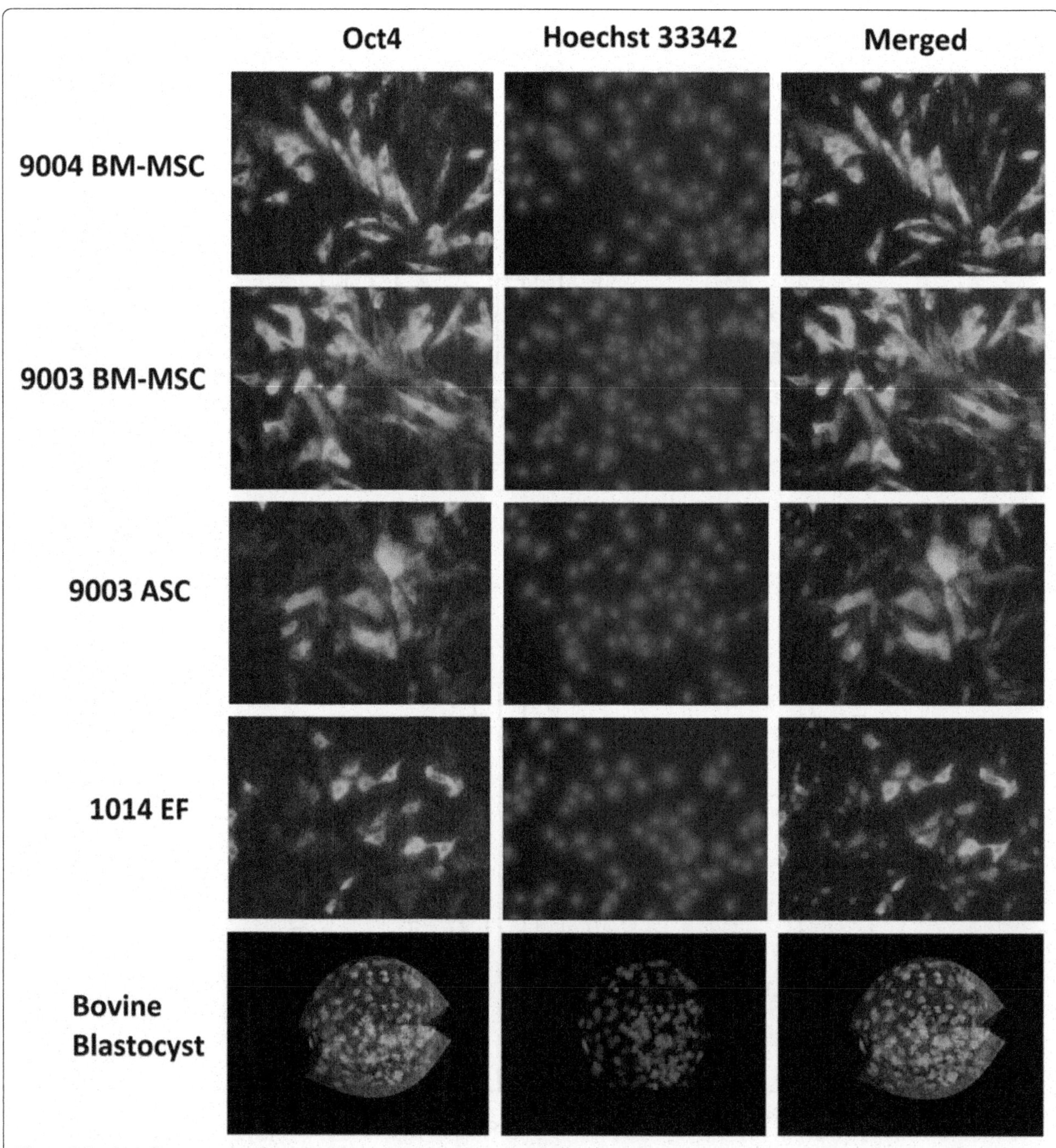

Figure 9 Immunofluorescent staining of OCT4 in MSCs and fibroblasts. P5 cells were stained with antibodies against OCT4. OCT4 was localized in the cytoplasm of MSCs and fibroblasts, whereas staining for OCT4 was specific to the nuclei in the bovine blastocyst. Representative images are shown at 200X magnification.

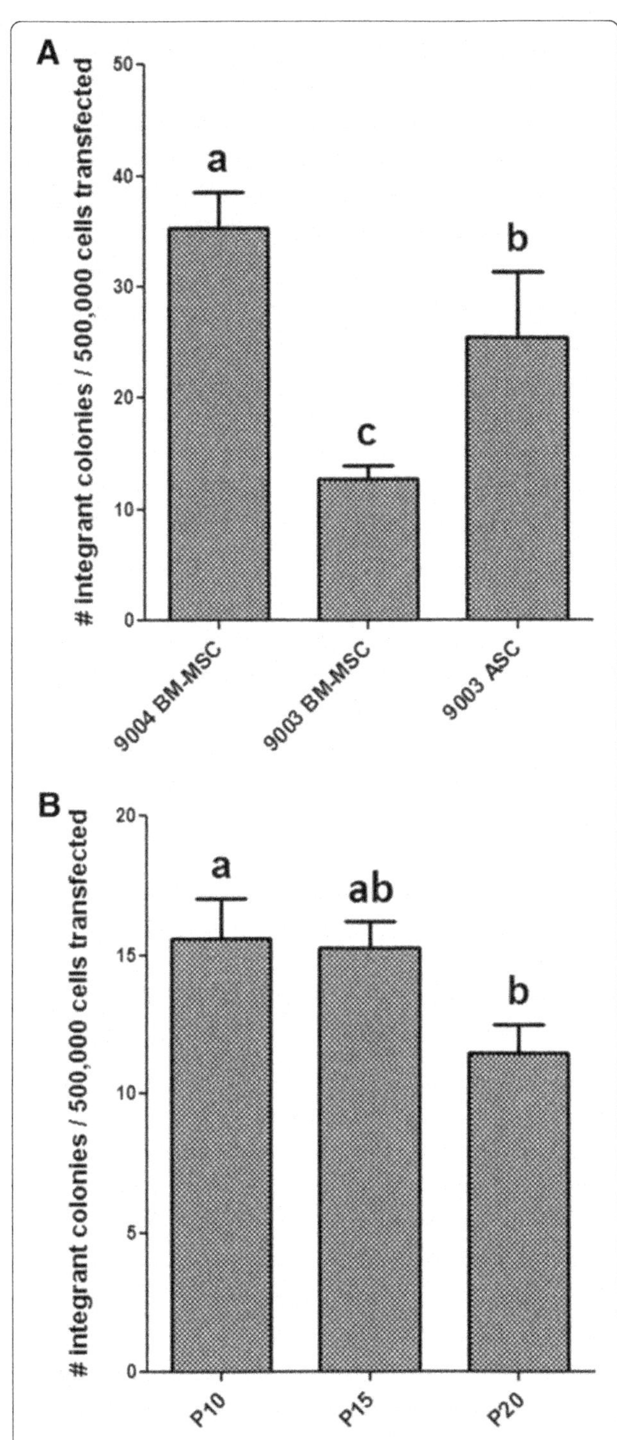

Figure 10 Transfection and integration efficiency in MSCs. P5 MSCs **(A)** and 9004 BM-MSC at passages 10, 15 and 20 **(B)** were transfected with pEGFP-N1. Integrant number were expressed as the number of discrete colonies that were resistant to G418 selection. Statistically significant groups by Tukey's test is indicated by letters ($P < 0.05$). 9004 BM-MSC showed the highest number of integrants, followed by 9003 ASC and 9003 BM-MSC. 9004 BM-MSC showed a decreasing trend in number of integrants with increasing passages.

Goat ASCs were more competent at adipogenesis than BM-MSCs, as reported for other species [8,9]. This could be explained by intrinsic epigenetic differences that primes resident MSCs in different source tissues to differentiate into the surrounding tissue. This is consistent with the finding that MSCs derived from adipose, bone marrow and muscle have similar but not identical promoter methylation profiles [46]. According to a review by Boquest et al. [47], studies showed that undifferentiated ASCs are epigenetically primed for adipogenesis. We also saw significant differences between two lines of BM-MSCs isolated from two donor goats. Large donor-to-donor variability in adipogenic differentiation of BM-MSCs also was seen in a study by Aldridge et al. [48]. Even though observations on our goat MSCs are consistent with what has been seen in MSCs of other species, it is difficult to draw the same conclusions based on only one ASC line and two BM-MSC lines. Future studies should aim to solidify these findings with more MSC isolates.

In this study, fibroblasts derived from ear cartilage were observed to be capable of adipogenic differentiation at a low capacity, as seen with other isolated fibroblast lines [25,48] and suggesting a rare subpopulation of mesenchymal progenitors reside in fibroblast preparations [49]. Consistent with our results, Alt et al. [25] also observed significantly more Oil Red O-staining adipocytes in human ASCs compared to skin fibroblasts. Stress-induced lipid accumulation in the fibroblasts is also a possibility, such as lipid formation due to overconfluence [50] and not adipogenesis.

After 21 d of adipogenic differentiation, the BM-MSCs appeared to exhibit a low adipogenic potential, and in some cases, even lower than the fibroblast cell line. However, from visual observations throughout the 21-d period of adipogenic differentiation via the continuous induction method (Adipo I), lipid-positive cells in BM-MSCs were seen to increase until approximately 10 d and appear to subsequently decrease in number. Adipogenic cell counts in both BM-MSC lines at Day 10 were found to be higher than the numbers acquired at Day 21 [Additional file 1: Figure S3]. Thus, the adipocyte count in BM-MSCs after 21 d of differentiation do not represent the maximal number that could be achieved in these cell lines, and although Adipo I did indeed yield greater numbers, this differentiation method was not able to sustain adipocyte numbers past two weeks of culture. The loss of adipocytes in BM-MSCs could either be explained by external aspects, such as composition of culture media, or intrinsic factors, such as the nature of the cells themselves. Bosch et al. [51] showed a plateau in number of lipid-positive cells after 10 d in porcine BM-MSCs. It is possible that the media compositions used are not sufficient to maintain BM MSCs differentiating down the adipogenic lineage, and needs to be further defined for less studied species such as goats and pigs.

In our study, we observed an increase in adipogenic differentiation capacity as passage number increased in 9004 BM-MSCs, as shown by adipocyte counts and mRNA expression of *PPARG* and *FABP4*. This is concordant with the observed increased adipogenesis in late-passage porcine BM-MSCs [52]. On the other hand, studies in rhesus macaque and human BM-MSCs show a decrease in adipogenic potential at higher passages [29,30,53]. These opposing results could be due to species-specific differences in the biology of MSCs. Also, an increase in CFE with passaging was observed, suggesting that the passage-related increase in adipogenic potential could also be attributed to the increased proportion of progenitor cells within the MSC culture. If this were the case, overall enhanced trilineage differentiation should be observed. Alternatively, should the increase in adipogenicity be due to shifts in the multipotentiality of the progenitor cells and not their numbers, an inverse effect between adipogenesis and osteogenesis is possible [54-57]. A concomitant increase in osteogenic potential has been demonstrated with the decrease in adipogenic potential [30,53], or vice versa [52]. Further research should investigate the effects of passaging on the osteochondrogenic potential of goat BM-MSCs, and whether the same trend occurs in goat ASCs. This is especially relevant as goat MSCs are often utilized in bone and cartilage tissue engineering applications.

MSC cultures are known to be heterogeneous [58]. Therefore, each MSC isolation will have a heterogeneous mixture of cells – each may have a different proportion of mesenchymal progenitor cells, which is demonstrated by the difference in differentiation capacity and low density CFE shown in this work. The colony-forming unit (CFU) assay, or colony-forming efficiency assay, is an assay that measures the ability of a cell line or isolate to propagate and form colonies originating from single cells after plating at a very low density. CFU assays are useful in estimating the proportion of early progenitors in different preparations of MSCs [59]. A positive correlation between CFE and differentiation capacity have been demonstrated in other studies [29,60,61]. In this study, between the BM-MSC lines, the cell line with the higher CFE (9004 BM-MSC) was also observed to have higher differentiation capacity. When cultured through higher passages, 9004 BM-MSC exhibited an increase in CFE, suggesting that long-term *in vitro* culture, at least up to 20 passages, selects for progenitor cells as they can self-renew.

Goat MSC lines express CD90, CD105 and CD73, but do not express CD45, concordant with the minimal criteria set for defining human MSCs [5]. However, thus far, trilineage differentiation appears to be a more reliable method for identifying MSCs as it is consistent across species, whereas the expression of cell surface markers is more variable. For example, CD90 expression is variable in the mouse [62,63] and is not expressed in rabbit MSCs [64]. Furthermore, the positive cell surface markers defined for MSCs are also expressed in fibroblasts [25,31,65], as shown in our study as well. Thus, these positive cell surface markers are not ideal for characterizing MSCs, given that MSCs share morphological and other characteristics with fibroblasts. Finding markers that distinguish MSCs from fibroblasts is important, as like MSCs [1], fibroblasts can also be found in virtually all tissues of the body and are morphologically similar to MSCs. As such, other research groups have endeavored to identify markers that distinguish MSCs from fibroblasts [31,66], and also use fibroblasts as a control cell line in MSC characterization experiments [44,67].

Data from qRT-PCR of cell surface markers showed that CD90 and CD73 mRNA levels did not change between passages 5 and 20. CD105 appeared to increase between passages 5 and 10, but did not change in further passages. CD105 was defined as a positive marker for its prevalence in MSCs at isolation [5], and was used to select for a more 'pure' MSC population [68]. Since no difference in CD105 expression was detected in 9004 BM-MSC between passages 10 and 20, while significant difference in adipogenesis was observed, it is unlikely that CD105 expression is related to adipogenic potential in MSCs. A study by Lysy et al. [69] also showed no link between CD105 expression and adipogenic capacity in MSCs and fibroblasts, whereas lung fibroblasts that also express CD105 were not able to undergo adipogenesis [25]. The fact that CFE in 9004 BM-MSC at passage 5 and 10 were comparable indicates that the increase in CD105 expression cannot be attributed to a growing number of progenitor cells. CD105, or endoglin, is a transmembrane protein involved in TGF-β signaling [70]. CD105 is required for angiogenesis [71] and is regulated during heart development [72]. Since MSCs are known to promote angiogenesis [73,74], perhaps the increase in CD105 expression is indicative of an increase in angiogenic potential. It has been shown that gene expression data of cell surface markers is representative of protein expression measured in flow cytometry [31]; thus, quantitating gene expression could be useful to estimate the percentage of cells expressing certain markers, when the use of antibodies prove to be difficult due to lack of availability and cross-reactivity in certain species, such as the goat. Nevertheless, there is still a need for antibodies with specificity to goats, as flow cytometry can provide information on marker expression on a per cell basis and has the potential to reflect heterogeneity in the MSC population.

Expression of CD34, a negative cell surface marker for MSCs defined by Dominici et al. [5] was detected in

9003 ASC at passage 5. This is consistent with a joint statement of the International Federation for Adipose Therapeutics and Science (IFATS) and the International Society for Cellular Therapy (ISCT) that CD34 expression in adipose-derived MSCs is variable and generally detected at early passages [10]. This could be due to heterogeneity in the early phase of ASC culture, as ASCs have been found to be more homogeneous in later passages [75]. This is consistent with the high standard deviation seen in the CFE of 9003 ASCs (which was due to differences between biological replicates, not technical replicates), suggesting a greater heterogeneity in the cell line. Detection of CD34 expression in early passage ASCs due to contamination of hematopoietic cells is also possible, and more population doublings in culture may be needed to attain homogeneity [63]. Another possible explanation is that CD34 is gradually lost in ASCs in *ex vivo* culture [75,76].

In this study, the expression of the pluripotency-associated transcription factor genes *OCT4, SOX2* and *NANOG* were detected in our goat MSCs by RT-PCR, and OCT4 protein was detected by immunofluorescence. In the immunofluorescence experiment, OCT4 was shown to be localized exclusively in the cytoplasm of MSCs. This is consistent with the OCT4B isoform, which is the isoform associated with cell stress response [77,78] and has been shown to be cytoplasmic [43,79]. Localization of OCT4A, the isoform that is associated with pluripotency, is restricted to the nuclei of pluripotent cells [43,79]. Cytoplasmic localization has been demonstrated in other reports of OCT4 detection in MSCs [29,80,81], confirming that the OCT4 isoform detected in MSCs is most likely OCT4B. A study on goat umbilical cord MSCs, which were reported to be more primitive, showed that OCT4 and SOX2 expression was more restricted to the nucleus with some expression in the cytoplasm, but with NANOG expression completely restricted to the cytoplasm [22]. When working with antibodies, one has to consider the possibility of non-specific staining, especially when using antibodies specific to other species and when no positive or negative control was used. One study reported that the detection of OCT4 and SOX2 expression in the nucleus and the cytoplasm varied with the different commercial antibodies used [82], demonstrating the questionable nature of the detection of these proteins in adult stem cells and the importance of using controls. In our study, we showed that the antibody used against OCT4 is specific to the nucleus in a bovine blastocyst, suggesting that the antibody was not isoform specific, and thus could be used to show differing localization that correlates with different isoforms. We also tried an antibody against NANOG in our study, which cross-reacted in our MSCs, but showed non-specific reactivity in bovine blastocysts (data not shown). *OCT4, NANOG* and *SOX2* were also expressed in the fibroblast line 1014 EF; it is then likely that these genes perform a different function in adult cells that is unrelated to pluripotency, or no function at all. This is supported by the observation that knocking out *OCT4* in MSCs as well as other somatic stem cells did not affect their self-renewal and multipotential abilities [83].

Investigating transfection and integration of introduced DNA constructs in MSCs is interesting for assessing their potential use in genetic engineering. The integration efficiencies observed in our goat MSCs are similar to the low efficiencies of random integration seen in other mammalian cells [84]. Haleem-Smith et al. saw no significant differences in transfection efficiency of human BM-MSCs from different donors [85], consistent with our findings. Nevertheless, they demonstrated higher Nucleofection efficiencies compared to ours, which could be explained by our use of a different plasmid [85], as well as the fact that the kits and protocols available were optimized for human MSCs and not goat MSCs. We also observed that, compared to 9004 BM-MSC, 9003 BM-MSC yielded significantly fewer integrant colonies and also lower CFE. This is unsurprising, as the subpopulation of cells that can attach and proliferate from low-density plating should represent the cells that are robust enough to stably integrate introduced DNA constructs, survive selection and proliferate into colonies of integrants.

Long-term *ex vivo* expansion of goat BM-MSCs may also have altered the proportion of MSC subpopulations. Higher passage BM-MSCs contained more enlarged, flattened and polygonal cells, which could be indicative of an increase in the large, flat and polygonal Type II subpopulation of MSCs [86,87], which proliferate slower and tend to increase in proportion in higher passage MSCs [86,88]. A large number of the lipid-filled adipocytes in P15 and P20 MSCs are rounded and appear less attached to the substrate, suggesting that the flatter cells may be less structurally supportive to adjacent cells. Additionally, a progressive decrease in G418-resistant integrant colonies from transfections was observed, even though DNA uptake ability of MSCs did not change with passaging. This suggests that genetic modification of higher passage MSCs may be less efficient due the increase of in the slow-growing Type II subpopulation that may be less robust against selection procedures. MSCs have been genetically modified *in vitro* for various applications, such as enhancing tissue regeneration [89,90] or utilizing MSCs as delivery vehicles for transgene products [91-93]. In exploring the application of in cell-based therapy and tissue engineering, it is useful to investigate whether *ex vivo* expansion alters their amenability to genetic modification.

Our results show that two different adipogenic differentiation protocols resulted in significantly different profiles of adipogenesis, as shown by the morphology and percentage of Oil Red O-positive cells. MSCs that were exposed to alternating induction and maintenance media produced larger fat droplets than those produced by MSCs that were exposed to the same adipogenic medium consistently. The different patterns of adipogenesis due to different induction method show that MSCs could potentially be used as an *in vitro* model for illustrating adipogenesis [94]. There have been studies conducted on more differentiated cell types, such as immortalized pre-adipocytic cell lines or mature adipocytes from the stromal-vascular fraction in digested adipose tissue, as *in vitro* models to elucidate mechanisms involved in adipogenesis [95,96]. Using adult stem cells like MSCs could bring another dimension to these studies; for instance, the mechanisms of differentiation and adipogenic lineage commitment could be examined as well.

Measuring adipogenic capacity by counting adipocytes is laborious, but is more informative as it takes into account the number of cells that actually differentiated. Methods that measure adipogenesis by solubilizing dyes and measuring absorbance do not provide information on the number of adipocytes or the average levels of lipids per cell. On the other hand, counting cells provides information on the number of cells that actually differentiated, and may be a better representation of adipogenic capacity and has been employed in other studies as well [25]. As demonstrated by our results, different methods of differentiation may yield more Oil Red O staining that do not necessarily equal to more adipocytes. It has been established that *FABP4* is one of the best markers of adipogenesis due to its significant upregulation after adipogenic induction [97]. Moreover, our results showed that *FABP4* expression is highly correlative to the number of lipid-positive cells in a given treatment. The large standard deviation seen in 9003 ASC's *FABP4* expression (but not in *PPARG*), which was due to a large difference between the two biological replicates, mirrors the large standard deviation seen in the adipocyte count as well. The expression of *FABP4* (but not *PPARG*) in 9003 BM-MSC was also concordant with the adipocyte count, demonstrating that *FABP4* expression could be a reliable representation of adipocyte count. Hence, measuring *FABP4* mRNA could be used as quick measurement of the extent of adipogenesis within a given cell line. It was also shown in another study that semi-quantitative scoring of lipid staining was highly correlative to FABP4 protein quantification via flow cytometry, which also provides information on a per cell basis [48]. Results from our study and others [48] showed that *PPARG* expression was not representative of the extent of adipogenic differentiation, suggesting that expression of the early marker *PPARG* is not indicative of the progression of adipogenic differentiation.

Conclusions

Our experiments demonstrate that 9004 BM-MSC, 9003 BM-MSC and 9003 ASC are indeed goat MSCs, as evidenced by their ability to undergo osteogenic, chondrogenic and adipogenic differentiation. Like human MSCs, goat MSCs also express CD90, CD105 and CD73, and do not express the hematopoietic marker CD45. As evidenced by CFE, different MSC isolations from the same tissue type can result in significantly different quantities of progenitor cells. Thus, performing a colony-forming assay early in a MSC culture may be useful to assess the purity of a MSC isolate and influence choice of cell lines for future experiments.

This study also sheds light on how the donor, source tissue and passage number of goat MSCs could lead to differences in adipogenic differentiation. Cytochemical staining and gene expression analysis showed that the goat ASC line had a greater capacity for adipogenic differentiation compared to the two BM-MSC lines. Goat fibroblasts were also able to undergo adipogenic differentiation, but at a low capacity. The differences in adipogenic potential between two BM-MSC lines demonstrate donor-to-donor variability between MSCs isolated from the same tissue, even when the donors were siblings from the same litter. Last but not least, passaging and prolonged *ex vivo* culture of BM-MSCs had an impact on adipogenicity as well.

In conclusion, this study provides further characterization information on goat MSCs, as well as an insight to how MSC characteristics change during *ex vivo* expansion. BM-MSCs expanded over increasing passages *in vitro* maintained karyotypic stability up to 20 passages in culture, exhibited an increase in adipogenic differentiation and CFE, but showed altered morphology and amenability to genetic modification by selection. Different methods of inducing adipogenesis also affect the extent and profile of adipogenic differentiation in MSCs, indicating the potential of goat MSCs as an *in vitro* model for adipogenesis for ruminant species.

Additional file

Additional file 1: This PDF file contains Supplementary Figures 1, 2 and 3, as mentioned in the manuscript text, along with the materials and methods involved to produce these figures.

Abbreviations

ASCs: Adipose-derived mesenchymal stem cells; BM-MSCs: Bone marrow-derived mesenchymal stem cells; CFE: Colony-forming efficiency; CFU: Colony-forming unit; FBS: fetal bovine serum; EGFP: Enhanced green fluorescent protein; MSCs: Mesenchymal stem cells; Neo^R: Neomycin resistance/resistant; P5: Passage 5; P10: Passage 10; P15: Passage 15; P20: Passage 20; PBS: Phosphate-buffered saline.

Competing interests

The authors declare that they have no competing interests.

Authors' contributions

NM conceived of the study, carried out all the experiments and drafted the manuscript. PJR assisted with the immnuofluorescence experiment and the interpretation of its results. EAM and JDM participated in the study's design and coordination. All authors contributed to, read, and approved the final manuscript.

Acknowledgements

The authors thank Drs. Matthew Wheeler and Elisa Monaco for the isolation of the goat MSCs, and Jan Carlson for her assistance in the goat ear sample collection. NM wishes to thank Jenna Robles and Kira Yoshimura for their assistance in optimizations and data collection, as well as Barbara Jean Nitta-Oda, Jennifer Hughes, James Chitwood, Angela Canovas, Juan Reyes, Lydia Garas and Erica Scott for their advice and technical assistance. NM was funded by Majlis Amanah Rakyat (Malaysia) throughout the duration of this study. The study was funded by Jastro Shields Research Fellowship and USDA-CREES W2171 Regional Research Project.

Author details

[1]Department of Animal Science, University of California, Davis, California 95616, USA. [2]Department of Population Health and Reproduction, University of California, Davis, California 95616, USA. [3]Institute of Ocean and Earth Sciences, University of Malaya, 50603 Kuala Lumpur, Malaysia.

References

1. da Silva Meirelles L, Chagastelles PC, Nardi NB. Mesenchymal stem cells reside in virtually all post-natal organs and tissues. J Cell Sci. 2006;119:2204–13.
2. Caplan AI. Mesenchymal stem cells. J Orthop Res. 1991;9:641–50.
3. Pittenger MF, Mackay AM, Beck SC, Jaiswal RK, Douglas R, Mosca JD, et al. Multilineage potential of adult human mesenchymal stem cells. Science. 1999;284:143–7.
4. Zuk PA, Zhu M, Mizuno H, Huang J, Futrell JW, Katz AJ, et al. Multilineage cells from human adipose tissue: implications for cell-based therapies. Tissue Eng. 2001;7:211–28.
5. Dominici M, Le Blanc K, Mueller I, Slaper-Cortenbach I, Marini F, Krause D, et al. Minimal criteria for defining multipotent mesenchymal stromal cells. The International Society for Cellular Therapy position statement. Cytotherapy. 2006;8:315–7.
6. Koga H, Muneta T, Nagase T, Nimura A, Ju YJ, Mochizuki T, et al. Comparison of mesenchymal tissues-derived stem cells for in vivo chondrogenesis: suitable conditions for cell therapy of cartilage defects in rabbit. Cell Tissue Res. 2008;333:207–15.
7. Im GI, Shin YW, Lee KB. Do adipose tissue-derived mesenchymal stem cells have the same osteogenic and chondrogenic potential as bone marrow-derived cells? Osteoarthritis Cartilage. 2005;13:845–53.
8. Monaco E, Sobreira de Lima A, Bionaz M, Maki A, Wilson SM, Hurley WL, et al. Morphological and transcriptomic comparison of adipose and bone marrow derived porcine stem cells. Open Tissue Eng Regen Med J. 2009;2:20–33.
9. Spencer ND, Chun R, Vidal MA, Gimble JM, Lopez MJ. In vitro expansion and differentiation of fresh and revitalized adult canine bone marrow-derived and adipose tissue-derived stromal cells. Vet J. 2012;191:231–9.
10. Bourin P, Bunnell BA, Casteilla L, Dominici M, Katz AJ, March KL, et al. Stromal cells from the adipose tissue-derived stromal vascular fraction and culture expanded adipose tissue-derived stromal/stem cells: a joint statement of the International Federation for Adipose Therapeutics and Science (IFATS) and the International Society for Cellular Therapy (ISCT). Cytotherapy. 2013;15:641–8.
11. De Ugarte DA, Alfonso Z, Zuk PA, Elbarbary A, Zhu M, Ashjian P, et al. Differential expression of stem cell mobilization-associated molecules on multi-lineage cells from adipose tissue and bone marrow. Immunol Lett. 2003;89:267–70.
12. Proffen BL, McElfresh M, Fleming BC, Murray MM. A comparative anatomical study of the human knee and six animal species. Knee. 2012;19:493–9.
13. Liu X, Li X, Fan Y, Zhang G, Li D, Dong W, et al. Repairing goat tibia segmental bone defect using scaffold cultured with mesenchymal stem cells. J Biomed Mater Res B Appl Biomater. 2010;94:44–52.
14. Vertenten G, Lippens E, Girones J, Gorski T, Declercq H, Saunders J, et al. Evaluation of an injectable, photopolymerizable, and three-dimensional scaffold based on methacrylate-endcapped poly(D, L-lactide-co-epsilon-caprolactone) combined with autologous mesenchymal stem cells in a goat tibial unicortical defect model. Tissue Eng A. 2009;15:1501–11.
15. Coburn JM, Gibson M, Monagle S, Patterson Z, Elisseeff JH. Bioinspired nanofibers support chondrogenesis for articular cartilage repair. Proc Natl Acad Sci U S A. 2012;109:10012–7.
16. Toh WS, Lim TC, Kurisawa M, Spector M. Modulation of mesenchymal stem cell chondrogenesis in a tunable hyaluronic acid hydrogel microenvironment. Biomaterials. 2012;33:3835–45.
17. Murphy JM, Fink DJ, Hunziker EB, Barry FP. Stem cell therapy in a caprine model of osteoarthritis. Arthritis Rheum. 2003;48:3464–74.
18. Zhu S, Zhang B, Man C, Ma Y, Hu J. NEL-like molecule-1-modified bone marrow mesenchymal stem cells/poly lactic-co-glycolic acid composite improves repair of large osteochondral defects in mandibular condyle. Osteoarthritis Cartilage. 2011;19:743–50.
19. Sun XD, Jeng L, Bolliet C, Olsen BR, Spector M. Non-viral endostatin plasmid transfection of mesenchymal stem cells via collagen scaffolds. Biomaterials. 2009;30:1222–31.
20. Ren Y, Wu H, Zhou X, Wen J, Jin M, Cang M, et al. Isolation, expansion, and differentiation of goat adipose-derived stem cells. Res Vet Sci. 2012;93:404–11.
21. Knippenberg M, Helder MN, Doulabi BZ, Semeins CM, Wuisman PI, Klein-Nulend J. Adipose tissue-derived mesenchymal stem cells acquire bone cell-like responsiveness to fluid shear stress on osteogenic stimulation. Tissue Eng. 2005;11:1780–8.
22. Qiu P, Bai Y, Liu C, He X, Cao H, Li M, et al. A dose dependent function of follicular fluid on the proliferation and differentiation of umbilical cord mesenchymal stem cells (MSCs) of goat. Histochem Cell Biol. 2012;138:593–603.
23. Azari O, Babaei H, Derakhshanfar A, Nematollahi-Mahani SN, Poursahebi R, Moshrefi M. Effects of transplanted mesenchymal stem cells isolated from Wharton's jelly of caprine umbilical cord on cutaneous wound healing; histopathological evaluation. Vet Res Commun. 2011;35:211–22.
24. Hematti P. Mesenchymal stromal cells and fibroblasts: a case of mistaken identity? Cytotherapy. 2012;14:516–21.
25. Alt E, Yan Y, Gehmert S, Song YH, Altman A, Gehmert S, et al. Fibroblasts share mesenchymal phenotypes with stem cells, but lack their differentiation and colony-forming potential. Biol Cell. 2011;103:197–208.
26. Potchoiba MJ, Lu CD, Pinkerton F, Sahlu T. Effects of all-milk diet on weight gain, organ development, carcass characteristics and tissue composition, including fatty acids and cholestrol contents, of growing male goats. Small Rumin Res. 1990;3:583–92.
27. Hausman GJ, Dodson MV, Ajuwon K, Azain M, Barnes KM, Guan LL, et al. Board-invited review: the biology and regulation of preadipocytes and adipocytes in meat animals. J Anim Sci. 2009;87:1218–46.
28. Du M, Yin J, Zhu MJ. Cellular signaling pathways regulating the initial stage of adipogenesis and marbling of skeletal muscle. Meat Sci. 2010;86:103–9.
29. Izadpanah R, Trygg C, Patel B, Kriedt C, Dufour J, Gimble JM, et al. Biologic properties of mesenchymal stem cells derived from bone marrow and adipose tissue. J Cell Biochem. 2006;99:1285–97.
30. Wagner W, Horn P, Castoldi M, Diehlmann A, Bork S, Saffrich R, et al. Replicative senescence of mesenchymal stem cells: a continuous and organized process. PLoS One. 2008;3:e2213.
31. Halfon S, Abramov N, Grinblat B, Ginis I. Markers distinguishing mesenchymal stem cells from fibroblasts are downregulated with passaging. Stem Cells Dev. 2011;20:53–66.
32. Tsai CC, Su PF, Huang YF, Yew TL, Hung SC. Oct4 and Nanog directly regulate Dnmt1 to maintain self-renewal and undifferentiated state in mesenchymal stem cells. Mol Cell. 2012;47:169–82.
33. Kretlow JD, Jin YQ, Liu W, Zhang WJ, Hong TH, Zhou G, et al. Donor age and cell passage affects differentiation potential of murine bone marrow-derived stem cells. BMC Cell Biol. 2008;9:60.
34. Pittenger MF. Mesenchymal stem cells from adult bone marrow. Methods Mol Biol. 2008;449:27–44.
35. Behboodi E, Bondareva A, Begin I, Rao K, Neveu N, Pierson JT, et al. Establishment of goat embryonic stem cells from in vivo produced blastocyst-stage embryos. Mol Reprod Dev. 2011;78:202–11.
36. He S, Pant D, Schiffmacher A, Bischoff S, Melican D, Gavin W, et al. Developmental expression of pluripotency determining factors in caprine

embryos: novel pattern of NANOG protein localization in the nucleolus. Mol Reprod Dev. 2006;73:1512–22.

37. Li JW, Guo XL, He CL, Tuo YH, Wang Z, Wen J, et al. In vitro chondrogenesis of the goat bone marrow mesenchymal stem cells directed by chondrocytes in monolayer and 3-dimetional indirect co-culture system. Chin Med J. 2011;124:3080–6.
38. Wang YH, Bower NI, Reverter A, Tan SH, De Jager N, Wang R, et al. Gene expression patterns during intramuscular fat development in cattle. J Anim Sci. 2009;87:119–30.
39. Wei S, Zan LS, Wang HB, Cheng G, Du M, Jiang Z, et al. Adenovirus-mediated interference of FABP4 regulates mRNA expression of ADIPOQ, LEP and LEPR in bovine adipocytes. Genet Mol Res. 2013;12:494–505.
40. Livak KJ, Schmittgen TD. Analysis of relative gene expression data using real-time quantitative PCR and the 2(–Delta Delta C(T)) Method. Methods. 2001;25:402–8.
41. Schmittgen TD, Livak KJ. Analyzing real-time PCR data by the comparative C(T) method. Nat Protoc. 2008;3:1101–8.
42. Khan DR, Dube D, Gall L, Peynot N, Ruffini S, Laffont L, et al. Expression of pluripotency master regulators during two key developmental transitions: EGA and early lineage specification in the bovine embryo. PLoS One. 2012;7:e34110.
43. Cauffman G, Liebaers I, Van Steirteghem A, Van de Velde H. POU5F1 isoforms show different expression patterns in human embryonic stem cells and preimplantation embryos. Stem Cells. 2006;24:2685–91.
44. Kern S, Eichler H, Stoeve J, Kluter H, Bieback K. Comparative analysis of mesenchymal stem cells from bone marrow, umbilical cord blood, or adipose tissue. Stem Cells. 2006;24:1294–301.
45. Lorenz K, Sicker M, Schmelzer E, Rupf T, Salvetter J, Schulz-Siegmund M, et al. Multilineage differentiation potential of human dermal skin-derived fibroblasts. Exp Dermatol. 2008;17:925–32.
46. Sorensen AL, Jacobsen BM, Reiner AH, Andersen IS, Collas P. Promoter DNA methylation patterns of differentiated cells are largely programmed at the progenitor stage. Mol Biol Cell. 2010;21:2066–77.
47. Boquest AC, Noer A, Collas P. Epigenetic programming of mesenchymal stem cells from human adipose tissue. Stem Cell Rev. 2006;2:319–29.
48. Aldridge A, Kouroupis D, Churchman S, English A, Ingham E, Jones E. Assay validation for the assessment of adipogenesis of multipotential stromal cells–a direct comparison of four different methods. Cytotherapy. 2013;15:89–101.
49. Chen FG, Zhang WJ, Bi D, Liu W, Wei X, Chen FF, et al. Clonal analysis of nestin(–) vimentin(+) multipotent fibroblasts isolated from human dermis. J Cell Sci. 2007;120:2875–83.
50. Delikatny EJ, Chawla S, Leung DJ, Poptani H. MR-visible lipids and the tumor microenvironment. NMR Biomed. 2011;24:592–611.
51. Bosch P, Pratt SL, Stice SL. Isolation, characterization, gene modification, and nuclear reprogramming of porcine mesenchymal stem cells. Biol Reprod. 2006;74:46–57.
52. Vacanti V, Kong E, Suzuki G, Sato K, Canty JM, Lee T. Phenotypic changes of adult porcine mesenchymal stem cells induced by prolonged passaging in culture. J Cell Physiol. 2005;205:194–201.
53. Izadpanah R, Joswig T, Tsien F, Dufour J, Kirijan JC, Bunnell BA. Characterization of multipotent mesenchymal stem cells from the bone marrow of rhesus macaques. Stem Cells Dev. 2005;14:440–51.
54. Jaiswal RK, Jaiswal N, Bruder SP, Mbalaviele G, Marshak DR, Pittenger MF. Adult human mesenchymal stem cell differentiation to the osteogenic or adipogenic lineage is regulated by mitogen-activated protein kinase. J Biol Chem. 2000;275:9645–52.
55. Beresford JN, Bennett JH, Devlin C, Leboy PS, Owen ME. Evidence for an inverse relationship between the differentiation of adipocytic and osteogenic cells in rat marrow stromal cell cultures. J Cell Sci. 1992;102(Pt 2):341–51.
56. Zhao LJ, Jiang H, Papasian CJ, Maulik D, Drees B, Hamilton J, et al. Correlation of obesity and osteoporosis: effect of fat mass on the determination of osteoporosis. J Bone Miner Res. 2008;23:17–29.
57. Lecka-Czernik B, Moerman EJ, Grant DF, Lehmann JM, Manolagas SC, Jilka RL. Divergent effects of selective peroxisome proliferator-activated receptor-gamma 2 ligands on adipocyte versus osteoblast differentiation. Endocrinology. 2002;143:2376–84.
58. Schellenberg A, Stiehl T, Horn P, Joussen S, Pallua N, Ho AD, et al. Population dynamics of mesenchymal stromal cells during culture expansion. Cytotherapy. 2012;14:401–11.
59. Pochampally R. Colony forming unit assays for MSCs. Methods Mol Biol. 2008;449:83–91.
60. Russell KC, Phinney DG, Lacey MR, Barrilleaux BL, Meyertholen KE, O'Connor KC. In vitro high-capacity assay to quantify the clonal heterogeneity in trilineage potential of mesenchymal stem cells reveals a complex hierarchy of lineage commitment. Stem Cells. 2010;28:788–98.
61. Digirolamo CM, Stokes D, Colter D, Phinney DG, Class R, Prockop DJ. Propagation and senescence of human marrow stromal cells in culture: a simple colony-forming assay identifies samples with the greatest potential to propagate and differentiate. Br J Haematol. 1999;107:275–81.
62. Eslaminejad MB, Nadri S. Murine mesenchymal stem cell isolated and expanded in low and high density culture system: surface antigen expression and osteogenic culture mineralization. In Vitro Cell Dev Biol Anim. 2009;45:451–9.
63. Meirelles Lda S, Nardi NB. Murine marrow-derived mesenchymal stem cell: isolation, in vitro expansion, and characterization. Br J Haematol. 2003;123:702–11.
64. Lapi S, Nocchi F, Lamanna R, Passeri S, Iorio M, Paolicchi A, et al. Different media and supplements modulate the clonogenic and expansion properties of rabbit bone marrow mesenchymal stem cells. BMC Res Notes. 2008;1:53.
65. Kisselbach L, Merges M, Bossie A, Boyd A. CD90 Expression on human primary cells and elimination of contaminating fibroblasts from cell cultures. Cytotechnology. 2009;59:31–44.
66. Ishii M, Koike C, Igarashi A, Yamanaka K, Pan H, Higashi Y, et al. Molecular markers distinguish bone marrow mesenchymal stem cells from fibroblasts. Biochem Biophys Res Commun. 2005;332:297–303.
67. Ramkisoensing AA, Pijnappels DA, Askar SF, Passier R, Swildens J, Goumans MJ, et al. Human embryonic and fetal mesenchymal stem cells differentiate toward three different cardiac lineages in contrast to their adult counterparts. PLoS One. 2011;6:e24164.
68. Roura S, Farre J, Soler-Botija C, Llach A, Hove-Madsen L, Cairo JJ, et al. Effect of aging on the pluripotential capacity of human CD105+ mesenchymal stem cells. Eur J Heart Fail. 2006;8:555–63.
69. Lysy PA, Smets F, Sibille C, Najimi M, Sokal EM. Human skin fibroblasts: From mesodermal to hepatocyte-like differentiation. Hepatology. 2007;46:1574–85.
70. Duff SE, Li C, Garland JM, Kumar S. CD105 is important for angiogenesis: evidence and potential applications. FASEB J. 2003;17:984–92.
71. Li C, Hampson IN, Hampson L, Kumar P, Bernabeu C, Kumar S. CD105 antagonizes the inhibitory signaling of transforming growth factor beta1 on human vascular endothelial cells. FASEB J. 2000;14:55–64.
72. Vincent EB, Runyan RB, Weeks DL. Production of the transforming growth factor-beta binding protein endoglin is regulated during chick heart development. Dev Dyn. 1998;213:237–47.
73. Li L, Chopp M, Ding GL, Qu CS, Li QJ, Lu M, et al. MRI measurement of angiogenesis and the therapeutic effect of acute marrow stromal cell administration on traumatic brain injury. J Cereb Blood Flow Metab. 2012;32:2023–32.
74. Quertainmont R, Cantinieaux D, Botman O, Sid S, Schoenen J, Franzen R. Mesenchymal stem cell graft improves recovery after spinal cord injury in adult rats through neurotrophic and pro-angiogenic actions. PLoS One. 2012;7:e39500.
75. Mitchell JB, McIntosh K, Zvonic S, Garrett S, Floyd ZE, Kloster A, et al. Immunophenotype of human adipose-derived cells: temporal changes in stromal-associated and stem cell-associated markers. Stem Cells. 2006;24:376–85.
76. Varma MJ, Breuls RG, Schouten TE, Jurgens WJ, Bontkes HJ, Schuurhuis GJ, et al. Phenotypical and functional characterization of freshly isolated adipose tissue-derived stem cells. Stem Cells Dev. 2007;16:91–104.
77. Guo CL, Liu L, Jia YD, Zhao XY, Zhou Q, Wang L. A novel variant of Oct3/4 gene in mouse embryonic stem cells. Stem Cell Res. 2012;9:69–76.
78. Wang X, Zhao Y, Xiao Z, Chen B, Wei Z, Wang B, et al. Alternative translation of OCT4 by an internal ribosome entry site and its novel function in stress response. Stem Cells. 2009;27:1265–75.
79. Lee J, Kim HK, Rho JY, Han YM, Kim J. The human OCT-4 isoforms differ in their ability to confer self-renewal. J Biol Chem. 2006;281:33554–65.
80. Violini S, Ramelli P, Pisani LF, Gorni C, Mariani P. Horse bone marrow mesenchymal stem cells express embryo stem cell markers and show the ability for tenogenic differentiation by in vitro exposure to BMP-12. BMC Cell Biol. 2009;10:29.

81. Ock SA, Jeon BG, Rho GJ. Comparative characterization of porcine mesenchymal stem cells derived from bone marrow extract and skin tissues. Tissue Eng Part C Methods. 2010;16:1481–91.
82. Zuk PA. The intracellular distribution of the ES cell totipotent markers OCT4 and Sox2 in adult stem cells differs dramatically according to commercial antibody used. J Cell Biochem. 2009;106:867–77.
83. Lengner CJ, Camargo FD, Hochedlinger K, Welstead GG, Zaidi S, Gokhale S, et al. Oct4 expression is not required for mouse somatic stem cell self-renewal. Cell Stem Cell. 2007;1:403–15.
84. Vasquez KM, Marburger K, Intody Z, Wilson JH. Manipulating the mammalian genome by homologous recombination. Proc Natl Acad Sci U S A. 2001;98:8403–10.
85. Haleem-Smith H, Derfoul A, Okafor C, Tuli R, Olsen D, Hall DJ, et al. Optimization of high-efficiency transfection of adult human mesenchymal stem cells in vitro. Mol Biotechnol. 2005;30:9–20.
86. Mets T, Verdonk G. In vitro aging of human bone marrow derived stromal cells. Mech Ageing Dev. 1981;16:81–9.
87. Colter DC, Sekiya I, Prockop DJ. Identification of a subpopulation of rapidly self-renewing and multipotential adult stem cells in colonies of human marrow stromal cells. Proc Natl Acad Sci U S A. 2001;98:7841–5.
88. Neuhuber B, Swanger SA, Howard L, Mackay A, Fischer I. Effects of plating density and culture time on bone marrow stromal cell characteristics. Exp Hematol. 2008;36:1176–85.
89. Mizrahi O, Sheyn D, Tawackoli W, Kallai I, Oh A, Su S, et al. BMP-6 is more efficient in bone formation than BMP-2 when overexpressed in mesenchymal stem cells. Gene Ther. 2013;20:370–7.
90. Sheyn D, Kallai I, Tawackoli W, Cohn Yakubovich D, Oh A, Su S, et al. Gene-modified adult stem cells regenerate vertebral bone defect in a rat model. Mol Pharm. 2011;8:1592–601.
91. Dembinski JL, Wilson SM, Spaeth EL, Studeny M, Zompetta C, Samudio I, et al. Tumor stroma engraftment of gene-modified mesenchymal stem cells as anti-tumor therapy against ovarian cancer. Cytotherapy. 2013;15:20–32.
92. Benabdallah BF, Allard E, Yao S, Friedman G, Gregory PD, Eliopoulos N, et al. Targeted gene addition to human mesenchymal stromal cells as a cell-based plasma-soluble protein delivery platform. Cytotherapy. 2010;12:394–9.
93. Li H, Zhang B, Lu Y, Jorgensen M, Petersen B, Song S. Adipose tissue-derived mesenchymal stem cell-based liver gene delivery. J Hepatol. 2011;54:930–8.
94. Janderova L, McNeil M, Murrell AN, Mynatt RL, Smith SR. Human mesenchymal stem cells as an in vitro model for human adipogenesis. Obes Res. 2003;11:65–74.
95. Shimba S, Wada T, Hara S, Tezuka M. EPAS1 promotes adipose differentiation in 3 T3-L1 cells. J Biol Chem. 2004;279:40946–53.
96. Poulos SP, Dodson MV, Hausman GJ. Cell line models for differentiation: preadipocytes and adipocytes. Exp Biol Med. 2010;235:1185–93.
97. Boucher S, Lakshmipathy U, Vemuri M. A simplified culture and polymerase chain reaction identification assay for quality control performance testing of stem cell media products. Cytotherapy. 2009;11:761–7. 767 e761-762.

Conceptus elongation in ruminants: roles of progesterone, prostaglandin, interferon tau and cortisol

Kelsey Brooks, Greg Burns and Thomas E Spencer*

Abstract

The majority of pregnancy loss in ruminants occurs during the first three weeks after conception, particularly during the period of conceptus elongation that occurs prior to pregnancy recognition and implantation. This review integrates established and new information on the biological role of ovarian progesterone (P4), prostaglandins (PGs), interferon tau (IFNT) and cortisol in endometrial function and conceptus elongation. Progesterone is secreted by the ovarian corpus luteum (CL) and is the unequivocal hormone of pregnancy. Prostaglandins (PGs) and cortisol are produced by both the epithelial cells of the endometrium and the trophectoderm of the elongating conceptus. In contrast, IFNT is produced solely by the conceptus trophectoderm and is the maternal recognition of pregnancy signal that inhibits production of luteolytic pulses of $PGF_{2\alpha}$ by the endometrium to maintain the CL and thus production of P4. Available results in sheep support the idea that the individual, interactive, and coordinated actions of P4, PGs, IFNT and cortisol regulate conceptus elongation and implantation by controlling expression of genes in the endometrium and/or trophectoderm. An increased knowledge of conceptus-endometrial interactions during early pregnancy in ruminants is necessary to understand and elucidate the causes of infertility and recurrent early pregnancy loss and provide new strategies to improve fertility and thus reproductive efficiency.

Keywords: Conceptus, Cortisol, Endometrium, Interferon, Prostaglandin, Ruminant

Introduction

This review integrates established and new information on the biological role of ovarian progesterone (P4), prostaglandins (PGs), interferon tau (IFNT) and cortisol in endometrial function and conceptus elongation during the peri-implantation period of pregnancy in ruminants. Our knowledge of the complex biological and genetic mechanisms governing conceptus elongation and implantation remains limited in domestic ruminants [1], but is essential to ameliorate early pregnancy losses and increase fertility of ruminants.

Establishment of pregnancy in domestic ruminants (i.e., sheep, cattle, goats) begins at the conceptus stage and includes pregnancy recognition signaling, implantation, and placentation [2-5]. The morula-stage embryo enters the uterus on days 4 to 6 post-mating and then forms a blastocyst that contains an inner cell mass and a blastocoele or central cavity surrounded by a monolayer of trophectoderm. After hatching from the zona pellucida (days 8 to 10), the blastocyst slowly grows into a tubular or ovoid form and is then termed a conceptus (embryo-fetus and associated extraembryonic membranes) [5,6]. In sheep, the ovoid conceptus of about 1 mm in length on day 11 begins to elongate on day 12 and forms a filamentous conceptus of 15 to 19 cm or more in length by day 15 that occupies the entire length of the uterine horn ipsilateral to the corpus luteum (CL) with extraembryonic membranes extending into the contralateral uterine horn. In cattle, the hatched blastocyst forms an ovoid conceptus between days 12 to 14 and is only about 2 mm in length on day 13. By day 14, the conceptus is about 6 mm, and the elongating bovine conceptus reaches a length of about 60 mm (6 cm) by day 16 and is 20 cm or more by day 19. Thus, the bovine blastocyst/conceptus doubles in length every day between days 9 and 16 with a significant increase (~30-fold) in length between days 12 and 15 [7,8]. In both

* Correspondence: thomas.spencer@wsu.edu
Department of Animal Science and Center for Reproductive Biology, Washington State University, Pullman, WA 99164, USA

sheep and cattle, conceptus elongation involves exponential increases in length and weight of the trophectoderm [9] and onset of extraembryonic membrane differentiation, including gastrulation of the embryo and formation of the yolk sac and allantois that are vital for embryonic survival and formation of a functional placenta [5,6]. Trophoblast elongation observed in ruminants appears to not involve geometrical changes in cell shape but rather occurs from cell proliferation [10]. Successively, the elongated conceptus begins the process of central implantation and placentation around day 16 in sheep and day 19 in cattle [11].

Blastocyst growth into an elongated conceptus does not occur *in vitro*, as it requires secretions supplied by the endometrium of the uterus [12-14]. The uterine luminal fluid (ULF) contains substances, collectively termed histotroph, that govern elongation of the conceptus via effects on trophectoderm proliferation and migration as well as attachment and adhesion to the endometrial luminal epithelium (LE) [4,15,16]. Histotroph is derived primarily from transport and (or) synthesis and secretion of substances by the endometrial LE and glandular epithelia (GE), and it is a complex and rather undefined mixture of proteins, lipids, amino acids, sugars (glucose, fructose), ions and exosomes/microvesicles [17-21]. The recurrent early pregnancy loss observed in uterine gland knockout (UGKO) ewes established the importance of uterine epithelial-derived histotroph for support of conceptus elongation and implantation [14]. Available evidence supports the idea that ovarian P4 induces expression of a number of genes, specifically in the endometrial epithelia, that are then further stimulated by factors from the conceptus (e.g., IFNT, PGs, cortisol) as well as the endometrium itself (e.g., PGs and cortisol) [22]. The genes and functions regulated by these hormones and factors in the endometrial epithelia elicit specific changes in the intrauterine histotrophic milieu necessary for conceptus elongation [4,15,16,22,23].

Progesterone regulation of endometrial function and conceptus elongation

Progesterone stimulates and maintains endometrial functions necessary for conceptus growth, implantation, placentation, and development to term. In cattle, concentrations of P4 during early pregnancy clearly affect embryonic survival [13,24]. In both lactating dairy cows and heifers, there is a strong positive association between the postovulatory rise in P4 and embryonic development. Increasing concentrations of P4 after ovulation enhanced conceptus elongation in beef heifers [25,26], dairy cows [27], and sheep [28], while lower P4 concentrations in the early luteal phase retarded embryonic development in sheep and cattle [24,29,30]. Supplementation of cattle with P4 during early pregnancy has equivocal effects to increase embryonic survival [31] and is unlikely to rescue development of embryos with inherent genetic defects or in high-producing dairy cows [27,32,33].

Progesterone predominantly exerts an indirect effect on the conceptus via the endometrium to regulate blastocyst growth and conceptus elongation [28,30,34-36]. Similar to humans [37,38], the endometria of both cyclic and pregnant sheep and cattle express genes implicated in uterine receptivity, which can be defined as a physiological state of the uterus when conceptus growth and implantation for establishment of pregnancy is possible. The absence of a sufficiently developed conceptus to signal pregnancy recognition results in those genes being turned 'off' as luteolysis ensues and the animal returns to estrus for another opportunity to mate. The outcome of the P4-induced changes in the cyclic and pregnant uterus is to modify the intrauterine milieu, such as an increase in select amino acids, glucose, cytokines and growth factors in histotroph, for support of blastocyst growth into an ovoid conceptus and elongation to form a filamentous conceptus [4,15,22,23].

Sheep

Actions of ovarian P4 on the uterus are essential for conceptus survival and growth in sheep [28]. Between days 10 and 12 after onset of estrus or mating in cyclic and pregnant ewes, P4 induces the expression of many conceptus elongation- and implantation-related genes (Figure 1 and Table 1). The initiation of expression of those genes requires P4 action and is temporally associated with the loss of progesterone receptors (PGR) between days 10 and 12 in the endometrial LE and between days 12 and 14 to 16 in the GE after onset of estrus; however, PGR remain present in the stroma and myometrium in the ovine uterus throughout pregnancy [39]. In the endometrial LE and superficial GE (sGE), P4 induces genes that encode secreted attachment and migration factors (galectin-15 [LGALS15] and insulin-like growth factor binding protein one [IGFBP1]), intracellular enzymes (prostaglandin G/H synthase and cyclooxygenase 2 [PTGS2] and hydroxysteroid (11-beta) dehydrogenase 1 [HSD11B1]), secreted proteases (cathepsin L1 [CTSL1]), secreted protease inhibitors (cystatin C [CST] 3 and 6), a secreted candidate cell proliferation factor (gastrin releasing peptide [GRP]), glucose transporters (SLC2A1, SLC2A5, SLC5A1), and a cationic amino acid (arginine, lysine and ornithine) transporter (SLC7A2) [3,4,15]. In the endometrial GE, P4 induces genes that encode for a secreted cell proliferation factor (GRP), a glucose transporter (SLC5A11), secreted adhesion protein (secreted phosphoprotein one or SPP1), a candidate regulator of calcium/phosphate homeostasis (stanniocalcin one or STC1), and a potential immunomodulatory factor (SERPINA14, also known as uterine milk proteins or uterine

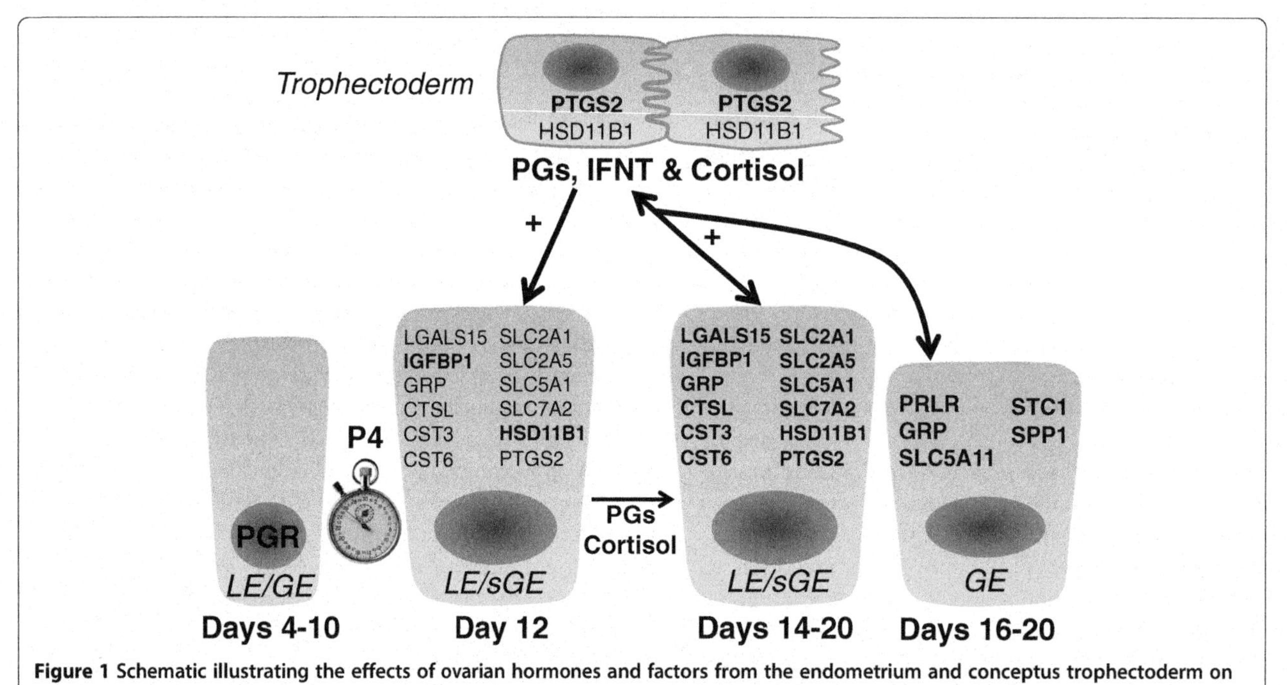

Figure 1 Schematic illustrating the effects of ovarian hormones and factors from the endometrium and conceptus trophectoderm on expression of elongation- and implantation-related genes in the endometrial epithelia of the ovine uterus during early pregnancy. Progesterone action for 8–10 days down-regulate expression of the progesterone receptor (PGR). The loss of PGR is correlated with the induction of many genes in the endometrial LE and sGE, including PTGS2 and HSD11B1 involved in prostaglandin (PG) and cortisol production, respectively, in both cyclic and pregnant ewes. If the ewe is pregnant, the trophectoderm synthesizes and secretes PGs, interferon tau (IFNT), and cortisol that act on the endometrium in a cell-specific manner to up-regulate the expression of many P4-induced genes that govern endometrial functions and/or elongation of the conceptus. Legend: GE, glandular epithelia; IFNT, interferon tau; LE, luminal epithelium; PG, prostaglandins; PGR, progesterone receptor; sGE, superficial glandular epithelia.

serpins) [3,4,15]. Several of those P4-induced genes in the epithelia are further stimulated by the actions of PGs, IFNT and/or cortisol, resulting in changes in components of the uterine luminal fluid histroph that regulate conceptus elongation via effects on trophectoderm proliferation and migration (Figures 1 and 2).

Cattle

Comparisons of the endometrial transcriptome in cyclic and pregnant heifers (days 5, 7, 12 and 13) found no difference prior to pregnancy recognition (days 15 or 16) [41,42]. Indeed, the major changes required to drive conceptus elongation and establish uterine receptivity to implantation occur between days 7 and 13 in response to ovarian P4, irrespective of whether an appropriately developed embryo/conceptus is present or not [23,30,41,43-46]. Similar to sheep, PGR protein is lost from the LE by day 13 and in the GE by day 16, and PGR loss is associated with the down- and up-regulation of genes expressed in the endometrial epithelia [47]. Using a global gene profiling approach, studies have identified the temporal changes that occur in endometrial gene expression in both cyclic [30] and pregnant [43] heifers following an elevation or diminution of post-ovulatory P4 during metestrus that promotes or delays conceptus elongation, respectively [30,34,48]. As summarized in a recent review [23], the expression of several genes is lost in the LE and GE, including PGR and a protease (alanyl (membrane) aminopeptidase [ANPEP]), and in the GE, a lipase (lipoprotein lipase [LPL]), protease (matrix metallopeptidase 2 [MMP2]) and immunomodulatory protein with antimicrobial activity (lactotransferrin [LTF]), between days 7 and 13 after onset of estrus or mating in cyclic and pregnant heifers. As expected, many conceptus elongation- and implantation-related genes appear in the endometrial epithelia between days 7 and 13 in cyclic and pregnant heifers. Genes up-regulated in the LE encode a mitogen (connective tissue growth factor [CTGF]) and in the GE encode a transport protein (retinol binding protein 4 [RBP4]), a glucose transporter (SLC5A1), and a protein involved in transport and cell proliferation (fatty acid binding protein 3 [FABP3]). Further, some genes are up regulated in both the LE and GE that encode secreted attachment and migration factors (lectin, galactoside-binding, soluble, 9 [LGALS9] and insulin-like growth factor binding protein one [IGFBP1]) as well as an intracellular enzyme (PTGS2). Those gene expression changes in the endometrium elicit changes in the ULF histotroph that are hypothesized to support conceptus elongation [23,49,50]. It is quite clear that substantial differences in gene expression

Table 1 Effects of ovarian progesterone (P4) and intrauterine infusion of interferon tau (IFNT), prostaglandins (PGs) or cortisol on elongation- and implantation-related genes expressed in the endometrial epithelia of the ovine uterus[1]

Gene symbol	P4	IFNT	PGs[2]	Cortisol
Transport of glucose				
SLC2A1	↑	+	+	+
SLC2A5	n.d.	n.e.	+	+
SLC2A12	n.d.	+	n.e. or +	+
SLC5A1	↑	+	n.e. or +	+
SLC5A11	↑	+	n.e. or +	n.e.
Transport of amino acids				
SLC1A5	n.d.	n.d.	+	+
SLC7A2	↑	+	n.e.	n.e.
Cell proliferation, migration and (or) attachment				
GRP	↑	+	+	+
IGFBP1	↑	+	++	n.e.
LGALS15	↑	++	++	++
SPP1	↑	+	n.d.	++
Proteases and their inhibitors				
CTSL1	↑	++	n.e. or +	+
CST3	↑	+	n.e. or +	n.e.
CST6		+	n.e.	+
Enzymes				
HSD11B1	↑	+	++	+
PTGS2	↑	n.e. (+ activity)	n.e. (+ activity)	n.e. (+ activity)

[1]Effect of hormone or factor denoted as induction (↑), stimulation (+ or ++), no effect (n.e.), decrease (–) or not determined (n.d.). [2]Summary data for infusion of PGE2, PGF2α, or PGI2 [40].

occur between the receptive endometrium of sheep and cattle, as one of the most abundant genes (LGALS15) induced by P4 and stimulated by IFNT in the endometrium of sheep is not expressed in cattle [51]. However, PTGS2 and IGFBP1 are common uterine receptivity markers and regulators of conceptus elongation in both sheep and cattle [46,52]. Of note, *in vitro* produced bovine embryos will elongate when transferred into a receptive ovine uterus [53].

Interferon tau regulation of endometrial function and conceptus elongation

Maternal recognition of pregnancy is the physiological process whereby the conceptus signals its presence to the maternal system and prolongs the lifespan of the ovarian CL [54]. In ruminants, IFNT is the pregnancy recognition signal secreted by the elongating conceptus that acts on the endometrium to inhibit development of the luteolytic mechanism [15,16,55,56]. Interferon tau is secreted predominantly by the elongating conceptus before implantation [57,58]. The antiluteolytic effects of IFNT inhibit transcription of the *estrogen receptor alpha* (*ESR1*) gene in sheep and *oxytocin receptor* (*OXTR*) gene in both sheep and cattle specifically in the endometrial LE. The absence of OXTR in the endometrium prevents the release of luteolytic pulses of PGF2α, thereby ensuring maintenance of the CL and continued production of P4 [3,59]. Although IFNT inhibits *OXTR* expression, it does not inhibit expression of *PTGS2*, which is important for the generation of PGs that are critical regulators of conceptus elongation during early pregnancy [60]. In addition to antiluteolytic effects, IFNT acts in a paracrine manner on the endometrium to induce or enhance expression of IFN-stimulated genes (ISGs) that are hypothesized to regulate uterine receptivity and conceptus elongation and implantation [4,40,61,62].

Classical type I IFN-stimulated genes in the endometrium

A number of transcriptional profiling experiments conducted with human cells, ovine endometrium, bovine endometrium, and bovine peripheral blood lymphocytes have elucidated classical ISG induced by IFNT during pregnancy [3,4,41,42,63]. In cattle, comparisons of days 15 to 18 pregnant and non-pregnant or cyclic endometria revealed conceptus effects on endometrial gene expression, particularly the induction or up-regulation of classical ISGs [23,41-43,64,65]. In sheep, *ISG15* (ISG15 ubiquitin-like modifier) is expressed in LE of the ovine uterus on days 10 or 11 of the estrous cycle and pregnancy, but is not detected in the LE by days 12 to 13 of pregnancy [66]. In response to IFNT from the elongating conceptus, *ISG15* is induced in the stratum compactum stroma and GE by days 13 to 14, and expression extends to the stratum spongiosum stroma, deep glands, and myometrium as well as resident immune cells of the ovine uterus by days 15 to 16 of pregnancy [66,67]. As IFNT production by the conceptus trophectoderm declines, expression of ISG in the stroma and GE also declines, but some remain abundant in endometrial stroma and GE on days 18 to 20 of pregnancy. Similar temporal and spatial alterations in *ISG15* expression occur in the bovine uterus during early pregnancy [68,69].

In vivo studies revealed that the majority of classical ISG (*B2M, GBP2, IFI27, IFIT1, ISG15, IRF9, MIC, OAS, RSAD2, STAT1,* and *STAT2*) are not induced or up-regulated by IFNT in endometrial LE or sGE of the ovine uterus during early pregnancy [66,70-73]. This finding was initially surprising, because all endometrial cell types express *IFNAR1* (interferon [alpha, beta and omega] receptor 1) and *IFNAR2* subunits of the common Type I IFN receptor [74]. Further, bovine endometrial, ovine endometrial, and human 2fTGH fibroblast cells were used to determine that IFNT activates the canonical janus kinase-signal transducer and activator of transcription-

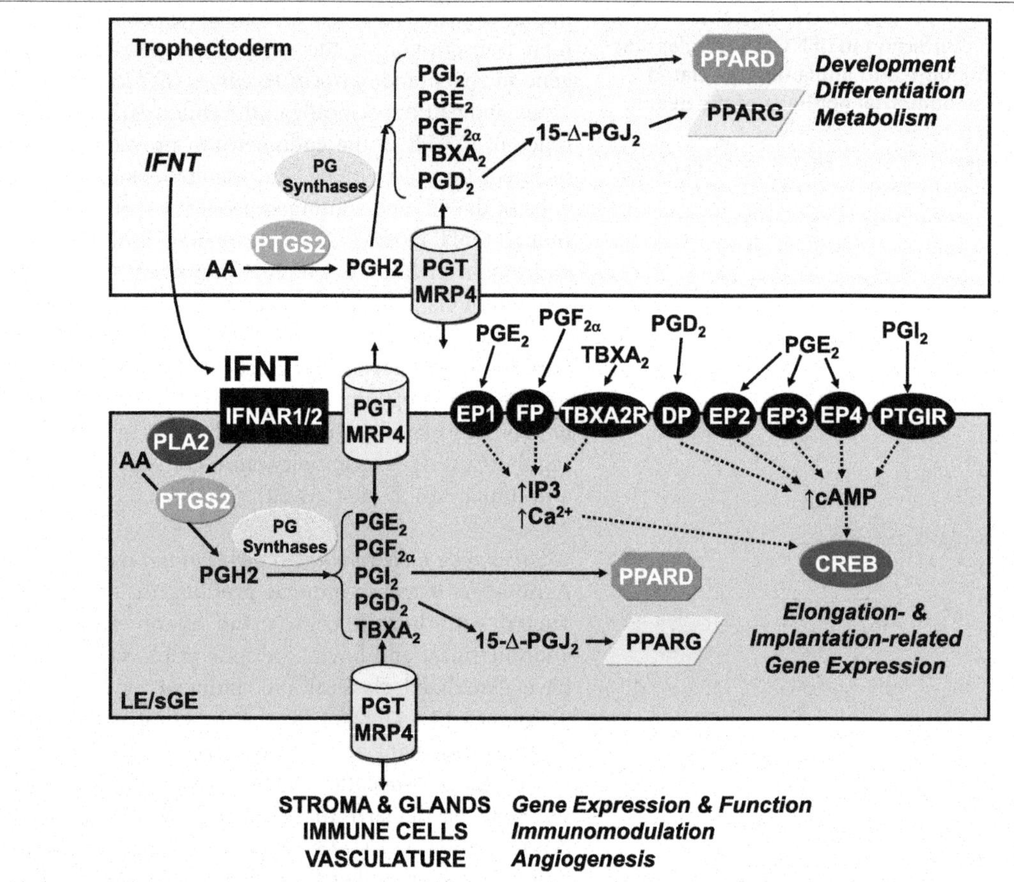

Figure 2 Schematic illustrating working hypothesis of the biological role of interferon tau (IFNT) and prostaglandins (PGs) in uterine function and conceptus elongation during early pregnancy in sheep. See text for detailed description. Legend: ABCC4, ATP-binding cassette, sub-family C (CFTR/MRP), member 4; CREB, cAMP responsive element binding protein; IFNAR, interferon (alpha, beta and omega) receptor; DP, prostaglandin D receptor (PTGDR); EP, prostaglandin E receptor (PTGER); FP, prostaglandin F receptor (PTGFR); IP, prostaglandin I receptor (PTGIR); PLA2, phospholipase A2; PPARD, peroxisome proliferator-activated receptor delta; PPARG, peroxisome proliferator-activated receptor gamma; PTGS2, prostaglandin-endoperoxide synthase 2 (prostaglandin G/H synthase and cyclooxygenase); PG Synthases, prostaglandin synthases (AKR1C3, PTGDS, PTGES, PTGFS, PTGIS, TBXAS); SLCO2A1, solute carrier organic anion transporter family, member 2A1 (prostaglandin transporter); TBXA2R, thromboxane A2 receptor.

interferon regulatory factor (JAK-STAT-IRF) signaling pathway used by other Type I IFNs [75]. About the same time, it was discovered that IRF2, a potent transcriptional repressor of ISGs [76], is expressed specifically in the endometrial LE and sGE and represses transcriptional activity of genes containing IFN-stimulated response element (ISRE)-containing promoters [70,77]. In fact, all components of the ISGF3 transcription factor complex (*STAT1, STAT2, IRF9*) and other classical ISGs (*B2M, GBP2, IFI27, IFIT1, ISG15, MIC, OAS*) contain one or more ISRE in their promoters. Thus, IRF2 in LE appears to restrict IFNT induction of most classical ISG to stroma and GE of the ovine uterus. The silencing of *MIC* and *B2M* genes in endometrial LE or sGE during pregnancy may be a critical mechanism preventing immune rejection of the semi-allogeneic conceptus [71]. As IRF2 is not expressed in other uterine cell types, classical ISGs are substantially increased in the endometrial stroma, GE and immune cells by IFNT from the conceptus during early pregnancy. Of particular note, several reports indicate induction or increases in ISGs in peripheral blood lymphocytes and the CL during pregnancy of sheep and cattle or in ewes receiving intrauterine injections of IFNT [61,63]. Recent evidence indicates that IFNT exits the uterus to exert systemic effects that alter maternal physiology, including function of the CL [61,78-80].

One challenge has been to determine which of the large number of classical ISGs induced in the endometrium by IFNT have a biological role in conceptus-endometrial interactions, as traditionally the main function of Type I IFN is to inhibit viral infection and has primarily been associated with cellular antiviral responses [81]. One classical ISG with reported biological effects on trophectoderm growth and adhesion in ruminants is *CXCL10* [chemokine (C-X-C motif) ligand 10; alias IP-10], a member of the C-X-C chemokine family that regulates multiple aspects of inflammatory and immune responses primarily through chemotactic activity toward subsets of leukocytes [82,83]. ISG15 conjugates to intracellular proteins through a ubiquitin-like mechanism [40], and deletion of *Isg15* in mice results in 50% pregnancy loss manifest during early placentation [84]. In addition, MX proteins are thought to regulate secretion through an unconventional secretory pathway [85]. The enzymes which comprise the 2′,5′-oligoadenylate synthetase (OAS) family regulate ribonuclease L antiviral responses and may play additional roles in control of cellular growth and differentiation [72].

Non-classical IFNT-stimulated genes in the endometrium

Although IFNT is the only known IFN to act as the pregnancy recognition signal, IFNs appear to have a biological role in uterine receptivity, decidualization, and placental growth and development in primates, ruminants, pigs, and rodents [40,62]. Transcriptional profiling of human U3A (STAT1 null) cells and ovine endometrium, as well as candidate gene analyses were used to discover novel 'non-classical' ISG in the endometrial LE during pregnancy such as *CST3, CTSL, HSD11B1, IGFBP1, LGALS15* and *WNT7A* (wingless-type MMTV integration site family, member 7A) [28,86-90]. Subsequently, a series of transcriptomic and candidate gene studies found that IFNT stimulates expression of a number of elongation- and implantation-related genes that are initially induced by P4 (*CST3, CST6, CTSL, GRP, HSD11B1, IGFBP1, LGALS15, SLC2A1, SLC2A5, SLC5A11, SLC7A2*) specifically in the endometrial LE, sGE, and (or) GE [3,4,62,90] (Figure 1). None of these genes are classical Type I ISG, and are referred to as 'non-classical or novel' ISG. Indeed, IFNT stimulation of these non-classical ISG requires induction by P4 and loss of PGR in the epithelia. Importantly, all of the non-classical ISG encode factors that have actions on the trophectoderm (proliferation, migration, attachment and (or) adhesion, nutrient transport) important for conceptus elongation (Table 1). For example, knockdown of an arginine transporter (SLC7A1) in the conceptus trophectoderm and inhibition of PTGS2 or HSD11B1 activity in utero compromised conceptus elongation in sheep [60,91,92]. The effects of IFNT in the bovine endometrium are not as well understood in terms of non-classical ISGs, but recent studies have started to unravel those effects in cattle [41,42,93].

Given that the critical signaling components of the JAK-STAT signaling system (STAT1, STAT2, IRF9) are not expressed in endometrial LE or sGE [70], IFNT must utilize a noncanonical, STAT1-independent signaling pathway to regulate expression of genes in endometrial LE and sGE of the ovine uterus. The noncanonical pathway mediating IFNT stimulation of genes in the endometrial LE and sGE has not been entirely elucidated, but other Type I IFN utilize mitogen-activated protein kinase (MAPK) and phosphatidylinositol 3-kinase (PI3K) cascades [94]. Available evidence suggests that IFNT activates distinct epithelial and stromal cell-specific JAK, epidermal growth factor receptor, MAPK (ERK1/2), PI3K-AKT, and (or) Jun N-terminal kinase (JNK) signaling modules to regulate expression of PGE_2 receptors in the endometrium of the ovine uterus or in ovine uterine LE cells *in vitro* [95,96]. As discussed subsequently, recent evidence indicates that PTGS2-derived PGs and HSD11B1-derived cortisol are part of the noncanonical pathway of IFNT action on the endometrium in sheep [60,97].

Prostaglandin regulation of endometrial function and conceptus elongation

Results of recent studies in sheep support the concept that PGs regulate expression of elongation- and implantation-related genes in the endometrial epithelia of ruminants during early pregnancy and are involved in conceptus elongation [46,60,98] (Figures 1 and 2). The conceptus and endometria synthesize a variety of PGs during early pregnancy in both sheep and cattle [99-104]. The endometrium produces and uterine lumen contains substantially more PGs during early pregnancy than during the estrous cycle [105-107]. The dominant cyclooxygenase expressed in both the endometrium and trophectoderm of the elongating conceptus is PTGS2 [104-106]. Although the antiluteolytic effects of IFNT are to inhibit expression of the *OXTR* in the endometrial LE and sGE of early pregnant ewes, it does not impede up-regulation of PTGS2, a rate-limiting enzyme in PG synthesis [102,107]. In sheep, PTGS2 activity in the endometrium is stimulated by IFNT, and PTGS2-derived PG were found to mediate, in part, the effects of P4 and IFNT on the endometrium of the ovine uterus. In those studies, the abundance of *HSD11B1* and *IGFBP1* mRNA in the endometrium was considerably reduced by intrauterine infusion of meloxicam, a selective PTGS2 inhibitor. As illustrated in Figure 1, *PTGS2* expression appears between days 10 and 12 post-estrus and mating in the endometrial LE and sGE and is induced by ovarian P4 [98,102]. In the bovine uterus, PTGS2 is also not down-regulated in endometria of early pregnant cattle, but rather is up-regulated by IFNT [108,109]. Thus, IFNT acts as a molecular switch that stimulates PGE_2 production in the bovine endometrium [110]. Indeed, Type I IFNs were found to stimulate phospholipase A2 (PLA_2) and synthesis of PGE_2 and $PGF_{2\alpha}$ in several different cell types over 25 years ago [111,112].

Prostaglandins are essential for conceptus elongation, as intrauterine infusions of meloxicam prevented conceptus elongation in early pregnant sheep [60,98]. The elongating conceptuses of both sheep and cattle synthesize and secrete more PG than the underlying endometrium [99,100,113]. Thus, PG levels are much greater in the uterine lumen of pregnant as compared with cyclic or non-pregnant cattle [106]. In sheep, Charpigny and coworkers

[103] found that PTGS2 was abundant in day 8 to 17 blastocysts/conceptuses, whereas PTGS1 was undetectable. PTGS2 protein increased in the conceptus trophectoderm between days 8 and 14 and was maximal between days 14 and 16. In fact, there was a 30-fold increase in PTGS2 content per protein extract between days 10 and 14, corresponding to a 50,000-fold increase in the whole conceptus, and PTGS2 protein in the conceptus then declined substantially after Day 16 to become undetectable by day 25 of pregnancy. Other studies found that Day 14 sheep conceptuses *in vitro* release mainly cyclooxygenase metabolites including PGF2α, 6-keto-PGF1α (i.e., a stable metabolite of PGI_2), and PGE_2 [103], and day 16 conceptuses produce substantially more of those PGs than day 14 conceptuses [101]. Given that membrane and nuclear receptors for PGs are present in all cell types of the ovine endometrium and conceptus during early pregnancy [60,114], PTGS2-derived PGs from the conceptus likely have paracrine, autocrine, and perhaps intracrine effects on endometrial function and conceptus development during early pregnancy (Figure 2). Indeed, expression of *PTGS2* in biopsies of day 7 bovine blastocysts is a predictor of the successful development of that blastocyst to term and delivery of a live calf [114]. Further, pregnancy rates were substantially reduced in heifers that received meloxicam, a partially selective inhibitor of PTGS2, on day 15 after insemination [115]. Thus, PGs are critical regulators of conceptus elongation and implantation in ruminants, as they are for blastocyst implantation and decidualization during pregnancy in mice, rats, hamsters, mink and likely humans [116-118].

Recently, Dorniak and coworkers [52] infused PGE_2, $PGF_{2\alpha}$, PGI_2, or IFNT, at the levels produced by the day 14 conceptus, into the uterus of cyclic ewes. In that study, expression of *GRP, IGFBP1*, and *LGALS15* were increased by PGE_2, PGI_2, and IFNT, but only IFNT increased *CST6* (Table 1). Differential effects of PG were also observed for *CTSL1* and its inhibitor *CST3*. For glucose transporters, IFNT and all PG increased *SLC2A1*, but only PG increased *SLC2A5* expression, whereas *SLC2A12* and *SLC5A1* were increased by IFNT, PGE_2, and $PGF_{2\alpha}$. Infusions of all PGs and IFNT increased the amino acid transporter *SLC1A5*, but only IFNT increased *SLC7A2*. In the uterine lumen, only IFNT increased glucose levels, and only PGE_2 and $PGF_{2\alpha}$ increased total amino acids [52]. Thus, available results support the idea that PG and IFNT from the conceptus coordinately regulate endometrial functions important for growth and development of the conceptus during the peri-implantation period of pregnancy [22] (Figures 1 and 2).

Prostaglandins also have intracrine effects within cells. Both PGI_2 and PGJ_2 can activate nuclear peroxisome proliferator-activating receptors (PPARs) [119]. PGI_2 is a ligand for PPARD, and PGD_2 spontaneously forms 15-deoxy-Δ12,14-PGJ_2 within cells that is a ligand for PPARG [120-123]. PPARs dimerize with retinoid X receptors (RXRs) and regulate transcription of target genes. Although PGs are lipid-derived, their efflux out and influx into cells depends on specific PG transporters (PGT) termed solute carrier organic anion transporter family, member 20A1 (SLC20A1) and ATP-binding cassette, subfamily C (CFTR/MRP), member 4 (ABCC4 or MRP4). PGJ_2 and PGI_2 are not as efficiently transported as other PGs (PGE_2, $PGF_{2\alpha}$, $TBXA_2$). Expression of prostacyclin (PGI2) synthase (PTGIS), PGI2 receptors (PTGIR), PPARs and RXRs in uteri and conceptuses of sheep during early pregnancy has been well documented [124]. In the endometrium, *PTGIS* mRNA and protein were expressed mainly in the endometrial LE/sGE as early as day 9 of pregnancy, but levels declined from days 12 to 17. Expression of PTGIR, PPARs (PPARA, PPARD, PPARG) and RXRs (RXRA, RXRB, RXRG) was detected in the endometrium, and PPARD and PPARG were particularly abundant in the endometrial LE and sGE. In the conceptus trophectoderm, *PTGIS* expression increased and then peaked at day 17. *PTGIR* and *PPARA* mRNAs peaked before day 12 and then declined and were nearly undetectable by Day 17, whereas *PPARD* and *PPARG* mRNAs increased from Days 12 to 17 in the conceptus. These results suggest that PPARG may also regulate conceptus trophectoderm development and differentiation due to intrinsic actions of PGJ_2, which is spontaneously formed within cells from PGD_2.

Unexpectedly, genetic studies in mice found that *Pparg* is essential for placental development, as null mutation of *Pparg* in mice resulted in placentae with poor differentiation and vascular anomalies, leading to embryonic death by gestational day 10 [123]. In mink, treatment of trophoblast cells with PGJ_2 attenuated cell proliferation, increased PPARG expression, elicited the appearance of enlarged and multinuclear cells, and increased the expression of adipophilin or ADRP (adipose differentiation-related protein), a protein involved in lipid homeostasis, and SPP1 [125]. PPARs alter the transport, cellular uptake, storage, and use of lipids and their derivatives [119]. In extravillous cytotrophoblasts of human placentae, PPARG stimulates synthesis of chorionic gonadotrophin (hCG) and increases free fatty acid (FFA) uptake. PPARG-regulated genes include fatty acid binding proteins (FABP) and fatty acid transport proteins [FATP or SLC27As] required for lipid uptake and triacylglycerol synthesis, which is undoubtedly important in rapidly growing and elongating conceptuses producing large amounts of PGs.

Mice deficient in *Ppard* also exhibit placental defects and reduced or inhibited trophoblast giant cell differentiation [126,127]. PPARD is activated by PGI_2, and treatment of rat trophoblast cells with a specific PPARD agonist triggered early differentiation of giant cells that expressed

CSH1 (chorionic somatomammotropin hormone one or placental lactogen) and reduced expression of inhibitor of differentiation two (ID2), which is an inhibitor of several basic helix-loop-helix (bHLH) transcription factors, such as HAND1, that promote giant cell differentiation. Further, PPARD increases expression of ADRP, and PPARD potentiates cell polarization and migration in the skin [128], which are all cellular activities implicated in conceptus elongation. Thus, PTGS2-derived PGs and PPARG may impact conceptus elongation via effects on trophectoderm growth and survival as well as expression of elongation- and implantation-related genes in the endometrial epithelia.

Cortisol regulation of endometrial function and conceptus elongation

Initially identified as a candidate P4-regulated gene in the endometrium that potentially governed conceptus elongation [36,52], *HSD11B1* was found to be expressed specifically in the endometrial LE and sGE and is induced by P4 and stimulated by IFNT and PGs in the endometrium of the ovine uterus [98] (Table 1). Expression of *HSD11B1* is also up-regulated in the endometrium of cattle between days 7 and 13 of pregnancy [43]. One of two isoforms of hydroxysteroid (11-beta) dehydrogenases that regulate intracellular levels of bioactive glucocorticoids within key target tissues [129], HSD11B1 is a low affinity NADP(H)-dependent bidirectional dehydrogenase/reductase for glucocorticoids, and the direction of HSD11B1 activity is determined by the relative abundance of $NADP^+$ and NADPH co-factors. The endometrium of the ovine uterus as well as conceptus generates active cortisol from inactive cortisone [97]. Cortisol regulates gene expression via the nuclear receptor subfamily 3, group C, member 1 (NR3C1 or glucocorticoid receptor [GR]), a transcriptional regulator that modulates expression of primary target genes that either directly affect cellular physiology or alter the expression of other secondary target genes, which then confer hormonal responses [130,131].

Recent findings support the idea that PGs mediate, in part, P4 induction and IFNT stimulation of *HSD11B1* expression in the ovine endometrium [60,97]. Similarly, PG regulate activity of HSD11B1 in bovine endometria [132], and $PGF_{2\alpha}$ stimulates the activity of HSD11B1 in human fetal membranes [133,134]. Whereas PG stimulate HSD11B1 activity, glucocorticoids enhance PG synthesis by up-regulating expression and activity of PLA_2 and PTGS2 in the ovine placenta, thereby establishing a positive feed-forward loop implicated in the timing of parturition [135]. This tissue-specific stimulatory role of glucocorticoids on PG synthesis contradicts the classical concept that glucocorticoids exert anti-inflammatory effects on immune cells [136].

Available results support the idea that cortisol from the endometrium as well as conceptus regulates endometrial functions important for conceptus elongation during early pregnancy in sheep. The day 14 conceptus expresses both *HSD11B1* and *HSD11B2* as well as *NR3C1* [60,98]. Indeed, the elongating sheep conceptus generates cortisol from cortisone via HSD11B1, and elevated levels of cortisol are found in the uterine lumen of early pregnant sheep [97]. Thus, cortisol may have paracrine and intracrine effects on the endometrium and conceptus trophectoderm during early pregnancy (Figure 1). As summarized in Table 1, intrauterine infusions of cortisol at early pregnancy levels into the uterus of cyclic ewes from day 10 to 14 post-estrus increased the expression of several elongation- and implantation-related genes expressed in the endometrial epithelia of the ovine uterus and increased endometrial PTGS2 and HSD11B1 expression and/or activity [92]. Similar to IFNT actions, PTGS2-derived PGs mediated some effects of cortisol. In order to determine if HSD11B1-derived cortisol is important for conceptus elongation, PF915275, a selective HSD11B1 inhibitor, was infused into the uterine lumen of bred ewes from days 8 to 14 post-mating [92]. Inhibition of HSD11B1 activity in utero prevented conceptus elongation. Thus, HSD11B1-derived cortisol is an essential regulator of conceptus elongation via effects on trophectoderm growth and survival as well as expression of elongation- and implantation-related genes in the endometrial epithelia.

The effect of knocking out NR3C1 in the elongating conceptus has not been reported in ruminants. Indeed, NR3C1 targets hundreds of genes, including those involved in lipid metabolism and triglyceride homeostasis, in other organs and cell types [130,137]. In humans, the proposed positive roles of HSD11B1-generated cortisol at the conceptus-maternal interface include stimulation of hormone secretion by the trophoblast, promotion of trophoblast growth/invasion, and stimulation of placental transport of glucose, lactate and amino acids. Indeed, glucocorticoids can have positive as well as negative effects during pregnancy [138]. Administration of synthetic glucocorticoids to women during pregnancy can alter normal development of the fetus and compromise pregnancy success by inhibiting cytokine-PG signaling, restricting trophoblast invasion, and inducing apoptosis in placenta. Similarly, administration of synthetic glucocorticoids to pregnant ewes reduced placental growth and development, numbers of trophoblast giant binucleate cells in the placenta, and circulating levels of placental lactogen [139]. On the other hand, natural glucocorticoids are hypothesized to have positive effects during early pregnancy [138]. Interestingly, administration of glucocorticoids increased pregnancy rates in women undergoing assisted reproductive technologies and pregnancy outcomes in women with a history of recurrent miscarriage [140,141].

Conclusions

The individual, additive and synergistic actions of P4, IFNT, PGs and cortisol regulate expression of elongation- and implantation-related genes in the endometrial epithelia in ruminants. The outcome of the carefully orchestrated changes in gene expression is secretion or transport of substances (e.g., glucose, amino acids, proteins) from the endometrium into the uterine lumen that govern conceptus survival and elongation via effects on trophectoderm proliferation, migration, attachment, and adhesion. Moreover, conceptus elongation is also likely governed by intracrine factors and pathways such as PGs and PPARs. A systems biology approach is necessary to fully understand conceptus elongation and the multifactorial phenomenon of early pregnancy loss. Such information is critical to provide a basis for new strategies to improve the fertility and reproductive efficiency in ruminant livestock.

Ethical approval

This is a review paper; however, all results reported based on research by the authors was approved by the Washington State University Institutional Animal Care and Use Committee.

Competing interests

The authors declare that they have no competing interests.

Authors' contributions

TES, GB and KB contributed to the writing of this review paper. All authors read and approved the manuscript.

Acknowledgments

Support for the work described in this review paper was supported, in part, by AFRI competitive grants 2009-01722 and 2012-67015-30173 from the USDA National Institute of Food and Agriculture.

References

1. Ulbrich SE, Groebner AE, Bauersachs S: **Transcriptional profiling to address molecular determinants of endometrial receptivity–lessons from studies in livestock species.** *Methods* 2013, **59**:108–115.
2. Spencer TE, Johnson GA, Bazer FW, Burghardt RC: **Implantation mechanisms: insights from the sheep.** *Reproduction* 2004, **128**:657–668.
3. Spencer TE, Johnson GA, Bazer FW, Burghardt RC: **Fetal-maternal interactions during the establishment of pregnancy in ruminants.** *Soc Reprod Fertil Suppl* 2007, **64**:379–396.
4. Spencer TE, Sandra O, Wolf E: **Genes involved in conceptus-endometrial interactions in ruminants: insights from reductionism and thoughts on holistic approaches.** *Reproduction* 2008, **135**:165–179.
5. Guillomot M: **Cellular interactions during implantation in domestic ruminants.** *J Reprod Fertil Suppl* 1995, **49**:39–51.
6. Hue I, Degrelle SA, Turenne N: **Conceptus elongation in cattle: genes, models and questions.** *Anim Reprod Sci* 2012, **134**:19–28.
7. Berg DK, van Leeuwen J, Beaumont S, Berg M, Pfeffer PL: **Embryo loss in cattle between Days 7 and 16 of pregnancy.** *Theriogenology* 2010, **73**:250–260.
8. Betteridge KJ, Eaglesome MD, Randall GC, Mitchell D: **Collection, description and transfer of embryos from cattle 10–16 days after oestrus.** *J Reprod Fertil* 1980, **59**:205–216.
9. Wales RG, Cuneo CL: **Morphology and chemical analysis of the sheep conceptus from the 13th to the 19th day of pregnancy.** *Reprod Fertil Dev* 1989, **1**:31–39.
10. Wang J, Guillomot M, Hue I: **Cellular organization of the trophoblastic epithelium in elongating conceptuses of ruminants.** *C R Biol* 2009, **332**:986–997.
11. Guillomot M, Flechon JE, Wintenberger-Torres S: **Conceptus attachment in the ewe: an ultrastructural study.** *Placenta* 1981, **2**:169–182.
12. Betteridge KJ, Flechon JE: **The anatomy and physiology of pre-attachment bovine embryos.** *Theriogenology* 1988, **29**:155–187.
13. Lonergan P: **Influence of progesterone on oocyte quality and embryo development in cows.** *Theriogenology* 2011, **76**:1594–1601.
14. Gray CA, Taylor KM, Ramsey WS, Hill JR, Bazer FW, Bartol FF, Spencer TE: **Endometrial glands are required for preimplantation conceptus elongation and survival.** *Biol Reprod* 2001, **64**:1608–1613.
15. Bazer FW, Wu G, Spencer TE, Johnson GA, Burghardt RC, Bayless K: **Novel pathways for implantation and establishment and maintenance of pregnancy in mammals.** *Mol Hum Reprod* 2010, **16**:135–152.
16. Spencer TE, Johnson GA, Bazer FW, Burghardt RC, Palmarini M: **Pregnancy recognition and conceptus implantation in domestic ruminants: roles of progesterone, interferons and endogenous retroviruses.** *Reprod Fertil Dev* 2007, **19**:65–78.
17. Bazer FW: **Uterine protein secretions: relationship to development of the conceptus.** *J Anim Sci* 1975, **41**:1376–1382.
18. Gray CA, Bartol FF, Tarleton BJ, Wiley AA, Johnson GA, Bazer FW, Spencer TE: **Developmental biology of uterine glands.** *Biol Reprod* 2001, **65**:1311–1323.
19. Bazer FW, Song G, Kim J, Erikson DW, Johnson GA, Burghardt RC, Gao H, Carey Satterfield M, Spencer TE, Wu G: **Mechanistic mammalian target of rapamycin (MTOR) cell signaling: effects of select nutrients and secreted phosphoprotein 1 on development of mammalian conceptuses.** *Mol Cell Endocrinol* 2012, **354**:22–33.
20. Koch JM, Ramadoss J, Magness RR: **Proteomic profile of uterine luminal fluid from early pregnant ewes.** *J Proteome Res* 2010, **9**:3878–3885.
21. Burns G, Brooks K, Wildung M, Navakanitworakul R, Christenson LK, Spencer TE: **Extracellular vesicles in luminal fluid of the ovine uterus.** *PLoS One* 2014, **9**:e90913.
22. Dorniak P, Bazer FW, Spencer TE: **Physiology and endocrinology symposium: biological role of interferon tau in endometrial function and conceptus elongation.** *J Anim Sci* 2013, **91**:1627–1638.
23. Forde N, Lonergan P: **Transcriptomic analysis of the bovine endometrium: what is required to establish uterine receptivity to implantation in cattle?** *J Reprod Dev* 2012, **58**:189–195.
24. Mann GE, Lamming GE: **Relationship between maternal endocrine environment, early embryo development and inhibition of the luteolytic mechanism in cows.** *Reproduction* 2001, **121**:175–180.
25. Garrett JE, Geisert RD, Zavy MT, Gries LK, Wettemann RP, Buchanan DS: **Effect of exogenous progesterone on prostaglandin F2 alpha release and the interestrous interval in the bovine.** *Prostaglandins* 1988, **36**:85–96.
26. Carter F, Forde N, Duffy P, Wade M, Fair T, Crowe MA, Evans AC, Kenny DA, Roche JF, Lonergan P: **Effect of increasing progesterone concentration from Day 3 of pregnancy on subsequent embryo survival and development in beef heifers.** *Reprod Fertil Dev* 2008, **20**:368–375.
27. Mann GE, Fray MD, Lamming GE: **Effects of time of progesterone supplementation on embryo development and interferon-tau production in the cow.** *Vet J* 2006, **171**:500–503.
28. Satterfield MC, Bazer FW, Spencer TE: **Progesterone regulation of preimplantation conceptus growth and galectin 15 (LGALS15) in the ovine uterus.** *Biol Reprod* 2006, **75**:289–296.
29. Nephew KP, McClure KE, Ott TL, Dubois DH, Bazer FW, Pope WF: **Relationship between variation in conceptus development and differences in estrous cycle duration in ewes.** *Biol Reprod* 1991, **44**:536–539.
30. Forde N, Beltman ME, Duffy GB, Duffy P, Mehta JP, O'Gaora P, Roche JF, Lonergan P, Crowe MA: **Changes in the endometrial transcriptome during the bovine estrous cycle: effect of low circulating progesterone and consequences for conceptus elongation.** *Biol Reprod* 2011, **84**:266–278.
31. Beltman ME, Lonergan P, Diskin MG, Roche JF, Crowe MA: **Effect of progesterone supplementation in the first week post conception on embryo survival in beef heifers.** *Theriogenology* 2009, **71**:1173–1179.

32. Wiltbank MC, Souza AH, Carvalho PD, Bender RW, Nascimento AB: **Improving fertility to timed artificial insemination by manipulation of circulating progesterone concentrations in lactating dairy cattle.** *Reprod Fertil Dev* 2011, **24**:238–243.
33. Lonergan P, Woods A, Fair T, Carter F, Rizos D, Ward F, Quinn K, Evans A: **Effect of embryo source and recipient progesterone environment on embryo development in cattle.** *Reprod Fertil Dev* 2007, **19**:861–868.
34. Clemente M, de La Fuente J, Fair T, Al Naib A, Gutierrez-Adan A, Roche JF, Rizos D, Lonergan P: **Progesterone and conceptus elongation in cattle: a direct effect on the embryo or an indirect effect via the endometrium?** *Reproduction* 2009, **138**:507–517.
35. Larson JE, Krisher RL, Lamb GC: **Effects of supplemental progesterone on the development, metabolism and blastocyst cell number of bovine embryos produced in vitro.** *Reprod Fertil Dev* 2011, **23**:311–318.
36. Satterfield MC, Song G, Kochan KJ, Riggs PK, Simmons RM, Elsik CG, Adelson DL, Bazer FW, Zhou H, Spencer TE: **Discovery of candidate genes and pathways in the endometrium regulating ovine blastocyst growth and conceptus elongation.** *Physiol Genomics* 2009, **39**:85–99.
37. Kao LC, Tulac S, Lobo S, Imani B, Yang JP, Germeyer A, Osteen K, Taylor RN, Lessey BA, Giudice LC: **Global gene profiling in human endometrium during the window of implantation.** *Endocrinology* 2002, **143**:2119–2138.
38. Giudice LC, Ferenczy A: **The Endometrial Cycle.** In *Reproductive Endocrinology, Surgery and Technology.* Edited by Adashi EY, Rock J, Rosenwaks Z. Philadelphia: Lippencott-Raven; 1996:271–300.
39. Spencer TE, Bazer FW: **Biology of progesterone action during pregnancy recognition and maintenance of pregnancy.** *Front Biosci* 2002, **7**:d1879–1898.
40. Hansen TR, Austin KJ, Perry DJ, Pru JK, Teixeira MG, Johnson GA: **Mechanism of action of interferon-tau in the uterus during early pregnancy.** *J Reprod Fertil* 1999, **54**:329–339.
41. Forde N, Carter F, Spencer TE, Bazer FW, Sandra O, Mansouri-Attia N, Okumu LA, McGettigan PA, Mehta JP, McBride R, O'Gaora P, Roche JF, Lonergan P: **Conceptus-induced changes in the endometrial transcriptome: how soon does the cow know she is pregnant?** *Biol Reprod* 2011, **85**:144–156.
42. Bauersachs S, Ulbrich SE, Reichenbach HD, Reichenbach M, Buttner M, Meyer HH, Spencer TE, Minten M, Sax G, Winter G, Wolf E: **Comparison of the effects of early pregnancy with human interferon, alpha 2 (IFNA2), on gene expression in bovine endometrium.** *Biol Reprod* 2012, **86**:46.
43. Forde N, Carter F, Fair T, Crowe MA, Evans AC, Spencer TE, Bazer FW, McBride R, Boland MP, O'Gaora P, Lonergan P, Roche JF: **Progesterone-regulated changes in endometrial gene expression contribute to advanced conceptus development in cattle.** *Biol Reprod* 2009, **81**:784–794.
44. Forde N, Spencer TE, Bazer FW, Song G, Roche JF, Lonergan P: **Effect of pregnancy and progesterone concentration on expression of genes encoding for transporters or secreted proteins in the bovine endometrium.** *Physiol Genomics* 2010, **41**:53–62.
45. Forde N, Mehta JP, Minten M, Crowe MA, Roche JF, Spencer TE, Lonergan P: **Effects of low progesterone on the endometrial transcriptome in cattle.** *Biol Reprod* 2012, **87**:124.
46. Simmons RM, Erikson DW, Kim J, Burghardt RC, Bazer FW, Johnson GA, Spencer TE: **Insulin-like growth factor binding protein one in the ruminant uterus: potential endometrial marker and regulator of conceptus elongation.** *Endocrinology* 2009, **150**:4295–4305.
47. Okumu LA, Forde N, Fahey AG, Fitzpatrick E, Roche JF, Crowe MA, Lonergan P: **The effect of elevated progesterone and pregnancy status on mRNA expression and localisation of progesterone and oestrogen receptors in the bovine uterus.** *Reproduction* 2010, **140**:143–153.
48. Beltman ME, Roche JF, Lonergan P, Forde N, Crowe MA: **Evaluation of models to induce low progesterone during the early luteal phase in cattle.** *Theriogenology* 2009, **72**:986–992.
49. Forde N, Mehta JP, McGettigan PA, Mamo S, Bazer FW, Spencer TE, Lonergan P: **Alterations in expression of endometrial genes coding for proteins secreted into the uterine lumen during conceptus elongation in cattle.** *BMC Genomics* 2013, **14**:321.
50. Forde N, McGettigan PA, Mehta JP, O'Hara L, Mamo S, Bazer FW, Spencer TE, Lonergan P: **Proteomic analysis of uterine fluid during the pre-implantation period of pregnancy in cattle.** *Reproduction* 2014, **147**:575–587.
51. Lewis SK, Farmer JL, Burghardt RC, Newton GR, Johnson GA, Adelson DL, Bazer FW, Spencer TE: **Galectin 15 (LGALS15): a gene uniquely expressed in the uteri of sheep and goats that functions in trophoblast attachment.** *Biol Reprod* 2007, **77**:1027–1036.
52. Dorniak P, Bazer FW, Wu G, Spencer TE: **Conceptus-derived prostaglandins regulate endometrial function in sheep.** *Biol Reprod* 2012, **87**(9):1–7.
53. Black SG, Arnaud F, Burghardt RC, Satterfield MC, Fleming JA, Long CR, Hanna C, Murphy L, Biek R, Palmarini M, Spencer TE: **Viral particles of endogenous betaretroviruses are released in the sheep uterus and infect the conceptus trophectoderm in a transspecies embryo transfer model.** *J Virol* 2010, **84**:9078–9085.
54. Bazer FW, Thatcher WW, Hansen PJ, Mirando MA, Ott TL, Plante C: **Physiological mechanisms of pregnancy recognition in ruminants.** *J Reprod Fertil Suppl* 1991, **43**:39–47.
55. Spencer TE, Bazer FW: **Conceptus signals for establishment and maintenance of pregnancy.** *Reprod Biol Endocrinol* 2004, **2**:49.
56. Spencer TE, Ott TL, Bazer FW: **tau-Interferon: pregnancy recognition signal in ruminants.** *Proc Soc Exp Biol Med* 1996, **213**:215–229.
57. Robinson RS, Fray MD, Wathes DC, Lamming GE, Mann GE: **In vivo expression of interferon tau mRNA by the embryonic trophoblast and uterine concentrations of interferon tau protein during early pregnancy in the cow.** *Mol Reprod Dev* 2006, **73**:470–474.
58. Roberts RM, Ezashi T, Rosenfeld CS, Ealy AD, Kubisch HM: **Evolution of the interferon tau genes and their promoters, and maternal-trophoblast interactions in control of their expression.** *Reprod Suppl* 2003, **61**:239–251.
59. Thatcher WW, Hansen PJ, Gross TS, Helmer SD, Plante C, Bazer FW: **Antiluteolytic effects of bovine trophoblast protein-1.** *J Reprod Fertil Suppl* 1989, **37**:91–99.
60. Dorniak P, Bazer FW, Spencer TE: **Prostaglandins regulate conceptus elongation and mediate effects of interferon tau on the ovine uterine endometrium.** *Biol Reprod* 2011, **84**:1119–1127.
61. Hansen TR, Henkes LK, Ashley RL, Bott RC, Antoniazzi AQ, Han H: **Endocrine actions of interferon-tau in ruminants.** *Soc Reprod Fertil Suppl* 2010, **67**:325–340.
62. Bazer FW, Spencer TE, Johnson GA: **Interferons and uterine receptivity.** *Seminars Reproductive Medicine* 2009, **27**:90–102.
63. Ott TL, Gifford CA: **Effects of early conceptus signals on circulating immune cells: lessons from domestic ruminants.** *Am J Reproductive Immunol* 2010, **64**:245–254.
64. Bauersachs S, Ulbrich SE, Gross K, Schmidt SE, Meyer HH, Wenigerkind H, Vermehren M, Sinowatz F, Blum H, Wolf E: **Embryo-induced transcriptome changes in bovine endometrium reveal species-specific and common molecular markers of uterine receptivity.** *Reproduction* 2006, **132**:319–331.
65. Cerri RL, Thompson IM, Kim IH, Ealy AD, Hansen PJ, Staples CR, Li JL, Santos JE, Thatcher WW: **Effects of lactation and pregnancy on gene expression of endometrium of Holstein cows at day 17 of the estrous cycle or pregnancy.** *J Dairy Sci* 2012, **95**:5657–5675.
66. Johnson GA, Spencer TE, Hansen TR, Austin KJ, Burghardt RC, Bazer FW: **Expression of the interferon tau inducible ubiquitin cross-reactive protein in the ovine uterus.** *Biol Reprod* 1999, **61**:312–318.
67. Johnson GA, Spencer TE, Burghardt RC, Joyce MM, Bazer FW: **Interferon-tau and progesterone regulate ubiquitin cross-reactive protein expression in the ovine uterus.** *Biol Reprod* 2000, **62**:622–627.
68. Johnson GA, Austin KJ, Collins AM, Murdoch WJ, Hansen TR: **Endometrial ISG17 mRNA and a related mRNA are induced by interferon-tau and localized to glandular epithelial and stromal cells from pregnant cows.** *Endocrine* 1999, **10**:243–252.
69. Austin KJ, Carr AL, Pru JK, Hearne CE, George EL, Belden EL, Hansen TR: **Localization of ISG15 and conjugated proteins in bovine endometrium using immunohistochemistry and electron microscopy.** *Endocrinology* 2004, **145**:967–975.
70. Choi Y, Johnson GA, Burghardt RC, Berghman LR, Joyce MM, Taylor KM, Stewart MD, Bazer FW, Spencer TE: **Interferon regulatory factor-two restricts expression of interferon- stimulated genes to the endometrial stroma and glandular epithelium of the ovine uterus.** *Biol Reprod* 2001, **65**:1038–1049.
71. Choi Y, Johnson GA, Spencer TE, Bazer FW: **Pregnancy and interferon tau regulate MHC class I and beta-2-microglobulin expression in the ovine uterus.** *Biol Reprod* 2003, **68**:1703–1710.
72. Johnson GA, Stewart MD, Gray CA, Choi Y, Burghardt RC, Yu-Lee LY, Bazer FW, Spencer TE: **Effects of the estrous cycle, pregnancy, and interferon tau on 2',5'- oligoadenylate synthetase expression in the ovine uterus.** *Biol Reprod* 2001, **64**:1392–1399.

73. Song G, Bazer FW, Spencer TE: **Pregnancy and interferon tau regulate RSAD2 and IFIH1 expression in the ovine uterus.** *Reproduction* 2007, **133**:285–295.
74. Rosenfeld CS, Han CS, Alexenko AP, Spencer TE, Roberts RM: **Expression of interferon receptor subunits, IFNAR1 and IFNAR2, in the ovine uterus.** *Biol Reprod* 2002, **67**:847–853.
75. Stark GR, Kerr IM, Williams BR, Silverman RH, Schreiber RD: **How cells respond to interferons.** *Annu Rev Biochem* 1998, **67**:227–264.
76. Taniguchi T, Ogasawara K, Takaoka A, Tanaka N: **IRF family of transcription factors as regulators of host defense.** *Annu Rev Immunol* 2001, **19**:623–655.
77. Spencer TE, Ott TL, Bazer FW: **Expression of interferon regulatory factors one and two in the ovine endometrium: effects of pregnancy and ovine interferon tau.** *Biol Reprod* 1998, **58**:1154–1162.
78. Bott RC, Ashley RL, Henkes LE, Antoniazzi AQ, Bruemmer JE, Niswender GD, Bazer FW, Spencer TE, Smirnova NP, Anthony RV, Hansen TR: **Uterine vein infusion of interferon tau (IFNT) extends luteal life span in ewes.** *Biol Reprod* 2010, **82**:725–735.
79. Romero JJ, Antoniazzi AQ, Smirnova NP, Webb BT, Yu F, Davis JS, Hansen TR: **Pregnancy-associated genes contribute to antiluteolytic mechanisms in ovine corpus luteum.** *Physiol Genomics* 2013, **45**:1095–1108.
80. Antoniazzi AQ, Webb BT, Romero JJ, Ashley RL, Smirnova NP, Henkes LE, Bott RC, Oliveira JF, Niswender GD, Bazer FW, Hansen TR: **Endocrine delivery of interferon tau protects the corpus luteum from prostaglandin F2 alpha-induced luteolysis in ewes.** *Biol Reprod* 2013, **88**:144.
81. Pestka S: **The interferons: 50 years after their discovery, there is much more to learn.** *J Biol Chem* 2007, **282**:20047–20051.
82. Nagaoka K, Sakai A, Nojima H, Suda Y, Yokomizo Y, Imakawa K, Sakai S, Christenson RK: **A chemokine, interferon (IFN)-gamma-inducible protein 10 kDa, is stimulated by IFN-tau and recruits immune cells in the ovine endometrium.** *Biol Reprod* 2003, **68**:1413–1421.
83. Nagaoka K, Nojima H, Watanabe F, Chang KT, Christenson RK, Sakai S, Imakawa K: **Regulation of blastocyst migration, apposition, and initial adhesion by a chemokine, interferon gamma-inducible protein 10 kDa (IP-10), during early gestation.** *J Biol Chem* 2003, **278**:29048–29056.
84. Ashley RL, Henkes LE, Bouma GJ, Pru JK, Hansen TR: **Deletion of the Isg15 gene results in up-regulation of decidual cell survival genes and down-regulation of adhesion genes: implication for regulation by IL-1beta.** *Endocrinology* 2010, **151**:4527–4536.
85. Toyokawa K, Leite F, Ott TL: **Cellular localization and function of the antiviral protein, ovine Mx1 (oMx1): II. The oMx1 protein is a regulator of secretion in an ovine glandular epithelial cell line.** *Am J Reproductive Immunol* 2007, **57**:23–33.
86. Song G, Spencer TE, Bazer FW: **Cathepsins in the ovine uterus: regulation by pregnancy, progesterone, and interferon tau.** *Endocrinology* 2005, **146**:4825–4833.
87. Kim S, Choi Y, Bazer FW, Spencer TE: **Identification of genes in the ovine endometrium regulated by interferon tau independent of signal transducer and activator of transcription 1.** *Endocrinology* 2003, **144**:5203–5214.
88. Gray CA, Abbey CA, Beremand PD, Choi Y, Farmer JL, Adelson DL, Thomas TL, Bazer FW, Spencer TE: **Identification of endometrial genes regulated by early pregnancy, progesterone, and interferon tau in the ovine uterus.** *Biol Reprod* 2006, **74**:383–394.
89. Song G, Spencer TE, Bazer FW: **Progesterone and Interferon Tau Regulate Cystatin C (CST3) in the Endometrium.** *Endocrinology* 2006, **147**:3478–3483.
90. Bazer FW, Spencer TE, Johnson GA, Burghardt RC, Wu G: **Comparative aspects of implantation.** *Reproduction* 2009, **138**:195–209.
91. Wang X, Frank JW, Little DR, Dunlap KA, Carey Satterfield M, Burghardt RC, Hansen TR, Wu G, Bazer FW: **Functional role of arginine during the peri-implantation period of pregnancy. I. Consequences of loss of function of arginine transporter SLC7A1 mRNA in ovine conceptus trophectoderm.** *FASEB J* 2014, **28**:2852–2863.
92. Dorniak P, Welsh TH Jr, Bazer FW, Spencer TE: **Cortisol and interferon tau regulation of endometrial function and conceptus development in female sheep.** *Endocrinology* 2013, **154**:931–941.
93. Forde N, Duffy GB, McGettigan PA, Browne JA, Mehta JP, Kelly AK, Mansouri-Attia N, Sandra O, Loftus BJ, Crowe MA, Fair T, Roche JF, Lonergan P, Evans AC: **Evidence for an early endometrial response to pregnancy in cattle: both dependent upon and independent of interferon tau.** *Physiol Genomics* 2012, **44**:799–810.
94. Platanias LC: **Mechanisms of type-I- and type-II-interferon-mediated signalling.** *Nat Rev Immunol* 2005, **5**:375–386.
95. Banu SK, Lee J, Stephen SD, Nithy TK, Arosh JA: **Interferon tau regulates PGF2alpha release from the ovine endometrial epithelial cells via activation of novel JAK/EGFR/ERK/EGR-1 pathways.** *Mol Endocrinol* 2010, **24**:2315–2330.
96. Lee J, Banu SK, Nithy TK, Stanley JA, Arosh JA: **Early pregnancy induced expression of prostaglandin E2 receptors EP2 and EP4 in the ovine endometrium and regulated by interferon tau through multiple cell signaling pathways.** *Mol Cell Endocrinol* 2012, **348**:211–223.
97. Dorniak P, Welsh TH Jr, Bazer FW, Spencer TE: **Endometrial HSD11B1 and cortisol regeneration in the ovine uterus: effects of pregnancy, interferon tau, and prostaglandins.** *Biol Reprod* 2012, **86**:124.
98. Simmons RM, Satterfield MC, Welsh TH Jr, Bazer FW, Spencer TE: **HSD11B1, HSD11B2, PTGS2, and NR3C1 expression in the peri-implantation ovine uterus: effects of pregnancy, progesterone, and interferon tau.** *Biol Reprod* 2010, **82**:35–43.
99. Lewis GS, Thatcher WW, Bazer FW, Curl JS: **Metabolism of arachidonic acid in vitro by bovine blastocysts and endometrium.** *Biol Reprod* 1982, **27**:431–439.
100. Lewis GS, Waterman RA: **Effects of endometrium on metabolism of arachidonic acid by bovine blastocysts in vitro.** *Prostaglandins* 1983, **25**:881–889.
101. Lewis GS, Waterman RA: **Metabolism of arachidonic acid in vitro by ovine conceptuses recovered during early pregnancy.** *Prostaglandins* 1985, **30**:263–283.
102. Charpigny G, Reinaud P, Tamby JP, Creminon C, Martal J, Maclouf J, Guillomot M: **Expression of cyclooxygenase-1 and −2 in ovine endometrium during the estrous cycle and early pregnancy.** *Endocrinology* 1997, **138**:2163–2171.
103. Charpigny G, Reinaud P, Tamby JP, Creminon C, Guillomot M: **Cyclooxygenase-2 unlike cyclooxygenase-1 is highly expressed in ovine embryos during the implantation period.** *Biol Reprod* 1997, **57**:1032–1040.
104. Ellinwood WE, Nett TM, Niswender GD: **Maintenance of the corpus luteum of early pregnancy in the ewe. II. Prostaglandin secretion by the endometrium in vitro and in vivo.** *Biol Reprod* 1979, **21**:845–856.
105. Marcus GJ: **Prostaglandin formation by the sheep embryo and endometrium as an indication of maternal recognition of pregnancy.** *Biol Reprod* 1981, **25**:56–64.
106. Ulbrich SE, Schulke K, Groebner AE, Reichenbach HD, Angioni C, Geisslinger G, Meyer HH: **Quantitative characterization of prostaglandins in the uterus of early pregnant cattle.** *Reproduction* 2009, **138**:371–382.
107. Kim S, Choi Y, Spencer TE, Bazer FW: **Effects of the estrous cycle, pregnancy and interferon tau on expression of cyclooxygenase two (COX-2) in ovine endometrium.** *Reprod Biol Endocrinol* 2003, **1**:58.
108. Arosh JA, Banu SK, Kimmins S, Chapdelaine P, Maclaren LA, Fortier MA: **Effect of interferon-tau on prostaglandin biosynthesis, transport, and signaling at the time of maternal recognition of pregnancy in cattle: evidence of polycrine actions of prostaglandin E2.** *Endocrinology* 2004, **145**:5280–5293.
109. Emond V, MacLaren LA, Kimmins S, Arosh JA, Fortier MA, Lambert RD: **Expression of cyclooxygenase-2 and granulocyte-macrophage colony-stimulating factor in the endometrial epithelium of the cow is up-regulated during early pregnancy and in response to intrauterine infusions of interferon-tau.** *Biol Reprod* 2004, **70**:54–64.
110. Krishnaswamy N, Chapdelaine P, Tremblay JP, Fortier MA: **Development and characterization of a simian virus 40 immortalized bovine endometrial stromal cell line.** *Endocrinology* 2009, **150**:485–491.
111. Fuse A, Mahmud I, Kuwata T: **Mechanism of stimulation by human interferon of prostaglandin synthesis in human cell lines.** *Cancer Res* 1982, **42**:3209–3214.
112. Fitzpatrick FA, Stringfellow DA: **Virus and interferon effects on cellular prostaglandin biosynthesis.** *J Immunol* 1980, **125**:431–437.
113. Lewis GS: **Prostaglandin secretion by the blastocyst.** *J Reprod Fertil Suppl* 1989, **37**:261–267.
114. El-Sayed A, Hoelker M, Rings F, Salilew D, Jennen D, Tholen E, Sirard M-A, Schellander K, Tesfaye D: **Large-scale transcriptional analysis of bovine embryo biopsies in relation to pregnancy success after transfer to recipients.** *Physiol Genomics* 2006, **28**:84–96.
115. Erdem H, Guzeloglu A: **Effect of meloxicam treatment during early pregnancy in Holstein heifers.** *Reprod Domest Anim* 2010, **45**:625–628.
116. Dey SK, Lim H, Das SK, Reese J, Paria BC, Daikoku T, Wang H: **Molecular cues to implantation.** *Endocr Rev* 2004, **25**:341–373.

117. Kennedy TG, Gillio-Meina C, Phang SH: **Prostaglandins and the initiation of blastocyst implantation and decidualization.** *Reproduction* 2007, **134**:635–643.
118. Wang H, Dey SK: **Roadmap to embryo implantation: clues from mouse models.** *Nat Rev Genet* 2006, **7**:185–199.
119. Desvergne B, Wahli W: **Peroxisome proliferator-activated receptors: nuclear control of metabolism.** *Endocr Rev* 1999, **20**:649–688.
120. Forman BM, Tontonoz P, Chen J, Brun RP, Spiegelman BM, Evans RM: **15-Deoxy-delta 12, 14-prostaglandin J2 is a ligand for the adipocyte determination factor PPAR gamma.** *Cell* 1995, **83**:803–812.
121. Kliewer SA, Lenhard JM, Willson TM, Patel I, Morris DC, Lehmann JM: **A prostaglandin J2 metabolite binds peroxisome proliferator-activated receptor gamma and promotes adipocyte differentiation.** *Cell* 1995, **83**:813–819.
122. Lim H, Gupta RA, Ma WG, Paria BC, Moller DE, Morrow JD, DuBois RN, Trzaskos JM, Dey SK: **Cyclo-oxygenase-2-derived prostacyclin mediates embryo implantation in the mouse via PPARdelta.** *Genes Dev* 1999, **13**:1561–1574.
123. Lim H, Dey SK: **PPAR delta functions as a prostacyclin receptor in blastocyst implantation.** *Trends Endocrinol Metab* 2000, **11**:137–142.
124. Cammas L, Reinaud P, Bordas N, Dubois O, Germain G, Charpigny G: **Developmental regulation of prostacyclin synthase and prostacyclin receptors in the ovine uterus and conceptus during the peri-implantation period.** *Reproduction* 2006, **131**:917–927.
125. Desmarais JA, Lopes FL, Zhang H, Das SK, Murphy BD: **The peroxisome proliferator-activated receptor gamma regulates trophoblast cell differentiation in mink (Mustela vison).** *Biol Reprod* 2007, **77**:829–839.
126. Barak Y, Liao D, He W, Ong ES, Nelson MC, Olefsky JM, Boland R, Evans RM: **Effects of peroxisome proliferator-activated receptor delta on placentation, adiposity, and colorectal cancer.** *Proc Natl Acad Sci U S A* 2002, **99**:303–308.
127. Nadra K, Anghel SI, Joye E, Tan NS, Basu-Modak S, Trono D, Wahli W, Desvergne B: **Differentiation of trophoblast giant cells and their metabolic functions are dependent on peroxisome proliferator-activated receptor beta/delta.** *Mol Cell Biol* 2006, **26**:3266–3281.
128. Tan NS, Icre G, Montagner A, Bordier-ten-Heggeler B, Wahli W, Michalik L: **The nuclear hormone receptor peroxisome proliferator-activated receptor beta/delta potentiates cell chemotactism, polarization, and migration.** *Mol Cell Biol* 2007, **27**:7161–7175.
129. Michael AE, Thurston LM, Rae MT: **Glucocorticoid metabolism and reproduction: a tale of two enzymes.** *Reproduction* 2003, **126**:425–441.
130. Wang JC, Derynck MK, Nonaka DF, Khodabakhsh DB, Haqq C, Yamamoto KR: **Chromatin immunoprecipitation (ChIP) scanning identifies primary glucocorticoid receptor target genes.** *Proc Natl Acad Sci U S A* 2004, **101**:15603–15608.
131. Yamamoto KR: **Steroid receptor regulated transcription of specific genes and gene networks.** *Annu Rev Genet* 1985, **19**:209–252.
132. Lee HY, Acosta TJ, Skarzynski DJ, Okuda K: **Prostaglandin F2alpha stimulates 11Beta-hydroxysteroid dehydrogenase 1 enzyme bioactivity and protein expression in bovine endometrial stromal cells.** *Biol Reprod* 2009, **80**:657–664.
133. Alfaidy N, Xiong ZG, Myatt L, Lye SJ, MacDonald JF, Challis JR: **Prostaglandin F2alpha potentiates cortisol production by stimulating 11beta-hydroxysteroid dehydrogenase 1: a novel feedback loop that may contribute to human labor.** *J Clin Endocrinol Metab* 2001, **86**:5585–5592.
134. Alfaidy N, Li W, MacIntosh T, Yang K, Challis J: **Late gestation increase in 11beta-hydroxysteroid dehydrogenase 1 expression in human fetal membranes: a novel intrauterine source of cortisol.** *J Clin Endocrinol Metab* 2003, **88**:5033–5038.
135. Challis JRG, Matthews SG, Gibb W, Lye SJ: **Endocrine and paracrine regulation of birth at term and preterm.** *Endocr Rev* 2000, **21**:514–550.
136. Goppelt-Struebe M: **Molecular mechanisms involved in the regulation of prostaglandin biosynthesis by glucocorticoids.** *Biochem Pharmacol* 1997, **53**:1389–1395.
137. Yu CY, Mayba O, Lee JV, Tran J, Harris C, Speed TP, Wang JC: **Genome-wide analysis of glucocorticoid receptor binding regions in adipocytes reveal gene network involved in triglyceride homeostasis.** *PLoS One* 2010, **5**:e15188.
138. Michael AE, Papageorghiou AT: **Potential significance of physiological and pharmacological glucocorticoids in early pregnancy.** *Hum Reprod Update* 2008, **14**:497–517.
139. Braun T, Li S, Moss TJ, Newnham JP, Challis JR, Gluckman PD, Sloboda DM: **Maternal betamethasone administration reduces binucleate cell number and placental lactogen in sheep.** *J Endocrinol* 2007, **194**:337–347.
140. Boomsma CM, Keay SD, Macklon NS: **Peri-implantation glucocorticoid administration for assisted reproductive technology cycles.** *Cochrane Database Syst Rev* 2007, **1**:CD005996.
141. Quenby S, Kalumbi C, Bates M, Farquharson R, Vince G: **Prednisolone reduces preconceptual endometrial natural killer cells in women with recurrent miscarriage.** *Fertil Steril* 2005, **84**:980–984.

15

Developmental programming: the role of growth hormone

Anita M Oberbauer

Abstract

Developmental programming of the fetus has consequences for physiologic responses in the offspring as an adult and, more recently, is implicated in the expression of altered phenotypes of future generations. Some phenotypes, such as fertility, bone strength, and adiposity are highly relevant to food animal production and *in utero* factors that impinge on those traits are vital to understand. A key systemic regulatory hormone is growth hormone (GH), which has a developmental role in virtually all tissues and organs. This review catalogs the impact of GH on tissue programming and how perturbations early in development influence GH function.

Keywords: Epigenetic, Growth hormone, Programming, Transgeneration

Introduction

Broadly considered, development has two distinct yet interrelated facets: alterations in gene expression associated with the normal developmental profile of an organism and alterations associated with developmental plasticity permitting adaptation to environment perturbations. The fundamental role of normal fetal development is best exemplified by pattern formation in the embryo. "Programming" was a term coined in 1986 to reflect that events occurring *in utero* alter adult phenotypic and metabolic traits [1]. The influence of fetal and neonatal nutritional environments on metabolic programming was first conceptualized by Lucas [2] and the concept of developmental programming was later expanded to include other environmental perturbations [3,4]. The impact of environmental challenges experienced in the neonatal period emphasizes the need to understand both inherent development as well as the mechanistic processes by which developmental programming is achieved.

Recent data highlight that the metabolic environment experienced by the fetus influences phenotypic expression and disease susceptibility in later life [5,6]. Small for gestational age (SGA) offspring, often arising as a consequence of malnourished mothers, is correlated with adult hypertension, glucose intolerance, insulin resistance, type 2 diabetes, dyslipidemia, and diminished measures of bone strength. Metabolic bone disease is also associated with prenatal nutrient deprivation [7]. Adult-onset disorders associated with SGA neonates represent programming occurring at the gene level by methylation, gene silencing, and other epigenetic modifications established during fetal development (reviewed in [8]). It is worth noting that the definition of epigenetics used in this review is "the structural adaptation of chromosomal regions so as to register, signal or perpetuate altered activity states" first proposed by Bird in 2007 to encompass the many aspects of epigenetic alterations [9].

Although not developmental programming, the most familiar illustration of developmental epigenetic imprinting is the insulin-like growth factor-2 (IGF-2) pathway. The methylation status differs between the paternally and maternally inherited *Igf-2* gene with the paternal copy as the sole source of IGF-2 during development and the maternally inherited copy silenced. It is hypothesized that the maternal and paternal expression are balanced to maximally promote fetal growth while preventing excessive depletion of the dam's resources [10]. Another example of the role of epigenetic imprinting occurs in Prader-Willi Syndrome, a syndrome characterized by growth hormone (GH) insufficiency [11]. The genetic cause is the deletions of paternal chromosome 15; the maternal region is silenced by epigenetic factors. While IGF-2's role is predominantly in prenatal development, the key postnatal hormone is GH. Growth hormone effects are either direct or mediated through the

Correspondence: amoberbauer@ucdavis.edu
Department of Animal Science, University of California, One Shields Ave, Davis, CA 95616, USA

induction of IGF-1 [12] to regulate overall body growth through its effects on adipose, bone, and muscle, is growth hormone (GH) [13].

Importantly, developmental programming may have multigenerational consequences transmitted through epigenetic modifications of the genome. Transgenerational epigenetic programming was initially recognized as a result of malnourishment. The Dutch famine, sometimes referred to as the "Hunger winter", is most often cited as the initial epidemiological human case study illustrating the impacts of nutritional stress upon both physical and behavioral traits in subsequent generations (reviewed in [14]). More recent evidence of transgenerational consequences has been associated with environmental perturbations including toxicants and hormonal manipulations [15-17]. Although many effects are termed "transgenerational," Heard and Martienssen [18] recommend a more parsimonious use of the term and define much of what has been observed as "intergenerational" which will be employed in this review. Nevertheless, understanding the link between fetal and neonatal programming and subsequent phenotypic expression is important for human health and has value for sustainable food animal production systems. Nutritional, physical, and psychological stress experienced by livestock influences immediate productivity and health [19-21] and yet notably can also affect future generations' production through developmental programming.

Review

Role of GH in normal development

The accepted role of GH and the GH-IGF axis in tissue development is predominantly postnatal with other hormones assuming importance *in utero* [5]. Yet GH is known to influence fetal development. For example, transgenic mice overexpressing GH, produce offspring with a 12% reduction in birth weight [22] and calves born to dams given exogenous bovine somatotropin (GH) have reduced birth weights [23,24]. These findings indicate that GH, or its downstream regulators, plays a significant role in normal fetal development. The intimate link between GH and generalized growth has led to the speculation that GH may also affect the expression of IGF-2 though the evidence does not support this supposition to any great extent [25].

Evidence that fetal perturbations program GH signaling is also emerging. A common model used to define the role of GH in developmental programming is the neonate that experienced impaired fetal growth, specifically SGA infants, who are characterized by compromised bone growth and reduced body mass. Infants having low birth weights trend to lower circulating GH as young adults [26] indicating persistent effects by early developmental experiences on GH secretion. Despite prenatal growth impairment, approximately 90% of SGA births have accelerated growth in the first two postnatal years achieving heights in the normal range [27,28]; however the remaining 10% do not exhibit catch-up growth giving evidence of fetal programming of the GH-IGF endocrine axis as a consequence of early growth dysfunction [29]. Also associated with SGA is reduced bone mass and strength in adulthood [30] indicative of developmental programming of bone remodeling. GH, with its role in bone mineral density [31], may mediate the adult bone characteristics arising from *in utero* growth restriction. Furthermore, adult height is correlated with placental GH expression. Genetic variation within the sequence of the placental-form of GH is significantly associated with mature height confirming that altered expression of GH *in utero* governs longitudinal bone growth potential [32]. Immune function is depressed chronically in SGA rat pups, although the condition can be reversed by provision of exogenous GH during the pre-weaning growth stage [33] indicating GH is important in the ontogeny of immune competence.

Birth weight and neonatal growth is predictive of adult circulating GH levels suggesting that the intrauterine environment programs GH secretion [34] or tissue responsiveness to GH. Waldman and Chia postulate that idiopathic short stature reflects epigenetic changes to the GH receptor or to IGF-1 genes [35]. Research exploring the chromatin landscape suggests a tissue specificity in the accessiblity of the IGF-1 promoter indicating epigenetic regulation of the GH-IGF-1 axis [36,37]. This view is supported by developmental profiling of IGF-1 mRNA expression in cattle; as female calves mature, IGF-1 gene expression in the pituitary is reduced, particularly the exon most responsive to GH, whereas IGF-1 expression in the uterus increased [38].

Elevated GH models

In utero exposure

The role of GH in normal development was established decades ago through ablation studies and assessing the physiological consequences of the absence of GH. Additional models to characterize GH action elevate the hormone either through exogenous administration of GH, GH-transgene expression, or chemical compounds that induce GH. Elevated GH *in utero* reduces birth weights of calves [23], lambs [39], and rodents [22]. In sheep, fetal adipogenesis is enhanced when dams are exposed to GH indicating the direct effect of GH on adipose metabolism [39]. For rodent models, neonates are both lighter in weight and have reduced skeletal size at birth. In contrast, pregnant sows given hydroxyl beta methylbutyrate (HMB), a compound that enhances fetal GH and IGF-1, give birth to piglets with increased birth weights [40]. Furthermore, the piglet data indicate that fetal exposure to elevated GH has persistent effects on growth parameters into adulthood; they had increased growth rate to slaughter, lean carcass content and indices of bone strength [40]. Similar to the rodents, piglets from HMB treated sows had shorter femurs although in a different study, provision of GH directly to pregnant sows did not affect bone length in the neonates [41]. Ewes treated with extended -release GH during gestation delivered lambs having heavier birth weights and a sustained postnatal growth advantage over lambs born of untreated ewes as well as a blunted response to stimulation of the GH axis [42,43]. Taken together, the evidence suggests that exposure to elevated GH *in utero* exerts long term consequences on the offspring possibly through alteration of maternal metabolic pathways and placental function.

Postnatal exposure

Elevated GH postnatally has significant effects on bone, muscle, and adipose. Human SGA infants given GH respond with increased bone growth velocity for the duration of GH treatment [8]. In rodents, provision of GH during the postnatal period can reverse the *in utero* growth inhibition and restore overall bone length [22]. GH exerts stimulatory effects on linear growth rates in rodents with transient elevation of GH increasing bone growth rate. At the cellular level, GH accelerates bone growth by hyperplasia with little impact on hypertrophy [44]. Provision of elevated GH in a GH-transgenic mouse model also increases the duration of bone elongation though the degree of responsiveness is sensitive to the chronological age of exposure with early neonatal tissue more responsive than slightly older ages.

Lambs given HMB during the first 3 postnatal weeks show elevated circulating levels of GH, IGF-1, and biochemical markers of bone turnover. However once the HMB treatment concludes, these indices all fall to control values indicating resistance to long term bone programming by HMB when provided postnatally [45]. The failure of persistent HMB treatment effects in lambs, in contrast to the direct GH effects seen in rodents and pigs, may reflect the precocial development of sheep when compared to that of rodents and pigs suggesting that the period responsive to GH programming may be *in utero* for lambs.

The enhanced gain accompanying elevated GH has a greater proportion of protein than normal tissue accrual in rodents. Rodent development is characterized by each unit of gain being composed of the same proportions of water, lipid, protein and ash across the entire growth phase, despite the speed of accrual [46]. In contrast, under conditions of elevated GH, protein comprises a larger proportion of gain [47]. This preferential repartitioning by GH of nutrients to muscle is also seen in pigs transgenic for GH [48], broilers supplemented with HMB [49], fish given exogenous GH [50], and GH-transgenic fish [51]. Exogenous GH increases protein anabolism by enhancing protein synthesis [52,53].

Therapeutic GH treatment of SGA children offers growth advantages as noted above, yet the elevated GH also reduces insulin sensitivity potentially increasing the risk of diabetes. Recent studies of children born SGA and treated with GH would suggest this is not a valid concern as they have normal insulin post-GH treatment [8]. However, these children have not aged to a point wherein the long term consequences of GH treatment can be evaluated; thus, adult onset diabetes remains a concern. In a rodent model, elevating GH in adult mice increases circulating insulin with levels remaining elevated even after reduction of GH consistent with an insulin resistant state, though persistence of this condition was not determined [54].

The role of GH in adipogenesis is complex. *In utero* or postnatal exposure to elevated GH in a GH-transgenic mouse model increases adipose storage by increasing adipocyte hyperplasia, differentiation, and cellular lipid content [39,54,55]. Artificially elevating GH induces lipoprotein lipase which in turn stimulates adipogenesis; the elevated GH also increases circulating insulin which then can induce lipoprotein lipase further promoting adipose storage [56]. Postnatal exposure to GH acts on membrane lipids to program response to cellular perturbations. Elevated GH in rodents creates a more unsaturated lipid profile in cell membranes by activating membrane desaturase pathways [57]. This net flux through the lipid metabolism pathways generates eicosanoids important in the inflammation process [58].

The interplay with leptin adds more complexity to the influence of GH on adipose. Leptin, synthesized by adipose cells, regulates energy metabolism (reviewed in [59]) and influences fetal brain and bone development [60]. Leptin also defines energy storage adequacy during neonatal life and can program the neuronal regulation of adult food intake and satiety with leptin programming in early development altering adiposity at later ages [61,62]. Newborn rodents and piglets born SGA experience adipogenesis however neonatal provision of exogenous leptin can reverse the adipocyte proliferation induced by intrauterine growth restriction [61,63]. Leptin levels correlate with birth weight with SGA infants having low leptin levels while large for gestational age babies have high leptin concentrations [64]. Although the correlation is largely independent of the GH-IGF-1 axis (reviewed by [62]), GH levels influence maternal and fetal leptin levels. In sheep, fetal adipogenesis is enhanced when dams are exposed to GH and leptin levels are reduced in the dam and the fetus, demonstrating a direct effect of GH on adipose *in utero* [39] with potential of long term programming impacts.

Early dysregulation of GH promotes adipogenesis which in turn elevates leptin that influences adipose function at later ages. Elevated GH in a GH-transgenic mouse model increases plasma leptin [65] while mice transiently exposed to elevated GH during early postnatal development become obese having increased circulating leptin once the excess GH is withdrawn [54]. In these GH-transgenic mice, although each adipocyte expresses less leptin, the leptin in circulation is greater due to the increased overall adipocity. This suggests that prolonged elevation of GH disrupts normal physiological responses to elevated plasma leptin, creating leptin resistance and obesity; in the case of elevated GH, the most likely site of leptin signaling impairment is at the hypothalamic axis. Supporting this view of GH generating leptin resistance is that the elevated GH in the GH- transgenic mouse increases plasma leptin as well as the leptin receptor (NPY) gene and protein expression [65]. It is proposed that leptin resistance can be programmed during fetal and neonatal life with long term impacts on body energy stores potentiating adult onset metabolic disorders [63].

Models of *in utero* programming

Studies in multiple mammalian species demonstrate that aberrant programming of adult tissue response is associated with stress-induced glucocorticoid secretion during fetal and perinatal development. Maternal stress in humans is known to impact neurological circuitry of the offspring: infants have an increased risk of neuropsychiatric disease when the mother experienced psychological stress during the first trimester (reviewed in [66]). Similarly, prospective stress research with rodents and non-human primates have identified gestational periods of increased susceptibility to developmental disruption [66]. Importantly the maternal stress response is characterized by altered methylation patterns in the fetal DNA [67] thereby programming future gene expression patterns that may also be transmitted to the next generation. Similar changes in methylation patterns have been detected in response to maternal exposure to toxins and hormones [68] resulting in intergenerational epigenetic changes having significant impact on future physiological performance and significant implications on selection for production species.

Early research demonstrated that adult GH secretory patterns are influenced by perturbations during the perinatal period. For example, male rats normally express a high amplitude secretory pattern of GH whereas females have low amplitude pulses within a higher basal background. Transient manipulation of sex steroids in the neonatal period can modify the GH secretory pattern to that of the opposite sex [69]. Glucocorticoids are viewed as critical modulators of GH in a wide range of species including chickens, rodents, and humans with impacts on the hypothalamus and pituitary regulation of GH release, somatotrope development, and peripheral tissue responsivity [70-74]. In turn, GH mediates gene regulation by increasing accessibility of the transcriptional machinery by chromatin remodeling of promoters through histone acetylation [36]. Taken together, stress has profound impacts on the development of the neonate that may transcend a single generation and GH plays a role in programming the consequences of the stress.

Nutritional stress

Mothers who are undernourished during pregnancy give birth to SGA infants and when adult those children exhibit adult-onset obesity, insulin resistance, hypertension, and metabolic dysfunction. Maternal nutritional deprivation stress also is implicated in intergenerational epigenetic programming to alter future generations' growth and metabolic phenotypes. This phenomenon is suggested to account for the rising cases of human obesity, diabetes, and coronary disease (reviewed in [66]).

Levels of placental GH found in maternal circulation are positively correlated with fetal birth weight and in times of maternal nutrient deprivation and SGA pregnancies, placental GH secretion is reduced [32]. A recent study of ewes undernourished during gestation reported that although maternal circulating GH and placental weight were not reduced [75], birth weights and crown rump lengths were [76]. Newbern and Freemark [77] review the evidence that supports a direct role of GH and placental GH in the programming of fetal growth and long-term metabolic function. Some evidence is based upon transgenic mice that overexpress placental GH and their high birth weight pups with enhanced growth capacity relative to their non-transgenic littermates [78].

As noted above, there are generalized consequences of SGA on bone characteristics but SGA due to nutritional deprivation also programs bone through the GH- IGF axis. The necessity of adequate fetal and neonatal nutrition for normal bone growth and adult bone integrity was underscored by Eastell and Lambert in their review [79]. In rats, post-weaning dietary protein restriction alters the GH secretory profile, blunting the peak GH pulses [80] which alters growth and metabolic responses. It is known that neonatal nutrition programs adult bone mass [81] and bone density is positively associated with peak GH and IGF-1 concentrations but elevated median GH concentrations, as seen with dietary restriction, is negatively associated with bone density [34]. Presumably the effects on bone are due to the role that GH plays in opening the chromatin surrounding the IGF-1 gene. Moreover, maternal malnutrition alters methylation of the GH responsive promoter 2 of the IGF-1 gene (reviewed in [12]) and intrauterine growth restriction is associated with sustained hypermethlyation of the IGF-1 promoter in rats [82] thereby amplifying the signaling impairment of GH in the neonate.

Another mechanism of nutritional programming on bone is by its effects on the formation of the embryonic skeletal anlage. Fetal and postnatal bone growth depends upon chondrocyte hypertrophy and hyperplasia. Fibroblast growth factors (FGF) and their cognate fibroblast growth factor receptors (FGFR) are key signaling molecules of chondrocyte function in developing bone. Activation of different FGFR family members both promote and inhibit chondrocyte proliferation [83]. The FGF21 ligand directly inhibits chondrocyte proliferation and antagonizes the proliferative effects of GH on growth plate chondrocytes [84] thereby impairing chondrogenesis and bone growth. FGF21 is increased during food restriction and attenuates the action of GH at the chondrocytic level and at the systemic level by impairing IGF-1 expression [84]. Antagonizing GH action by elevated FGF as result of food restriction has implications for many of the body systems because of GH's extensive broad range of target tissues.

In rats, restricting maternal dietary protein to induce intrauterine growth restriction enlarges abdominal adipose depots when animals are adult [85]. The enlarged adipose depots upregulate expression of genes associated with enhanced lipid metabolism and adipose accrual. Provision of exogenous GH postnatally can minimize those adult conditions [33,86]. Specifically, adult-onset obesity seen in pups born to undernourished rat dams can be prevented with preweaning GH treatment [87] (reviewed in [88]). This provides compelling evidence of early GH programming the adult physiological response.

Intrauterine growth restricted babies, piglets, and rodent pups have reduced leptin at birth and then propensity for obesity at adulthood [89]; provision of leptin reverses this tendency [61]. At birth infants considered SGA have greater methylation of the leptin promoter correlating with the reduced circulating leptin levels [90]. In contrast, using a mouse model, early nutritional programming of the fetus by low protein diets is correlated with hypomethylation of the leptin gene that is maintained throughout life [91]. In the latter study, the authors acknowledge their results differ from the pattern of leptin expression, and methylation, observed in other species in response to maternal malnutrition. Adipocyte hypertrophy results in hypomethylation of the leptin gene and Increases leptin expression which increases adipose hyperplasia [89]. Clearly the specifics of the nutritional deprivation, timing, and species evaluated have significant impact on the manifestation of maternal malnutrition on adult susceptibility to metabolic disruption.

Physical stress

Physical stressors, often modeled with induced hypoxia, have been used to assess the epigenetic consequences of stress. Hypoxia-inducible factor 1 (HIF-1), induced under conditions of hypoxia to adapt to reduced oxygen supply, is also required for normal fetal tissue and skeletal development [92]. The transcription factor HIF-I coordinates with additional factors to induce chromatin remodeling of target genes [93]. In addition to the role of HIF-1 in hypoxia adaptation, in rodents the GH axis modulates the hypoxic stress response [94]. In periods of intermittent hypoxia, GH is neuroprotective [95]. However, hypoxia elevates glucocorticoid expression from the fetal adrenal gland [96] and elevated glucocorticoids can lead to the programming of the fetal GH axis as detailed earlier in this article [73,74]. Furthermore, hypoxia inhibits GH release by enhancing the expression of somatostatin, the hypothalamic GH inhibitor [97]. Varvarigou et al [98] propose that severe hypoxia permanently damages the fetal hypothalamic-hypophyseal axis leading to greatly diminished GH secretory profiles. Despite the knowledge that fetal programming is a consequence of hypoxia, defining the long term and intergenerational effects of physical stressors are yet to be fully explored.

Conclusions

Growth hormone has a pivotal role in pre and postnatal development although its effects on intergenerational programming are as yet under studied. Given the plasticity of postnatal development, defining fetal and neonatal programming and the effect upon later phenotypes is important to maximize the expression of desirable traits in food animals. This is particularly important on the impact of more permanent epigenetic alterations of gene expression that may have future consequences. Historically, classical genetic selection has permitted extremely favorable phenotypic advances. The implementation of quantitative trait loci selection schemes for livestock based upon genomic signatures will accelerate genetic improvement. Overlaid upon genetic selection one must be mindful of the epigenetic programming that environmental perturbations can exert on future trait expression and how that may factor into selection schemes for agricultural livestock production in a changing environment. Better defined knowledge of the mechanisms of programming will facilitate incorporation of epigenetic factors into selection.

Abbreviations

GH: Growth hormone; Hlf-1: Hypoxia-inducible factor 1; HMB: Hydroxyl beta methylbutyrate; IGF-1: Insulin-like growth factor 1; FGF: Fibroblast growth factors; FGFR: Fibroblast growth factor receptors; SGA: Small for gestational age.

Competing interests

The author declares that she has no competing interests.

Authors' contributions

AMO reviewed the literature and drafted and edited the manuscript.

References

1. Barker DJ, Osmond C. Infant mortality, childhood nutrition, and ischaemic heart disease in England and Wales. Lancet. 1986;327:1077–81.
2. Lucas A. Programming by early nutrition in man. Childhood Environ Adult Dis. 1991;1991:38–55.
3. Myatt L. Placental adaptive responses and fetal programming. J Physiol. 2006;572:25–30.
4. Langley-Evans SC. Developmental programming of health and disease. Proc Nutr Soc. 2006;65:97–105.
5. Fowden AL, Forhead AJ. Endocrine regulation of feto-placental growth. Horm Res Paediatr. 2009;72:257–65.
6. Gluckman PD, Hanson MA, Cooper C, Thornburg KL. Effect of in utero and early-life conditions on adult health and disease. N Engl J Med. 2008;359:61–73.
7. Cooper C, Javaid M, Taylor P, Walker-Bone K, Dennison E, Arden N. The fetal origins of osteoporotic fracture. Calcif Tissue Int. 2002;70:391–4.
8. Jung H, Rosilio M, Blum WF, Drop SL. Growth hormone treatment for short stature in children born small for gestational age. Adv Ther. 2008;25:951–78.
9. Bird A. Perceptions of epigenetics. Nature. 2007;447:396–8.
10. Bergman D, Halje M, Nordin M, Engström W. Insulin-like growth factor 2 in development and disease: a mini-review. Gerontology. 2012;59:240–9.
11. Cassidy SB, Schwartz S, Miller JL, Driscoll DJ. Prader-willi syndrome. Genet Med. 2011;14:10–26.
12. Oberbauer A. The regulation of IGF-1 gene transcription and splicing during development and aging. Front Endocrinol. 2013;4:1–9. doi:10.3389/fendo.2013.00039.
13. Bartke A, Sun LY, Longo V. Somatotropic Signaling: Trade-Offs Between Growth, Reproductive Development, and Longevity. Physiol Rev. 2013;93:571–98.
14. Schulz LC. The Dutch Hunger Winter and the developmental origins of health and disease. Proc Natl Acad Sci. 2010;107:16757–8.
15. Goerlich VC, Nätt D, Elfwing M, Macdonald B, Jensen P. Transgenerational effects of early experience on behavioral, hormonal and gene expression responses to acute stress in the precocial chicken. Horm Behav. 2012;61:711–8.
16. Guerrero-Bosagna C, Skinner MK. Environmentally induced epigenetic transgenerational inheritance of male infertility. Curr Opin Genet Dev. 2014;26:79–88.
17. Baker TR, Peterson RE, Heideman W. Using Zebrafish as a Model System for Studying the Transgenerational Effects of Dioxin. Toxic Sci. 2014:kfu006.
18. Heard E, Martienssen RA. Transgenerational Epigenetic Inheritance: Myths and Mechanisms. Cell. 2014;157:95–109.
19. Zulkifli I. Review of human-animal interactions and their impact on animal productivity and welfare. J Anim Sci Biotechnol. 2013;4:25.
20. Reynolds L, Borowicz P, Caton J, Vonnahme K, Luther J, Hammer C, et al. Developmental programming: the concept, large animal models, and the key role of uteroplacental vascular development. J Anim Sci. 2010;88:E61–72.
21. Reynolds LP, Caton JS. Role of the pre-and post-natal environment in developmental programming of health and productivity. Mol Cell Endocrinol. 2012;354:54–9.
22. Oberbauer A, Cruickshank J, Thomas A, Stumbaugh A, Evans K, Murray J, et al. Effects of pre and antenatal elevated and chronic oMt1a-oGH transgene expression on adipose deposition and linear bone growth in mice. Growth Dev Aging. 2001;65:3–13.
23. Oldenbroek J, Garssen G, Jonker L, Wilkinson J. Effects of treatment of dairy cows with recombinant bovine somatotropin over three or four lactations. J Dairy Sci. 1993;76:453–67.
24. Gallo GF, Block E. Effects of recombinant bovine somatotropin on nutritional status of dairy cows during pregnancy and of their calves. J Dairy Sci. 1990;73:3266–75.
25. Holly JM. The IGF-II enigma. Growth Horm IGF Res. 1998;8:183–4.
26. Jensen RB, Vielwerth S, Frystyk J, Veldhuis J, Larsen T, Mølgaard C, et al. Fetal growth velocity, size in early life and adolescence, and prediction of bone mass: association to the GH–IGF axis. J Bone Miner Res. 2008;23:439–46.
27. Karlberg J, Albertsson-Wikland K. Growth in full-term small-for-gestational-age infants: from birth to final height. Pediatr Res. 1995;38:733–9.
28. Hokken-Koelega A, De Ridder M, Lemmen R, Den Hartog H, Keizer-Schrama SDM, Drop S. Children Born Small for Gestational Age: Do They Catch Up? Pediatr Res. 1995;38:267–71.
29. Jensen RB, Chellakooty M, Vielwerth S, Vaag A, Larsen T, Greisen G, et al. Intrauterine growth retardation and consequences for endocrine and cardiovascular diseases in adult life: does insulin-like growth factor-I play a role? Horm Res Paediatr. 2003;60:136–48.
30. Sayer AA, Cooper C. Fetal programming of body composition and musculoskeletal development. Early Hum Dev. 2005;81:735–44.
31. Doga M, Bonadonna S, Gola M, Mazziotti G, Nuzzo M, Giustina A. GH deficiency in the adult and bone. J Endocrinol Invest. 2004;28:18–23.
32. Timasheva Y, Putku M, Kivi R, Kožich V, Männik J, Laan M. Developmental programming of growth: Genetic variant in *GH2* gene encoding placental growth hormone contributes to adult height determination. Placenta. 2013;34:995–1001.
33. Reynolds CM, Li M, Gray C, Vickers MH. Pre-weaning growth hormone treatment ameliorates bone marrow macrophage inflammation in adult male rat offspring following maternal undernutrition. PLoS ONE. 2013;8:e68262.
34. Fall C, Hindmarsh P, Dennison E, Kellingray S, Barker D, Cooper C. Programming of Growth Hormone Secretion and Bone Mineral Density in Elderly Men: A Hypothesis 1. J Clin Endocrinol Metab. 1998;83:135–9.
35. Waldman LA, Chia DJ. Towards identification of molecular mechanisms of short stature. Int J Pediatr Endocrinol. 2013;2013:19.
36. Chia DJ, Rotwein P. Defining the epigenetic actions of growth hormone: acute chromatin changes accompany GH-activated gene transcription. Mol Endocrinol. 2010;24:2038–49.
37. Chia DJ, Young JJ, Mertens AR, Rotwein P. Distinct alterations in chromatin organization of the two IGF-I promoters precede growth hormone-induced activation of IGF-I gene transcription. Mol Endocrinol. 2010;24:779–89.
38. Oberbauer A, Belanger J, Rincon G, Cánovas A, Islas-Trejo A, Gularte-Mérida R, et al. Bovine and murine tissue expression of insulin like growth factor-I. Gene. 2014;535:101–5.
39. Wallace JM, Matsuzaki M, Milne J, Aitken R. Late but not early gestational maternal growth hormone treatment increases fetal adiposity in overnourished adolescent sheep. Biol Reprod. 2006;75:231–9.
40. Tatara MR, Śliwa E, Krupski W. Prenatal programming of skeletal development in the offspring: effects of maternal treatment with β-hydroxy-β-methylbutyrate (HMB) on femur properties in pigs at slaughter age. Bone. 2007;40:1615–22.
41. Rehfeldt C, Kuhn G, Nürnberg G, Kanitz E, Schneider F, Beyer M, et al. Effects of exogenous somatotropin during early gestation on maternal performance, fetal growth, and compositional traits in pigs. J Anim Sci. 2001;79:1789–99.
42. Koch J, Wilmoth T, Wilson M. Periconceptional growth hormone treatment alters fetal growth and development in lambs. J Anim Sci. 2010;88:1619–25.
43. Costine B, Inskeep E, Wilson M. Growth hormone at breeding modifies conceptus development and postnatal growth in sheep. J Anim Sci. 2005;83:810–5.
44. Oberbauer A, Pomp D, Murray J. Dependence of increased linear bone growth on age at oMT1a-oGH transgene expression in mice. Growth Dev Aging. 1994;58:83–93.
45. Tatara MR. Neonatal programming of skeletal development in sheep is mediated by somatotrophic axis function. Exp Physiol. 2008;93:763–72.
46. Thonney ML, Oberbauer AM, Duhaime DJ, Jenkins TC, Firth NL. Empty body component gain of rats grown at different rates to a range of final weights. J Nutr. 1984;114:1777–86.
47. Pomp D, Narcarrow CD, Ward KA, Murray JD. Growth, feed efficiency and body composition of transgenic mice expressing a sheep metallothionein 1a-sheep growth hormone fusion gene. Livest Prod Sci. 1992;31:335–50.
48. Pursel VG, Hammer R, Bolt D, Palmiter R, Brinster R. Integration, expression and germ-line transmission of growth-related genes in pigs. J Reprod Fertil Suppl. 1989;41:77–87.
49. Qiao X, Zhang H, Wu S, Yue H, Zuo J, Feng D, et al. Effect of β-hydroxy-β-methylbutyrate calcium on growth, blood parameters, and carcass qualities of broiler chickens. Poult Sci. 2013;92:753–9.
50. Gahr SA, Vallejo RL, Weber GM, Shepherd BS, Silverstein JT, Rexroad III CE. Effects of short-term growth hormone treatment on liver and muscle transcriptomes in rainbow trout (Oncorhynchus mykiss). Physiol Genomics. 2008;32:380–92.
51. Johnston IA, Devlin RH: Muscle fibre size optimisation provides flexibility to energy budgeting in calorie-restricted Coho salmon transgenic for growth hormone. J Experi Biol 2014. jeb. 107664.
52. Bush JA, Burrin DG, Suryawan A, O'Connor PM, Nguyen HV, Reeds PJ, et al. Somatotropin-induced protein anabolism in hindquarters and portal-drained viscera of growing pigs. Am J Phy Endocrin Metab. 2003;284:E302–12.

53. López-Oliva M, Agis-Torres A, Muñoz-Martínez E. Growth hormone administration produces a biphasic cellular muscle growth in weaning mice. J Physiol Biochem. 2001;57:255–63.
54. Oberbauer A, Stern J, Johnson P, Horwitz B, German J, Phinney S, et al. Body composition of inactivated growth hormone (oMt1a-oGH) transgenic mice: generation of an obese phenotype. Growth Dev Aging. 1997;61:169–79.
55. Oberbauer AM, Runstadler JA, Murray JD, Havel PJ. Obesity and Elevated Plasma Leptin Concentration in oMT1A-o Growth Hormone Transgenic Mice. Obes Res. 2001;9:51–8.
56. Oberbauer A, Stiglich C, Murray J, Keen C, Fong D, Smith L, et al. Dissociation of body growth and adipose deposition effects of growth hormone in oMt1a-oGH transgenic mice. Growth Dev Aging. 2003;68:33–45.
57. Murray J, Oberbauer A, Sharp K, German J. Expression of an ovine growth hormone transgene in mice increases archidonic acid in cellular membranes. Transgenic Res. 1994;3:241–8.
58. Oberbauer A, German J, Murray J. Growth Hormone Enhances Arachidonic Acid Metabolites in a Growth Hormone Transgenic Mouse. Lipids. 2011;46:495–504.
59. Forhead AJ, Fowden AL. The hungry fetus? Role of leptin as a nutritional signal before birth. J Physiol. 2009;587:1145–52.
60. Hoggard N, Haggarty P, Thomas L, Lea R. Leptin expression in placental and fetal tissues: does leptin have a functional role? Biochem Soc Trans. 2001;29:57–62.
61. Vickers M, Gluckman P, Coveny A, Hofman P, Cutfield W, Gertler A, et al. Neonatal leptin treatment reverses developmental programming. Endocrinology. 2005;146:4211–6.
62. Alexe D-M, Syridou G, Petridou ET. Determinants of early life leptin levels and later life degenerative outcomes. Clin Med Res. 2006;4:326–35.
63. Attig L, Djiane J, Gertler A, Rampin O, Larcher T, Boukthir S, et al. Study of hypothalamic leptin receptor expression in low-birth-weight piglets and effects of leptin supplementation on neonatal growth and development. Am J Phy Endocrin Metab. 2008;295:E1117–25.
64. Lea R, Howe D, Hannah L, Bonneau O, Hunter L, Hoggard N. Placental leptin in normal, diabetic and fetal growth-retarded pregnancies. Mol Hum Reprod. 2000;6:763–9.
65. Thomas A, Murray J, Oberbauer A. Leptin modulates fertility under the influence of elevated growth hormone as modeled in oMt1a-oGH transgenic mice. J Endocrinol. 2004;182:421–32.
66. Dunn GA, Morgan CP, Bale TL. Sex-specificity in transgenerational epigenetic programming. Horm Behav. 2011;59:290–5.
67. Mueller BR, Bale TL. Sex-specific programming of offspring emotionality after stress early in pregnancy. J Neurosci. 2008;28:9055–65.
68. Anway MD, Cupp AS, Uzumcu M, Skinner MK. Epigenetic transgenerational actions of endocrine disruptors and male fertility. Science. 2005;308:1466–9.
69. Jansson J-O, Ekberg S, Isaksson O, Mode A, Gustafsson J-Å. Imprinting of Growth Hormone Secretion, Body Growth, and Hepatic Steroid Metabolism by Neonatal Testosterone*. Endocrinology. 1985;117:1881–9.
70. Giustina A, Wehrenberg WB. The role of glucocorticoids in the regulation of growth hormone secretion mechanisms and clinical significance. Trends Endocrinol Metab. 1992;3:306–11.
71. Mazziotti G, Giustina A. Glucocorticoids and the regulation of growth hormone secretion. Nat Rev Endocrinol. 2013;9:265–76.
72. Giustina A, Mazziotti G: Impaired growth hormone secretion associated with low glucocorticoid levels: an experimental model for the Giustina effect. Endocrine 2014:1-3
73. Nogami H, Hisano S. Functional maturation of growth hormone cells in the anterior pituitary gland of the fetus. Growth Horm IGF Res. 2008;18:379–88.
74. Dean CE, Morpurgo B, Porter TE. Induction of somatotroph differentiation in vivo by corticosterone administration during chicken embryonic development. Endocrine. 1999;11:151–6.
75. Lemley C, Meyer A, Neville T, Hallford D, Camacho L, Maddock-Carlin K, et al. Dietary selenium and nutritional plane alter specific aspects of maternal endocrine status during pregnancy and lactation. Domest Anim Endocrinol. 2014;46:1–11.
76. Meyer A, Reed J, Neville T, Taylor J, Hammer C, Reynolds L, et al. Effects of plane of nutrition and selenium supply during gestation on ewe and neonatal offspring performance, body composition, and serum selenium. J Anim Sci. 2010;88:1786–800.
77. Newbern D, Freemark M. Placental hormones and the control of maternal metabolism and fetal growth. Curr Opin Endocrinol Diabetes Obes. 2011;18:409–16.
78. Barbour LA, Shao J, Qiao L, Pulawa LK, Jensen DR, Bartke A, et al. Human placental growth hormone causes severe insulin resistance in transgenic mice. Am J Obstet Gynecol. 2002;186:512–7.
79. Eastell R, Lambert H. Diet and healthy bones. Calcif Tissue Int. 2002;70:400–4.
80. Harel Z, Tannenbaum GS. Long-term alterations in growth hormone and insulin secretion after temporary dietary protein restriction in early life in the rat. Pediatr Res. 1995;38:747–53.
81. Cooper C, Walker Bone K, Arden N, Dennison E. Novel insights into the pathogenesis of osteoporosis: the role of intrauterine programming. Rheumatology. 2000;39:1312–5.
82. Fu Q, Yu X, Callaway CW, Lane RH, McKnight RA. Epigenetics: intrauterine growth retardation (IUGR) modifies the histone code along the rat hepatic IGF-1 gene. FASEB J. 2009;23:2438–49.
83. Smith LB, Belanger JM, Oberbauer AM: Fibroblast growth factor receptor 3 effects on proliferation and telomerase activity in sheep growth plate chondrocytes. J Anim Sci Biotec. 2012, 3:doi: 10.1186/2049-1891-1183-1139.
84. Wu S, Levenson A, Kharitonenkov A, De Luca F. Fibroblast growth factor 21 (FGF21) inhibits chondrocyte function and growth hormone action directly at the growth plate. J Biol Chem. 2012;287:26060–7.
85. Guan H, Arany E, van Beek JP, Chamson-Reig A, Thyssen S, Hill DJ, et al. Adipose tissue gene expression profiling reveals distinct molecular pathways that define visceral adiposity in offspring of maternal protein-restricted rats. Am J Phy Endocrin Metab. 2005;288:E663–73.
86. Gray C, Li M, Reynolds CM, Vickers MH. Pre-weaning growth hormone treatment reverses hypertension and endothelial dysfunction in adult male offspring of mothers undernourished during pregnancy. PLoS ONE. 2013;8:e53505.
87. Reynolds C, Li M, Gray C, Vickers M. Pre-weaning growth hormone treatment ameliorates adipose tissue insulin resistance and inflammation in adult male offspring following maternal undernutrition. Endocrinology. 2013;154:2676–86.
88. Vickers MH, Sloboda DM. Strategies for reversing the effects of metabolic disorders induced as a consequence of developmental programming. Front Physiol. 2012;3:342.
89. Sarr O, Yang K, Regnault TR: In utero programming of later adiposity: the role of fetal growth restriction. J Preg. 2012, 2012:doi:10.1155/2012/134758.
90. Lesseur C, Armstrong DA, Paquette AG, Koestler DC, Padbury JF, Marsit CJ. Tissue-specific Leptin promoter DNA methylation is associated with maternal and infant perinatal factors. Mol Cell Endocrinol. 2013;381:160–7.
91. Jousse C, Parry L, Lambert-Langlais S, Maurin A-C, Averous J, Bruhat A, et al. Perinatal undernutrition affects the methylation and expression of the leptin gene in adults: implication for the understanding of metabolic syndrome. FASEB J. 2011;25:3271–8.
92. Shao Y, Zhao F-Q: Emerging evidence of the physiological role of hypoxia in mammary development and lactation. J Anim Sci Biotec. 2014, 5:doi:10.1186/2049-1891-1185-1189.
93. Melvin A, Mudie S, Rocha S. The chromatin remodeler ISWI regulates the cellular response to hypoxia: role of FIH. Mol Biol Cell. 2011;22:4171–81.
94. Nair D, Ramesh V, Li RC, Schally AV, Gozal D. Growth hormone releasing hormone (GHRH) signaling modulates intermittent hypoxia-induced oxidative stress and cognitive deficits in mouse. J Neurochem. 2013;127:531–40.
95. Li RC, Guo SZ, Raccurt M, Moudilou E, Morel G, Brittian KR, et al. Exogenous growth hormone attenuates cognitive deficits induced by intermittent hypoxia in rats. Neuroscience. 2011;196:237–50.
96. Braems G, Han V. Gestational age-dependent changes in the levels of mRNAs encoding cortisol biosynthetic enzymes and IGF-II in the adrenal gland of fetal sheep during prolonged hypoxemia. J Endocrinol. 1998;159:257–64.
97. Chen X-Q, Du J-Z. Increased somatostatin mRNA expression in periventricular nucleus of rat hypothalamus during hypoxia. Regul Pept. 2002;105:197–201.
98. Varvarigou A, Vagenakis A, Makri M, Beratis N. Growth hormone, insulin-like growth factor-I and prolactin in small for gestational age neonates. Neonatology. 1994;65:94–102.

16

Alternatives to antibiotics as growth promoters for use in swine production: a review

Philip A Thacker

Abstract

In the past two decades, an intensive amount of research has been focused on the development of alternatives to antibiotics to maintain swine health and performance. The most widely researched alternatives include probiotics, prebiotics, acidifiers, plant extracts and neutraceuticals such as copper and zinc. Since these additives have been more than adequately covered in previous reviews, the focus of this review will be on less traditional alternatives. The potential of antimicrobial peptides, clay minerals, egg yolk antibodies, essential oils, eucalyptus oil-medium chain fatty acids, rare earth elements and recombinant enzymes are discussed. Based on a thorough review of the literature, it is evident that a long and growing list of compounds exist which have been tested for their ability to replace antibiotics as feed additives in diets fed to swine. Unfortunately, the vast majority of these compounds produce inconsistent results and rarely equal antibiotics in their effectiveness. Therefore, it would appear that research is still needed in this area and that the perfect alternative to antibiotics does not yet exist.

Keywords: Antimicrobial peptides, Clay minerals, Egg yolk antibodies, Essential oils, Eucalyptus oil-medium chain fatty acids, Rare earth elements, Recombinant enzymes

Background

Antibiotics have played a major role in the growth and development of the swine industry for more than 50 years. Their efficiency in increasing growth rate, improving feed utilization and reducing mortality from clinical disease is well documented [1]. However, consumers are becoming increasingly concerned about drug residues in meat products [2]. In addition, it has been suggested that the continuous use of antibiotics may contribute to a reservoir of drug-resistant bacteria which may be capable of transferring their resistance to pathogenic bacteria in both animals and humans [3]. As a result, many countries have banned or are banning the inclusion of antibiotics in swine diets as a routine means of growth promotion.

In the past two decades, an intensive amount of research has been focused on the development of alternatives to antibiotics to maintain swine health and performance and many excellent reviews have already been published on this subject. The most widely researched alternatives include probiotics [4-6], prebiotics [4,7], enzymes [8-10], acidifiers [11-14], plant extracts [4,15, 16] and neutraceuticals such as copper and zinc [17,18]. Since these additives have been more than adequately covered, the focus of this review will be on less traditional alternatives.

Antimicrobial peptides

Antimicrobial peptides, as the name implies, are peptides with antimicrobial properties. They have been isolated and characterized from virtually all living organisms ranging from prokaryotes to humans [19]. They are important components of the host's defense system and are effector molecules of innate immunity with direct antimicrobial and mediator function [20]. Most antimicrobial peptides contain between 30 and 60 amino acids and are polar molecules with spatically separated hydrophobic and charged regions. Antimicrobial peptides have been identified that have activity against Gram-positive and Gram-negative bacteria as well as against fungi and enveloped viruses [20].

More than 700 antimicrobial peptides are known to exist [20]. Bioscreening, cloning strategies and computer-based database searches have been used to identify antimicrobial peptides which have potential to be used

Correspondence: phil.thacker@usask.ca
Department of Animal and Poultry Science, University of Saskatchewan, 51 Campus Drive, Saskatoon, Saskatchewan S7N 5A8, Canada

as alternatives to antibiotics [20]. Once identified, it is possible to chemically synthesize most antimicrobial peptides but the high cost of this process precludes the production of peptides through this method for use as feed additives. However, several research groups have developed recombinant systems for expression of antimicrobial peptides.

Antimicrobial proteins produced by bacteria are called bacteriosins. These proteins have several characteristics that make them desirable alternatives to conventional antibiotics for use in swine production. Most importantly, bacteria have difficulty in developing resistance against these peptides [21]. Peptides have a narrow spectrum of activity so they can be used to target specific pathogenic bacteria without affecting the normal native flora. There is almost no risk of residues in meat because they are proteins and therefore will not be absorbed as an intact molecule. In addition, antimicrobial peptides can tolerate a wide range of pH and temperatures [22].

The antimicrobial activity of peptides is based on several mechanisms. In most cases, interactions between the peptide and the surface membranes of the target bacteria are thought to be responsible for their killing activity [20]. These interactions are proposed to lead to a loss of membrane function including breakdown of membrane potential, leakage of metabolites and ions, and alteration of membrane permeability [19]. These alterations in the bacterial membrane can result in cell lysis or, alternatively, can lead to the formation of transient pores and the transport of peptides inside the cell bringing them into contact with intracellular targets. Other mechanisms of antimicrobial activity include the inhibition of protein and RNA synthesis [20].

To date, the most prevalent use of antimicrobial peptides has been in the preservation of foods and few studies have been conducted using antimicrobial peptides with swine. One promising research area has been in the use of the antimicrobial peptide colicin. Colicins are a class of bacteriocin produced by and effective against *Escherichia coli* (*E. coli*) and closely related species. They have been shown to be effective against many pathogenic *E. coli* strains including those responsible for post-weaning diarrhea and edema disease in pigs [23,24].

A chemically synthesized antimicrobial peptide A3 has been shown to have beneficial effects on weanling pig performance, nutrient digestibility, intestinal morphology as well as intestinal and fecal microflora [25,26]. In addition, an antimicrobial peptide isolated from the intestine of the Rongchang pig improved performance but had no effect on diarrhea incidence in weanling pigs [27]. However, the antimicrobial peptide appeared to act synergistically with zinc as the two additives in combination were superior to either additive fed separately.

The results of a feeding trial in which the antimicrobial peptide cecropin, originally isolated from the silkworm *Hyalophora cecropia*, was fed to weanling pigs challenged with enterotoxigenic *E. coli K88* are shown in Table 1. Use of the antimicrobial peptide cecropin resulted in similar performance to pigs fed a combination of antibiotics [21]. The improvement in performance appeared to be related to improvements in nutrient digestibility and intestinal morphology. Cecropin treatment decreased total aerobes while increasing total anaerobes in the ileum compared with the control (Table 2). Cecropin also increased the numbers of beneficial lactobacillus in the cecum. Cecropin increased serum IgA and IgG and the inflammatory cytokines interleukin-1β and interleukin 6 indicating that cecropin activates both systemic and local immune systems in response to *E. coli* challenge.

Although there is little research on these compounds, the use of antimicrobial peptides appears to have considerable potential as a replacement for antibiotics in rations fed to swine. A commercial entity (Beijing Longkefangzhou Biological Engineering Technology Company, Beijing, China) has started to market cecropin for use in swine rations in China.

Clay minerals

Clay minerals are formed by a net of stratified tetrahedral and octahedral layers [2]. They contain molecules of silicon, aluminum and oxygen. The natural extracted clays (bentonites, zeolite, kaolin) are a mixture of various clays differing in chemical composition. The best known are montmorillonite, smectite, illite, kaolinite, biotic and clinoptilolite [2].

Clays added to the diet can bind and immobilize toxic materials in the gastrointestinal tract of animals and thereby reduce their biological availability and toxicity [2]. Clay minerals can bind aflatoxins, plant metabolites, heavy metals, and toxins. The extent of adsorption is determined by the chemistry of the clay minerals, exchangeable ions, surface properties and the fine structure of the clay particles [2]. An important role is played by pH, dosage and exposure time. As a result of their binding properties, clays have been widely used in swine diets to improve pig performance when diets containing mycotoxins are fed [28,29].

Clays have also been shown to prevent diarrhea in weaned pigs [2,30,31]. Based on this fact, several research groups have attempted to determine whether or not the inclusion of various clays in swine diets can improve pig performance. The results have been inconclusive with some trials demonstrating positive results particularly for younger pigs [30], but the vast majority of the experiments have failed to show improvements

Table 1 Effects of antibiotics or an antimicrobial peptide cecropin on the performance of four week old weaned pigs after challenge with *E. coli* as well as nutrient digestibility before challenge

Items	Control	Antibiotics[1]	Cecropin	SEM	*P*-value
Performance (day 13-19)					
Weight gain, g/d	312[a]	367[b]	358[b]	6.4	<0.01
Feed intake, g/d	566	597	592	9.8	0.08
Feed efficiency	0.55[a]	0.62[b]	0.61[b]	0.01	<0.01
Diarrhea incidence, %	37.50	17.86	19.64		
Nutrient digestibility					
Nitrogen retention, g/d	10.1[a]	11.5[b]	10.7[ab]	0.38	0.04
Nitrogen digestibility, %	73.2	76.9	75.0	1.20	0.17
Energy retention, MJ/kg/d)	2.5[a]	3.0[b]	2.8[ab]	0.13	0.04
Energy digestibility, %	84.6	88.2	86.4	1.71	0.14

Wu et al. [21].
[1]Kitasamycin and colistin sulfate.
[a,b]Within row, means followed by same or no letter do not differ ($P>0.05$).

Table 2 Effects of antibiotics or the antimicrobial peptide cecropin on intestinal morphology and intestinal microflora of four week old weaned pigs after challenge with *E. coli*

Items	Control	Antibiotic[1]	Cecropin	SEM	*P*-value
Intestinal Morphology					
Duodenum					
Villus height, μm	418	439	431	10.7	0.53
Crypt depth, μm	233	227	232	5.3	0.41
Villus height to crypt depth ratio	1.83	1.96	1.89	0.24	0.18
Jejunum					
Villus height, μm	401	448	420	18.4	0.37
Crypt depth, μm	212[b]	233[a]	220[b]	6.8	0.04
Villus height to crypt depth ratio	1.89[b]	1.97[a]	1.91[ab]	0.01	0.03
Ileum					
Villus height, μm	357[b]	396[a]	384[a]	12.4	0.04
Crypt depth, μm	211	217	213	5.6	0.37
Villus height to crypt depth ratio	1.74[b]	1.85[a]	1.82[ab]	0.04	0.04
Intestinal Microflora (log_{10} CFU/g of digesta)					
Ileum					
E. coli	4.37	4.14	4.25	0.18	0.85
Lactobacillus	9.38	10.00	9.62	0.20	0.42
Total aerobes	6.69[a]	6.60[ab]	6.43[b]	0.08	0.04
Total anaerobes	9.36[b]	9.87[ab]	10.12[a]	0.23	0.03
Cecum					
E. coli	3.37[a]	3.09[b]	3.22[ab]	0.12	0.04
Lactobacillus	8.89[b]	9.47[a]	9.23[a]	0.14	0.03
Total aerobes	3.88	3.77	3.49	0.44	0.63
Total anaerobes	8.79	9.37	9.26	0.28	0.38

Wu et al. [21].
[1]Kitasamycin and colistin sulfate.
[a,b]Within row, means followed by same or no letter do not differ ($P>0.05$).

[32-35]. It would appear that clay minerals are not viable alternatives to antibiotics as growth promoters.

Egg yolk antibodies

One technique that appears to have considerable potential as an alternative to antibiotics for growth promotion in the presence of disease causing organisms is the use of egg yolk antibodies generally referred to as IgY [36]. In order to produce these antibodies, laying hens are injected with organisms that cause specific diseases in swine. The injection of these antigens induces an immune response in the hen which results in the production of antibodies. These antibodies are typically deposited in the egg yolk. Booster immunizations are given to ensure continued transfer of antibodies from the hen to the egg yolk. These antibodies are then extracted from the egg yolk and processed. Antibodies can be administered in the feed in several forms including whole egg powder, whole yolk powder, water-soluble fraction powder or purified IgY [37]. Details concerning IgY production including choice of adjuvant, route of immunization, dose, immunization frequency and techniques for IgY extraction from the yolk have been reviewed by Chalghoumi et al. [37] and Kovacs-Nolan and Mine [38].

Compared with the use of mammals such as rabbits or sheep for antibody production, the immunization of chickens for antibody production is an attractive approach. Chicken housing is inexpensive, egg collection is non-invasive, the IgY antibodies are concentrated in egg yolk and isolation is fast and simple. In addition, chicken immunnoglobin does not react with mammalian IgG or IgM and also it does not activate mammalian complement factors [38]. Finally, the use of IgY elicits no undesirable side effects, disease resistance or toxic residues [36].

IgY antibodies have been tested against a number of enteric pathogens in swine including *E. coli*, *Salmonella* and *Rotavirus* with varying degrees of success [39-43]. Table 3 shows the results of an experiment where the performance of pigs fed egg yolk antibodies was compared with that of pigs fed diets supplemented with zinc oxide, fumaric acid or antibiotics. All four feed additives successfully increased pig performance compared with unsupplemented pigs with significant reductions observed in scour score and piglet mortality. In this experiment, egg yolk antibody was equal to antibiotics in enhancing pig performance.

Unfortunately, there are several reports where egg yolk antibody failed to improve pig performance [42,44]. The most likely explanation for the failure of egg yolk antibody to improve performance is that the antibody failed to survive passage through the gastrointestinal tract [45]. It appears that the IgY molecule is less stable than the IgG molecule due to its higher molecular weight, lower percentage of β-sheet structure and reduced flexibility [45]. It has been reported that the activity of IgY was decreased at pH 3.5 or lower and almost completely lost activity with irreversible change at pH 3 [37]. In addition, IgY is fairly sensitive to pepsin digestion [45]. Therefore, a recent avenue of research has been to use microencapsulation techniques to protect IgY from gastric inactivation [46,47].

Table 4 shows the results of an experiment where chitosan-alginate microcapsules were used for oral delivery of egg yolk immunoglobulin in weaned pigs challenged with enterotoxigenic *E. coli* C83903 [46]. The percentage of pigs with diarrhea 24 h after treatment and the diarrhea score were improved in pigs receiving encapsulated IgY compared with non-encapsulated IgY. In addition, weight gain over the three day period was significantly higher in pigs receiving encapsulated IgY compared with non-encapsulated IgY. Both encapsulated and non-encapsulated IgY treatments were numerically superior to an aureomycin treated group.

The mechanism through which IgY counteracts pathogen activity has not been determined. However, several mechanisms were proposed by Xu et al. [36] including agglutination of bacteria, inhibition of adhesion, opsoni-

Table 3 Effect of egg yolk antibody, zinc oxide, fumaric acid and antibiotic on the performance and intestinal morphology of 10 to 24 day old pigs fed diets based on pea protein concentrate

Items	Control	Egg yolk antibody	Zinc oxide	Fumaric acid	Carbadox	SEM
Weight gain, g/d	100.9	151.2	158.9	155.4	152.6	16.6
Feed intake, g/d	141.0	208.1	214.7	211.6	222.4	15.3
Feed conversion	1.39	1.38	1.35	1.36	1.45	0.04
Scour score	2.7	1.3	1.4	1.3	1.1	-
Mortality, %	40.0	6.6	13.3	6.6	13.3	-
Villus height, m	355	564	488	573	570	20.0
Crypt depth, m	204	183	190	207	204	10.1
Villous height:crypt depth	1.7	3.1	2.6	2.8	2.8	0.11

Owusu-Asiedu et al. [41].

Table 4 Effect of encapuslation of IgY on performance and the incidence of diarrhea in pigs challenged with *E. coli*

Items	Percentage of pigs with diarrhea after specific times (Fecal score in brackets)[1]				Weight gain (g/d)	Recovery rate (%)
	9 h	24 h	48 h	72 h		
Negative control, unchallenged	0% (0.5)	0% (0.0)	0% (0.4)	0% (0.0)	116.6[a]	-
Positive control	75% (2.5)	75% (2.5)	75% (2.0)	75% (2.0)	13.5[d]	0%
Non-encapsulated IgY	100% (2.0)	75% (1.3)	25% (1.0)	0% (0.0)	78.1[b]	100%
Microencapsulated IgY	75% (2.0)	0% (0.0)	0% (0.0)	0% (0.0)	110.4[a]	100%
Aureomycin	100% (2.0)	50% (2.0)	75% (1.5)	50% (1.5)	54.1[c]	50%

Li et al. [46].
[1]Fecal score is the mean fecal consistency score where 0 = normal, 1 = soft feces, 2 = mild diarrhea, 3 = severe diarrhea.
[a,b,c,d]Within column, means followed by same or no letter do not differ (P>0.05).

zation followed by phagocytosis and toxin neutralization. Further research is necessary to determine the exact mechanism for the growth promoting activity of IgY.

Essential oils

Essential oils are aromatic oily liquids obtained from plant material and usually have the characteristic odor or flavor of the plant from which they are obtained [48]. They are typically mixtures of secondary plant metabolites and may contain phenolic compounds (i.e. thymol, carvacrol and eugenol), terpenes (i.e. citric and pinapple extracts), alkaloids (capsaicine), lectins, aldehydes (i.e. cinnamaldehyde), polypeptides or polyacetylenes [49]. They can be extracted from plants with organic solvents or steam distillation [49]. An estimated 3000 essential oils are known to exist but cinnamaldehyde, carvacrol, eugenol and thymol have received the most interest for use in swine production.

Interest in the use of essential oils as a potential replacement for antibiotics in swine rations has been generated as a result of *in vitro* studies showing that essential oils have antimicrobial activity against microflora commonly present in the pig gut [50]. The exact mode of action of essential oils has not been established but the activity may be related to changes in lipid solubility at the surface of the bacteria [48]. The hydrophobic constituents of essential oils allow them to disintegrate the outer membrane of *E. coli* and *Salmonella* and thus inactivate these pathogens [48]. This would result in a shift in the microbial ecology in favor of lactic acid producing bacteria and reducing the number of pathogenic bacteria [50]. Essential oils containing phenolic compounds tend to have greater antimicrobial activity than oils containing other compounds [51].

Based on the fact that essential oils appear to control pathogenic bacteria, several research groups have attempted to determine whether or not the inclusion of essential oils in swine diets can improve pig performance [52]. The results have been inconclusive with some trials demonstrating positive results [53-55] while others have reported no beneficial effects [56,57]. The most compelling evidence for including essential oils in diets fed to swine can be obtained from the results of Li et al. [55]. This trial compared the performance of pigs fed an unsupplemented control diet with that of pigs fed a diet supplemented with antibiotics or a combination of thymol and cinnamaldehyde (Table 5). Weight gain, feed conversion and fecal consistency of pigs fed essential oils was essentially equal to that of pigs fed antibiotics. The improved performance appeared to be mediated by improvements in dry matter and protein digestibility arising from improvements in intestinal morphology. In addition, total antioxidant capacity and levels of the cytokines interleukin-6 and tumor necrosis factor-α were altered by inclusion of essential oils (Table 6).

The reason for the variability in results when essential oils are fed is likely due to differences in the type of essential oils used and the dose provided [55]. As noted previously, oils containing phenolic compounds tend to have greater antimicrobial activity than those based on other compounds. In addition, if the dose used is too high, the strong smell can reduce feed intake and thereby limit pig performance [48]. Another important consideration is the stability of essential oils during pelleting. Maenner et al. [54] reported considerable loss

Table 5 Effect of essential oils on weanling pig performance, nutrient digestibility and fecal consistency

Items	Control	Antibiotic[1]	Essential oil	SEM
Performance				
Weight gain, g/d	442[a]	505[b]	493[b]	15
Feed intake, g/d	783	846	789	24
Feed conversion	1.79	1.67	1.62	0.06
Fecal consistency	1.53[a]	1.22[b]	1.30[b]	0.06
Nutrient digestibility				
Dry matter	84.33[a]	87.03[b]	86.92[b]	0.65
Crude protein	76.51[a]	83.53[b]	81.34[b]	1.25

Li et al. [55].
[1]Chlortetracycline, colistin sulfate and kitasamycin.
[a,b] Within row, means followed by same or no letter do not differ (P>0.05).

Table 6 Effect of essential oils on intestinal morphology, antioxidant capacity, and cytokine levels in weanling pigs

Items	Control	Antibiotic[1]	Essential oil	SEM
Villus height, μm	466	509	535	24
Crypt depth, μm	164	156	162	8
Villus height:crypt depth	2.96^{a}	3.41^{b}	3.38^{b}	0.09
Total antioxidant capacity, U/mL	10.46^{a}	11.97^{ab}	12.37^{b}	0.52
Interleukin-6, ng/L	44.21^{a}	40.39^{a}	27.40^{b}	2.76
Tumor necrosis factor-α, ng/L	208^{a}	237^{ab}	260^{b}	13

Li et al. [55].
[1]Chlortetracycline, colistin sulfate and kitasamycin.
[a,b]Within row, means followed by same or no letter do not differ ($P>0.05$).

of activity of essential oils when a pelleting temperature of 58°C was applied.

Eucalyptus oil-medium chain fatty acids

Eucalyptus oil is obtained from the leaves of the eucalyptus, a tree which belongs to the plant family *Myrtaceae* and is cultivated worldwide. In humans, eucalyptus oil has been shown to have antibacterial effects on pathogenic bacteria in the respiratory tract [58]. Eucalyptus oil has also been shown to stimulate the immune system by affecting the phagocytic ability of monocyte-derived macrophages [59]. In poultry, dietary inclusion of eucalyptus has been shown to improve production performance and stimulate the immunity of commercial laying hens [60].

Medium-chain fatty acids have been suggested as an alternative feed additive to antibiotics for piglets [61-63]. Medium chain fatty acids have been shown to have antimicrobial activity against *Salmonella* [64] and *E. coli* [61]. Hong et al. [63] reported that feeding a blend of caprylic and caproic acids improved performance and nutrient digestibility in 3 and 4 week old weaned pigs during the first two weeks following weaning.

Micro-encapsulation of medium chain fatty acids is a process in which medium chain fatty acids are nano-micronized to extremely small particles and then encapsulated. Han et al. [65] tested a product where eucalyptus extract was mixed with caprylic and carpric acids and encapsulated with palm oil in comparison with antibiotics or zinc oxide (Table 7). The performance of pigs fed the eucalyptus-medium chain fatty acid blend was essentially equal to that of antibiotics or zinc oxide. The performance enhancing effects of the blend appeared to be mediated through improvements in nutrient digestibility (Table 8). The process used to produce the micro-encapsualted eucalyptus-medium chain fatty acid blend has been patented by the Korean Intellectual Property Office under patent number 10-2009-0025329.

Rare earth elements

Rare earth elements comprise the elements scandium, yttrium, lanthanum and the 14 chemical elements following lanthanum in the periodic table called lanthanoids [66]. The application of rare earth elements as feed additives for livestock has been practiced in China for decades [66]. There are many articles in the Chinese literature concerning the performance enhancing effects of rare earth elements for swine [67,68] and many more have been reviewed by Rambeck and Wehr [69] and Redling [66]. In the Chinese literature, body weight gain was shown to be improved by 5 to 23% and feed conversion between 4 and 19% under the influence of rare earth elements.

Research concerning the effect of rare earth elements on swine performance have been published in the Western literature since about the year 2000 with some reports indicating significant improvements in pig performance [70,71] while others have observed no change [72]. Table 9 shows the results of a recent trial in which the performance of weaned pigs fed a lanthanum-yeast mixture was similar to that of pigs fed diets supplemented with antibiotics or zinc oxide [73].

The products commonly used as feed additives for swine are typically mixtures of rare earth elements

Table 7 Effects of antibiotics, zinc oxide, and eucalyptus-medium chain fatty acids (MCFA) on nursery pig performance

Items	Control	Antibiotics[1]	ZnO (1,500 ppm)	ZnO (2,500 ppm)	Eucalyptus-MCFA	SEM	*P*
Weight gain, g/d	243^{a}	315^{b}	298^{b}	308^{b}	310^{b}	13.6	<0.01
Feed intake, g/d	361^{a}	431^{b}	426^{b}	429^{b}	448^{b}	18.1	<0.01
Feed conversion	1.53	1.41	1.44	1.41	1.46	0.05	0.35

Han et al. [65].
[1]Tiamulin and lincomycin.
[a,b]Within row, means followed by same or no letter do not differ ($P>0.05$).

Table 8 Effects of antibiotics, zinc oxide, and eucalyptus-medium chain fatty acids (MCFA) on nutrient digestibility for weaned pigs

Items	Antibiotics[1]	ZnO (1,500 ppm)	ZnO (2,500 ppm)	Eucalyptus-MCFA	SEM	*P*
Dry matter	91.74[a]	90.58[b]	90.44[b]	92.17[a]	0.26	< 0.01
Crude protein	74.18[a]	72.01[a]	71.23[a]	78.93[b]	1.13	< 0.01
Calcium	56.31[a]	48.26[b]	46.75[b]	65.93[c]	1.56	<0.01
Phosphorus	54.48[a]	38.25[b]	42.77[b]	66.10[b]	2.01	<0.01
Energy	82.92[a]	81.60[b]	81.00[b]	86.00[c]	0.61	< 0.01
Lysine	79.13[a]	80.25[b]	78.25[a]	83.80[b]	0.88	< 0.01
Methionine	83.94[a]	80.95[b]	80.78[b]	84.23[a]	0.63	<0.01
Threonine	73.56[b]	73.57[b]	73.43[b]	79.40[a]	1.40	0.02

Han et al. [65].
[1]Tiamulin and lincomycin.
[a,b]Within row, means followed by same or no letter do not differ ($P>0.05$).

mainly containing lanthanum, cerium and praseodymium [73]. Both inorganic and organic rare earth compounds have been used as feed additives but it is believed that best results are obtained with organic compounds [66].

Several mechanisms have been proposed for the growth promoting effects of rare earth elements. It has been suggested that rare earth elements may promote growth by influencing the development of undesirable bacterial species within the gastrointestinal tract. For example, lanthanum has been shown to bind to the surface of bacteria [69]. This reduces the surface charge and retards electrophoretic migration. When the surface charge is completely neutralized, flocculation occurs. In addition, bacterial respiration has been shown to be strongly inhibited by lanthanides [69].

Another explanation for the growth promoting effects of rare earth elements is due to improvements in nutrient digestibility and availability as was observed by Han and Thacker [73; Table 10]. It has been suggested that rare earth elements may influence the permeability of the intestines thereby enhancing the absorption of different nutrients [66]. Enhanced secretion of digestive fluids and increased gastrointestinal motility have also been proposed as explanations for the enhanced digestibility of nutrients following dietary inclusion of rare earth elements [66].

Rare earth elements have several properties that make them attractive alternatives to antibiotics. Generally, absorption of orally applied rare earths is low with more than 95% being recovered in the feces of animals [66]. As a result, the chances of residues being present in meat are low with studies reporting no higher levels of rare earth elements in the muscle tissue of supplemented animals than those fed commercial diets [66]. In addition, there have been no reports of the development of bacterial resistance in treated animals [66].

Recombinant enzymes

Enzymes are biologically active proteins that break specific chemical bonds to release nutrients for further digestion and absorption. They accelerate chemical reactions in the body which would otherwise proceed very slowly or not at all [74]. Enzymes used in the feed industry are commonly produced by bacteria (i.e. *Bacillus subtilis*), fungus (i.e. *Trichoderma reesei, Aspergillus niger*) or yeast (*Saccharomyces cerevisiae*).

The supplementation of swine diets with exogenous enzymes to enhance performance is not a new concept and research articles in this field date back to the 1950's [10]. The most common reasons for enzyme supplementation include degrading feed components resistant to endogenous enzymes (i.e. β-glucanase, xylanase, mannanase, pectinase and galactosidase), inactivating antinutritional factors (i.e. phytase) and supplementing endogenous enzymes that may be present in insufficient amounts (i.e. proteases, lipases and amylases). This

Table 9 Effects of zinc oxide, antibiotic, or lanthanum-yeast on the performance of weanling pigs (day 0 to 28)

Items	Control	Antibiotic[1]	Zinc (1,500 ppm)	Zinc (2,500 ppm)	Lanthanum-yeast	SEM	*P* Values
Weight gain, g/d	302[b]	353[a]	352[a]	369[a]	359[a]	14.0	0.02
Feed intake, g/d	467[b]	518[ab]	530[ab]	558[a]	501[ab]	22.6	0.10
Feed conversion	1.55[a]	1.47[ab]	1.50[ab]	1.52[ab]	1.41[b]	0.04	0.31

Han and Thacker [73].
[1]Tiamulin and chlortetraccycline.
[a,b] Within row, means followed by same or no letter do not differ ($P>0.05$).

Table 10 Effects of antibiotics, zinc oxide or lanthanum-yeast on nutrient digestibility

Items	Antibiotics[1]	Zinc(1,500 ppm)	Zinc (2,500 ppm)	Lanthanum-yeast	SEM	*P*-value
Dry matter	95.19[a]	93.83[b]	93.98[b]	95.46[a]	0.30	<0.01
Crude protein	74.51[ab]	71.55[b]	72.33[b]	78.34[a]	1.38	0.01
Calcium	56.59[b]	46.98[c]	48.50[c]	65.10[a]	1.69	<0.01
Phosphorus	54.87[b]	43.07[c]	38.52[c]	66.11[a]	2.09	<0.01
Energy	83.51[b]	81.42[b]	81.33[b]	86.89[a]	0.80	<0.01
Lysine	81.45[b]	79.42[b]	80.32[b]	85.15[a]	0.95	<0.01
Methionine	83.49[b]	83.67[b]	86.76[a]	87.32[a]	0.79	<0.01
Phenylalanine	74.21[b]	73.75[b]	75.41[ab]	78.96[a]	1.32	0.05
Threonine	76.19[b]	75.13[b]	75.28[b]	81.19[a]	1.58	0.04

Han and Thacker [73].
[1]Tiamulin and chlortetraccycline.
[a,b,c]Within row, means followed by same or no letter do not differ ($P>0.05$).

review will focus on the use of enzymes to degrade feed components resistant to endogenous enzymes.

The cell walls of cereal grains, legumes and oilseed meals are comprised of complex carbohydrates commonly referred to as non-starch polysaccharides [75]. Non-starch polysaccharides consist of a wide range of polymers which include cellulose, hemicellulose, pectins, β-glucans, α-galctosides (raffinose, stachnyose and verbascose) and xylans [8]. These non-starch polysaccharides reduce the nutritional value of feed ingredients in a number of ways [74]. Firstly, they are indigestible by mammalian enzymes and therefore dilute the energy and nutrient content of the feed. Secondly, non-starch polysaccharides exhibit a so called "cage effect" whereby normally highly digestible nutrients such as starch, fat and protein are entrapped in a coating of non-starch polysaccharides preventing access of the endogenous enzymes to these substrates [76]. In addition, certain non-starch polysaccharides may increase intestinal viscosity. It has also been suggested that non-starch polysaccharides allow microbial populations to assimilate a greater proportion of the nutrients contained in the feed into their own system thereby reducing the availability of these nutrients to the host [8].

Carbohydrases include all enzymes that catalyze a reduction in the molecular weight of polymeric carbohydrate but more than 80% of the global carbohydrase market is accounted for by xylanase and β-glucanase [10]. Other commercially available carbohydrases include α-amylase, β-mannanase, α-galactosidase and pectinase. These carbohydrases have widespread application in the poultry industry but are used less commonly in feeds for swine.

The effect of carbohydrase supplementation on the performance of pigs is inconsistent. There are reports of positive responses to carbohydrase supplementation [77,78], whereas others have reported no improvement in weight gain in response to enzymes [79-81]. Where positive effects on performance are observed, they are commonly associated with increases in nutrient digestibility likely as a result of increased accessibility of endogenous enzymes to nutrients as a result of inhibition of the "cage effect" as well as hydrolysis or partial hydrolysis of the non-starch polysaccharide. There also seems to be an influence on th e composition of the microflora in the digestive tract [76]. Hydrolysis of non-starch polysaccharides results in increased sugar release in the large and small intestine and thereby stimulates the growth of lactobacilli which produce lactic acid. Increased proportions of lactic acid promote gut health by suppressing the growth of coliforms such as pathogenic *E. coli.*

Based on a review of the literature, it is clear that the response of pigs to supplementation with carbohydrases is less consistent than has been observed with poultry. The question is why? What differences are there in the physiology of the pig and the chicken that might account for the differences in the magnitude of the results obtained. One clear difference is the pH in the gut. In the pig, the duration that feed is exposed to a low pH is significantly longer than in the chicken [82]. Therefore,

Table 11 Comparison of the effects of a β-mannase produced using normal fermentation technology with that of a recombinant β-mannase on the performance of growing-finishing pigs

Items	Control	β-mannase	% Improvement
Traditional fermentation[1]			
Weight gain, g/d	0.84	0.87	3.4
Feed intake, g/d	2.50	2.48	-
Feed efficiency	0.337	0.351	3.9
Recombinant technology[2]			
Weight gain, g/d	0.66	0.79	16.4
Feed intake, g/d	1.66	1.61	3.0
Feed efficiency	0.404	0.491	17.7

[1]Pettey et al. [87].
[2]Lv et al. [85].

it is possible that exposure to the low pH in the stomach of the pig is either partially or totally denaturing the enzyme accounting for the lower magnitude of responses obtained when carbohydrases are fed to pigs compared with poultry.

Many of the enzyme preparations used in the past were unsuitable for use in the harsh environment of the pig's gastrointestinal tract. The pH in the stomach of the pig is usually between 2 and 3.5 and substantial reductions in β-glucanase [82] and xylanase [83] activity were reported when ten commercially available enzyme products were exposed *in vitro* to a pH of 2.5 or 3.5 for 30 min.

The application of genetic engineering in the process of enzyme production allows the development of enzymes targeted for specific purposes [84-86]. Recently, several carbohydrases have been developed by molecular directed evolution which have considerable potential for animal feed application [84-86]. Enzymes have been developed which are active over a broad pH range, exhibit thermostability, are resistant to pepsin and trypsin, and viable under simulated gastric conditions.

Inclusion of a recombinant β-mannanase in corn soybean meal diets fed to growing pigs increased weight gain by 16.1% and feed efficiency by 17.7% compared with an unsupplemented diet (Table 11). The magnitude of the improvement was notably greater than previous experiments using β-mannanase produced by traditional fermentation techniques. For example, Pettey et al. [87] reported that weight gain was only increased 3.4% and feed efficiency 3.9% in their experiment in which growing–finishing pigs were fed diets supplemented with β-mannanase.

Enzymes added to feed are broken down in the digestive tract in the same way as other proteins [74]. Therefore, there are not any issues with residues and it is not necessary to observe any withdrawal periods before animals fed enzymes can be slaughtered [74]. For this reason, the amount of enzyme required is very small compared with the amount of substrate and therefore only small quantities are needed when using enzymes in ration formulation.

Miscellaneous compounds

Many additional compounds have been tested for their potential to replace antibiotics as growth promoters for use in swine production. They are too numerous to be able to go into much detail regarding their effectiveness. Some of the more promising include spray-dried porcine plasma [88,89], yeast culture [90-92], bacteriophages [93], lysozyme [94], bovine colostrum [95], lactoferrin [96-98], conjugated fatty acids [99,100], chito-oligosaccarides [101,102] and seaweed extract [103].

Conclusions

Clearly, a long and growing list of compounds exist which have been tested for their ability to replace antibiotics as feed additives to maintain swine health and performance. Unfortunately, the vast majority of these compounds produce inconsistent results and rarely equal antibiotics in their effectiveness. Therefore, it would appear that research is still needed in this area and that the perfect alternative does not exist as yet.

Competing interests
The author declares they have no competing interests.

References

1. Cromwell GL: **Why and how antibiotics are used in swine production.** *Anim Biotechnol* 2002, **13**:7–27.
2. Vondruskova H, Slamova R, Trckova M, Zraly Z, Pavli I: **Alternatives to antibiotic growth promotors in prevention of diarrhea in weaned piglets: a review.** *Vet Med* 2010, **55**:199–224.
3. Van der Fels-Klerx HJ, Puister-Jansen LF, Van Asselt ED, Burgers SL: **Farm factors associated with the use of antibiotics in pig production.** *J Anim Sci* 2011, **89**:1922–1929.
4. Jacela JY, DeRouchey JM, Tokach MD, Goodband RD, Nelssen JL, Renter DG, Dritz SS: **Feed additives for swine: fact sheets-prebiotics and probiotics, and phytogenics.** *J Swine Health Prod* 2010, **18**:132–136.
5. Simon O: **An interdisciplinary study on the mode of action of probiotics in pigs.** *J Anim Feed Sci* 2010, **19**:230–243.
6. Cho JH, Zhao PY, Kim IH: **Probiotics as a dietary additive for pigs: a review.** *J Anim Vet Adv* 2011, **10**:2127–2134.
7. Halas V, Nochta I: **Mannan oligosaccharides in nursery pig nutrition and their potential mode of action.** *Animals* 2012, **2**:261–274.
8. Thacker PA: **Recent advances in the use of enzymes with special reference to β-glucanases and pentosanases in swine rations.** *Asian-Aust J Anim Sci* 2000, **13**:376–385 (Special Issue).
9. Jacela JY, DeRouchey JM, Tokach MD, Goodband RD, Nelssen JL, Renter DG, Dritz SS: **Feed additives for swine: fact sheets-carcass modifers, carbohydrate-degrading enzymes and proteases, and anthelmintics.** *J Swine Health Prod* 2009, **17**:325–332.
10. Adeola O, Cowieson AJ: **Opportunities and challenges in using exogenous enzymes to improve nonruminant animal production.** *J Anim Sci* 2011, **89**:3189–3218.
11. Jacela JY, DeRouchey JM, Tokach MD, Goodband RD, Nelssen JL, Renter DG, Dritz SS: **Feed additives for swine: fact sheets-acidifiers and antibiotics.** *J Swine Health Prod* 2009, **17**:270–275.
12. Kil DY, Kwon WB, Kim BG: **Dietary acidifiers in weanling pig diets: a review.** *Revista Colombian de Diencias Pecuarias* 2011, **24**:1–22.
13. Suruanarayana MV, Suresh J, Rajasekhar MV: **Organic acids in swine feeding: a review.** *Agric Sci Res J* 2012, **2**:523–533.
14. Papatsiros VG, Billinis C: **The prophylactic use of acidifiers as antibacterial agents in swine.** In *Antimicrobial agents.* Edited by Bobbarala V; 2012:295–310. InTech, DOI:10.5772/32278. Available from: http://www.intechopen.com/books/antimicrobial-agents/the-prophylactic-use-of-acidifiers-as-anti bacterial-agents-in-swine. ISBN 978-953-51-0723-1.
15. Windisch W, Schedle K, Plitzner C, Kroismayr A: **Use of phytogenic products as feed additives for swine and poultry.** *J Anim Sci* 2008, **86**(E. Suppl):E140–E148.
16. Liu HW, Tong JM, Zhou DW: **Utilization of Chinese herbal feed additives in animal production.** *Agric Sci China* 2011, **10**:1262–1272.
17. Pettigrew JE: **Reduced use of antibiotic growth promoters in diets fed to weanling pigs: dietary tools, part 1.** *Anim Biotechnol* 2006, **17**:207–215.
18. Jacela JY, DeRouchey JM, Tokach MD, Goodband RD, Nelssen JL, Renter DG, Dritz SS: **Feed additives for swine: fact sheets-high dietary levels of copper and zinc for young pigs, and phytase.** *J Swine Health Prod* 2010, **18**:87–91.

19. Li YM, Xiang Q, Zhang QH, Huang YD, Su ZJ: **Overview on the recent study of antimicrobial peptides: origins, functions, relative mechanisms and application.** *Peptides* 2012, **37:**207–215.
20. Koczulla AR, Bals R: **Antimicrobial peptides: current status and therapeutic potential.** *Drugs* 2003, **63:**389–406.
21. Wu SD, Zhang FR, Huang ZM, Liu H, Xie CY, Zhang J, Thacker PA, Qiao S: **Effect of the antibacterial peptide cecropin AD on performance and intestinal health in weaned piglets challenged with Escherichia coli.** *Peptides* 2012, **35:**225–230.
22. Yusuf MA, Hamid TH: **Lactic acid bacteria: bacteriocin producer: a mini review.** *IOSR J Pharm* 2013, **3:**44–50.
23. Stahl CH, Callaway TR, Lincoln LM, Lonergan SM, Genovese KJ: **Inhibitory activities of colicins against *Esherichia coli* strains responsible for postweaning diarrhea and edema disease in swine.** *Antimicrob Agents Chemother* 2004, **48:**3119–3121.
24. Cutler SA, Lonergan SM, Cornick N, Johnson AK, Stahl CH: **Dietary inclusion of colicin E1 is effective in preventing postweaning diarrhea caused by F18-positive Esherichia coli in pigs.** *Antimicrob Agents Chemother* 2007, **51:**3830–3835.
25. Yoon JH, Ingale SL, Kim JS, Kim KH, Lohakare J, Park YK, Park JC, Kwon LK, Chae BJ: **Effects of dietary supplementation with antimicrobial peptide-P5 on growth performance, apparent total tract digestibility, faecal and intestinal microflora and intestinal morphology of weanling pigs.** *J Sci Food Agric* 2013, **93:**587–592.
26. Yoon JH, Ingale SL, Kim JS, Kim KH, Lee SH, Park YK, Kwon IK, Chae BJ: **Effects of dietary supplementation of antimicrobial peptide-A3 on growth performance, nutrient digestibility, intestinal and fecal microflora and intestinal morphology in weanling pigs.** *Anim Feed Sci Technol* 2012, **177:**98–107.
27. Wang JH, Wu CC, Feng J: **Effect of dietary antibacterial peptide and zinc-methionine on performance and serum biochemical parameters in piglets.** *Czech J Anim Sci* 2011, **56:**30–36.
28. Schell TC, Lindemann MD, Kornegay ET, Blodgett DJ: **Effects of feeding aflatoxin-contained diets with and without clay to weanling and growing pigs on performance, liver function and mineral metabolism.** *J Anim Sci* 1993, **71:**1209–1218.
29. Schell TC, Lindemann MD, Kornegay ET, Blodgett DJ: **Effects of different types of clay for reducing the detrimental effects of aflatoxin-contained diets on performance and serum profiles of weanling pigs.** *J Anim Sci* 1993, **71:**1226–1231.
30. Trckova M, Vondruskova H, Zraly Z, Alexa P, Kummer V, Maskova J, Mrlik V, Krizova K, Slana I, Leva L, Pavlik I: **The effect of kaolin feeding on efficiency, health status and course of diarrheoal infections caused by enterotoxigenic *Esherichia coli* strains in weaned piglets.** *Vet Med* 2009, **54:**47–63.
31. Song M, Liu Y, Soares JA, Che TM, Osuna O, Maddox CW, Pettigrew JE: **Dietary clays alleviate diarrhea of weaned pigs.** *J Anim Sci* 2012, **90:**345–360.
32. Thacker PA: **Performance of growing-finishing pigs fed diets containing graded levels of Biotite, and alumninosilicate clay.** *Asian-Aust J Anim Sci* 2003, **16:**1666–1672.
33. Chen YJ, Kwon OS, Min BJ, Son KS, Cho JH, Hong JW, Kim IH: **The effects of dietary Biotite V supplementation as an alternative substance to antibiotics in growing pigs.** *Asian-Aust J Anim Sci* 2005, **18:**1642–1645.
34. Prvulovic D, Jovanovic-Galovic A, Stanic B, Popovic M, Grubor-Lajsic G: **Effects of a clinoptilolite supplement in pig diets on performance and serum parameters.** *Czech J Anim Sci* 2007, **52:**159–164.
35. Yan L, Hong SM, Kim IH: **Effect of bacteriophage supplementation on the growth performance, nutrient digestibility, blood characteristics, and fecal microbial shedding in growing pigs.** *Asian-Aust J Anim Sci* 2012, **25:**1451–1456.
36. Xu Y, Li X, Jin L, Zhen Y, Lu Y, Li S, You J, Wang L: **Application of chicken egg yolk immunoglobulins in the control of terrestrial and aquatic animal diseases: a review.** *Biotechnol Adv* 2011, **29:**860–868.
37. Chalghoumi R, Beckers Y, Portetelle D, Thewis A: **Hen egg yolk antibodies (IgY) production and use for passive immunization against bacterial enteric infections in chicken: a review.** *Biotechnol Agron Soc Environ* 2009, **13:**295–308.
38. Kovacs-Nolan J, Mine Y: **Egg yolk antibodies for passive immunity.** *Annu Rev Food Sci Technol* 2012, **3:**163–182.
39. Marquardt RR, Jin LZ, Kim JW, Fang L, Frohlich AA, Baidoo SK: **Passive protective effect of egg-yolk antibodies against enterotoxigenic *Esherichia coli* K88+ infection in neonatal and early-weaned piglets.** *FEMS Immunol Med Microbiol* 1999, **23:**283–288.
40. Owusu-Asiedu A, Nyachoti CM, Baidoo SK, Marquardt RR, Yang X: **Response of early-weaned pigs to an enterotoxigenic *Esherichia coli* (K88) challenge when fed diets containing spray-dried porcine plasma or pea protein isolate plus egg yolk antibody.** *J Anim Sci* 2003, **81:**1781–1789.
41. Owusu-Asiedu A, Nyachoti CM, Marquardt RR: **Response of early-weaned pigs to an enterotoxigenic *Esherichia coli* (K88) challenge when fed diets containing spray-dried porcine plasma or pea protein isolate plus egg yolk antibody, zinc oxide, fumaric acid or antibiotic.** *J Anim Sci* 2003, **81:**1790–1798.
42. Hong JW, Kwon OS, Min BJ, Lee WB, Shon KS, Kim IH, Kim JW: **Evaluation effects of spray-dried egg protein containing specific egg yolk antibodies as a substitute for spray-dried plasma protein or antibiotics in weaned pigs.** *Asian-Aust J Anim Sci* 2004, **17:**1139–1144.
43. Zhang ZF, Kim IH: **Effects of egg yolk immunoglobulin on growth performance, diarrhea score, diarrhea incidence and serum antibody titer in pre-and post-weaned pigs.** *Wayamba J Anim Sci* 2013, **578X:**590–597.
44. Chernysheva LV, Friendship RM, Dewey CE, Gyles CL: **The effect of dietary chicken egg-yolk antibodies on the clinical response in weaned pigs challenged with a K88+ *Esherichia coli* isolate.** *J Swine Health Prod* 2003, **12:**119–122.
45. Kovacs-Nolan J, Mine Y: **Microencapsulation for the gastric passage and controlled intestinal release of immunoglobulin Y.** *J Immunol Methods* 2005, **296:**199–209.
46. Li XY, Jin LJ, Uzonna JE, Li SY, Liu JJ, Li HQ, Lu YN, Zhen YH, Xu YP: **Chitosan-alginate microcapsules for oral delivery of egg yolk immunoglobulin (IgY): in vivo evaluatin in a pig model of enteric colibacillosis.** *Vet Immunol Immunopathol* 2009, **129:**132–136.
47. Li XY, Jin LJ, McAllister TA, Stanford K, Xu JY, Lu YN, Zhen YH, Sun YX, Xu YP: **Chitosan-alginate microcapsules for oral delivery of egg yolk immunoglobulin (IgY).** *J Agric Food Chem* 2007, **55:**2911–2917.
48. Stein HH, Kil DY: **Reduced use of antibiotic growth promoters in diets fed to weanling pigs: dietary tools, Part 2.** *Anim Biotechnol* 2006, **17:**217–231.
49. Gatnau R: *Use of plant extracts in swine*; 2009. Available at http://www.pig333.com/nutrition/use-of-plant-extracts-in-swine_957/.
50. Michiels J, Missotten JA, Fremaut D, De Smet S, Dierick NA: **In vitro characterization of the antimicrobial activity of selected essential oil components and binary combinations against the pig gut flora.** *Anim Feed Sci Technol* 2009, **151:**111–127.
51. Brenes A, Roura E: **Essential oils in poultry nutrition: main effects and modes of action.** *Anim Feed Sci Technol* 2010, **158:**1–14.
52. Ragland D, Stevenson D, Hill MA: **Oregano oil and multi-component carbohydrases as alternatives to antimicrobials in nursery diets.** *Swine Health Prod* 2008, **16:**238–243.
53. Cho JH, Chen YJ, Min BJ, Kim HJ, Kwon OS, Shon KS, Kim IH, Kim SJ, Asamer A: **Effects of essential oils supplementation on growth performance, IgG concentration and fecal noxious gas concentration of weaned pigs.** *Asian-Aust J Anim Sci* 2006, **19:**80–85.
54. Maenner K, Vahjen W, Simon O: **Studies on the effects of essential-oil based feed additives on performance, ileal nutrient digestibility, and selected bacterial groups in the gastrointestinal tract of piglets.** *J Anim Sci* 2011, **89:**2106–2112.
55. Li PF, Piao XS, Ru YJ, Han X, Xue LF, Zhang HY: **Effects of adding essential oil to the diet of weaned pigs on performance, nutrient utilization, immune response and intestinal health.** *Asian-Aust J Anim Sci* 2012, **25:**1617–1626.
56. Ahmed ST, Hossain ME, Kim GM, Hwang JA, Ji H, Yang CJ: **Effect of resveratrol and essential oils on growth performance, immunity, digestibility and fecal microbial shedding in challenged piglets.** *Asian-Aust J Anim Sci* 2013, **26:**683–690.
57. Huang Y, Yoo JS, Kim HJ, Wang Y, Chen YJ, Cho JH, Kim IH: **Effects of dietary supplementation with blended essential oils on growth performance, nutrient digestibility, blood profiles and fecal characteristics in weanling pigs.** *Asian-Aust J Anim Sci* 2010, **23:**607–613.
58. Salari MH, Amine G, Shirazi MH, Hafezi R, Mohammadypour M: **Antibacterial effects of Eucalyptus globulus leaf extract on pathogenic bacteria isolated from specimens of patients with respiratory tract disorders.** *Clin Microbiol Infect* 2006, **12:**194–196.
59. Serafino A, Vallebona PS, Andreola F, Zonfrillo M, Mercuri L, Federici M, Rasi G, Garaci E, Pierimarchi P: **Stimulatory effect of Eucalyptus essential oil on innate cell-mediated immune response.** *BMC Immunol* 2008, **9:**17.

60. Abd El-Motaal AM, Ahmed AMH, Bahakaim ASA, Fathi MM: **Productive performance and immunocompetence of commercial laying hens given diets supplemented with Eucalyptus.** *Int J Poult Sci* 2008, **7**:445–449.
61. Dierick NA, Decuypere JA, Molly K, Van Beek E, Vanderbecke E: **The combined use of triacylglycerols containing medium-chain fatty acids and exogenous lipolytic enzymes as an alternative for nutritional antibiotics in piglet nutrition: I: in vitro screening of the release of MCFAs from selected fat sources by selected exogenous lipolytic enzymes under simulated pig gastric conditions and their effects on the gut flora of piglets.** *Livest Prod Sci* 2002, **75**:129–142.
62. Dierick NA, Decuypere JA, Molly K, Van Beek E, Vanderbecke E: **The combined use of triacylglycerols containing medium-chain fatty acids and exogenous lipolytic enzymes as an alternative for nutritional antibiotics in piglet nutrition: II: in vivo release of MCFAs in gastric cannulated and slaughtered piglets by endogenous and exogenous lipases: effects on the luminal gut flora and growth performance.** *Livest Prod Sci* 2002, **76**:1–16.
63. Hong SM, Hwang JH, Kim IH: **Effect of medium-chain triglyceride (MCT) on growth performance, nutrient digestibility, blood characteristics in weanling pigs.** *Asian-Aust J Anim Sci* 2012, **25**:1003–1008.
64. Rossi R, Pastorelli G, Cannata S, Corino C: **Recent advances in the use of fatty acids as supplements in pig diets: a review.** *Anim Feed Sci Technol* 2010, **162**:1–11.
65. Han YK, Hwang UH, Thacker PA: **Use of a micro-encapsulated medium chain fatty acid product as an alternative to zinc oxide and antibiotics for weaned pigs.** *J Swine Health Prod* 2011, **19**:34–43.
66. Redling K: *Rare earth elements in agriculture with emphasis on animal husbandry.* Muenchen: Diss Ludwig-Maximilians-Universitaet; 2006:325.
67. Zhu X, Li D, Yang W, Xiao C, Chen H: **Effects of rare earth elements on the growth and nitrogen balance of piglets.** *Feed Industry* 1994, **15**:23–25.
68. He R, Xia Z: **Effects of rare earth elements on growing and fattening of pigs.** *Guangxi Agric Sci* 1998, **5**:243–245.
69. Rambeck WA, Wehr U: **Rare earth elements as alternative growth promoters in pig production.** *Arch Tierernahr* 2000, **53**:323–334.
70. He ML, Rambeck WA: **Rare earth elements: a new generation of growth promoters for pigs.** *Arch Anim Nutr* 2000, **53**:323–334.
71. He ML, Ranz D, Rambeck WA: **Study on the performance enhancing effect of rare earth elements in growing and finishing pigs.** *J Anim Physiol Anim Nutr* 2001, **85**:263–270.
72. Kraatz M, Taras D, Manner K, Simon O: **Weaning pig performance and faecal microbiota with and without in-feed addition of rare earth elements.** *J Anim Physiol Anim Nutr* 2006, **90**:361–368.
73. Han YK, Thacker PA: **Effect of antibiotics, zinc oxide and rare earth mineral yeast on performance, nutrient digestibility and blood parameters in weaned pigs.** *Asian-Aust J Anim Sci* 2010, **23**:1057–1065.
74. Buhler M, Limper J, Muller A, Schwarz G, Simon O, Sommer M, Spring W: *Enzymes in animal nutrition.* German Feed Additives Association Fact Sheet; 2013. Available at http://www.awt-feedadditives.de/Publikationen/Enzymbro-engl.pdf.
75. Choct M: *Feed non-starch polysaccharides: chemical structures and nutritional significance.* Singapore: Proceedings of the Feed Ingredients Asia 97 Conference; 1997.
76. Metzler B, Bauer B, Mosenthin R: **Microflora management in the gastrointestinal tract of piglets.** *Asian-Aust J Anim Sci* 2005, **18**:1353–1362.
77. Kiarie E, Nyachoti CM, Slominski BA, Blank G: **Growth performance, gastrointestinal microbial activity, and nutrient digestibility in early-weaned pigs fed diets containing flaxseed and carbohydrase enzyme.** *J Anim Sci* 2007, **85**:2982–2993.
78. Emiola IA, Opapeju FO, Slominski BA, Nyachoti CM: **Growth performance and nutrient digestibility in swine fed wheat distillers dried grains with solubles-based diets supplemented with a multi-carbohydrase enzyme.** *J Anim Sci* 2009, **87**:2315–2322.
79. Thacker PA: **Effect of enzyme supplementation on the performance of growing-finishing pigs fed barley based diets supplemented with soybean meal or canola meal.** *Asian-Aust J Anim Sci* 2001, **14**:1008–1013.
80. Olukosi OA, Sands JS, Adeola O: **Supplementation of carbohydrases or phytase individually or in combination to diets for weanling and growing-finishing pigs.** *J Anim Sci* 2007, **85**:1702–1711.
81. Jones CK, Bergstrom JR, Tokach MD, DeRouchey JM, Goodband RD, Nelssen JL, Dritz SS: **Efficacy of commercial enzymes in diets containing various concentrations and sources of dried distillers grains with solubles for nursery pigs.** *J Anim Sci* 2010, **88**:2084–2091.
82. Baas TC, Thacker PA: **Impact of gastric pH on dietary enzyme activity and survivability in swine fed β-glucanase supplemented diets.** *Can J Anim Sci* 1996, **76**:245–252.
83. Thacker PA, Baas TC: **Effect of gastric pH on the activity of exogenous pentosanase and the effect of pentosanase supplementation of the diet on the performance of growing-finishing pigs.** *Anim Feed Sci Technol* 1996, **63**:187–200.
84. He J, Yin J, Wang L, Yu B, Chen D: **Functional characterization of a recombinant xylanase from Pichia pastoris and effect of the enzyme on nutrient digestibility in weaned pigs.** *Brit J Nutr* 2010, **103**:1507–1513.
85. Lv JN, Chen YQ, Guo XJ, Piao XS, Cao YH, Dong B: **Effects of supplementation of β-mannanase in corn-soybean meal diets on performance and nutrient digestibility in growing pigs.** *Asian-Aust J Anim Sci* 2013, **26**:579–587.
86. Cai H, Shi P, Luo H, Bai Y, Huang H, Yang P, Yao B: **Acidic β-mannanase from penicillium pinophilum C1: cloning, characterization and assessment of its potential for animal feed application.** *J Biosci Bioeng* 2011, **112**:551–557.
87. Pettey LA, Carter SD, Senne BW, Shriver JA: **Effect of beta-mannanase adding to corn-soybean meal diets on growth performance, carcass traits and apparent nutrient digestibility in growing-finishing pigs.** *J Anim Sci* 2002, **80**:1012–1019.
88. Van Dijk AJ, Everts H, Nabuurs MJA, Margry RJ, Beynen AC: **Growth performance of weanling pigs fed spray-dried animal plasma: a review.** *Livest Prod Sci* 2001, **68**:263–274.
89. Torrallardona D: **Spray dried animal plasma as an alternative to antibiotics in weanling pigs: a review.** *Asian-Aust J Anim Sci* 2010, **23**:131–148.
90. Bontempo V, Di Giancamillo A, Savoini G, Dell'Orto V, Domeneghini C: **Live yeast supplementation acts upon intestinal morpho-functional aspects and growth in weanling piglets.** *Anim Feed Sci Technol* 2006, **129**:224–236.
91. Li JY, Li DF, Gong LM, Ma YX, He YH, Zhai HX: **Effects of live yeast on the performance, nutrient digestibility, gastrointestinal microbiota and concentration of volatile fatty acids in weanling pigs.** *Arch Anim Nutr* 2006, **60**:277–288.
92. Shen YB, Piao XS, Kim SW, Wang L, Liu P, Yoon I, Zhen YG: **Effects of yeast culture supplementation on growth performance, intestinal health, and immune response of nursery pigs.** *J Anim Sci* 2009, **87**:2614–2624.
93. Yan L, Han DL, Meng QW, Lee JH, Park CJ, Kim IH: **Effects of anion supplementation on growth performance, nutrient digestibility, meat quality and fecal noxious gas content in growing-finishing pigs.** *Asian-Aust J Anim Sci* 2010, **23**:1073–1079.
94. Nyachoti CM, Kiarie E, Bhandari SK, Zhang G, Krause DO: **Weaned pig responses to Esherichia coli K88 oral challenge when receiving a lysozyme supplement.** *J Anim Sci* 2012, **90**:252–260.
95. Huguet A, Le Dividich J, Le Huerou-Luron I: **Improvement of growth performance and sanitary status of weaned piglets fed a bovine colostrum-supplemented diet.** *J Anim Sci* 2012, **90**:1513–1520.
96. Wang Y, Shan T, Xu Z, Liu J, Feng J: **Effect of lactoferrin on the growth performance, intestinal morphology and expression of PR-39 and protegrin-1 genes in weaned piglets.** *J Anim Sci* 2006, **84**:2636–2641.
97. Shan T, Wang Y, Liu J, Xu Z: **Effect of dietary lactoferrin on the immune functions and serum iron level of weanling piglets.** *J Anim Sci* 2007, **85**:2140–2146.
98. Garcia-Montoya IA, Cendon TS, Arevalo-Gallegos S, Rascon-Cruz Q: **Lactoferrin a multiple bioactive protein: an overview.** *Biochim Biophys Acta* 1820, **2012**:226–236.
99. Lai CH, Yin JD, Li DF, Zhao LD, Qiao SY, Xing JJ: **Conjugated linoleic acid attenuates the production and gene expression of pro-inflammatory cytokines in weaned pigs challenged with lipopolysaccharide.** *J Nutr* 2005, **135**:239–244.
100. Lai CH, Yin JD, Li DF, Zhao LD, Chen XJ: **Effects of dietary conjugated linoleic acid supplementation on the performance and immunological responses of weaned pigs after an escherichia coli lipopolysaccharide challenge.** *J Anim Vet Adv* 2005, **2**:299–305.
101. Liu P, Piao XS, Thacker PA, Zheng ZK, Wong D, Kim SW: **Chito-oligosaccharide reduces diarrhea incidence and attenuates the immune response of weanling pigs challenged with E. coli K88.** *J Anim Sci* 2010, **88**:3871–3879.
102. Liu P, Piao XS, Kim SW, Li XJ, Wang L, Shen YB, Lee HS, Li SY: **Effects of chito-oligosaccharide supplementation on the growth performance, nutrient digestibility, intestinal morphology, and fecal shedding of**

Escherichia coli and Lactobacilli in weaning pigs. *J Anim Sci* 2008, **86:**2609–2618.

103. O'Doherty JV, Dillon S, Figat S, Callan JJ, Sweeney T: **The effects of lactose inclusion and seaweed extract derived from *Laminaria spp.* on performance, digestibility of diet components and microbial populations in newly weaned pigs.** *Anim Feed Sci Technol* 2010, **157:**173–180.

Effects of ensiling processes and antioxidants on fatty acid concentrations and compositions in corn silages

Liying Han and He Zhou*

Abstract

Background: Corn silage is the main dietary component used for ruminant breeding in China and is an important dietary source of fatty acids for these animals. However, little is known regarding effective means to protect the fatty acid (FA) contents in silages. In this study, we examined the changes in FA contents and compositions during corn ensiling and screened several antioxidants for their inhibition of lipid oxidation during corn ensiling.

Methods: We conducted two different experiments. In Experiment 1, corn was ensiled in 30 polyethylene bottles (bottle volume: 1 L, silage density: 600 g/dm^3) and three bottles were opened at 0.5 d, 1 d, 1.5 d, 2 d, 2.5 d, 3 d, 5 d, 7 d, 14 d, and 28 d after ensiling. In Experiment 2, corn was treated with various antioxidants: (1) No additives (CK); (2) BHA (Butylated hydroxyanisole); (3) TBHQ (Tertiary butyl hydroquinone); (4) TPP (Tea polyphenols); and (5) VE (Vitamin E). These treatments were applied at 50 mg/kg and 100 mg/kg of fresh weight with each treatment replicated 3 times.

Results: During ensiling in Experiment 1, saturated fatty acids (SFA; C16:0 and C18:0) and malondialdehyde (MDA) contents tended to increase, whereas unsaturated fatty acids (UFA; C18:1, C18:2 and C18:3) tended to decrease. However, these changes were only significant on the first 2 days of ensiling. In Experiment 2, all of the antioxidants tested affected the total FA contents and those of unsaturated fatty acids (C18:1, C18:2 and C18:3) and MDA. The effects of TBHQ and TPP were greater than those of the other antioxidants.

Conclusions: The reduced total FA contents in corn silages were due to unsaturated fatty acids' oxidation during the early stages of ensiling. Adding an antioxidant could prevent fatty acids' oxidation in corn silages.

Keywords: Antioxidant, Fatty acid, Silage

Background

Corn silage is the main dietary component used for ruminant breeding in China and is an important dietary source of fatty acids for these animals. However, the ensiling process may reduce the positive effects of herbage lipids on the fatty acid (FA) composition of milk due to oxidation during the period between plant cutting and ensiling [1,2]. During harvesting and the early stages of ensiling, enzymatic hydrolysis of triacylglycerols yield free fatty acids (FFAs) from damaged tissues after cutting and non-esterified polyunsaturated fatty acids (PUFA) from damaged membranes are rapidly converted to hydroperoxy PUFA by the actions of lipoxygenases (LOX) [3]. The most common substrates for plant lipoxygenases are linolenic and linoleic acids. These are abundant in plant membranes and can be further broken down into aldehydes and ketones [4], which may affect feed preferences, palatability, and ingestion by animals [5,6].

Numerous studies have been done on chemically characterizing silages and their nutritional value in order to obtain high quality silages [7-9]. Dewhurst et al. [10] concluded that plant species and cutting intervals affected the FA compositions of grasses. The effects of additives, such as formalin, formic acid, inoculants, and enzymes, on the FA compositions of grass silages have also been investigated [11-14], although these results suggested that they only had minimal effects on the FA contents.

* Correspondence: zhouhe@cau.edu.cn
College of Animal Science and Technology, China Agricultural University, Yuanmingyuan West Road 2#, Haidian District Beijing, PRC, 100193, China

Lourenco et al. [15] and Lee et al. [16] attributed reduced lipid oxidation in red or white clover silages to their high polyphenol oxidase (PPO) contents. However, due to wilting and ensiling, ensiled forages contain fewer antioxidants as compared to fresh pasture forage [13,17]. Some antioxidant phenolic compounds have been widely used to inhibit lipid oxidation in the food industry. Thus, it might be possible to use these as silo lipid oxidation inhibitors.

Thus, in this study, we examined the changes in FA contents and compositions during corn ensiling and screened several antioxidants for their inhibition of lipid oxidation during corn ensiling.

Methods

Corn and ensiling

We used corn for our experiments (*Zea mays* L., JKN928), which was sown on April 25, 2012. The temperature range and total precipitation during the growing season were 16.2–27.4°C and 94.5 mm, respectively. We selected a high cutting height for corn (1 m above ground) in order to increase the FA contents during ensiling. To determine their compositions, 10 plants from 10 randomly selected sites were sampled, chopped, and stored at -80°C. Corn, including the ear, was harvested at the one-half milk line stage (August 5, 2012) and chopped into 10 mm lengths using a conventional forage harvester. Then, we used two different experiments.

In Experiment 1, corn was ensiled in 30 polyethylene bottles (bottle volume: 1 L, silage density: 600 g/dm^3) in the dark at 25 ± 2°C. Each treatment was replicated 3 times. Three bottles were opened after having been ensiled for 0.5 d, 1 d, 1.5 d, 2 d, 2.5d, 3 d, 5 d, 7 d, 14d, and 28 d. About 400 g samples from each bottle were removed and vacuum packed at −18°C to determine the fermentation quality, FA contents and compositions, and malondialdehyde (MDA) contents.

In Experiment 2, corn was divided into equal portions for different treatments. These treatments were: (1) No additives (CK); (2) Butylated hydroxyanisole (BHA, synthetic antioxidant); (3) Tertiary butyl hydroquinone (TBHQ, synthetic antioxidant); (4) Tea polyphenols (TPP, natural antioxidant); and (5) Vitamin E (VE, natural antioxidant). These treatments were applied at 50 mg/kg and 100 mg/kg of fresh weight. All antioxidants were purchased from Beijing Sky Bamboo Bird Food Additives Co., Ltd. (Beijing, China). The antioxidants were diluted with distilled water to obtain the designated application concentrations and sprayed onto fresh corn. For a control, the same amount of distilled water was sprayed onto corn samples. About 200 g of each treated or untreated corn sample was frozen immediately in liquid nitrogen and used to determine LOX activity.

Treated and untreated corn samples were ensiled in polyethylene bottles (bottle volume: 1 L, silage density: 600 g/dm^3) in the dark at 25 ± 2°C for 60 d. Each treatment was replicated 3 times. The FA contents and compositions (C16:0, C18:0, C16:1, C18:1, C18:2, C18:3), and the contents of MDA, fermentation quality, dry matter (DM), water soluble carbohydrates (WSC), crude proteins (CP), neutral detergent fiber (NDF), and acid detergent fiber (ADF) in these silages were determined when the silage bottles were opened.

Chemical analyses

Using 250 μmol/L linolenic acid as the substrate, lipoxygenase (LOX) activity was determined as the increase in absorbance at 234 nm due to the formation of conjugated dienes using a spectrophotometer over 5–10 min [18]. A sample (1 g) was diluted in 50 mmol/L Na phosphate buffer (pH 7.0) and incubated for 30 min on ice with occasional vortexing. The sample was then centrifuged (10,000 rpm) at 4°C for 30 min and the supernatant was used as a crude enzyme solution. Protein concentrations in the enzyme solutions were determined using the Bradford method (Bio-Rad, Hercules, CA, USA).

To initiate the assay, 0.05 mL of an enzyme extract was mixed with 0.25 mL of substrate stock solution, followed by incubation at 30°C for 4 min. After incubation, 1 mol/L NaOH (0.7 mL) was added to stop the reaction. Hydroperoxides produced by LOX were monitored using a spectrophotometer (Thermo Electron Co., PA, USA) at 234 nm. One unit of enzyme activity was defined as an increase in absorbance of 0.001 at 234 nm per mg of protein per minute (Units/mg protein/min). Protein concentrations were determined using the Coomassie Brilliant Blue method [19].

FA compositions were determined by gas chromatography (GC) after methylation [20]. GC analyses were done using a Shimadzu GC-2010 chromatograph equipped with an Agilent chromatography column for FA methyl esters (FAME) (100 m × 0.25 mm × 0.2 μm). The temperature program was: starting temperature of 180°C for 10 min, which was then increased by 4°C/min until the temperature reached 200°C; the injector temperature was set at 250°C; and the detector temperature was 280°C. Malondialdehyde (MDA) was determined spectrophotometrically as thiobarbituric acid reactive substances (TBARS) after reaction with thiobarbituric acid (TBA) at 100°C in acidic media; the absorbance of a reaction mixture was measured at 532 nm [21].

Fermentation indices were determined using the following methods. A sample silage (20 g) was homogenized in 180 mL of distilled water for 1 min at high speed (12,000 rpm). The resulting suspension was filtered through four layers of cheese cloth and then centrifuged for 20 min at 27,500 × g, after which the pellet discarded.

Supernatant samples were used for pH, lactic acid, acetic acid, propionic acid, butyric acid, and NH_3-N analyses. pH was determined with a pH meter (PHS-3C). Lactic acid, acetic acid, propionic acid, and butyric acid were determined by HPLC (SHIMADZE-10A, Shimadze, Japan) as described [22]. The HPLC system included a Shimadzu system controller (SCL-10A) and a Shodex Rspak KC-811 S-DVB gel column (300 mm × 8 mm) at a column temperature of 50°C. The mobile phase was a solution of 3 mmol perchloric acid at a rate of 1 mL/min. The injection volume was 50 μL. A UV detector (SPD-10A) was used and analyses were made at 210 nm.

Ammonia-N (NH_3-N) was determined by the Phenol-Hypochlorite colorimetric method as described [23]. DM was determined by oven drying at 65°C for 48 h. Crude protein (CP) was determined using the Kjeldahl method [24]. NDF and ADF were analyzed as described [25]. Water-soluble carbohydrate (WSC) was determined by the Deriaz method [26].

Statistical analysis

Statistical comparisons were made by one-way analysis of variance followed by Duncan's new multiple range test [27]. Results for similar treatments at different time points were compared using paired t-tests. Statistical analyses were done using SAS 9.1.3 software (SAS Institute, Cary, NC, USA). $P < 0.05$ was considered significant.

Table 1 DM, WSC, CP, NDF, ADF, FA contents and compositions in fresh whole plant corn

Items	N[1]	Mean	SD[2]
DM, g/kg	10	283.3	12.6
WSC, g/kg DM	10	130.4	4.0
CP, g/kg DM	10	115.4	7.5
NDF, g/kg DM	10	384.1	4.5
ADF, g/kg DM	10	271.2	2.1
Total FA content, mg/kg DM	10	24.74	0.61
Proportion in DM, mg/kg DM			
C16:0	10	4.16	0.16
C18:0	10	1.61	0.11
C18:1	10	3.86	0.27
C18:2	10	12.97	0.14
C18:3	10	1.85	0.08
Proportion in total FA,%			
C16:0	10	16.81	1.56
C18:0	10	6.51	0.73
C18:1	10	14.87	1.03
C18:2	10	52.43	3.45
C18:3	10	7.48	0.74

[1]*N*, observation numbers.
[2]*SD*, standard deviation.

Results

Material compositions

Dry matter (DM) contents and the chemical compositions (WSC, CP, NDF, ADF, and FA contents and compositions) of fresh chopped whole corn plants before ensiling are shown in Table 1. The DM contents and WSC of the corn used in this study were higher than the recommended contents to ensure successful ensiling and to obtain good fermentation rates [28,29]. High CP and total FA contents and low NDF and ADF contents were found because of the high cutting height used in this study (1 m above ground). More than one half of the FA in fresh corn was C18:2.

Fatty acid changes during ensiling

The FA contents and compositions and the MDA contents during ensiling are shown in Table 2. No differences were found for DM during ensiling, whereas significant decreases in WSC and pH had occurred, as was expected. Total FA contents decreased markedly during the first two days of ensiling. Significant changes in C16:0, C18:0, C18:1, C18:2, and C18:3 compositions were also found during the first two or three days of ensiling. Saturated fatty acid (SFA; C16:0 and C18:0) compositions tended to increase, whereas unsaturated fatty acid (UFA; C18:1, C18:2, and C18:3) compositions tended to decrease. The greatest decrease in the proportion of C18:2 among total FA occurred on the 28th day of ensiling. The proportion of MDA increased as the number of days of ensiling increased, although this was only significant during the first two days.

Lipoxygenase (LOX) activity

LOX activity results are shown in Table 3. The LOX activities of treated corn were lower than that of the control. LOX activity decreased as the concentration of additives increased for most treatments used. However, LOX activities were not detected when the added concentrations of TPP and BTHQ were 100 mg/kg.

Effects of antioxidants

The fermentation quality and chemical compositions of the corn silages after 60 d of conservation are shown in Table 4. The pH values of all the silages when bottles were opened were < 4.0. The lactic acid (LA) contents were high, and there was little butyric acid (BA) (< 0.1 g/kg DM). All of the treatments used affected ammoniacal nitrogen/total nitrogen (NH_3-N/TN) and WSC contents. There were no significant differences between treated and untreated silages for DM, CP, NDF, and ADF contents.

The effects of the different antioxidants on the FA contents and compositions and the MDA contents are shown in Table 5. All of these antioxidants affected the

Table 2 Effect of the ensiling process on total FA content and composition, MDA , DM and WSC content in corn silage

Treatments	0.5d	1d	1.5d	2d	3d	5d	7d	14d	28d	SEM[1]
pH	4.61^{a}	4.25^{b}	4.08^{c}	3.85^{d}	3.76^{de}	3.73^{de}	3.70^{e}	3.69^{e}	3.68^{c}	0.06
Total FA, mg/kg DM	23.68^{a}	23.04^{a}	21.73^{b}	19.56^{c}	19.43^{c}	19.45^{c}	19.51^{c}	19.57^{c}	19.75^{c}	0.76
Proportion in DM, mg/kg DM										
C16:0	4.19^{c}	4.27^{b}	4.36^{b}	4.55^{a}	4.56^{a}	4.58^{a}	4.59^{a}	4.62^{a}	4.66^{a}	0.18
C18:0	1.73^{c}	1.78^{c}	1.88^{b}	1.85^{b}	1.91^{a}	2.04^{a}	2.02^{a}	2.05^{a}	2.09^{a}	0.09
C18:1	3.34^{a}	3.01^{b}	2.99^{b}	2.57^{c}	2.45^{c}	2.48^{c}	2.46^{c}	2.45^{c}	2.56^{c}	0.21
C18:2	11.92^{a}	10.64^{b}	9.74^{c}	8.53^{d}	8.34^{d}	8.23^{d}	8.18^{d}	8.19^{d}	8.21^{d}	0.26
C18:3	1.78^{a}	1.59^{b}	1.44^{c}	1.32^{d}	1.31^{d}	1.27^{d}	1.23^{d}	1.19^{d}	1.18^{d}	0.20
Proportion in total FA,%										
C16:0	17.69^{d}	18.53^{c}	20.06^{b}	23.26^{a}	23.46^{a}	23.54^{a}	23.52^{a}	23.60^{a}	23.59^{a}	0.41
C18:0	7.31^{d}	7.73^{d}	8.65^{c}	9.46^{b}	9.83^{a}	10.49^{a}	10.35^{a}	10.47^{a}	10.58^{a}	0.52
C18:1	14.10^{a}	13.06^{b}	12.84^{b}	13.14^{b}	12.61^{b}	12.24^{b}	12.61^{b}	12.52^{b}	12.96^{b}	0.35
C18:2	50.33^{a}	46.18^{b}	44.82^{c}	43.61^{d}	42.92^{d}	42.31^{d}	41.93^{d}	41.84^{d}	41.56^{d}	0.62
C18:3	7.52^{a}	6.90^{b}	6.63^{b}	6.74^{b}	6.74^{b}	6.53^{b}	6.30^{b}	6.08^{b}	5.97^{b}	0.24
MDA, μmol/g FW	35.12^{d}	47.33^{c}	60.82^{b}	72.11^{a}	72.18^{a}	73.39^{a}	73.69^{a}	74.60^{a}	74.89^{a}	0.52
DM, g/kg FW	28.83	28.45	27.72	27.69	28.71	28.16	29.80	28.00	27.72	1.69
WSC, g/kg DM	9.93^{a}	9.62^{a}	9.24^{a}	6.73^{b}	6.71^{b}	5.27^{bc}	4.76^{c}	2.29^{d}	2.01^{d}	0.86

^{a,b,c,d,e}Within a row, means without a common superscript differ ($P < 0.05$).
[1]*SEM*, standard error mean.

Table 3 Effect of different antioxident additives on the LOX activity in corn silage

Additives	Amount	LOX activity,%
BHA	0, Ck[1]	100^{a}
	50 mg/kg	46.32^{b}
	100 mg/kg	23.71^{c}
SE[2]		9.15
TBHQ	0, Ck	100^{a}
	50 mg/kg	26.15^{b}
	100 mg/kg	0.00^{c}
SE		8.47
TPP	0, Ck	100^{a}
	50 mg/kg	22.04^{b}
	100 mg/kg	0.00^{c}
SE		10.26
VE	0, Ck	100^{a}
	50 mg/kg	37.69^{b}
	100 mg/kg	15.33^{c}
SE		9.15

^{a,b,c}Within a column, means without a common superscript differ ($P < 0.05$).
[1]The LOX activity in CK was considered as 100%.
[2]*SE*, standard error.

total FA contents and treatments with TBHQ100 and TPP100 were better than the other antioxidants. There were no significant differences in saturated fatty acids (C16:0 and C18:0) between treated and untreated silages, whereas the unsaturated fatty acid (C18:1, C18:2 and C18:3) compositions of treated silages were higher than those of the control. The MDA contents in treated silages were lower than in the control.

Discussion

Corn that is ensiled along with the ear can improve the nutritional value of feed because the high WSC contents should increase bacterial activity. In this study, high WSC and CP contents and low NDF and ADF contents resulted because of the high cutting height we used. Linoleic acid (C18:2) in corn increases as the plant matures, whereas the C18:3 concentration progressively declines [30]. Thus, C18:2 corresponded to more than 50% of the total FA in this study, which was in agreement with Shingfield et al. [13].

Lipid oxidation depends on the activity of lipoxygenases. Lipoxygenase activity is found in a plant at all growth stages and this activity increases when a plant enters the mature and senescence stages or after tissue injury. UFA's that are hydrolyzed from lipids are oxidized by LOX to form hydroperoxides, which are further decomposed into aldehydes and ketones. Malondialdehyde (MDA) is primarily produced via the lipoxygenase pathway of fatty acid oxidation and is widely used in food science as an

Table 4 Effect of different antioxidants on the formation of corn silage

Treatments	pH	LA g/kg DM	AA g/kg DM	BA g/kg DM	NH_3-N g/kg TN	DM g/kg FW	WSC g/kg DM	CP g/kg DM	NDF g/kg DM	ADF g/kg DM
CK	3.75	46.3	24.1	<0.1	36.1^{a}	281.7	10.3c	132.4	394.1	221.7
BHA50	3.62	47.1	23.4	<0.1	24.1^{b}	285.3	13.7b	139.8	381.5	218.5
BHA100	3.60	46.9	22.5	<0.1	21.2^{c}	290.7	14.3a	145.2	378.6	209.5
TBHQ50	3.71	46.4	22.0	<0.1	25.4^{b}	282.4	13.2b	138.6	385.3	219.6
TBHQ100	3.70	46.3	21.3	<0.1	20.1^{c}	291.1	14.1a	140.9	382.6	205.4
TPP50	3.72	48.2	24.8	<0.1	20.3^{c}	292.1	14.5a	141.5	379.5	212.5
TPP100	3.70	47.8	23.9	<0.1	19.7^{c}	296.9	15.2a	146.3	375.1	200.1
VE50	3.74	47.6	25.0	<0.1	21.1^{c}	288.6	12.9b	136.6	388.9	214.7
VE100	3.71	47.5	24.5	<0.1	20.5^{c}	294.3	14.7a	142.5	379.1	203.5
[1]SEM	0.42	4.13	2.51	–	1.33	6.72	1.72	4.12	7.25	6.57

[a,b,c]Within a column, means without a common superscript differ ($P < 0.05$).
[1]*SEM*, standard error mean.

index of lipid oxidation and rancidity in foods and food products [31]. Thus, for this study, MDA was determined as a product of lipid peroxidation and used as an index of lipid oxidation.

The differences in FA compositions could be related to various factors, such as plant species, cutting date, wilting, and the ensiling process [10,11,32-34]. Plant enzymes can remain functional in silages, although the activity of plant enzymes generally declines during ensiling. According to Elgersma et al [35], almost all of the total fat in fresh grass is in the form of esterified fatty acids, whereas in silage, a large proportion is in the form of free fatty acids (FFAs). These FFAs are then further oxidized by LOX.

UFAs are more susceptible to oxidation than are SFAs. Thus, the decrease in total FA contents together with the increase in SFA contents and the decreases in UFA contents on 28 d of corn ensiling were due to UFA oxidation, particularly that of C18:2. Malondialdehyde (MDA), one of the final products of lipid oxidation, increased during ensiling, which further indicated FA oxidation during ensiling. The rapid changes in FA contents and compositions mainly occurred during the first two days of ensiling in this study. This indicated that enzyme activity was higher during the first two days than during the other ensiling periods. This may have been related to the pH changes that occurred during ensiling.

Table 5 Effect of diierent antioxidants on the FA content and composition and MDA content of corn silages (mg/kg DM)

Treatments	Total FA	C16:0	C18:0	C18:1	C18:2	C18:3	MDA, µmol/g FW
CK	19.37^{c}	3.34	1.62	2.89^{c}	9.06^{d}	1.52^{c}	68.99^{a}
BHA50	21.81^{b}	3.36	1.65	3.06^{b}	10.73^{c}	1.65^{b}	46.69^{b}
BHA100	22.12^{a}	3.29	1.56	3.15^{b}	11.65^{b}	1.67^{b}	34.45^{c}
TBHQ50	22.74^{a}	3.41	1.58	2.99^{b}	11.68^{b}	1.79^{a}	32.75^{c}
TBHQ100	23.04^{a}	3.30	1.51	3.11^{b}	11.95^{a}	1.82^{a}	32.31^{c}
TPP50	22.56^{a}	3.35	1.55	3.22^{b}	11.62^{b}	1.80^{a}	32.44^{c}
TPP100	23.05^{a}	3.22	1.50	3.56^{a}	12.02^{a}	1.85^{a}	31.82^{c}
VE50	21.95^{b}	3.38	1.60	3.02^{b}	10.69^{c}	1.63^{b}	45.68^{b}
VE100	22.74^{a}	3.25	1.57	3.23^{b}	11.33^{b}	1.72^{b}	34.12^{c}
[1]SEM	0.65	0.17	0.09	0.27	1.33	0.12	2.41

[a,b,c,d]Within a column, means without a common superscript differ ($P < 0.05$).
[1]*SEM*, standard error mean.

A high LOX activity would be plausible in high pH silages, whereas these enzymes would be inhibited by low pH [15,36]. In addition, lipid or FA oxidation requires oxygen, whereas ensiling is a process that progresses from aerobic conditions to anaerobic conditions. Thus, LOX activity would decline as oxygen was consumed during ensiling.

Many additives, such as formalin, formic acid, inoculants, and enzymes, have been used to alter the FA contents and compositions of silages [11-14]. However, their effects were either minimal or they had no effects. Among the various methods that have recently been used to prevent lipid peroxidation, adding antioxidants has received the most attention. Most of the commonly used antioxidants are phenolics that can be divided into synthetic compounds and natural ingredients.

Synthetic compounds, such as BHA, BHT, and TBHQ, are chemically stable, inexpensive, and readily available. However, the safety of these synthetic antioxidants has been questioned due to their potential risks to human health [37]. Thus, there is a growing interest in natural ingredients because they are more acceptable to consumers, more palatable, stable. and improve the shelf-lives of food products. They have also shown beneficial health effects against degenerative diseases and certain cancers.

Polyphenols are known to have important protective roles during lipoperoxidation. There is considerable data for the use of polyphenols as natural antioxidants and an interest in the antioxidant properties of polyphenols from pomegranate has recently emerged [38,39]. Vitamin E is commonly added to animal and human diets because it can inhibit lipoperoxidation [40]. It is also an important nutrient that aids in stabilizing unsaturated fatty acids in milk [41,42].

Antioxidants prevent enzyme catalysis by disrupting fatty acid peroxidation chain reactions or chelating Cu and Fe ions to form stable chelation compounds [43,44]. The lipoperoxidase (LOX) activities found in this study were lower after adding antioxidants as compared to the control. These results were consistent with those in a previous report [45], which indicated that antioxidants considerably inhibited lipid oxygenation.

All of the silages in this study were of high quality due to their high lactic acid contents, low pH values, and NH_3-N/TN contents. Furthermore, as compared to the control, all of the treatments we used affected the NH_3-N/TN contents. The reason may have been that the antioxidants used in this study inhibited protease activity. For the same reason, the treated silages had higher CP contents than the control, although these differences were not significant. Aerobic microorganism metabolism would also be restricted because of the lower oxygen tension. Thus, the WSC contents in treated silages were higher than they were in the control on day 28 of this study. The antioxidants used in this study could not break down cellulose, so there were no significant differences between treated and untreated silages for NDF and ADF contents.

We found that all of the antioxidants we used successfully protected fatty acids. Higher total FA and UFA contents, and lower MDA contents were found in all of the treated silages. These results were consistent with the predicted loss of antioxidants in silage during ensiling [13,17]. Furthermore, TPP and TBHQ were better than BHA and VE for inhibiting FA oxidation. These results were consistent with the inhibitory effects on LOX activity shown in Table 3, which indicated that the FA losses in silages were mainly due to the effects on LOX activity and that higher FA concentrations could be obtained by inhibiting LOX activity.

Conclusions

A reduction in total FA contents in corn silages was due to the oxidation of unsaturated fatty acids by LOX during the early stages of ensiling. However, adding an antioxidant could prevent fatty acids' oxidation in corn silages.

Abbreviations

DM: Dry matter; FW: Fresh weight; WSC: Water-soluble carbohydrate; CP: Crude protein; NDF: Neutral detergent fiber; ADF: Acid detergent fiber; FA: Fatty acid; MDA: Malondialdehyde; LOX: Lipoperoxidation; LA: Lactic acid; AA: Acetic acid; BA: Butyric acid; NH3-N/TN: Ammoniacal nitrogen/total nitrogen; CK: No additives; BHA: Butylated hydroxyanisole; TBHQ: Tertiary butyl hydroquinone; TPP: Teapolyphenols; VE: Vitamin E; BHA50: 50 mg butylated hydroxyanisole/kg fresh corn; BHA100: 100 mg butylated hydroxyanisole/kg fresh corn; TBHQ50: 50 mg tertiary butyl hydroquinone/kg fresh corn; TBHQ100: 100 mg tertiary butyl hydroquinone/kg fresh corn; TPP50: 50 mg teaPolyphenols /kg fresh corn; TPP100: 100 mg teaPolyphenols/kg fresh corn; VE50: 50 mg vitamin E/kg fresh corn; VE100: 100 mg vitamin E/kg fresh corn.

Competing interests

The authors declare that they have no competing interests related to this study.

Authors' contributions

LYH did the chemical analyses, statistical analyses, and drafted the manuscript. HZ conceived the study, participated in its design and coordination, and helped draft the manuscript. All authors read and approved the final manuscript.

Acknowledgments

Funding for this research by the National Science and Technology Foundation (2011BAD17B02) and the Research Fund for the Doctoral Program of Higher Education of China (20120008110003) is gratefully acknowledged.

References

1. Eriksson SF, Pickova J: **Fatty acids and tocopherol levels in M.Longissimus dorsi of beef cattle in Sweden - a comparison between seasonal diets.** *Meat Sci* 2007, **76**:746–754.
2. Chilliard Y, Glasser F, Ferlay A, Bernard L, Rouel J, Doreau M: **Diet, rumen biohydrogenation and nutritional quality of cow and goat milk fat.** *Eur J Lipid Sci Technol* 2007, **109**:828–855.
3. Feussner I, Wasternack C: **The lipoxygenase pathway.** *Annu Rev Plant Biol* 2002, **53**:275–297.
4. Noci F, Monahan FJ, Scollan ND, Moloney AP: **The fatty acid composition of muscle and adipose tissue of steers offered unwilted or wilted grass silage supplemented with sunflower oil and fish oil.** *Br J Nutr* 2007, **97**:502–513.
5. Mo M, Selmer-Olsen I, Randby AT, Aakre SE, Asmyhr A: *New fermentation products in grass silage and their effects on feed intake and milk taste,* Proceedings of the 10th International Symposium on Forage Conservation. Brno, CZ: NutriVet Ltd; 2001.
6. Krizsan SJ, Westad F, Adnoy T, Odden E, Aakre SE, Randby AT: **Effect of volatile compounds in grass silage on voluntary intake by growing cattle.** *Anim* 2007, **1**:283–292.
7. Umana R, Staples CR, Bates DB, Wilcox CJ, Mahanna WC: **Effects of a microbial inoculant and (or) sugarcane molasses on the fermentation aerobic stability, and digestibility of bermudagrass ensiled at two moisture contents.** *J Anim Sci* 1991, **69**:4588–4601.
8. Sheperd AC, Maslank M, Quinn D, Kung L: **Additives containing bacteria and enzymes for alfalfa silage.** *J Dairy Sci* 1995, **78**:565–572.
9. Bureenok S, Namihira T, Kawamoto Y, Nakada T: **Additive effects of fermented juice of epiphytic lactic acid bacteria on the fermentative quality of guinea grass (Panicum maximum Jacq.) silage.** *Grassl Sci* 2005, **51**:243–248.
10. Dewhurst RJ, Scollan ND, Youell SJ, Tweed JKS, Humphreys MO: **Influence of species, cutting date and cutting interval on the fatty acid composition of grasses.** *Grass Forage Sci* 2001, **56**:68–74.
11. Dewhurst RJ, King PJ: **Effects of extended wilting, shading and chemical additives on the fatty acids in laboratory grass silages.** *Grass Forage Sci* 1998, **53**:219–224.

12. Boufaued H, Chouinard PY, Tremblay GF, Petit HV, Michaud R, Belanger G: **Fatty acids in forages. I. Factors affecting concentrations.** *Can J Anim Sci* 2003, **83**:501–511.
13. Shingfield KJ, Reynolds CK, Lupoli B, Toivonen V, Yurawecz MP, Delmonte P, Griinari JM, Grandison AS, Beever DE: **Effect of forage type and proportion of concentrate in the diet on milk fatty acid composition in cows given sunflower oil and fish oil.** *Anim Sci* 2005, **80**:225–238.
14. Arvidsson K, Gustavsson AM, Martinsson K: **Effects of conservation method on fatty acid composition of silage.** *Anim Feed Sci Technol* 2009, **148**:241–252.
15. Lourenco M, Van Ranst G, Fievez V: **Difference in extent of lipolysis in red or white clover and ryegrass silages in relation to polyphenol oxidase activity.** *Comm Agric Appl Biol Sci* 2005, **70**:169–172.
16. Lee MRF, Scott MB, Tweed JKS, Minchin FR, Davies DR: **Effects of polyphenol oxidase on lipolysis and proteolysis of red clover silage with and without a silage inoculant (Lactobacillus plantarum L54).** *Anim Feed Sci Technol* 2008, **144**:125–136.
17. Noziere P, Graulet B, Lucas A, Martin B, Grolier P, Doreau M: **Carotenoids for ruminants: from foragrs to dairy products.** *Anim Feed Sci Technol* 2006, **131**:418–450.
18. Surrey K: **Spectrophotometric determination of lipoxygenase activity.** *Plant Physiol* 1964, **38**:65–70.
19. Bradford MM: **A rapid and sensitive method for the quantitation of microgram quantities of protein utilizing the principle of protein-dye binding.** *Anal Biochem* 1976, **72**:248–254.
20. Raes K, de Smet S, Demeyer D: **Effect of double-muscling in Belgian blue young bulls on the intramuscular fatty acid composition with emphasis on conjugated linoleic acid and polyunsaturated fatty acids.** *Anim Sci* 2001, **73**:253–260.
21. Bird RP, Draper HH: **Comparative studies on different methods of malonaldehyde determination.** *Method Enzymol* 1984, **105**:299–305.
22. Owens VN, Albrecht KA, Muck RE, Duke SH: **Protein degradation and fermentation characteristics of red clover and alfalfa silage harvested with varying levels of total nonstructural carbohydrates.** *Crop Sci* 1999, **39**:1873–1880.
23. Broderica GA, Kang JH: **Automated simultaneous determination of ammonia and total amino acids in ruminal fluid and in vitro media.** *J Dairy Sci* 1980, **63**:64–75.
24. Association of Official Analytical Chemists: *Official Methods of Analysis of the Association of Official Analytical Chemists.* 16th edition. Arlington, VA: Association of Analytical Communities; 1995.
25. Van Soest PJ, Robortson JB, Lewis BS: **Methods for dietary fiber, neutral detergent fiber and non starch polysaccharides in relation to animal nutrition.** *J Dairy Sci* 1991, **73**:2583–3593.
26. Deriaz RE: **Routine analysis of carbohydrates and lignin in herbage.** *J Sci Food Agric* 1961, **12**:152–160.
27. Institute Inc SAS: *SAS/Genetics™ 9.1.3 User's Guide.* Cary, NC: SAS Institute Inc; 2005.
28. Castle ME, Watson JN: **The relationship between the DM content of herbage for silage making and effluent production.** *Grass Forage Sci* 1973, **28**:135–138.
29. Rooke JA, Hatfield RD: **Biochemistry of ensiling.** In *Silage Science and Technology.* Edited by Buxton DR, Muck RE, Harrison JH. Madison: ASA Inc; 2003.
30. Thompson JE, Froese CD, Madey E, Smith MD, Hong Y: **Lipid metabolism during plant senescence.** *Prog Lipid Res* 1998, **37**:119–141.
31. Kanner J, Hazan B, Doll L: **Catalytic 'free' iron ions in muscle foods.** *J Agric Food Chem* 1991, **36**:412–415.
32. Elgersma A, Ellen G, van der Horst H, Muuse BG, Boer H, Tamminga S: **Influence of cultivar and cutting date on the fatty acid composition of perennial ryegrass (Lolium perenne L.).** *Grass Forage Sci* 2003, **58**:323–331.
33. Lee MRF, Winters AL, Scollan ND, Dewhurst RJ, Theodorou MK, Minchin FR: **Plant-mediated lipolysis and proteolysis in red clover with different polyphenol oxidase activities.** *J Sci Food Agric* 2004, **84**:1639–1645.
34. Van Ranst G, Fievez V, De Riek J, Van Bockstaele E: **Influence of ensiling forages at different dry matters and silage additives on lipid metabolism and fatty acid composition.** *Anim Feed Sci Technol* 2009, **150**:62–74.
35. Elgersma A, Ellen G, van der Horst H, Muuse BG, Boer H, Tamminga S: **Comparison of the fatty acid composition of fresh and ensiled perennial ryegrass (Lolium perenne L.), affected by cultivar and regrowth interval.** *Anim Feed Sci Tech* 2003, **108**:191–205.
36. Zhong QX, Glatz CE: **Enzymatic assay method for evaluating the lipase activity in complex extracts from transgenic corn seed.** *J Agric Food Chem* 2006, **54**:3181–3185.
37. Kubow S: **Toxicity of dietary lipid peroxidation products.** *Trends Food Sci Technol* 1990, **1**:67–71.
38. Bozkurt H: **Utilization of natural antioxidants: green tea extract and Thymbraspicata oil in Turkish dry-fermented sausage.** *Meat Sci* 2006, **73**:442–450.
39. Naveena B, Sen A, Kingsly R, Singh D, Kondaiah N: **Antioxidant activity of pomegranate rind powder extract in cooked chicken patties.** *International J Food Sci and Technol* 2008, **43**:1807–1812.
40. Cuverlier ME, Berset C, Richard H: **Use of a new test for determining comparative antioxidant activity of butylated hydroxyanisole, butylated hydroxytoluene, alpha- and gamma-tocopherols and extract from rosemary and sage.** *Sci Aliment* 1990, **10**:797–806.
41. Charmley E, Nicholson JWG: **Influence of dietary fat source on oxidative stability and fatty acid composition of milk from cows receiving a low or high level of dietary vitamin E.** *Can J Anim Sci* 1994, **74**:657–664.
42. Focant M, Mignolet E, Marique M, Clabots F, Breyne T, Dalemans D, Larondelle Y: **The effect of vitamin E supplementation of cow diets containing rapeseed and linseed on the prevention of milk fat oxidation.** *J Dairy Sci* 1998, **81**:1095–1101.
43. Min DB, Boff JM: **Chemistry and reaction of singlet oxygen in foods.** *Comprehensive Reviews in Food Science and Food Satefy* 2002, **1**:28–72.
44. Gordon MH: **The development of oxidative rancidity in foods.** In *Antioxidants in food practical applications.* Edited by Pokorny J, Yanishlieva N, Gordon M. Cambridge: Woodhead Pubshing Limited; 2001.
45. Frankel EN: *Free radical oxidation. Lipid oxidation.* Scotland: The Oily Press Ltd; 1998.

Effect of genotype on duodenal expression of nutrient transporter genes in dairy cows

Sinéad M Waters[1*], Kate Keogh[1], Frank Buckley[2] and David A Kenny[1]

Abstract

Background: Studies have shown clear differences between dairy breeds in their feed intake and production efficiencies. The duodenum is critical in the coordination of digestion and absorption of nutrients. This study examined gene transcript abundance of important classes of nutrient transporters in the duodenum of non lactating dairy cows of different feed efficiency potential, namely Holstein-Friesian (HF), Jersey (JE) and their F_1 hybrid. Duodenal epithelial tissue was collected at slaughter and stored at -80°C. Total RNA was extracted from tissue and reverse transcribed to generate cDNA. Gene expression of the following transporters, namely nucleoside; amino acid; sugar; mineral; and lipid transporters was measured using quantitative real-time RT-PCR. Data were statistically analysed using mixed models ANOVA in SAS. Orthogonal contrasts were used to test for potential heterotic effects and spearman correlation coefficients calculated to determine potential associations amongst gene expression values and production efficiency variables.

Results: While there were no direct effects of genotype on expression values for any of the genes examined, there was evidence for a heterotic effect ($P < 0.05$) on *ABCG8*, in the form of increased expression in the F_1 genotype compared to either of the two parent breeds. Additionally, a tendency for increased expression of the amino acid transporters, *SLC3A1* ($P = 0.072$), *SLC3A2* ($P = 0.081$) and *SLC6A14* ($P = 0.072$) was also evident in the F_1 genotype. A negative ($P < 0.05$) association was identified between the expression of the glucose transporter gene *SLC5A1* and total lactational milk solids yield, corrected for body weight. Positive correlations ($P < 0.05$) were also observed between the expression values of genes involved in common transporter roles.

Conclusion: This study suggests that differences in the expression of sterol and amino acid transporters in the duodenum could contribute towards the documented differences in feed efficiency between HF, JE and their F_1 hybrid. Furthermore, positive associations between the expression of genes involved in common transporter roles suggest that these may be co-regulated. The study identifies potential candidates for investigation of genetic variants regulating nutrient transport and absorption in the duodenum in dairy cows, which may be incorporated into future breeding programmes.

Keywords: Bovine, Duodenum, Gene expression, Nutrient transporters

Background

In dairy cow systems feed is the single greatest variable cost, accounting for up to 80% of the costs of production [1]. As profitability is directly linked to the efficient conversion of feed into milk, the identification of feed efficient animals is critically important to the economic sustainability of the enterprise. In dairy cattle, residual milk solids production is a measure of feed efficiency and can be used to identify animals that produce higher amounts of milk solids but have a similar level of feed intake to their herd counterparts [2]. Indeed, studies [3,4] have also shown clear differences between dairy breeds and their feed intake and efficiencies. Schwerin *et al.* [5] reported differences in nutrient utilisation between dairy and beef breeds and, specifically, that the expression of genes involved in nutrient transportation in the liver and intestine differed between Charolais and Holstein bulls. Furthermore, recent data from an Irish study has shown that dairy cow genotype affects the

* Correspondence: Sinead.Waters@teagasc.ie
[1]Teagasc Animal and Bioscience Research Department, Animal and Grassland Research and Innovation Centre, Grange, Dunsany, Co. Meath, Ireland
Full list of author information is available at the end of the article

expression profiles of genes involved in energy homeostasis in duodenum and liver [6].

The duodenum plays a critical role in nutrient digestion and absorption and is the site of expression of key signalling molecules regulating energy homeostasis and feed efficiency in cattle [6]. A number of studies have previously examined the effect of diet type on the absorption of nutrients namely, sugar, nucleoside and amino acid in the duodenum of beef cattle [7-9]. However, there is a dearth of information detailing mineral and lipid transporter mRNA abundance in this tissue. Additionally, there is no information available on whether differences exist between contrasting dairy cow genotypes or animals of different feed efficiency potential for the absorption of nutrients in the small intestine. Thus the aim of this study was to determine the effect of dairy cow genotype on the expression profiles of a variety of genes involved in the transportation and absorption of nutrients and minerals in Holstein-Friesian (HF), Jersey (JE) and Holstein-Friesian Jersey cross (F_1). Gene transcript abundance of five important classes of nutrient transporters, namely nucleoside, amino acid, lipid, sugar and mineral transporters, was investigated.

Materials and methods

All procedures involving animals were carried out under a licence for the Irish Department of Health and Children in accordance with the European Community Directive 86/609/EC.

Experimental animals

This study was part of a larger experiment designed to evaluate the performance of three dairy genotypes, HF, JE and F_1 (JE × HF), on a pasture-based production system. All data were generated at the Ballydague research farm (52°8′N 8°26′W), Teagasc Moorepark Dairy Production Research Centre, Fermoy, Co. Cork, Ireland. Performance data were obtained from 110 animals, representing HF (n = 37), JE (n = 36) and F_1 (n = 37) cows and was calculated as described by Prendiville *et al.* [3].

A total of 6, 7 and 3 sires were represented in the HF, JE and F_1, respectively. All F_1 animals were sired by JE bulls and were born to HF cows. The HF sires were of North American (86%) and New Zealand (14%) origin. The mean predicted transmitting abilities (PTA) (across breed) and standard deviations for the HF sires used were: +163 kg (31.1), +13 kg (7.0), +10 kg (3.0), +0.12% (0.14) and +0.08% (0.06) for milk yield, fat yield, protein yield, fat and protein concentration, respectively (source www.ICBF.com, April 2009). Comparable PTAs for the JE sires were: -408 kg (193.5), +8 kg (6.6), -3 kg (7.4), +0.55% (0.27) and +0.24% (0.10). The JE sires were of New Zealand (56%) and Danish (44%) origin. Of the 7 JE sires, 1 was represented in both the JE and F_1 cows. This sire accounted for 14% and 50% of the JE and F_1 cows, respectively. All sires were representative of the sires commonly used through AI in Irish dairy herds.

Tissue sample collection

At the end of lactation, cows were dried off and subsequently fed grass silage *ad libitum* for two months. A sub group of 30 cows from the initial 110 were randomly selected for inclusion in this study representing 10 HF, 10 JE and 10 F_1. All 30 animals were slaughtered in a licensed abattoir (Dawn Meats, Charleville, Co. Cork, Ireland). Duodenal tissue (5 cm long) was harvested approximately 15 cm distal to the abomasal-duodenal juncture. Tissue samples were washed in DPBS. Epithelial tissue was then scraped from the underlying connective and muscular tissue using a glass microscope slide. The tissue was washed with sterile phosphate buffered saline (PBS), snap frozen in liquid nitrogen and subsequently stored at −80°C. All instruments used for tissue collection were sterilised and treated with RNA Zap (Ambion, Dublin, Ireland) before use.

RNA extraction and purification

Total RNA was isolated from approximately 40 mg of duodenal epithelial tissue using TRIzol reagent and chloroform (Sigma-Aldrich Ireland, Dublin, Ireland). Tissue samples were homogenised using a tissue lyser (Qiagen, UK), following which the RNA was precipitated using isopropanol. Samples were then treated with RQ1 RNase-free DNase (Promega UK, Southhampton, UK), according to the manufacturers instructions in order to remove any contaminating genomic DNA. The quantity of the RNA isolated was determined by measuring the absorbance at 260 nm using a Nanodrop spectrophotometer ND-1000 (NanoDrop Technologies, DE, USA). RNA quality was assessed on the Agilent Bioanalyser 2100 using the RNA 6000 Nano Lab Chip kit (Agilent Technologies Ireland Ltd., Dublin, Ireland). RNA quality was verified by ensuring all RNA samples had an absorbance ($A_{260/280}$) of between 1.8 and 2. RNA samples with 28S/18S ratios ranging from 1.8 and 2.0 and RNA integrity numbers (RINs), which is a measure of RNA quality based on the integrity of 18 and 28S ribosomal RNA, of between 8 and 10 were deemed high quality.

Complementary DNA synthesis

Total RNA (1 μg) was reverse transcribed into cDNA using a High Capacity cDNA Reverse Transcription Kit (Applied Biosystems, Foster City, CA,. USA) using the Multiscribe™ reverse transcriptase according to manufacturers instructions. Samples were stored at −20°C for subsequent analyses.

Primer design and reference gene selection

All the gene specific primers used in this study were designed using the web based software program Primer 3 (http://frodo.wi.mit.edu/primer3/). Potential primers were then subjected to BLAST analysis (http://www.ncbi.nlm.nih.gov/BLAST/), in order to confirm primer specificity and also to ensure that they were homologous to the bovine sequences. All primers for reference and specific target genes were obtained from a commercial supplier (Sigma-Aldrich Ireland, Dublin, Ireland). Details of primer sets used in this study are listed in Additional file 1: Table S1. All amplified PCR products were sequenced to verify their identity (Macrogen Europe, Meibergdreef 39, 1105AZ Amsterdam, The Netherlands).

In order to select stable reference genes relevant to duodenal tissue, analysis of putative reference genes was carried out using the geNorm version 3.4 Excel software package (Microsoft, Redmond, WA). Ct values were transformed to relative quantities using the comparative delta Ct method, to facilitate the calculation of the M value within geNorm software. The software calculates the intra- and intergroup CV and combines both coefficients to give a stability value minus a lower value implying a higher stability in gene expression. A gene was considered to be sufficiently stable within the duodenal tissue, if an M value of less than 1.5 was generated. Within this range of parameters, beta-actin (*ACTB*), glyceraldehydes 3-phosphate dehydrogenase (*GAPDH*) and ribosomal protein SP (*RPS9*) were selected as being suitable reference genes for this study.

Quantitative real time PCR (qPCR)

Following reverse transcription, cDNA quantity was determined and standardised to the required concentration for qPCR. Triplicate 20 μL reactions were carried out in 96-well optical reaction plates (Applied Biosystems, Warrington, UK), containing 1 μL cDNA (10–50 ng of RNA equivalents), 10 μL Fast SYBR® Green PCR Master Mix (Applied Biosystems, Warrington, UK), 8 μL nuclease-free H_2O, and 1 μL forward and reverse primers (250–1000 nM per primer). Assays were performed using the ABI 7500 Fast qPCR System (Applied Biosystems, Warrington, UK) with the following cycling parameters: 95°C for 20 s and 40 cycles of 95°C for 3 s, 60°C for 30 s followed by amplicon dissociation (95°C for 15 s, 60°C for 1 min, 95°C for 15 s and 60°C for 15 s). Amplification efficiencies were determined for all candidate and reference genes using the formula E = 10^(−1/slope), with the slope of the linear curve of cycle threshold (Ct) values plotted against the log dilution [10]. Only primers with PCR efficiencies between 90% and 110% were used. The software package GenEx 5.2.1.3 (MultiD Analyses AB, Gothenburg, Sweden) was used for efficiency correction of the raw cycle threshold (**Ct**) values, interplate calibration based on a calibrator sample included on all plates, averaging of replicates, normalization to the reference gene and the calculation of quantities relative to the greatest Ct. Expression of each target gene was normalised to the reference genes and relative differences in gene expression were calculated using the $2^{-\Delta\Delta CT}$ method [11].

Statistical analysis

All data were analysed using Statistical Analysis Systems (SAS Institute, Cary, NC; version 9.2). Data were tested for adherence to a normal distribution using the UNIVARIATE procedure of SAS. A Box-Cox transformation analysis was performed using the Transreg procedure in SAS to obtain appropriate lambda values for data which were not normally distributed. These data were then transformed by raising the variable to the power of lambda. A mixed model ANOVA (PROC MIXED, SAS) was conducted to determine the effect of genotype on the relative expression of each gene measured. The Tukey critical difference test was performed to determine the existence of statistical difference between the treatment groups. In an effort to determine whether there was any evidence for potential heterotic effects on the expression of genes of interest, orthogonal contrasts were used to examine differences between the combined mean of expression values for Holstein-Friesian and Jersey animals compared with their F_1 hybrid. Spearman partial correlation coefficients were calculated to determine associations among gene expression values for each gene in the duodenum in addition to associations amongst gene expression and production efficiency variables, including residual feed intake (RFI), total milk solid (kg) produced over a 305 day lactation period per 100 kg (SOLIDS_WGT), and milk solids produced (kg) per kg of total dry matter intake (SOLIDS_TDMI), using the CORR procedure of SAS. Data were corrected for the fixed effects of both cow genotype and parity.

Results

Effect of genotype on cow production efficiency

A more comprehensive explanation of the genotypes, experimental design, grazing management, sward composition, feed intake and production efficiency measurements has been reported [3]. In brief, genotype had a number of statistically significant effects on cow productive efficiency. For example, daily milk solids yield (MLKS; fat and protein yield) was similar for HF and JE but JE was lower than the F_1 cows (1.33 kg for HF, 1.28 kg for JE and 1.41 for F_1). Body weight was higher for HF (577 Kg compared to 435 kg for JE with the F_1 intermediate (520 kg; $P < 0.05$), whereas body condition score was highest ($P < 0.05$) for the F_1 cows (3.00 compared to 2.76 for HF and 2.93 for JE).

Dry matter intake (DMI) per unit body weight (3.99 kg for JE compared to 3.39 kg for HF and 3.63 kg for F_1)

and gross production efficiency (0.088 kg for JE compared to 0.087 Kg for F_1 and 0.079 Kg for HF) was highest in JE. Production efficiency, expressed as net energy intake per MLKS was highest for the F_1 cows (8.32 UFL compared to 8.11 UFL for HF and 7.45 UFL for JE). Animals were slaughtered at an average of 2.7 (s.d. 0.90), 2.4 (s.d. 0.52) and 2.8 (s.d. 0.46) lactations for HF, JE and F_1 respectively. In addition, at slaughter Jersey tissues internal organs (or components of the GIT) weighed less than tissues recovered from cows of the other two breed types with the exception of the omasum, which did not differ in size between breeds. On a proportion of metabolic liveweight basis, HF cows had a smaller rumen-reticulum, abomasum and total GIT than both J and F_1 cows. However, when expressed as a proportion of metabolic liveweight, the weight of these organs did not differ between the three breed types and were similar [12].

Effect of cow genotype on the expression of genes in duodenal tissue

The effect of cow genotype on the expression of genes involved in nutrient and mineral absorption in the duodenum is presented in Table 1. Out of 27 genes tested, 19 were found to be expressed in duodenal tissue. Of all the genes studied, only one *ABCG8* ($P = 0.042$), was identified as significantly differentially expressed between groups. However, there was a strong tendancy towards mRNA expression levels for *SLC3A1* ($P = 0.072$), *SLC3A2* ($P = 0.081$) and *SLC6A14* ($P = 0.072$) being different between the three genotypes. There was evidence for a heterotic effect ($P < 0.05$) on the duodenal expression of the lipid transporter *ABCG8*, with expression levels higher for the F_1 genotype, compared with the mean of the two parent breeds. There was no evidence for any potential heterotic effects ($P > 0.10$) for the expression of any other gene studied in duodenal tissue.

Associations between gene expression values within duodenal tissue samples

The results of a Spearman correlation analysis which was conducted to examine the associations between gene expression values in the duodenum is presented in Table 2. Gene expression of *SLC28A1* was positively associated with *SLC28A2* ($r = 0.89$; $P < 0.05$), *SLC3A2* ($r = 0.96$; $P < 0.01$), *SLC6A14* ($r = 0.95$; $P < 0.05$) and *SLC2A2* ($r = 0.97$; $P < 0.01$). Expression values of *SLC28A2* were positively correlated with *SLC3A2* ($r = 0.92$; $P < 0.05$), *SLC15A1* ($r = 0.94$; $P < 0.05$), *SLC2A2* ($r = 0.97$; $P < 0.01$) and *SLC2A5* ($r = 0.92$; $P < 0.05$). *SLC28A3* was positively associated with *SLC3A1* ($r = 0.93$; $P < 0.05$), *SLC31A1* ($r = 0.90$; $P < 0.01$) and *SLC39A4* ($r = 0.98$; $P < 0.01$). *SLC29A1* was positively associated with *SLC11A2* ($r = 0.94$; $P < 0.05$). *SLC3A1* was positively correlated with *SLC39A4* ($r = 0.91$; $P < 0.05$). *SLC3A2* was positively associated with *SLC7A6* ($r = 0.93$; $P < 0.05$), *SLC15A1* ($r = 0.93$: $P < 0.05$), *SLC2A2* ($r = 0.98$; $P < 0.01$), *SLC2A5* ($r = 0.89$; $P < 0.05$), and *SLC31A1* ($r = 0.91$; $P < 0.05$). *SLC6A14* was positively correlated with *SLC2A2* ($r = 0.86$; $P < 0.05$). *SLC7A1* was negatively associated with *SLC7A7* ($r = -0.95$; $P < 0.05$) and positively associated with *ABCG8* ($r = 0.98$; $P < 0.01$). *SLC7A6* was positively associated with *SLC15A1* ($r = 1.0$; $P < 0.001$), *SLC2A2* ($r = 0.93$; $P < 0.05$), and *SLC2A5* ($r = 0.94$; $P < 0.05$). *SLC7A7* was negatively associated with *ABCG8* ($r = -0.88$; $P < 0.05$). *SLC15A1* was positively correlated with *SLC2A2* ($r = 0.93$; $P < 0.05$), and *SLC2A5* ($r = 0.94$; $P < 0.05$). *SLC2A2* was positively associated with *SLC2A5* ($r = 0.88$; $P < 0.05$). *SLC31A1* was positively associated with *SLC39A4* ($r = 0.96$; $P < 0.01$).

Table 1 Effect of genotype and potential heterotic effects on the expression of nutrient and mineral transporter genes in the duodenal epithelium of dairy cows

	Genotype				Significance (P-value)	
Gene name	HF	J	F_1	SEM[1]	Genotype	HF + J vs. F_1
Nucleoside transporters						
SLC28A1	4.39	4.23	4.50	0.55	0.95	0.79
SLC28A2	4.49	4.48	4.49	0.62	0.86	0.59
SLC28A3	4.62	4.76	5.02	0.48	0.84	0.59
SLC29A1	0.85	0.87	0.86	0.03	0.95	0.99
Amino acid transporters						
SLC3A1	3.08	3.48	4.35	0.44	0.15	0.07
SLC3A2	0.98	1.11	1.28	0.10	0.13	0.08
SLC6A14	3.08	3.48	4.35	0.44	0.15	0.07
SLC7A1	1.96	2.04	2.69	0.45	0.48	0.24
SLC7A6	4.43	3.45	4.87	0.63	0.29	0.26
SLC7A7	0.95	0.96	1.17	0.15	0.54	0.27
SLC15A1	2.10	1.98	2.15	0.12	0.59	0.49
Sugar transporters						
SLC2A2	6.77	6.09	6.72	0.93	0.83	0.81
SLC2A5	2.76	2.35	2.97	0.55	0.73	0.57
SLC5A1	4.75	3.82	4.74	0.47	0.29	0.46
Lipid transporters						
ABCG8	4.81^a	4.66^a	8.17^b	1.24	**0.04***	**0.02***
Mineral transporters						
SLC11A2	2.82	2.16	2.60	0.08	0.69	0.93
SLC31A1	2.48	2.06	3.59	1.05	0.61	0.34
SLC39A4	2.51	2.69	4.17	1.12	0.22	0.10
TRPV6	4.02	3.91	4.57	0.09	0.92	0.68

HF = Holstein-Friesian; Je = Jersey; F_1 = Holstein-Friesian x Jersey. SEM[1] = pooled standard error, Genotype, *P < 0.05. Gene expression values are presented as ratios of cycle threshold (Ct) value for each gene normalized to that of the reference gene after adjustment for efficiencies and interplate variation.
[a,b]Means sharing the same superscript are not significantly different at P < 0.05.
Bold represents significant (P<0.05) results in terms of difference between means in Table 1 or correlations in Tables 2 and 3.

Table 2 Spearman partial correlation coefficients for the association between the expression of duodenal genes involved in common nutrient transport function

	Nucleoside transporters				Amino acid transporters						Sugar transporters			Mineral transporters				Lipid transporter
SLC	*28A2*	*28A3*	*29A1*	*3A1*	*3A2*	*6A14*	*7A1*	*7A6*	*7A7*	*15A1*	*2A2*	*2A5*	*5A1*	*11A2*	*31A1*	*39A4*	*TRPV6*	*ABCG8*
28A1	**0.89***	0.62	0.62	0.70	**0.96****	**0.95***	0.28	0.84	−0.01	0.84	**0.97****	0.78	0.36	0.75	0.84	0.71	0.83	0.45
28A2		0.35	0.75	0.34	**0.92***	0.73	0.36	0.94	−0.12	**0.94***	**0.97****	**0.92***	0.49	0.73	0.70	0.49	0.83	0.52
28A3			0.21	**0.93***	0.68	0.63	−0.24	0.51	0.47	0.51	0.54	0.46	0.42	0.44	**0.90***	**0.98****	0.61	−0.09
29A1				0.14	0.67	0.40	0.77	0.64	−0.58	−0.58	0.70	0.83	0.05	**0.94***	0.40	0.26	0.34	0.86
3A1					0.66	0.79	−0.16	0.43	0.38	0.43	0.56	0.33	0.23	0.42	0.85	**0.91***	0.57	−0.02
3A2						0.85	0.20	**0.93***	0.08	**0.93***	**0.98****	**0.89***	0.53	0.75	**0.91***	0.78	0.89	0.38
6A14							0.17	0.67	0.05	0.67	**0.86***	0.55	0.22	0.60	0.79	0.69	0.74	0.33
7A1								0.10	**−0.95***	0.10	0.31	0.33	−0.54	0.73	−0.13	−0.24	−0.18	**0.98****
7A6									0.15	**1.0*****	**0.93***	**0.94***	0.72	0.62	0.82	0.65	0.93	0.28
7A7										0.15	−0.04	−0.08	0.69	−0.50	0.41	0.50	0.43	**−0.88***
15A1											**0.93***	**0.94***	0.72	0.62	0.82	0.65	0.93	0.28
2A2												**0.88***	0.46	0.76	0.82	0.65	0.86	0.48
2A5													0.57	0.79	0.73	0.57	0.77	0.49
5A1														−0.01	0.63	0.54	0.80	−0.40
11A2															0.56	0.46	0.39	0.83
31A1																**0.96****	0.88	0.04
39A4																	0.74	−0.08
TRPV6																		−0.01

The probability of a coefficient not being statistically different from zero is denoted as follows: $^*P < 0.05$, $^{**}P < 0.01$ and $^{***}p < 0.001$.
Bold represents significant ($P<0.05$) results in terms of difference between means in Table 1 or correlations in Tables 2 and 3.

Associations between duodenal gene expression values and animal production efficiency variables

A Spearman partial correlation analysis was conducted to determine the association between the expression of genes involved in nutrient and mineral absorption in the duodenum and feed efficiency variables, previously reported by Prendiville *et al.* [3,13]. Correlation coefficients for these associations are presented in Table 3. Only one association was identified as reaching statistical significance, *viz.* the correlation between *SLC5A1* gene expression and total milk solids produced over a 305 day lactation period (kg) per 100 kg of body weight (r = 0.93; $P < 0.05$).

Table 3 Correlation of expression of genes involved in nutrient transporters in the duodenum and production efficiency variables

Traits	RFI	SOLIDS_WGT	SOLIDS_TDMI
Nucleoside transporters			
SLC28A1	−0.33	−0.27	0.31
SLC28A2	−0.51	−0.55	0.04
SLC28A3	0.05	−0.15	0.20
SLC29A1	0.10	−0.23	−0.49
Amino acid transporters			
SLC3A1	0.04	0.07	0.46
SLC3A2	−0.36	−0.46	0.14
SLC6A14	−0.28	−0.05	0.55
SLC7A1	0.38	0.31	−0.41
SLC7A6	−0.57	−0.72	0.02
SLC7A7	−0.45	−0.43	0.42
SLC15A1	−0.57	−0.72	0.02
Sugar transporters			
SLC2A2	−0.42	−0.43	0.17
SLC2A5	−0.33	−0.64	−0.26
SLC5A1	−0.73	**−0.93***	0.06
Lipid transporters			
ABCG8	0.29	0.20	−0.36
Mineral transporters			
SLC11A2	0.20	−0.05	−0.29
SLC31A1	−0.29	−0.43	0.23
SLC39A4	−0.10	−0.29	0.23
TRPV6	−0.69	−0.69	0.32

The probability of a coefficient not being statistically different from zero is denoted as follows: *P < 0.05, **P < 0.01 and ***p < 0.001.
RFI; Residual Feed Intake.
SOLIDS_WGT; Total milk solids produced over a 305 day lactation period (kg) per 100 kg body weight (kg).
SOLIDS_TDMI; Total milk solid produced over a 305 day lactation period (kg) total dry matter intake.
Bold represents significant (P<0.05) results in terms of difference between means in Table 1 or correlations in Tables 2 and 3.

Discussion

Heterosis, or hybrid vigour, where progeny show increased fitness relative to their parents [14] is of economic importance in livestock production [15]. Positive effects of heterosis on growth and BW traits [16,17] and feed efficiency [17] have been reported for beef cattle. In dairy cattle, crossbreeding programmes utilising divergent cow breeds such as HF and JE cows have been explored to address the demands of the dairy industry [18]. Differences in feed intake capacity and production efficiency in lactating HF, JE and their F_1 have previously been presented by Prendiville *et al.* [3]. The resulting F_1 progeny have demonstrated promise in improving several traits associated with milk production including feed efficiency [3]. Gozho and Mutsvangwa [19] showed that improved production performance with corn and barley diets appeared to be due to greater nutrient absorption in dairy cows fed oats and grass silage diets. It has been postulated that improvements in digestion or absorption of dietary energy and protein are a possible mechanism to explain variation in feed efficiency [20,21]. In the current study we hypothesised that the improvement in feed efficiency observed in the F_1 genotype is due to an enhancement in nutrient absorption in the GIT possibly mediated through a modification of gene expression in the nutrient transporters.

We have recently shown that key genes involved in energy homeostasis and appetite behaviour, including *POMC* and *GLP1R,* were differentially expressed in the duodenum and liver, between contrasting cow genotypes, in a tissue dependent fashion [6]. There is, however, a dearth of information regarding the effect of dairy cow genotype on the expression of genes involved in nutrient absorption and transport in the small intestine of cattle and their relationship with production efficiency variables. To uncover potential molecular mechanisms controlling the documented differences in production efficiencies between contrasting breeds, an investigation into the expression of nutrient transporter genes was employed. The duodenum, which is the first section of the small intestine, is a major site of nutrient absorption in all animals [22] and has also been shown to be sensitive to nutritional changes [9]. The current study focussed on examining duodenal gene expression profiles. To our knowledge, this is the first examination of the expression of nutrient transporters in the duodenal tissue of dairy cows.

Of all the nutrient transporter genes analysed, the lipid transporter *ABCG8,* was the only gene found to be differentially expressed across genotype. In addition, heterotic effects in the duodenal expression of *ABCG8* were also observed with mean expression higher in the F_1 animals compared with the mean of the two parent breeds. ABCG8 is a transporter of dietary cholesterol. While it is usually found co-expressed with *ABCG5*, there was no evidence for expression of this latter gene in duodenal tissue of dairy cows in the current study. Viturro *et al.* [23] examined the gene expression of sterol transporters *ABCG5* and *ABCG8* in a range of bovine tissues including the intestine. While expression was detected in the abomasum, jejunum and colon, the duodenum was not examined in that study. We have therefore shown expression of the *ABCG8* gene for the first time in the duodenum of the bovine. The protein encoded by this gene functions to exclude non-cholesterol sterol entry at the intestinal level, promote excretion of cholesterol and sterols into bile, and to facilitate transport of sterols back into the intestinal lumen. It is expressed in a tissue-specific manner in the liver, intestine, and gallbladder. As plant sterols are a major component of the ruminant diet [23] expression of this gene in the duodenum is not surprising. Therefore it is hypothesised that increased expression of this gene in the F_1 genotype may lead to enhanced transport of plant sterols, potentially lowering serum and milk cholesterol levels and contribute to improved feed efficiency compared to the parent breeds. Future studies should focus on the functional role of ABCG8 in the digestive tract of ruminants and how it may improve feed digestion and nutrient utilisation in cattle. The fatty acid transporter CD36 is frequently detected in tissues such as adipose [24] and mammary [25]. Recently, the expression of CD36 was shown to be dependent on diet and region of the intestine, with greater expression recorded in the upper jejunum compared to the ileum [26] however in the current study mRNA expression of this gene was not detected.

Amino acids are essential for optimal growth in cattle however there is little information available on amino acid transporter proteins expressed by the duodenum in dairy cattle. We observed a strong tendency towards increased expression of the amino acid transporter genes *SLC3A1, SLC3A2* and *SLC6A14* in the F_1 genotype, compared with the two parent breeds, consistent with the enhanced production efficiency reported for this genotype. SLC3A1 is involved in sodium independent transport of cystine, neutral and dibasic amino acids across the cell membrane. There is little published data available on this gene for cattle but it has been extensively studied in the human [27]. In the current study, expression of *SLC3A1* was strongly correlated with *SLC6A14, SLC7A6* and *SLC7A7* mRNA abundance. SLC7A7 is involved in the sodium dependent uptake of certain neutral amino acids and the sodium independent uptake of dibasic amino acids. It requires the co-expression of SLC3A2 to mediate the uptake of arginine, leucine and glutamine which probably explains the high correlation between these two genes. SLC3A2 is involved in light chain amino acid transport and functions as a sodium independent transporter of large neutral amino acids such as leucine, arginine, tyrosine and phenylalanine. SLC6A14 has a role involved in the sodium and chloride dependent transportation of neutral and basic amino acids. In a study by Liao *et al.* [8] the regulation of this gene amongst others was shown to be strongly regulated by diet. Furthermore gene expression of *SLC5A1* was negatively correlated with total milk solids produced over a 305 day lactation period (kg) per 100 kg of body weight. Liao *et al.* [8] found expression of SLC7A9 to be extremely low in the duodenum of beef steers and indeed, in our study, no expression of this gene was detected in the duodenal tissue of dairy cows. This could also be due to diet effects as animals in the current study were fed grass only while steers in the study of Liao *et al.* [8] were fed cornstarch, partially hydrolyzed by a heat-stable α-amylase. Chen *et al.* [28] found the gene *SLC15A1* to be expressed in the duodenum, jejunum and ileum of cattle while there was no expression detected in stomach, large intestine, liver, kidney and longissimus muscle tissue indicating that this gene is only expressed in the GIT.

In cattle, microbial-derived nucleic acids serve as a source of N and are absorbed as nucleosides through the small intestinal epithelia. Nucleosides are important nutrients for the development of gut and immune system function [29]. A supply of nucleosides is essential for many biological processes during animal development and growth, including DNA and RNA syntheses, energy (ATP) production, N and P recycling, cell signalling, and modulation of gene expression [29]. Liao *et al.* [30] showed that mRNA for nucleoside transporters are expressed throughout the small intestinal epithelia of growing beef steers and can be increased by augmenting the luminal supply of nucleotides. Nucleoside carriers bind to sodium ions as well as the nucleosides being transported. Consistent with our study, Liao *et al.* [30] found that *SLC28A1, SLC28A2, SLC28A3, SLC29A1* were expressed in the duodenum of beef steers. However their group also detected mRNA expression of SLC29A2 which we failed to detect in dairy cow duodenal tissue in our study potentially due to differences in the basal diet offered. SLC28A1 is a sodium coupled nucleoside transporter which has a higher affinity for binding to pyrimidines such as cytosine and thymine which enters a cell across a concentration gradient and uses the flow of sodium ions for transport into cells [31]. The expression of this gene was highly correlated with the expression of *SLC28A2,* which also codes for a sodium coupled nucleoside transporter and functions in the same manner. The high level of correlation could be due to the fact that SLC28A2 has a high affinity for purines such as adenosine and guanine and the expression of both of these genes are required for equal absorption of purines and pyrimidines. SLC28A3 is both purine and pyrimidine selective and functions in a similar fashion to both SLC28A1 and SLC28A2. Expression of *SLC28A3* is highly correlated with *SLC29A1* which is an equilibrative transporter. Unlike the other three nucleoside transporters studied, SLC29A1 is sodium independent and mediates the influx and efflux of nucleosides across a cell membrane. While there are studies published on the expression of this gene in cattle intestines [30], it has been extensively studied in humans due to its potential in aiding the uptake of chemotherapeutic drugs. None of the nucleoside transporters examined in the current study were differentially expressed between genotypes.

The absorption of monosaccharides from the small intestinal lumen of cattle involves sugar transporters, such as sodium-dependent glucose transporter 1 (encoded by the gene *SLC5A1*) which transports glucose and galactose; whereas glucose transporter (GLUT) 5 (GLUT5; encoded by the gene *SLC2A5*) transports fructose, across the apical membrane of enterocytes. Liao *et al.* [8] examined the expression profiles of glucose transporters along the intestinal tract. SLC5A1 is a sodium-glucose co-transporter and transcription of this gene has been extensively studied in humans. Work on the bovine *SLC5A1* gene has been conducted by Wood *et al.* [32] and Liao *et al.* [8]. Recently the expression of SLC5A1 in small intestinal epithelia was found to be influenced by the level of milk replacer fed to bull calves [26] and suggests that feeding high levels of milk replacer to calves can offer an advantage for greater uptake of lactose. SLC2A2 is a facilitated glucose transporter and is highly conserved among mammals such as humans, dogs, mice and rats. SLC2A5 is a cytochalasin B (a mycotoxin) sensitive fructose transporter. In our study expression of *SLC2A5* was highly correlated with that of *SLC2A2*, possibly due to the fact that they both transport sugars. While we failed to detect an effect of genotype on the expression of sugar transporter genes here, a negative association was observed between the expression of the glucose transporter gene SLC5A1 and total lactational milk solids corrected for body weight. Expression levels of SLC5A1, SLC2A2 and SLC2A5 were highly correlated in the current study. Similar to amino acid transporters, the expression of the sugar transporters is possibly co-regulated.

Conclusions

Taken together with the associated study of Alam *et al.* [6]. These data suggest a possible role for DEG in enhancing feed and production efficiency of dairy cows through improved facilitated absorptive capacity in the duodenum. There is evidence of enhanced expression of key genes involved nutrient transport in the F_1 genotype, compared with the two parent breeds, consistent with the enhanced production efficiency reported for this genotype. Expression of some genes involved in common nutrient transport roles are positively correlated, suggesting that these may be co-regulated. However a global gene expression approach, using tools such as microarrays or RNAseq, across regions of the GIT between breeds and individuals within breeds, is required to gain a greater understanding of the molecular control of feed efficiency and the contribution of GIT tissues in dairy cows. Furthermore, this study identifies potential candidates for investigation of genetic variants regulating nutrient transport and absorption in the duodenum in dairy cows, which may be incorporated into future breeding programmes.

Additional file

Additional file 1: Table S1. Bovine oligonucleotide primers used for qPCR.

Abbreviations

HF: Holstein-Friesian; JE: Jersey; cDNA: Complementary DNA; RT-PCR: Real time polymerase chain reaction; ANOVA: Analysis of variance; SAS: Statistical analysis systems; PTA: Predicted transmitting abilities; PBS: Phosphate buffered saline; Ct: Cycle threshold; DMI: Dry matter intake; GIT: Gastrointestinal tract; BW: Body weight; ME: Metabolisable energy; mRNA: Messenger RNA; N: Nitrogen; ATP: Adenosine tri-phosphate; P: Phosphate; DEG: Differentially expressed genes.

Competing interests

The authors declare that they have no competing interests.

Authors' contributions

SW and DK conceptualised the study and were responsible for experimental design. FB performed the animal study and collected data relating to animal performance. FB and DK collected the duodenal tissue. KK and SW performed real time PCR analysis of duodenal tissue. DK performed statistical analysis. SW, DK and KK participated in the data collection, data analysis and interpretation and drafted the manuscript. All authors approved the final version of the manuscript for publication.

Acknowledgements

This research was funded by Teagasc as part of the Irish National Development Plan. The technical assistance of Noel Byrne and the diligent work of the farm staff at the Ballydague Research Farm (Ballyhooly, Co. Cork, Ireland) are gratefully acknowledged. Authors would also like to acknowledge the advice of Dr. Eva Lewis, Teagasc Moorepark.

Author details

[1]Teagasc Animal and Bioscience Research Department, Animal and Grassland Research and Innovation Centre, Grange, Dunsany, Co. Meath, Ireland.
[2]Teagasc Animal and Bioscience Research Department, Animal and Grassland Research and Innovation Centre, Moorepark, Fermoy, Co. Cork, Ireland.

References

1. Shalloo L, Dillon P, Rath M, Wallace M: **Description and validation of the Moorepark dairy system model.** *J Dairy Sci* 2004, **87:**1945–1959.
2. Coleman J, Berry DP, Pierce KM, Brennan A, Horan B: **Dry matter intake and feed efficiency profiles of 3 genotypes of Holstein-Friesian within pasture-based systems of milk production.** *J Dairy Sci* 2010, **93:**4318–4331.
3. Prendiville R, Pierce KM, Buckley F: **An evaluation of production efficiencies among lactating Holstein-Friesian, Jersey, and Jersey x Holstein-Friesian cows at pasture.** *J Dairy Sci* 2009, **92:**6176–6185.
4. Olson KM, Cassell BG, Hanigan MD, Pearson RE: **Short communication: Interaction of energy balance, feed efficiency, early lactation health**

events, and fertility in first-lactation Holstein, Jersey, and reciprocal F1 crossbred cows. *J Dairy Sci* 2011, **94**:507–511.

5. Schwerin M, Kuehn C, Wimmers S, Walz C, Goldammer T: **Trait associated expressed hepatic and intestine genes in cattle of different metabolic type - putative functional candidates for nutrient utilization.** *J Anim Breed Genet* 2006, **123**:307–314.
6. Alam T, Kenny DA, Sweeney T, Buckley F, Prendiville R, McGee M, Waters SM: **Expression of genes involved in energy homeostasis in the duodenum and liver of Holstein-Friesian and Jersey cows and their F1 hybrid.** *Physiol Genomics* 2112, **44**:198–209.
7. Shellito SM, Ward MA, Lardy GP, Bauerand ML, Caton JS: **Effects of concentrated separator by-product (desugared molasses) on intake, ruminal fermentation, digestion, and microbial efficiency in beef steers fed grass hay.** *J Anim Sci* 2006, **84**:1535–1543.
8. Liao SF, Vanzant ES, Harmon DL, McLeod KR, Boling JA, Matthews JC: **Ruminal and abomasal starch hydrolysate infusions selectively decrease the expression of cationic amino acid transporter mRNA by small intestinal epithelia of forage-fed beef steers.** *J Dairy Sci* 2009, **92**:1124–1135.
9. Liao SF, Harmon DL, Vanzant ES, McLeod KR, Boling JA, Matthews JC: **The small intestinal epithelia of beef steers differentially express sugar transporter messenger ribonucleic acid in response to abomasal versus ruminal infusion of starch hydrolysate.** *J Anim Sci* 2010, **88**:306–314.
10. Higuchi R, Fockler C, Dollinger G, Watson R: **Kinetic PCR analysis: real-time monitoring of DNA amplification reactions.** *Biotechnology (NY)* 1993, **11**:1026–1030.
11. Livak KJ, Schmittgen TD: **Analysis of relative gene expression data using real-time quantitative PCR and the 2[–delta delta C(T)] method.** *Methods* 2001, **25**:402–408.
12. Lewis E, Thackaberry C, Buckley F: **Gastrointestinal tract size as a proportion of liveweight in Holstein, Jersey and Jersey-cross cows.** In *Proceedings from the Irish Agricultural Research Forum, Tullamore, Ireland.* 2011:104. 14th-15th March.
13. Prendiville R, Lewis E, Pierce KM, Buckley F: **Comparative grazing behavior of lactating Holstein-Friesian, Jersey, and Jersey x Holstein-Friesian dairy cows and its association with intake capacity and production efficiency.** *J Dairy Sci* 2010, **93**:764–774.
14. Shull GH: **The composition of a field of maize.** *American Breed Association Report* 1908, **4**:296–301.
15. Ren ZQ, Xiong YZ, Deng CY, Lei MG: **Cloning and identification of porcine SMPX differentially expressed in F1 crossbreds and their parents.** *Acta Biochimica et Biophysica Sinica (Shanghai)* 2006, **38**:753–758.
16. Chase CC Jr, Olson TA, Hammond AC, Menchaca MA, West RL, Johnson DD, Butts WT Jr: **Preweaning growth traits for Senepol, Hereford, and reciprocal crossbred calves and feedlot performance and carcass characteristics of steers.** *J Anim Sci* 1998, **76**:2967–2975.
17. Coleman SW, Chase CC Jr, Phillips WA, Riley DG, Olson TA: **Evaluation of tropically adapted straightbred and crossbred cattle: postweaning gain and feed efficiency when finished in a temperate climate.** *J Anim Sci* 2012, **90**:1955–1965.
18. Walsh S, Buckley F, Berry DP, Rath M, Pierce K, Byrne N, Dillon P: **Effects of breed, feeding system, and parity on udder health and milking characteristics.** *J Dairy Sci* 2007, **90**:5767–5779.
19. Gozho GN, Mutsvangwa T: **Influence of carbohydrate source on ruminal fermentation characteristics, performance, and microbial protein synthesis in dairy cows.** *J Dairy Sci* 2008, **91**:2726–2735.
20. Adams MW, Belyea RL: **Nutritional and energetic differences of dairy cows varying in milk yield.** *J Anim Sci* 1987, **70**:182.
21. Herd RM, Arthur PF: **Physiological basis for residual feed intake.** *J Anim Sci* 2009, **87**:64–71.
22. Latunde-Dada GO, Van der Westhuizen J, Vulpe CD, Anderson GJ, Simpson RJ, McKie AT: **Molecular and functional roles of duodenal cytochrome B (Dcytb) in iron metabolism.** *Blood Cells Mol Dis* 2002, **29**:356–360.
23. Viturro E, Farke C, Meyer HH, Albrecht C: **Identification, sequence analysis and mRNA tissue distribution of the bovine sterol transporters ABCG5 and ABCG8.** *J Dairy Sci* 2006, **89**:553–561.
24. Waters SM, Kenny DA, Killeen AP, Spellman SA, Fitzgerald A, Hennessy AA, Hynes AC: **Effect of level of eicosapentaenoic acid on the transcriptional regulation of D-9 desaturase using a novel in vitro bovine intramuscular adipocyte cell culture model.** *Animal* 2009, **3**:718–727.
25. Yonezawa T, Yonekura S, Sanosaka M, Hagino A, Katoh K, Obara Y: **Octanoate stimulates cytosolic triacylglycerol accumulation and CD36 mRNA expression but inhibits acetyl coenzyme a carboxylase activity in primary cultured bovine mammary epithelial cells.** *J Dairy Res* 2004, **71**:398–404.
26. Orihashi T, Mashiko T, Sera K, Roh SG, Katoh K, Obara Y: **Effects at early stage of life of elevated milk replacer feeding on growth rate, plasma IGF-I concentration and intestinal nutrient transporter expression in Holstein bull calves.** *Anim Sci J* 2012, **83**:77–82.
27. Desjeux JF: **The molecular and genetic base of congenital transport defects.** *Gut* 2000, **46**:585–587.
28. Chen H, Wong EA, Webb KE Jr: **Tissue distribution of a peptide transporter mRNA in sheep, dairy cows, pigs, and chickens.** *J Anim Sci* 1999, **77**:1277–1283.
29. Sánchez-Pozo A, Gil A: **Nucleotides as semiessential nutritional components.** *Br J Nutr* 2002, **87**:S135–S137.
30. Liao SF, Alman MJ, Vanzant ES, Miles ED, Harmon DL, McLeod KR, Boling JA, Matthews JC: **Basal expression of nucleoside transporter mRNA differs among small intestinal epithelia of beef steers and is differentially altered by ruminal or abomasal infusion of starch hydrolysate.** *J Dairy Sci* 2008, **91**:1570–1584.
31. Ritzel MW, Ng AM, Yao SY, Graham K, Loewen SK, Smith KM, Hyde RJ, Karpinski E, Cass CE, Baldwin SA, Young JD: **Recent molecular advances in studies of the concentrative Na + –dependent nucleoside transporter (CNT) family: identification and characterization of novel human and mouse proteins (hCNT3 and mCNT3) broadly selective for purine and pyrimidine nucleosides (system cib).** *Mol Membr Biol* 2001, **18**:65–72.
32. Wood IS, Dyer J, Hofmann RR, Shirazi-Beechey SP: **Expression of the Na+/glucose co-transporter (SGLT1) in the intestine of domestic and wild ruminants.** *Pflugers Arch* 2000, **441**:155–162.

Trans-10, *cis*-12 conjugated linoleic acid reduces neutral lipid content and may affect cryotolerance of *in vitro*-produced crossbred bovine embryos

Ribrio Ivan Tavares Pereira Batista[1,2], Nádia Rezende Barbosa Raposo[1], Paulo Henrique Almeida Campos-Junior[3], Michele Munk Pereira[1], Luiz Sergio Almeida Camargo[2], Bruno Campos Carvalho[2], Marco Antonio Sundfeld Gama[2] and João Henrique Moreira Viana[2*]

Abstract

Background: Due to high neutral lipids accumulation in the cytoplasm, *in vitro*-produced embryos from *Bos primigenius indicus* and their crosses are more sensitive to chilling and cryopreservation than those from *Bos primigenius taurus*. The objective of the present study was to evaluate the effects of *trans*-10, *cis*-12 conjugated linoleic acid (CLA) on the development and cryotolerance of crossbred *Bos primigenius taurus* x *Bos primigenius indicus* embryos produced *in vitro*, and cultured in the presence of fetal calf serum. Bovine zygotes (n = 1,692) were randomly assigned to one of the following treatment groups: 1) Control, zygotes cultured in Charles Rosenkrans 2 amino acid (CR2aa) medium (n = 815) or 2) CLA, zygotes cultured in CR2aa medium supplemented with 100 μmol/L of *trans*-10, cis-12 CLA (n = 877). Embryo development (cleavage and blastocyst rates evaluated at days 3 and 8 of culture, respectively), lipid content at morula stage (day 5) and blastocyst cryotolerance (re-expansion and hatching rates, evaluated 24 and 72 h post-thawing, respectively) were compared between groups. Additionally, selected mRNA transcripts were measured by Real–Time PCR in blastocyst stage.

Results: The CLA treatment had no effect on cleavage and blastocyst rates, or on mRNA levels for genes related to cellular stress and apoptosis. On the other hand, abundance of mRNA for the 1-acylglycerol-3-phosphate 0-acyltransferase-encoding gene (AGPAT), which is involved in triglycerides synthesis, and consequently neutral lipid content, were reduced by CLA treatment. A significant increase was observed in the re-expansion rate of embryos cultured with *trans*-10, *cis*-12 CLA when compared to control (56.3 vs. 34.4%, respectively, $P = 0.002$). However, this difference was not observed in the hatching rate (16.5 vs. 14.0%, respectively, $P = 0.62$).

Conclusions: The supplementation with *trans*-10, *cis*-12 CLA isomer in culture medium reduced the lipid content of *in vitro* produced bovine embryos by reducing the gene expression of 1-acylglycerol 3-phosphate 0-acyltransferase (AGPAT) enzyme. However, a possible improvement in embryo cryotolerance in response to CLA, as suggested by increased blastocyst re-expansion rate, was not confirmed by hatching rates.

Keywords: Blastocysts, CLA, Crossbred cattle, Cryopreservation, Lipid content

* Correspondence: henrique.viana@embrapa.br
[2]Embrapa Dairy Cattle Research Center, Juiz de Fora, MG 36038-330, Brazil
Full list of author information is available at the end of the article

Background

Dairy herds in tropical countries are often composed of crossbred *Bos primigenius taurus* x *Bos primigenius indicus* cows in order to take advantage of heterosis and characteristics from zebu breeds such as resistance to heat stress [1]. *In vitro* embryo production (IVEP) technique has been largely used in tropical countries like Brazil to improve the genetic value for milk traits in Zebu breeds and to obtain crossbred Holstein-Zebu dairy cows [2]. However, crossbred *Bos primigenius taurus* x *Bos primigenius indicus* embryos produced *in vitro* are less cryotolerant when compared to *Bos primigenius taurus* embryos, due to increased lipid accumulation in blastomeres' cytoplasm [3,4]. The low survival rate of those embryos after cryopreservation is a major obstacle to a wider use of IVEP and frozen embryos still represent less than 5% of all embryo transfers in Brazil, limiting the potential of *in vitro* embryo technology [2]. The high neutral lipid content, especially triglycerides stored in the lipid droplets present in the blastomeres [5], associated with the low concentration of polyunsaturated fatty acids in membrane phospholipids have been suggested as the major cause of low cryotolerance of *in vitro* produced embryos [6-8]. Excessive endogenous neutral lipid accumulation affects the equilibrium of dehydration and rehydration during embryo freezing and thawing [9], triggering ice crystals formation and thereby exacerbating the deleterious effects of cryopreservation.

The conjugated linoleic acid (CLA; generic term for a group of 18-carbon fatty acids with a conjugated double bond) has been shown to influence a range of biological processes, presenting anti-carcinogenic, antiatherogenic and anti-obesity properties. Specifically, the *trans*-10, *cis*-12 CLA isomer has been shown to alter membrane lipid composition [10], to inhibit lipid synthesis in different animals, and to regulate the expression of genes involved in the *de novo* fatty acids and triglycerides synthesis [11-13]. Additionally, supplementation of *in vitro* culture systems with *trans*-10, *cis*-12 CLA was suggested to improve bovine (*Bos primigenius taurus*) embryo survival after cryopreservation [14]. However, some studies with pre-neoplastic, neoplastic, and mammary gland cells have shown that *trans*-10, *cis*-12 CLA can also induce apoptosis, both by the endoplasmic reticulum (ER) pathway [15] and by the mitochondrial pathway [16], due to increased levels of free radicals and aldehydic products formed by the catabolism of this fatty acid [17]. Although these are different cell types, the early embryo cells show high metabolic activity, similarly to tumor cells [18], and it is possible that *trans*-10, *cis*-12 CLA causes some effects in the embryonic cells similar to those reported for somatic cells. The apparent contradiction between potentially beneficial and detrimental effects of *trans*-10, *cis*-12 CLA has not been addressed in previous studies evaluating embryos cultured in the absence of additional antioxidant protection, such as routinely used in the IVEP industry [17,19,20].

Eventual side effects of CLA can be evaluated by the analysis of the expression of genes related to cell stress and apoptosis, such as Peroxiredoxin (PRDX), heat shock proteins (HSP), and B-cell lymphoma 2 (Bcl-2) family. PRDXs are a ubiquitous family of thiol-specific antioxidant enzymes that control cytokine-induced peroxide levels, by reducing and detoxifying a wide range of organic hydroperoxides (ROOH) [21]. Hsp70s are an important part of the cell's machinery for protein folding, and help to protect cells in stress situation. The accumulation of unfolded or misfolded proteins caused by oxidative stress induces a cellular protective response called "unfolded protein response", characterized by increased expression of Hsp70 [22]. The B-cell lymphoma 2 family proteins control the release of cytochrome c during the apoptotic pathway. The proapoptotic proteins Bax and Bak form oligometric channels in the mitochondrial outer membrane, and cause the exit of the cytochrome c into the cytosol. In contrast, antiapoptotic family members Bcl-2 and Bcl-XL sequester the proapoptotic proteins, preventing the formation of protein-conducting channels and inhibiting apoptosis [23]. The proportion of the relative expression of Bax and Bcl-2 genes is commonly used as an indicator of apoptosis. However, Ip et al. [24] demonstrated that CLA induces apoptosis in mammary tumor cells by reducing the expression of Bcl-2, but without affecting the expression of proapoptotic members (Bak and Bax).

Thus, the aim of this study was to evaluate the effects of *trans*-10, *cis*-12 CLA in a conventional *in vitro* culture system on crossbred *Bos primigenius taurus* x *Bos primigenius indicus* embryos. The CLA effects were evaluated for embryo development rates, lipid accumulation, cryotolerance, abundance of mRNA for lipogenic enzymes such as acetyl-CoA carboxylase beta (ACACB), fatty acid synthase (FASN) and 1-acylglycerol 3-phosphate 0-acyltransferase (AGPAT1), as well as proteins involved in cell stress such as heat shock proteins 70.1 (Hsp70.1) and peroxiredoxin 1 (PRDX.1), and apoptosis, such as B-cell lymphoma 2 (Bcl-2).

Methods

Chemicals and reagents

All chemicals and reagents were purchased from Sigma Chemical Co. (St. Louis, MO, USA) unless otherwise stated.

In vitro embryo production

Ovaries of dairy crossbred cows (*Bos primigenius indicus* x *Bos primigenius taurus, primarily Holstein x Gir*) were collected from slaughterhouse and transported to the laboratory in saline solution (0.9% NaCl) at 35°C containing antibiotic (0.05 g/L of streptomycin sulfate). Follicles of 3–8 mm were aspirated and cumulus-oocyte complexes (COCs) of grades I and II [25] were selected, and washed twice in medium TALP-HEPES [26]. Groups of 30–40 COCs were then placed into four-well plates (NUNC, Roskilde, Denmark) for *in vitro* maturation (IVM). Each well contained 400 μL of tissue culture medium (TCM–199, Gibco Life Technologies, Inc., Grand Island, NY, USA) supplemented with 20 μg/mL follicle stimulating hormone (FSH; Pluset, Serono, Italy), 0.36 mmol/L sodium pyruvate, 10 mmol/L sodium bicarbonate, 50 mg/mL streptomycin/penicillin and 10% estrous cow serum. The COC were incubated for 22–24 h at 38.5°C in a humidified atmosphere containing 5% CO_2. For *in vitro* fertilization, frozen/thawed semen was centrifuged at 9,000 × g for 5 min in a Percoll (Nutricell Nutrientes Celulares, Campinas, SP, Brazil) discontinuous density gradient (45–90%) to obtain motile spermatozoa. The pellet was centrifuged again at 9,000 × g for 3 min in FERT-TALP medium [27]. Semen from the same bull (*Bos primigenius taurus*) was used throughout experiment. At the end of the maturation period, 25–30 COCs were co-incubated for 20 h with 2×10^6 spermatozoa/mL in 100 μL drops of FERT-TALP medium supplemented with heparin (20 μg/mL) and bovine serum albumin (BSA) fatty acid free (6 mg/mL) under mineral oil. For embryo culture, presumptive zygotes were partially denuded by mechanically pipetting in TALP-HEPES medium and then randomly allocated to the experimental groups and co-cultured with their respective cumulus cells in four-well plates (NUNC, Roskilde, Denmark). Embryo culture was performed under the same atmospheric conditions of fertilization. After 72 h post-insemination (hpi) (d 3) 50% of the medium (CR2aa) was renewed and cleavage rate assessed. Blastocyst rate was assessed at 192 hpi (d8).

Experimental design

After *in vitro* maturation and fertilization, presumptive zygotes (n = 1,692) were randomly assigned into two groups: Control – zygotes (n = 815) cultured in 500 μL CR2aa medium (modified from Rosenkrans Jr and First [28]) supplemented only with 10% FCS and 3 mg/mL BSA fraction V; or CLA – zygotes (n = 877) cultured in the same conditions of the control group, but with additional supplementation of the culture medium with 100 μmol/L *trans*–10, cis–12 CLA (Matreya, LLC, Pleasant Gap, PA, USA) as described by Pereira et al. [17]. Unlike Pereira et al. [17], however, in the present study no additional antioxidants were used in the culture medium. On the 3rd day of culture (d 3) cleavage rate was assessed. On d 5, samples of embryos at morula stage (n = 15/group) were fixed for analysis of lipids content. On d 7, another sample of Grade I and II embryos at blastocyst stage (3 pools of 10 embryos/group) were placed in Eppendorf tubes and stored at – 80°C until RNA extraction. For both lipid content and gene expression analysis, embryos were sampled from different batches, and the remaining d 7 and d 8 Grade I and II blastocysts from all batches underwent slow freezing for analysis of pos-thawing survival rate or were used as controls (93 and 49 from control group, and 103 and 44 from the CLA group, respectively). Blastocyst rate was assessed on d 8 considering all blastocysts produced in drops where embryos were not removed for lipids quantification.

Quantification of neutral lipid droplet content

Quantification of lipids was performed by Nile Red dye technique as described by Leroy et al. [29]. Briefly, d 5 embryos (n = 15 each groups) free of cumulus cells were fixed in a 500 μL of 2% glutaraldehyde and 2% formaldehyde solution at 4°C for at least 24 h. They were transferred to individual Eppendorf tubes (1 embryo/tube) containing 30 μL of a solution with 10 μg/mL Nile Red (Molecular Probes, Inc., Eugene, OR, USA) dissolved in saline solution (0.9% NaCl) with 1 mg/mL polyvinylpyrrolidone. Embryos were stained overnight in the dark and at room temperature. The Nile Red stock solution (1 mg/mL) was prepared previously with dimethyl sulfoxide (DMSO) and stored at room temperature in the dark. Final concentrations were obtained by diluting the stock with the saline solution. The amount of emitted fluorescent light of the whole embryo was evaluated at 582 ± 6 nm with an Olympus BX60 light microscope, equipped with an epifluorescence system using a 10 × objective. Images were captured separately using monochrome filters for FITC with a cooled charge-coupled device camera (Magnafire; Olympus). Images were imported into Adobe Photoshop (Adobe Systems, Mountain View, CA) as TIFF files. Fluorescence was quantified using the custom QUANTIPORO software (GNU Image Manipulation Program). The results were expressed in arbitrary units of fluorescence

RNA extraction and relative quantification by real-time PCR

Total RNA extraction was performed from 3 pools of 10 embryos/group using RNeasy Micro kit (Qiagen, Hilden, Germany) according to manufacturer's instructions and treated with DNase. Complementary DNA (cDNA) was synthesized using Superscript III First-strand supermix kit (Invitrogen, Carlsbad, CA, USA) and a random hexamer primer, according to manufacturer's instructions. The cDNA quantification from each pool per group was performed using 1 μL of sample and a spectrophotometer ND-100 (NanoDrop Products, Wilmington,

DE, USA). Relative quantification was performed in triplicate using Real-Time PCR (ABI Prisml 7300, Applied Biosystem, Forster City, CA, USA) and reactions using a mixture of Power SYBR Green PCR Master Mix (Applied Biosystem), 400 ng cDNA, nuclease-free water and specific primers for each reaction. Template cDNA was denatured at 95°C for 10 min, followed by 45 cycles of 95°C for 15 s; gene-specific primer annealing temperature for 30 s and elongation at 60°C for 30 s. After each PCR run, a melting curve analysis was performed to confirm that a single specific product was generated. Negative controls, comprised of the PCR reaction mix without nucleic acid, were also run for each group of samples. Amplicon size was confirmed by polycryalamide gel electrophoresis. Primer sequences and sizes of the amplified fragments for all transcripts are shown in Table 1. Primer efficiency was calculated for each reaction using Lin-RegPCR software [30]. The average efficiency of each set of primers was calculated and taking into account all groups. Expression of GAPDH gene was used as an endogenous reference. Relative abundance analyses were performed using REST software [31] and based on primer efficiency. Values found in embryos from CLA group are shown as n-fold differences relative to the control.

Cryopreservation and Survival of embryos after thawing

Blastocysts classified as grade I and II according to Kennedy et al. [32] from both Control and CLA groups were washed in a solution of Dulbecco's Phosphate-Buffered Saline (DPBS) with 0.4% BSA (Nutricell, Nutrientes Celulares, Campinas, SP, Brazil) and then dehydrated in a solution of 1.5 mol/L ethylene glycol in DPBS with 0.4% BSA (Nutricell, Nutrientes Celulares, Campinas, SP, Brazil), loaded into 0.25 mL straws and frozen in an automatic freezing device (Freeze Control, cryologic, Victoria, Australia) from – 6 to – 35°C at a 0.5°C per min rate. For thawing, each straw was kept 10 s in air and 20 s in a water bath at 35°C. Embryos were then co-cultured in CR2aa drops of 50 μL (15 embryos/drop) with a monolayer of granulosa cells for 72 h. The rate of re-expansion and hatching was assessed at 24 and 72 h of culture, respectively. The re-expansion and hatching rates were calculated based on the percentage of thawed embryos showing re-expansion of the blastocoel cavity and hatching of zona pellucida, respectively, after culture.

Statistical analysis

The results of cleavage and blastocyst rates from each routine of IVEP were considered as replicates, while the lipid content was analyzed for each embryo. Data from IVEP and intensity of fluorescence were tested for normality and homocedasticity by Lilliefors and Cochran & Bartlett tests, respectively; after that, analysis of variance (ANOVA) was used to assess statistical differences. Values of relative expression of target genes were performed using the software REST using the Pair Wise Fixed Reallocation Randomization Test tool. Data for embryo survival after cryopreservation (re-expansion and hatching) were analyzed by chi-square method. The statistical significance was determined based on a *P*-value of 0.05.

Results

Developmental competence

Addition of *trans*-10, *cis*-12 CLA in culture medium did not affect cleavage ($P = 0.06$) and blastocyst rates

Table 1 Sequence of primers and annealing temperature specific for each gene

Gene	Primer Sequence (5'→3')	Annealing Temperature	Fragment Size	GenBank Accession N°/Reference
GAPDH*	F: CCAACGTGTCTGTTGTGGATCTGA	58°C	237	[42]
	R: GAGCTTGACAAAGTGGTCGTTGAG			
FASN	F – 5' GCACCGGTACCAAGGTGGGC 3'	58°C	171	NM_001012669.1
	R – 5' CGTGCTCCAGGGACAGCAGC 3'			
ACACB	F – 5' CGGTGGTGCAGTGGCTGGAG 3'	58°C	254	XM_867921.3
	F – 5' CAGGAGGACCGGGGGTCAGG 3'			
AGPAT1	F – 5' CCGGAAGCGCACTGGGGATG 3'	58°C	170	NM_177518.1
	R – 5' TGGGAACCTGGGCCTGCACT 3'			
PRDX1	F – 5' ATGCCAGATGGTCAGTTCAAG 3'	53°C	224	[42]
	R – 5' CCTTGTTTCTTGGGTGTGTTG 3'			
HSP70.1	F – 5' AACAAGATCACCATCACCAACG 3'	59°C	275	NM_174550
	R – 5' TCCTTCTCCGCCAAGGTGTTG			
BCL-2	F– 5' TGGATGACCGAGTACCTGAA 3'	53°C	120	XM_586976
	R – 5' CAGCCAGGAGAAATCAAACA 3'			

*Reference gene.

Table 2 Development of bovine embryonic presumptive zygotes cultured in CR2aa medium supplemented with serum (Control) or with serum plus *trans*-10, *cis*-12 CLA

Treatment	Presumptive zygotes n	Cleavage d 3 n (% ± SD)	Blastocyst d 8 n (% ± SD)
Control	817	624 (74.5 ± 2.2)	276 (34.1 ± 2.6)
CLA	877	606 (70.3 ± 3.2)	280 (31.8 ± 2.4)

D3 = 72 h after fertilization, D8 = 192 h after fertilization, (n) = number of structures, SD = standard Deviation.

($P = 0.20$) as shown in Table 2. The developmental competence of cleaved embryos to reach blastocyst stage was also unaffected by CLA (44.2 and 46.2% for Control and CLA groups, respectively).

Lipids quantification

Figure 1 shows the result of the fluorescence intensity of control and CLA embryos after Nile red staining. Morulas that were cultured in the absence of *trans*-10, *cis*-12 CLA contained more lipid droplets expressed as a significantly higher amount of emitted fluorescence light ($P = 0.0001$), compared with the morulas that were cultured in presence *trans*-10, *cis*-12 CLA (130.4 ± 30.5 vs. 183.9 ± 17.9 arbitrary fluorescence units, Figure 2).

Analysis of target gene transcripts

Blastocysts cultured in CR2aa supplemented with 100 μM of *trans*-10, *cis*-12 CLA showed no effect ($P > 0.05$) on amount of mRNA expressed for Hsp70.1 (0.72 ± 0.17), PRDX1 (1.12 ± 0.27), and Bcl-2 (1.20 ± 0.26) genes when compared to the control group. Additionally, the analysis of transcripts of genes related to fatty acid synthesis showed a reduction ($P = 0.009$) in the amount of transcripts for AGPAT1 (0.16 ± 0.09) gene, but not for ACACB (0.6 ± 0.21, $P = 0.18$) and FASN (0.68 ± 0.30, $P = 0.30$) genes of blastocysts cultured with CLA (Figure 3).

Survival of embryos after cryopreservation

There was a higher ($P = 0.002$) re-expansion rate after thawing of embryos cultured in medium supplemented with CLA (Table 3). This difference, however, was not reflected in a higher hatching rate ($P = 0.62$). Hatching rate in relation to re-expanded embryos was 43.75 and 29.31% for the control and CLA group, respectively. Fresh embryos from Control and CLA groups had similar ($P = 0.16$) hatching rate, which was higher than that of frozen-thawed embryos of Control and CLA groups.

Discussion

In the present study, we evaluated the effect of the addition of *trans*-10, *cis*-12 CLA in the culture medium of a conventional *in vitro* embryo production system. *In vitro* development of crossbred *Bos primigenius taurus* x *Bos primigenius indicus* embryos was not affected by supplementation of *trans*-10, *cis*-12 CLA in the absence of any additional antioxidant agent other than serum. CLA was suggested to increase survival rate after cryopreservation [17,22]. In these studies, however, culture media were added with antioxidant glutathione and β-mercaptoethanol, respectively, to prevent fatty acids oxidation. Moreover, in the study of Pereira et al. [17] media was renewed in a daily basis, while in the current study we target the potential effects of CLA in a conventional culture system, i.e., only one renewal of medium 72 h.p.i. Interestingly, the expression of gene transcripts of the PRDX.1 peroxiredoxin family member, which protect against the deleterious effects of oxidation in the cell, were unaffected by the CLA treatment. The catabolism of *trans*-10, *cis*-12 CLA occurs preferentially in peroxisomes [33]. In these organelles, the first reaction of β-oxidation produces H_20_2 instead of FADH, as occurs in the mitochondria. Thus, the increase in PRDX1 expression by the embryo may occurs in response to the increased levels of H_20_2 induced by the CLA. The results from the present experiment, however, suggest a low

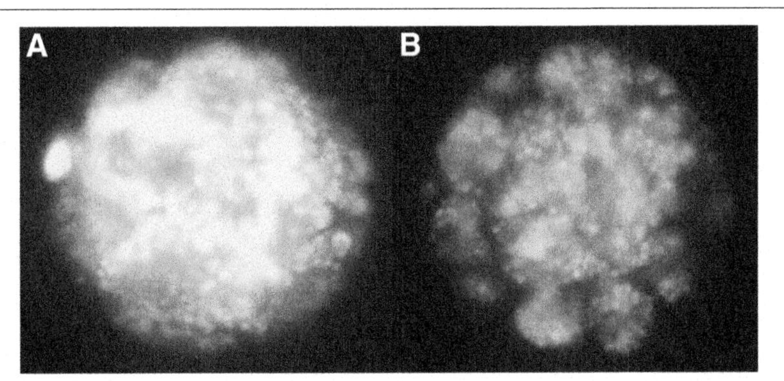

Figure 1 Lipid content in embryos cultured in the presence of *trans*-10, *cis*-12 CLA. Representative image of embryos produced *in vitro* and cultured in the absence **(A)** or presence **(B)** of *trans*-10, *cis*-12 CLA and stained with Nile Red die. The emitted fluorescence was restricted to lipid droplets (100 × magnification).

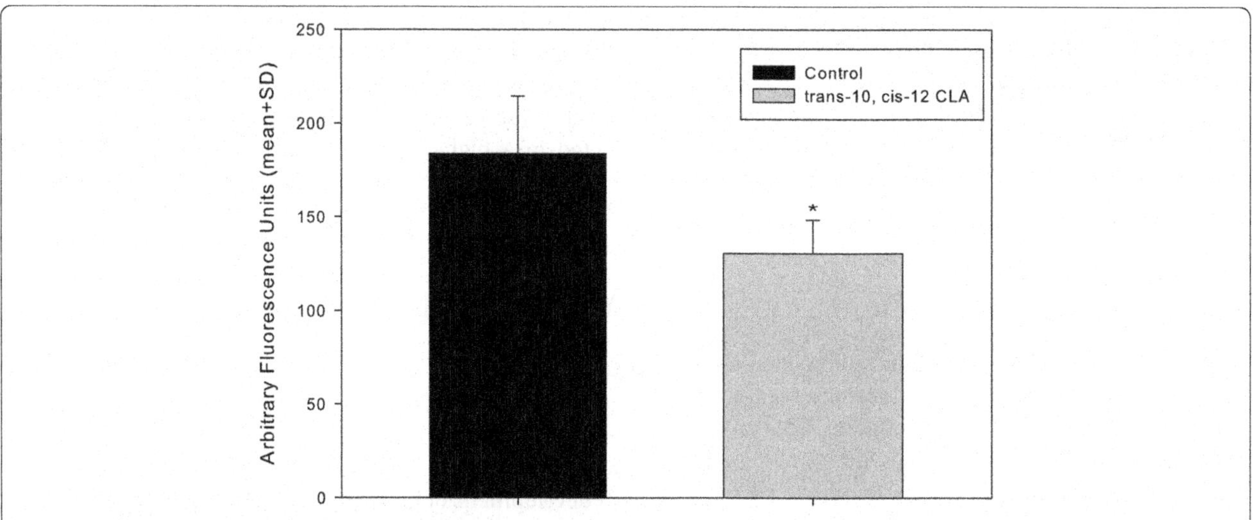

Figure 2 Effect of *trans*-10, *cis*-12 CLA in embryo lipid accumulation. Fluorescence emission (in arbitrary fluorescence units) of morula stage embryos cultured in the absence or presence of *trans*-10, *cis*-12 CLA. (*) Represents significant difference between the two groups ($P = 0.0001$).

fatty acid metabolism during this period of development, which may have minimized the deleterious effects associated with the production of free radicals. Coherently, Thompson et al. [34] reported a reduction in the amount of ATP produced by the blastomeres, possibly due to a reduction in the proportional contribution of oxidative phosphorylation. This strategy may be required by the embryo due to the low concentration of oxygen in the way from oviduct to the uterus [35].

Similarly, the expressions of PRDX.1, cellular stress gene (Hsp70.1) and anti-apoptotic (Bcl-2) were not affected by fatty acid supplementation during embryo culture. Hsp70.1 is over-expressed in stress situations, such as elevated temperature, hypoxia, presence of reactive oxygen species, abnormal pH, and others, and has been shown to attenuate apoptosis [36], being frequently used as an indicator of stress in bovine embryos [1,37]. The Bcl-2, when down-regulated, allows the activation of Bax and thus the formation of protein channels through which pro-apoptotic factors such as cytochrome C are released into the cytosol, initiating the apoptotic cascade [38,39]. Although no effect of *trans*-10, *cis*-12

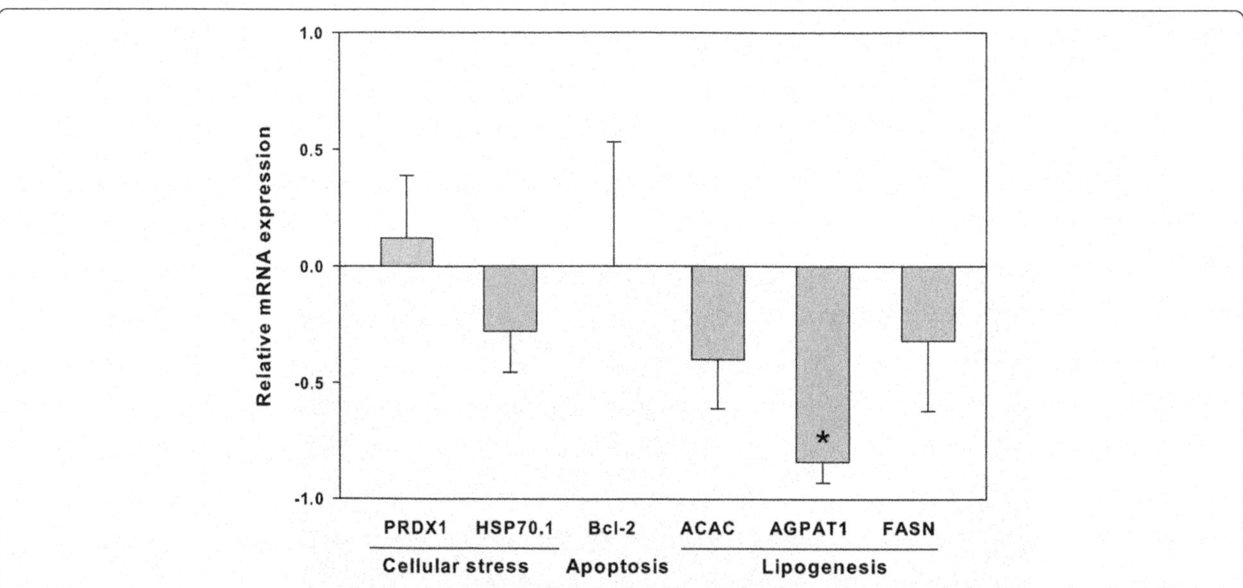

Figure 3 Gene expression in embryos cultured in the presence of *trans*-10, *cis*-12 CLA. Relative expression of transcripts of genes involved in cellular stress (PRDH1 and Hsp70.1), apoptotic process (Bcl-2), fatty acid synthesis (FASN and ACACB) and triglycerides synthesis (AGPAT1), through the Real - Time PCR in embryos cultured in the presence of *trans*-10, *cis*-12 CLA, in relation to the control group. (*) Represents significant difference between the two groups ($P = 0.009$).

Table 3 Survival and developmental capacity after cryopreservation (as indicated by rates of re-expansion and hatching) of embryos cultured in CR2aa medium supplemented with serum (Control) or with serum plus *trans*-10, *cis*-12 CLA (CLA), compared to non-frozen embryos from the same groups

Group	*In vitro* culture	Blastocyst n	Re-expansion % (n)	Hatching % (n)
Control	Fresh	49	-	77.6 (38)[a]
CLA	Fresh	44	-	88.6 (39)[a]
Control	Frozen-thawed	93	34.4 (32)[a]	14.0 (13)[b]
CLA	Frozen-thawed	103	56.3 (58)[b]	16.5 (17)[b]

[a,b]Different lower case letters in the same column indicate statistical difference by Chi-square test ($P < 0.01$).

CLA was reported on levels of Bax or on translocation of Bax to the mitochondria, significant down-regulation occurred in Bcl-2 expression in mouse mammary tumor cells cultured with CLA [18].

Trans-10, *cis*-12 CLA modulates the amount of intracellular lipids through transcription mediators, as transcription factors sterol-regulatory element binding protein (SREBP) [11]. In the present study, *trans*-10, *cis*-12 CLA supplementation during embryo culture down-regulated the expression of AGPAT1 gene in blastocysts, when compared to the control group. This result was associated with a reduction in lipid accumulation and with an increase in re-expansion rate after cryopreservation in the CLA group. The expressions of enzymes related to de novo synthesis of fatty acids (ACACB and FASN), however, were not affected. Previous studies suggested that the route by which the embryo synthesizes triglycerides was preferably by the assembly of fatty acids internalized by the cell [40,41]. Coherently, Gutgesell et al. [42] showed that dietary *trans*-10, *cis*-12 CLA down-regulates gene expression of the proteins responsible for fatty acids uptake by the cells. Altogether, the results of the present study bring additional evidences that the mechanism by which CLA affect lipid accumulation in the embryo is by a reduction in the incorporation and/or assembly of triglycerides, rather than by inhibiting de novo synthesis.

Hatching rate is largely used as an indicator of embryo developmental potential after cryopreservation. In the current study, we observed that *trans*-10, *cis*-12 CLA supplementation in culture medium did not increase hatching rate after cryopreservation. This result suggests that the reduction of intracytoplasmic fat caused by CLA, regardless a potential positive effect on re-expansion after cryopreservation, was not sufficient to protect the embryo from later detrimental effects of cryopresenvation. Alternatively, it is also possible that CLA supplementation in culture medium may have affected the mechanisms associated with the enzymatic digestion of the zona pellucida by the trofectoderm, impairing hatching. In fact, the reduction in intracytoplasmic fat content by delipidation does not disturb preimplantation development of *in vitro*-fertilized embryos and is beneficial for embryo survival after cryopreservation, but may impair further embryo development [43]. The apparent contradiction between the beneficial effects of CLA (reduced embryo neutral lipid content, increased re-expansion after thawing) and the lack of difference in hatching rates, as observed in the current, suggests that further adjustments in CLA concentration are needed to balance the its potential detrimental effects and improve embryo cryotolerance.

Conclusions

The use of *trans*-10, *cis*-12 CLA supplementation in a conventional *in vitro* embryo production system has no deleterious effects on blastocyst rate and thus can be used as an alternative to reduce embryo neutral lipid accumulation in crossbred zebu x *B. taurus* embryos. The mechanism by which *trans*-10, *cis*-12 CLA reduces the neutral lipid content of *in vitro* produced embryos involves a down-regulation in the expression of the 1-acylglycerol-3-phosphate 0-acyltransferase enzyme. However, a possible improvement in embryo cryotolerance in response to CLA, as suggested by increased blastocyst re-expansion rate, was not confirmed by hatching rates.

Abbreviations

ACACB: Lipogenic enzymes as acetyl-CoA carboxylase beta; AGPAT1: 1-acylglycerol 3-phosphate 0-acyltransferase; Bcl-2: B-cell lymphoma 2; BSA: Bovine serum albumin; CLA: Conjugated linoleic acid; COCs: Cumulus-oocyte complexes; DPBS: Dulbecco's Phosphate-Buffered Saline; ER: Endoplasmic reticulum; FASN: Fatty acid synthase; FCS: Fetal calf serum; HPI: After post-insemination; Hsp70.1: Heat shock proteins 70.1; IVEP: *In vitro* embryo production; PRDX.1: Peroxiredoxin 1.

Competing interests

The authors declare that they have no competing interests.

Authors' contributions

RTPB carried out the experiment, participated in the data collection, data analysis, and drafted the manuscript. PHAC-J helped in the *in vitro* embryo production, in embryo cryopreservation, and in quantification of lipid contents. MMP helped in gene expression analysis. NRBR, MASG and LSAC contributed in the conception and design of the experiment, preparation of the manuscript, and acquisition of funding. BCC helped in data analysis and interpretation. JHMV contributed in the conception and design of experiment, analysis and interpretation of data, preparation of the final version of the manuscript, critical revision of the content and acquisition of funding. All authors approved the final version of the manuscript for publication.

Acknowledgements

This study was supported by the Brazilian National Council for Scientific and Technological Development (CNPq), from the Minas Gerais State Research Foundation (FAPEMIG) and from Embrapa (Project 01.07.01.002). Batista received a grant from FAPEMIG. The authors thank Molecular Genetics Laboratory of Embrapa Dairy Cattle for technical assistance in Real-Time PCR.

Author details

[1]Federal University of Juiz de Fora, Juiz de Fora, MG 36036-900, Brazil.
[2]Embrapa Dairy Cattle Research Center, Juiz de Fora, MG 36038-330, Brazil.
[3]Federal University of Minas Gerais, Belo Horizonte, MG 31270-901, Brazil.

References

1. Camargo LSA, Viana JHM, Ramos AA, Serapião RV, de Sa F, Ferreira AM, Guimarães MFM, Do Vale Filho VR: **Developmental competence and expression of the Hsp 70.1 gene in oocytes obtained from Bos indicus and Bos taurus dairy cows in a tropical environment.** *Theriogenology* 2007, **68:**626–632.
2. Viana JHM, Siqueira LGB, Palhão MP, Camargo LSA: **Features and perspectives of the Brazilian *in vitro* embryo industry.** *Anim Reprod* 2012, **09:**12–18.
3. Visintin JA, Martins JF, Bevilacqua EM, Mello MR, Nicacio AC, Assumpcao ME: **Cryopreservation of Bos taurus vs. Bos indicus embryos: Are they really different?** *Theriogenology* 2002, **57:**345–359.
4. Steel R, Hasler JF: **Pregnancy rates resulting from transfer of fresh and frozen Holstein and Jersey embryos.** *Reprod Fertil Dev* 2004, **16:**182–183.
5. McEvoy TG, Coull GD, Broadbent PJ, Hutchinson JS, Speake BK: **Fatty acid composition of lipids in immature cattle, pig and sheep oocytes with intact zona pellucida.** *J Reprod Fertil* 2000, **118:**163–170.
6. Wilmut I: **The low temperature preservation of mammalian embryos.** *J Reprod Fert* 1972, **31:**513–514.
7. Polge C, Wilmut J, Rowson LEA: **Low temperature preservation of cow, sheep and pig embryos.** *Cryobiology* 1974, **11:**560.
8. Nagashima HH, Yamakawa H, Niemann H: **Freezability of porcine blastocyst at different peri-hatching stages.** *Theriogenology* 1992, **37:**839–850.
9. Vajta G, Rindom N, Peura TT, Holm P, Greve T, Callesen H: **The effect of media, serum and temperature on in vitro survival of bovine blastocysts after open pulled straw (OPS) vitrification.** *Theriogenology* 1999, **52**(Suppl 5):939–948.
10. Sampath H, Ntambi JM: **Polyunsaturated fatty acid regulation of genes of lipid metabolism.** *Annu Rev Nutr* 2005, **25:**317–340.
11. Pariza MH, Park Y, Cook ME: **The biologically active isomers of conjugated linoleic acid.** *Prog Lipid Res* 2001, **40:**283–298.
12. Kim JH, Hubbard NE, Ziboh V, Erickson KL: **Conjugated linoleic acid reduction of murine mammary tumor cell growth through 5-hydroxyeicosatetraenoic acid.** *Biochim Biophys Acta* 2005, **1687:**103–109.
13. Bauman DE, Perfield JW, Harvatine KJ, Baumgard LH: **Regulation of fat synthesis by conjugated linoleic acid: lactation and the ruminant model.** *J Nutr* 2008, **138:**403–409.
14. Pereira RM, Baptista MC, Vasques MI, Horta AEM, Portugal PV, Bessa RJB, Chagas e Silva J, Silva Pereira M, Marques CC: **Cryosurvival of bovine blastocysts is enhanced by culture with trans-10 cis-12 conjugated linoleic acid (10 t, 12c CLA).** *Anim Reprod Sci* 2007, **98:**293–301.
15. Ou L, Ip C, Lisafeld B, Ip MM: **Conjugated linoleic acid induces apoptosis of murine mammary tumor cells via Bcl-2 loss.** *Biochem Biophys Res Commun* 2007, **356:**1044–1049.
16. Ou L, Wu Y, Ip C, Meng X, Hsu Y-C, Ip MM: **Apoptosis induced by t10, c12-conjugated linoleic acid is mediated by an atypical endoplasmic reticulum stress response.** *J Lipid Res* 2008, **49:**985–994.
17. Hanada S, Harada M, Kumemura H, Bishr OM, Koga H, Kawaguchi T, Taniguchi E, Yoshida T, Hisamoto T, Yanagimoto C, Maeyama M, Ueno T, Sata M: **Oxidative stress induces the endoplasmic reticulum stress and facilitates inclusion formation in cultured cells.** *J Hepatol* 2007, **47:**93–102.
18. Krisher RL, Prather RS: **A Role for the Warburg Effect in Preimplantation Embryo Development: Metabolic Modification to Support Rapid Cell Proliferation.** *Mol Reprod Dev* 2012, **79**(5):311–320.
19. Pereira RM, Carvalhais I, Pimenta J, Baptista MC, Vasques MI, Horta AE, Santos IC, Marques MR, Reis A, Pereira MS, Marques CC: **Biopsied and vitrified bovine embryos viability is improved by trans10, cis12 conjugated linoleic acid supplementation during in vitro embryo culture.** *Anim Reprod Sci* 2008, **106:**322–332.
20. Darwich AA, Perreau C, Petit MH, Dupont PPJ, Guillaume D, Mermillod P, Guignot F: **Effect of PUFA on embryo cryoresistance, gene expression and AMPKα phosphorylation in IVF-derived bovine embryos.** *Prostaglandins Other Lipid Mediat* 2010, **93:**30–36.
21. Wood ZA, Schröder E, Harris JR, Poole LB: **Structure, mechanism and regulation of peroxiredoxins.** *Trends Biochem Sci* 2003, **28:**32–40.
22. Schröder M, Kaufman RJ: **ER stress and the unfolded protein response.** *Mutat Res* 2005, **569:**29–63.
23. Klener P Jr, Andera L, Klener P, Necas E, Zivný J: **Cell Death Signalling Pathways in the Pathogenesis and Therapy of Haematologic Malignancies: Overview of Apoptotic Pathways.** *Folia Biol (Praha)* 2006, **52**(1–2):34–44.
24. Ip C1, Ip MM, Loftus T, Shoemaker S, Shea-Eaton Shea-Eaton W: **Induction of apoptosis by conjugated linoleic acid in cultured mammary tumor cells and premalignant lesions of the rat mammary gland.** *Cancer Epidemiol Biomarkers Prev* 2000, **9:**689–696.
25. Viana JHM, Camargo LSA, Ferreira AM, Sa WF, Fernandes CAC, Marques-Junior AP: **Short intervals between ultrasonographically guided follicle aspiration improve oocyte quality but not prevent establishment of dominant follicle in the Gir breed (Bos indicus) of cattle.** *Anim Reprod Sci* 2004, **84:**1–12.
26. Bavister BD, Leibfried ML, Lieberman G: **Development of preimplantation embryos of the Golden hamster in a defined culture medium.** *Biol Reprod* 1983, **28:**235–243.
27. Gordon I: **Laboratory Production of Cattle Embryo.** In *CAB International.* London: Cambridge University; 1994:640.
28. Rosenkrans CF Jr, First NL: **Effect of free amino acids and vitamins on cleavage and developmental rate of bovine zygotes in vitro.** *J Anim Sci* 1994, **72:**434–437.
29. Leroy JL, Genicot G, Donnay I, Van Soom A: **Evaluation of the Lipid Content in Bovine Oocytes and Embryos with Nile Red: a Practical Approach.** *Reprod Domest Anim* 2005, **40:**76–78.
30. Ramakers C, Ruijter JM, Deprez RH, Moorman AFM: **Assumption-free analysis of quantitative real-time polymerase chain reaction (PCR) data.** *Neurosci Lett* 2003, **339:**62–66.
31. Pfaffl MW, Horgan GW, Dempfle L: **Relative expression software tool (REST©) for group-wise comparison and statistical analysis of relative expression results in real-time PCR.** *Nucleic Acids Res* 2002, **30:**26–36.
32. Kennedy LG, Boland MP, Gordon I: **The effects of embryo quality of freezing on subsequent development of thawed low embryo.** *Theriogenology* 1983, **19:**823–832.
33. Banni S, Petroni A, Blasevich M, Carta G, Angionia E, Murru E, Day BW, Melis MP, Spada S, Ip C: **Detection of conjugated C16 PUFAS in rat tissues as possible partial beta-oxidation products of naturally occurring conjugated linoleic acid and its metabolites.** *Biochim Biophys Acta* 2004, **1682:**120–127.
34. Thompson JG, Partridge RJ, Houghton FD, Cox CI, Leese HJ: **Oxygen uptake and carbohydrate metabolism by in vitro derived bovine embryos.** *J Reprod Fertil* 1996, **106:**299–306.
35. Fischer B, Bavister BD: **Oxygen tension in the oviduct and uterus of rhesus monkeys, hamsters and rabbits.** *J Reprod Fertil* 1993, **99:**673–679.
36. Kiang JG, Tsokos GC: **Heat shock protein 70 kda: molecular biology, biochemistry, and physiology.** *Pharmacol Ther* 1998, **80:**183–201.
37. Wrenzycki C, Herrmann D, Keskintepe L, Martins A-J, Sirisathien S, Brackett B, Niemann H: **Effects of culture system and protein supplementation on mRNA expression in pre-implantation bovine embryos.** *Hum Reprod* 2001, **16:**893–901.
38. Danial NN, Korsmeyer SJ: **Cell death: critical control points.** *Cell* 2004, **116:**205–219.
39. Adams JM, Cory S: **The Bcl-2 apoptotic switch in cancer development and therapy.** *Oncogene* 2007, **26:**1324–1337.
40. Ferguson EM, Leese HJ: **Triglyceride content of bovine oocytes and early embryos.** *J Reprod Fertil* 1999, **116:**373–378.
41. Sata R, Tsuji H, Abe H, Yamashita S, Hoshi H: **Fatty acid composition of bovine embryos cultured in serum-free and serum-containing medium during early embryonic development.** *J Reprod Dev* 1999, **45:**97–103.
42. Gutgesell A, Ringseis R, Eder K: **Short communication: Dietary conjugated linoleic acid down-regulates fatty acid transporters in the mammary glands of lactating rats.** *J Dairy Sci* 2009, **92:**1169–1173.
43. Diez C, Heyman Y, Bourhis D, Guyader-Joly C, Degrouard J, Renard JP: **Delipidating in vitro-produced bovine zygotes: effect on further development and consequences for freezability.** *Theriogenology* 2001, **55:**923–936.

Permissions

All chapters in this book were first published in JASB, by BioMed Central; hereby published with permission under the Creative Commons Attribution License or equivalent. Every chapter published in this book has been scrutinized by our experts. Their significance has been extensively debated. The topics covered herein carry significant findings which will fuel the growth of the discipline. They may even be implemented as practical applications or may be referred to as a beginning point for another development.

The contributors of this book come from diverse backgrounds, making this book a truly international effort. This book will bring forth new frontiers with its revolutionizing research information and detailed analysis of the nascent developments around the world.

We would like to thank all the contributing authors for lending their expertise to make the book truly unique. They have played a crucial role in the development of this book. Without their invaluable contributions this book wouldn't have been possible. They have made vital efforts to compile up to date information on the varied aspects of this subject to make this book a valuable addition to the collection of many professionals and students.

This book was conceptualized with the vision of imparting up-to-date information and advanced data in this field. To ensure the same, a matchless editorial board was set up. Every individual on the board went through rigorous rounds of assessment to prove their worth. After which they invested a large part of their time researching and compiling the most relevant data for our readers.

The editorial board has been involved in producing this book since its inception. They have spent rigorous hours researching and exploring the diverse topics which have resulted in the successful publishing of this book. They have passed on their knowledge of decades through this book. To expedite this challenging task, the publisher supported the team at every step. A small team of assistant editors was also appointed to further simplify the editing procedure and attain best results for the readers.

Apart from the editorial board, the designing team has also invested a significant amount of their time in understanding the subject and creating the most relevant covers. They scrutinized every image to scout for the most suitable representation of the subject and create an appropriate cover for the book.

The publishing team has been an ardent support to the editorial, designing and production team. Their endless efforts to recruit the best for this project, has resulted in the accomplishment of this book. They are a veteran in the field of academics and their pool of knowledge is as vast as their experience in printing. Their expertise and guidance has proved useful at every step. Their uncompromising quality standards have made this book an exceptional effort. Their encouragement from time to time has been an inspiration for everyone.

The publisher and the editorial board hope that this book will prove to be a valuable piece of knowledge for researchers, students, practitioners and scholars across the globe.

List of Contributors

Luis Gomez-Raya
Departamento de Mejora Genética Animal, Instituto Nacional de Investigación y Tecnología Agraria y Alimentaria (INIA), Ctra. de La Coruña km 7, 28040, Madrid, Spain

Amanda M Hulse
Interdisciplinary Program in Genetics, Texas A&M University, College Station, TX 77843, USA

David Thain
Department of Animal Biotechnology, University of Nevada, 1664 North Virginia Street, Reno, NV 89557, USA

Wendy M Rauw
Departamento de Mejora Genética Animal, Instituto Nacional de Investigación y Tecnología Agraria y Alimentaria (INIA), Ctra. de La Coruña km 7, 28040, Madrid, Spain

Hirwa Claire D'Andre
Rwanda Agriculture Board, Research Department, P. O Box 5016, Kigali, Rwanda
Department of Animal Genetics, Breeding and Reproduction, College of Animal Science, South China Agricultural University, Guangzhou, Guangdong 510642, China

Wallace Paul
Council for Scientific and Industrial Research (CSIR), Animal Research Institute, P. O. Box AH 20, Accra, Achimota, Ghana

Xu Shen
Department of Animal Genetics, Breeding and Reproduction, College of Animal Science, South China Agricultural University, Guangzhou, Guangdong 510642, China

Xinzheng Jia
Department of Animal Genetics, Breeding and Reproduction, College of Animal Science, South China Agricultural University, Guangzhou, Guangdong 510642, China

Rong Zhang
Department of Animal Genetics, Breeding and Reproduction, College of Animal Science, South China Agricultural University, Guangzhou, Guangdong 510642, China

Liang Sun
Department of Animal Genetics, Breeding and Reproduction, College of Animal Science, South China Agricultural University, Guangzhou, Guangdong 510642, China

Xiquan Zhang
Department of Animal Genetics, Breeding and Reproduction, College of Animal Science, South China Agricultural University, Guangzhou, Guangdong 510642, China

Caitlin A Cooper
Department of Animal Science, University of California, Davis, USA

Luis E Moraes
Department of Animal Science, University of California, Davis, USA

James D Murray
Department of Animal Science, University of California, Davis, USA
Department of Population Health and Reproduction, University of California, Davis, USA

Sean D Owens
Department of Pathology, Microbiology and Immunology, School of Veterinary Medicine, University of California, ne Shields Avenue, Davis, CA 95616, USA

Alison L Van Eenennaam
Department of Animal Science, 2113 Meyer Hall, University of California, One Shields Avenue, Davis, CA 95616, USA

Amanda Feeney
Center for Reproductive Biology, School of Biological Sciences, Washington State University, 99164-4236 Pullman, WA, USA

Eric Nilsson
Center for Reproductive Biology, School of Biological Sciences, Washington State University, 99164-4236 Pullman, WA, USA

Michael K Skinner
Center for Reproductive Biology, School of Biological Sciences, Washington State University, 99164-4236 Pullman, WA, USA

Bing Guo
Key Laboratory of Meat Processing and Quality Control, Synergetic Innovation Center of Food Safety and Nutrition, College of Food Science and Technology, Nanjing Agriculture University, Nanjing 210095, P.R. China
CSIRO Animal, Food and Health Sciences, St. Lucia QLD 4067, Australia

Kritaya Kongsuwan
CSIRO Animal, Food and Health Sciences, St. Lucia QLD 4067, Australia
National Institute of Animal Health, 50/2 Kasetklang, Ladyao, Bangkok 10900, Thailand

Paul L Greenwood
CSIRO Animal, Food and Health Sciences, Armidale NSW 2350, Australia
NSW Department of primary Industries, Armidale NSW 2350, Australia

Guanghong Zhou
Key Laboratory of Meat Processing and Quality Control, Synergetic Innovation Center of Food Safety and Nutrition, College of Food Science and Technology, Nanjing Agriculture University, Nanjing 210095, P.R. China

Wangang Zhang
Key Laboratory of Meat Processing and Quality Control, Synergetic Innovation Center of Food Safety and Nutrition, College of Food Science and Technology, Nanjing Agriculture University, Nanjing 210095, P.R. China

Brian P Dalrymple
CSIRO Animal, Food and Health Sciences, St. Lucia QLD 4067, Australia

Kai Xing
National Engineering Laboratory for Animal Breeding and MOA Key Laboratory of Animal Genetics and Breeding, Department of Animal Genetics and Breeding, China Agricultural University, 100193 Beijing, China

Feng Zhu
National Engineering Laboratory for Animal Breeding and MOA Key Laboratory of Animal Genetics and Breeding, Department of Animal Genetics and Breeding, China Agricultural University, 100193 Beijing, China

Liwei Zhai
National Engineering Laboratory for Animal Breeding and MOA Key Laboratory of Animal Genetics and Breeding, Department of Animal Genetics and Breeding, China Agricultural University, 100193 Beijing, China

Huijie Liu
National Engineering Laboratory for Animal Breeding and MOA Key Laboratory of Animal Genetics and Breeding, Department of Animal Genetics and Breeding, China Agricultural University, 100193 Beijing, China

Zhijun Wang
Tianjin Ninghe Primary Pig Breeding Farm, Ninghe 301500, Tianjin, China

Zhuocheng Hou
National Engineering Laboratory for Animal Breeding and MOA Key Laboratory of Animal Genetics and Breeding, Department of Animal Genetics and Breeding, China Agricultural University, 100193 Beijing, China

Chuduan Wang
National Engineering Laboratory for Animal Breeding and MOA Key Laboratory of Animal Genetics and Breeding, Department of Animal Genetics and Breeding, China Agricultural University, 100193 Beijing, China

Natalia Sevane
Departamento de Producción Animal, Facultad de Veterinaria, Universidad Complutense de Madrid, Madrid, Spain

Hubert Levéziel
INRA, UMR 1061, F-87000 Limoges, France
Université de Limoges, UMR 1061, F-87000 Limoges, France

Geoffrey R Nute
Division of Farm Animal Science, University of Bristol, Bristol BS40 5DU, UK

Carlos Sañudo
Departimento de Producción Animal y Ciencia de los Alimentos, Universidad de Zaragoza, 50013 Zaragoza, Spain

Alessio Valentini
Dipartimento di Produzioni Animali, Università della Tuscia, via De Lellis, 01100 Viterbo, Italy

John Williams
Parco Tecnologico Padano, Via Einstein, Polo Universitario, 26900 Lodi, Italy

Susana Dunner
Departamento de Producción Animal, Facultad de Veterinaria, Universidad Complutense de Madrid, Madrid, Spain

GeMQual Consortium

Jie Yang
National Engineering Laboratory for Animal Breeding; Key Laboratory of Animal Genetics, Breeding and Reproduction, Ministry of Agriculture of China; College of Animal Science and Technology, China Agricultural University, Beijing 100193, China

Xuan Liu
National Engineering Laboratory for Animal Breeding; Key Laboratory of Animal Genetics, Breeding and Reproduction, Ministry of Agriculture of China; College of Animal Science and Technology, China Agricultural University, Beijing 100193, China

Qin Zhang
National Engineering Laboratory for Animal Breeding; Key Laboratory of Animal Genetics, Breeding and Reproduction, Ministry of Agriculture of China; College of Animal Science and Technology, China Agricultural University, Beijing 100193, China

Li Jiang
National Engineering Laboratory for Animal Breeding; Key Laboratory of Animal Genetics, Breeding and Reproduction, Ministry of Agriculture of China; College of Animal Science and Technology, China Agricultural University, Beijing 100193, China

Kang Kang
College of Animal Science and Technology, Northwest A&F University, Yangling, Shaanxi 712100, China
College of Life Sciences, Shenzhen Key Laboratory of Microbial Genetic Engineering, Shenzhen University, Shenzhen 518060, China

Keli Yang
Hubei Key Laboratory of Animal Embryo and Molecular Breeding, Institute of Animal Husbandry and Veterinary, Hubei Academy of Agricultural Sciences, Wuhan 430064, China

Jiasheng Zhong
College of Life Sciences, Shenzhen Key Laboratory of Microbial Genetic Engineering, Shenzhen University, Shenzhen 518060, China

Yongxiang Tian
Hubei Key Laboratory of Animal Embryo and Molecular Breeding, Institute of Animal Husbandry and Veterinary, Hubei Academy of Agricultural Sciences, Wuhan 430064, China

Limin Zhang
College of Life Sciences, Shenzhen Key Laboratory of Microbial Genetic Engineering, Shenzhen University, Shenzhen 518060, China

Jianxin Zhai
Shenzhen Ao Dong Inspection and Testing Technology Co,. Ltd, Shenzhen 518000, China

Li Zhang
Shenzhen Ao Dong Inspection and Testing Technology Co,. Ltd, Shenzhen 518000, China

Changxu Song
Veterinary Medicine Institute, Guangdong Academy of Agricultural Sciences, Guangzhou 510640, China

Christine Yuan Gou
Northwestern University, Evanston, IL, USA

Jun Luo
College of Animal Science and Technology, Northwest A&F University, Yangling, Shaanxi 712100, China

Deming Gou
College of Life Sciences, Shenzhen Key Laboratory of Microbial Genetic Engineering, Shenzhen University, Shenzhen 518060, China

William Jon Meadus
AAFC-Lacombe, 6000 C&E Trail, Lacombe, AB, Canada T4L 1 W1

Pascale Duff
AAFC-Lacombe, 6000 C&E Trail, Lacombe, AB, Canada T4L 1 W1

Tanya McDonald
School of Innovation, Olds College, 4500-50th St., Olds, AB, Canada T4H 1R6

William R Caine
Caine Research Consulting, Box #1124, Nisku, AB, Canada T9E 8A8

Hongyan Sun
Department of Animal Genetics and Breeding, National Engineering Laboratory for Animal Breeding, College of Animal Science and Technology, China Agricultural University, Beijing 100193, China

Runshen Jiang
College of Animal Science and Technology, Anhui Agricultural University, Hefei 230036, China

Shengyou Xu
College of Animal Science and Technology, Anhui Agricultural University, Hefei 230036, China

Zebin Zhang
Department of Animal Genetics and Breeding, National Engineering Laboratory for Animal Breeding, College of Animal Science and Technology, China Agricultural University, Beijing 100193, China

Guiyun Xu
Department of Animal Genetics and Breeding, National Engineering Laboratory for Animal Breeding, College of Animal Science and Technology, China Agricultural University, Beijing 100193, China

Jiangxia Zheng
Department of Animal Genetics and Breeding, National Engineering Laboratory for Animal Breeding, College of Animal Science and Technology, China Agricultural University, Beijing 100193, China

Lujiang Qu
Department of Animal Genetics and Breeding, National Engineering Laboratory for Animal Breeding, College of Animal Science and Technology, China Agricultural University, Beijing 100193, China

Nuradilla Mohamad-Fauzi
Department of Animal Science, University of California, Davis, California 95616, USA
Institute of Ocean and Earth Sciences, University of Malaya, 50603 Kuala Lumpur, Malaysia

Pablo J Ross
Department of Animal Science, University of California, Davis, California 95616, USA

Elizabeth A Maga
Department of Animal Science, University of California, Davis, California 95616, USA

James D Murray
Department of Animal Science, University of California, Davis, California 95616, USA
Department of Population Health and Reproduction, University of California, Davis, California 95616, USA

Kelsey Brooks
Department of Animal Science and Center for Reproductive Biology, Washington State University, Pullman, WA 99164, USA

Greg Burns
Department of Animal Science and Center for Reproductive Biology, Washington State University, Pullman, WA 99164, USA

Thomas E Spencer
Department of Animal Science and Center for Reproductive Biology, Washington State University, Pullman, WA 99164, USA

Anita M Oberbauer
Department of Animal Science, University of California, One Shields Ave, Davis, CA 95616, USA

Philip A Thacker
Department of Animal and Poultry Science, University of Saskatchewan, 51 Campus Drive, Saskatoon, S7N 5A8, Canada

Liying Han
College of Animal Science and Technology, China Agricultural University, Yuanmingyuan West Road 2#, Haidian District Beijing, PRC, 100193, China

He Zhou
College of Animal Science and Technology, China Agricultural University, Yuanmingyuan West Road 2#, Haidian District Beijing, PRC, 100193, China

Sinéad M Waters
Teagasc Animal and Bioscience Research Department, Animal and Grassland Research and Innovation Centre, Grange, Dunsany, Co. Meath, Ireland

Kate Keogh
Teagasc Animal and Bioscience Research Department, Animal and Grassland Research and Innovation Centre, Grange, Dunsany, Co. Meath, Ireland

Frank Buckley
Teagasc Animal and Bioscience Research Department, Animal and Grassland Research and Innovation Centre, Moorepark, Fermoy, Co. Cork, Ireland

David A Kenny
Teagasc Animal and Bioscience Research Department, Animal and Grassland Research and Innovation Centre, Grange, Dunsany, Co. Meath, Ireland

Ribrio Ivan Tavares Pereira Batista
Federal University of Juiz de Fora, Juiz de Fora, MG 36036-900, Brazil
Embrapa Dairy Cattle Research Center, Juiz de Fora, MG 36038-330, Brazil

Nádia Rezende Barbosa Raposo
Federal University of Juiz de Fora, Juiz de Fora, MG 36036-900, Brazil

Paulo Henrique Almeida Campos-Junior
Federal University of Minas Gerais, Belo Horizonte, MG 31270-901, Brazil

Michele Munk Pereira
Federal University of Juiz de Fora, Juiz de Fora, MG 36036-900, Brazil

Luiz Sergio Almeida Camargo
Embrapa Dairy Cattle Research Center, Juiz de Fora, MG 36038-330, Brazil

Bruno Campos Carvalho
Embrapa Dairy Cattle Research Center, Juiz de Fora, MG 36038-330, Brazil

Marco Antonio Sundfeld Gama
Embrapa Dairy Cattle Research Center, Juiz de Fora, MG 36038-330, Brazil

João Henrique Moreira Viana
Embrapa Dairy Cattle Research Center, Juiz de Fora, MG 36038-330, Brazil

CPSIA information can be obtained at www.ICGtesting.com
Printed in the USA
BVOW05*1628231016

465803BV00004B/25/P